Quit

D0426143

1996

Broadband Telecommunications Technology

Second Edition

For a complete listing of the *Artech House Telecommunications Library,*
turn to the back of this book.

Broadband Telecommunications Technology

Second Edition

Byeong Gi Lee
Minho Kang
Jonghee Lee

Artech House
Boston • London

Library of Congress Cataloging-in-Publication Data
Lee, Byeong Gi.
    Broadband telecommunications technology/Byeong Gi Lee, Minho Kang,
Jonghee Lee. — 2nd ed.
        p.   cm.
    Includes bibliographical references and index.
    ISBN 0-89006-866-6 (alk. paper)
        1. Integrated services digital networks. 2. Broadband communication systems.
I. Kang, Minho. II. Lee, Jonghee. III. Title.
TK5103.7.L43 1996   96-24105
621.382—dc20   CIP

British Library Cataloguing in Publication Data
Lee, Byeong Gi
    Broadband telecommunications technology. —2nd ed.
    1. Broadband communication systems
    I. Title II. Kang, Minho III. Lee, Jonghee
    621.3'82

    ISBN 0-89006-866-6

Cover design by Kara Munroe-Brown and Lucia Colella

© **1996 ARTECH HOUSE, INC.**
685 Canton Street
Norwood, MA 02062

International Standard Book Number: 0-89006-866-6
Library of Congress Catalog Card Number: 96-24105

10 9 8 7 6 5 4 3 2 1

Contents

Preface

In very recent years we have seen proposals for an *information infrastructure* and *information superhighway* take shape and become reality. What distinguishes them from traditional proposals in the telecommunications field is that they were made by politicians, not by engineers. This fact signifies how important telecommunications is to society today. At the same time, we realize that we are just entering the *information age*. Although the telecommunications industry itself will grow enormously during this information age, its impact on other industries and society in general will be even more significant, especially in shaping the competitive direction of industrial activities and in enhancing the quality of individual lives. This leads us to the observation that the cumulative efforts of telecommunication engineers that have been exerted during the past century are now bearing fruit.

Technology generally progresses in a cyclical fashion, continuously repeating the research, development, practical applications, and feedback processes. In the case of telecommunications technology, the platform for practical applications is the *networks* and the objective of the applications is to provide *services*. Therefore, technology along with networks and services form the three elementary components of telecommunications. Neither one can be sacrificed for the other two: Technology without network is worthless, a network without services is useless, and services without technology are helpless. Therefore, it is of great importance to keep the three components balanced to realize the maximum benefit of services and technological developments, and this notion has been incorporated in this revised edition of *Broadband Telecommunications Technology*.

Broadband telecommunications is a main element of telecommunications evolution because it provides a means to realize the ideal of telecommunicating in a *direct* and *natural* manner. It is what enables visual communications and high-speed data transfer to pave the way toward the information age, and it is what communications engineers have strived for in the past couple of decades. The advancement of fiber-optic technologies has resolved the fundamental bandwidth problem, thus securing the physical medium for broadband transmission. The development of SDH/SONET technology has established the transmission convergence layer over the fiber-optic transmission medium, and the follow-up of ATM technology has enabled the consolidation of service adaptation and network functions on the SDH/SONET foundation. This, in effect, completes the formation of the transport platform for broadband telecommunications.

On the other hand, *video* processing technology has continued to evolve through decades to finally give birth to MPEG-2, a general-purpose video processing standard. In line with this, the digital version of HDTV has also been standardized. This, in fact, means the completion of a framework for broadband digital transmission and storage of all types of video information. Aside from this, progress has been made to convert the distributive CATV network into an interactive broadband communications network, in which televisions can be turned into broadband terminal equipment.

On the *computer* side, comparable evolutions have been made too. Supported by the rapid advancements in memory and processing devices, computers have been upgraded continuously with extended storage capacities and increased processing speeds. In addition, multimedia technology has been developed to diversify the functions of computers, and networking capability has been added to turn the computer into a network equipment. In the midst of this evolution, Internet services and the World Wide Web browser contributed a great deal to extensive diffusion of personal computers to the business and residential areas. Computers are now ready to communicate and process broadband interactive multimedia data.

Even with the trunk network turning into a broadband transport network and with televisions and personal computers changing into broadband and high-speed network terminals, broadband telecommunications services will yet be impossible unless the *subscriber network* carries the broadband transmission capability. For this reason, a considerable amount of research has concentrated on bandwidth broadening of the subscriber network in recent years, and consequently has yielded a profusion of enabling technologies. They can be categorized into twisted-pair-based technology such as HDSL, ADSL, and VDSL; cable-based technology such as HFC; fiber-based technology such as FTTO, FTTC, and FTTH; and wireless-based technology.

All core technologies that have contributed to the above evolutional process toward broadband telecommunications, together with the related networks and services, are discussed in this book. *SDH/SONET synchronous transmission* is discussed in Chapter 3; *BISDN and ATM technology* in Chapter 4; *high-speed data networks and services* in Chapter 5; *broadband video services and technology* in Chapter 6; the *broadband subscriber network* in Chapter 2; and the fundamental issues on broadband telecommunications are addressed in Chapter 1. In particular, this revised edition has expanded discussions on newly evolving topics, such as the broadband subscriber network and technologies; high-speed data networks and services, such as TCP/IP, Internet, high-speed and real-time protocols, and the information superhighway; and broadband video services and technologies such as MPEG-2 and HDTV.

More specifically, the revisions made in each chapter are as follows: In Chapter 2, the whole topic of discussion has been changed to the broadband subscriber network, and six of the seven sections are written afresh with only one section that explains the optical device technology being a revision from the first edition. In Chapter 3, the basics for SDH/SONET are simply updated, but the latter sections handling synchronous optical transmission and synchronous network survivability are new. Chapter 4 is entirely reorganized and updated, and the last three sections discussing the control and management of BISDN, traffic and congestion control in BISDN, and ATM switching are completely rewritten. Even more drastic changes appear in Chapter 5: The discussions on frame relay, SMDS, and other low-speed and miscellaneous topics have been eliminated. Instead, fundamental and emerging topics have been added, namely, TCP/IP protocol suites, wireless LANs, IEEE's high-speed LANs, data services in ATM networks, high-speed and real-time protocols, Internet, information infrastructure, and the information superhighway. In the case of Chapter 6, the fundamentals of video processing technology have merely been updated, but the texts for broadband video services and HDTV are thoroughly rewritten, and the section for MPEG-2 is new.

Acknowledgments

This revised version was made possible by numerous helpful people—graduate students in the *Telecommunications and Signal Processing* (TSP) Laboratory at *Seoul National University* (SNU), colleague faculty members at SNU, research people at *Korea Telecom Research Laboratories* (KTRL), *Electronics and Telecommunications Research Institute* (ETRI), and other colleague engineers in universities, research organizations, and industry.

We gratefully acknowledge the contributions of many people in writing, revising, translating, and reviewing sections and subsections, namely, Sung Hoon Hong (2.2), Man Seob Lee (2.3), Sun-Cheol Gweon (2.4), Byung Gon Chun (2.6, 5.4), Jae Geun Kim (3.7-3.10), Sang Hoon Lee (4.8), Jun Won Lee (4.8, 6.2), Chi Moon Han (4.10), Jeong Gyu Lee (4.10), Woo-June Kim (5.3, 5.10), Saewoong Bahk (5.4, 5.7), Chong Kwon Kim (5.5, 5.8, 5.9), Sang Jun Moon (5.7), Byoung Hoon Kim (6.2), Jechang Jeong (6.3, 6.4), Jun Hwan Kim (6.4), Kyung Soo Kim (6.5), Kyung Ho Yang (6.6), and Seung Joo Maeng (6.6).

We thank Byung Gon Chun and Deog Nyun Kim who previewed the revision and You Ze Cho, Doo Whan Choi, Jun Kyun Choi, Dong Wook Kang, Hwa Jong Kim, and Kiho Kim who reviewed the revised manuscript. We especially thank Tae Sub Yoon at UCLA for reviewing the revised manuscript line by line, refining the expressions.

We are also grateful to a number of people who provided references, discussions, and reviews of sections and chapters, including Hong Beom Chon, Ho Jae Lee, Hyung Jin Choi, and Yanghee Choi.

We deeply thank the members of two "processing engine" teams (the first team consisting of Myeong Kwang Byun, Kang Seok Seo, and Young Hwa Song; and the second team consisting of Dae Woo Lee, Ji Cheol Moon, and Jung Hyun Oh) that propelled the revision project, under the leadership of two "captains" Jun Won Lee and Sung Hoon Hong, by processing the text, drawing the figures, formatting the tables, and organizing the whole manuscript. Without the help these two "processing engine" teams, timely revision might have not been possible.

We are particularly indebted to Jun Won Lee who not only made technical contributions but even delayed the startup of a new job to help coordinate the revision project, and Sung Hoon Hong who took over the coordination job midway and exerted great efforts until its completion.

Last, but not least, we would like to thank our wives, Hyeon Soon Kang, Ae Soon Choi, and Iljoo Lee, who have supported us all through our careers by encouraging and helping us to concentrate on our jobs, which eventually enabled us to write and revise this book.

Byeong Gi Lee
Minho Kang
Jonghee Lee

October 1996

Introduction to Broadband Telecommunications

Modern communications can be said to be evolving toward a more direct and natural form of communication. To illustrate: Delivering messages via a third person was quickly replaced by the mail service, which was subsequently supplemented by the telephone, allowing direct, delay-free conversation between two remote parties. From this perspective, we can easily predict that the next evolutionary step following the telephone will be the video phone, which enables a more natural and realistic mode of conversing. Such an evolutionary scenario can also be applied to the field of broadcasting. A purely listening form of broadcasting was inevitably followed by television, with the added visual dimension, and black-and-white televisions became obsolete with the advent of color televisions. In the same context, ordinary color televisions of today will soon be transformed into *high-definition television* (HDTV), which can display even more realistic and natural-looking pictures.

As can be inferred from the examples of video phone and HDTV, the evolution of future communications will be via *broadband telecommunication* centered around video signals. The associated services make up a diverse set of high-speed and broadband services ranging from video services such as video phone, video conferencing, video surveillance, *cable television* (CATV) distribution, and HDTV distribution to the high-speed data services such as high-resolution image transmission, high-speed data transmission, and color facsimile. The means of standardizing these various broadband telecommunication services in an integrated manner is no other than the *broadband integrated services digital network* (BISDN). Simply put, therefore, the future communications network can be said to be a broadband telecommunication system based on the BISDN.

To realize the BISDN, the role of several broadband telecommunication technologies is crucial. Fortunately, the remarkable advances in the field of electronics and fiber optics have led to the maturation of broadband telecommunication technologies. As the BISDN becomes possible on the optical communication foundation, the relevant manufacturing technologies for light-source and passive devices and for optical fiber have advanced to considerable levels. Advances in high-speed device and integrated-circuit technologies for broadband signal processing are also worthy of close attention. There has also been notable progress in software, signal processing, and video equipment technologies. Hence, from the technological standpoint, the BISDN has finally reached a realization state.

On the other hand, standardization activities associated with broadband telecommunication have been progressing. The *synchronous optical network* (SONET) standardization attempts centered around the T1 Committee, which belongs to *American National Standards Institute* (ANSI), eventually bore fruit in the form of the *Synchronous Digital Hierarchy* (SDH) standards of the *International Telecommunication Union–Telecommunication* sector (ITU-T), paving the way for synchronous digital transmission based on optical communication. The standardization activities of the *integrated services digital network* (ISDN), which commenced in the early 1980s with the objective of integrating narrowband services, expanded in scope with the inclusion of broadband services, leading to the standardization of the BISDN in the late 1980s and also establishing the concept of *asynchronous transfer mode* (ATM) communication. In addition, standardization of various video signals has been finalized through cooperation among such organizations as ITU-T, *International Telecommunication Union–Radiocommunications* sector (ITU-R), the *International Standards Organization* (ISO), and the *International Electrotechnical Commission* (IEC), and reference protocols for high-speed packet communication have been standardized through ISO, ITU-T, and the *Institute of Electrical and Electronics Engineers* (IEEE). Recently, several young standards groups including the *Internet Engineering Task Force* (IETF), ATM Forum, *Digital Audio Visual Council* (DAVIC) have been active in the standardization of the Internet, ATM technology, audiovisual communications, and other related subjects.

Various factors mentioned above have made broadband telecommunication realizable. Therefore, the 1990s is the decade in which mature broadband telecommunication technologies will be used in conjunction with broadband standards to realize broadband telecommunication networks. In the broadband telecommunication network, the fiber-optic network will represent the physical medium for implementing broadband telecommunication, while synchronous transmission will make possible the transmission of broadband service signals over the optical medium. The BISDN will be the essential broadband telecommunication network established on the basis of synchronous transmission, and ATM is the communication mode that enables us to realize the BISDN. The most important broadband services to be provided through the BISDN are high-speed data communication services and video communication services.

The goal of this book is to provide a broad but detailed description of the technologies fundamental to constructing broadband telecommunication networks. Therefore, the five essential components described in Figure 1.1—the

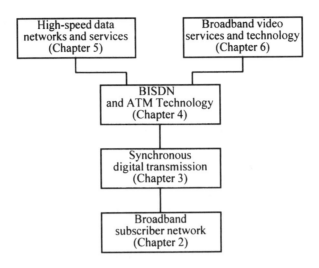

Figure 1.1 Networks, services, and technologies for broadband telecommunication systems.

broadband subscriber network, synchronous digital transmission, BISDN and ATM technology, high-speed data networks and services, and broadband video services and technology—will form the core contents of the book. These five topics are treated in order in Chapters 2 through 6. In this introductory chapter, the background of broadband telecommunications is reviewed briefly.

The organization of this chapter is as follows. The basic technologies underlying the communication network are reviewed first, and then the progress of telecommunication technologies is examined in terms of transmission, switching, signaling, packet communication, ISDN, and BISDN. Finally, the key technologies essential for the implementation of broadband telecommunication systems are discussed with reference to the subscriber network, broadband transmission, ATM communication, high-speed data network, and broadband video technologies.

1.1 TECHNOLOGIES IN TELECOMMUNICATION NETWORKS

The broadband telecommunication system takes the form of a communication network in which broadband services are provided in conjunction with existing narrowband services. The future broadband telecommunication system will thus be built on the foundation of the existing communication networks. In this context, it is desirable to examine the basic technologies underlying the existing communication networks, along with their developmental processes, in preparation for a full-fledged discussion of broadband telecommunication technologies. In this opening section, therefore, such telecommunication technologies are introduced, and their respective developmental processes are reviewed in the next section.

In this section, communication networks in general are surveyed first. Then the communication functions involved in the circuit-mode communication networks are reviewed in terms of transmission, switching, and signaling, followed by a review of the packet-mode network counterpart. Finally, we will review the technologies involved in the ISDN, in which the circuit-mode communication services meet the packet services.

1.1.1 Communication Networks

Defined in broad terms, a communication network can be said to be a system composed of the interconnection of basic communication components. Hence, a communication network can be represented in terms of nodes and links to interconnect the nodes. A communication network is required in providing communication services to a multiple number of users dispersed in a wide area. Communication services are represented in the form of traffic within the communication network. Here, traffic designates a flow of information or messages through the communication network. Consequently, within the context of the traffic concept, a communication network can be described as a system in which equipment is interconnected to transport traffic originating from various communication services.

When the communication network is depicted as a combination of nodes and links, the nodes represent communication equipment in the subscriber premises, as well as intraoffice and interoffice transmission equipment, while links represent the transmission facilities. So, if examined from the system standpoint, the basic components of a communication system can be divided into subscriber equipment, switching systems, and transmission facilities. Subscriber equipment is generally located within the subscriber premises and has the role of transmitting and receiving information, as well as controlling signals between subscribers and communication networks. Transmission facilities provide communication pathways for transporting information between subscribers. In general, transmission facilities consist of transmission media such as copper wire, waveguide, atmosphere, and optical fiber, and various electronic devices deployed along the transmission media. Here, electronic devices perform the function of amplifying, regenerating, and transforming transmitted signals. Also, transmission equipment in the central offices carries out the function of connecting transmission facilities to the switching systems. The switching system has the function of interconnecting transmission facilities at various locations and adjusting traffic pathways within the communication network. So the communication information generated from the subscriber equipment is transmitted to the switching systems via transmission facilities and interlinked via switching systems, thus accomplishing communication.

For the communication systems in the network to be comprehensible to one another, the transmission facilities and switching systems must be able to provide signaling functions in addition to the basic functions described above. *Signaling* refers to the process of transferring a variety of information for the purpose of controlling the setup of communication links within the network, as well as the related operations. For example, making a call involves the transmission of three types of signals: for indicating the beginning, the address, and the end of a call. The information capacity required for signaling is small compared to the general information transfer capacity; hence, signaling information has up to now been transmitted as part of the general communication channel, and signaling functions used to be performed by transmission equipment and switching systems.

The basic components of a communication system are as follows. The portion that links telephone offices is called the *interoffice transit network* or *trunk network*, and the portion that links telephone office and the subscriber is called the *subscriber network*. The transmission line that composes the interoffice network is called the *trunk*, and the transmission line inside the subscriber network is called the *subscriber loop* or *customer loop*. Also, the type of exchange that accommodates subscriber loops is called the *local exchange* (LE), and the exchange that links only the trunks is called the *interexchange* (IE) or *transit exchange* (TE). The corresponding signaling scheme is also divided into subscriber loop signaling and trunk signaling. If communication networks are classified in terms of traffic, they can be divided into *public switched telephone network* (PSTN), *packet-switched (public) data network* (PSPDN or PSDN), private data network, and telex network. Among them, PSTN is the largest and employs a circuit switching scheme, while PSPDN and private data network are data communication networks that are based on a packet-switching scheme.

In the following sections, we investigate the development and the future evolutionary direction of communication technologies. We first study the developmental process of transmission, switching, and signaling for public communication networks, and then we review packet communication networks. Last, we examine the ISD, which appeared in the process of integrating circuit-mode public communication networks with packet-mode communications, as well as the BISDN, which may be the final product of this integration process.

1.1.2 Transmission Technology

Transmission refers to the function of transferring subscriber information and control signals from one point inside the communication network to another. Transmission is classified into subscriber transmission, which joins subscribers to the central office, and interoffice transmission, which joins central offices. The transmitted information signals are in either the analog or digital form, and

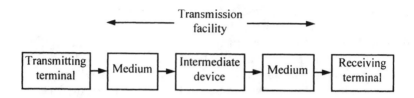

Figure 1.2 Block diagram of transmission systems (undirectional).

transmission is differentiated into analog transmission or digital transmission according to the transmission format. Also, depending on the transmission medium, transmission is divided into wired transmission via copper wire or coaxial cable, wireless transmission via terrestrial microwave links, satellite transmission via satellite links, and optical transmission via optical fibers.

A transmission system consists of transmission facilities and transmission terminal equipment. In the transmission system block diagram of Figure 1.2, the transmission medium and the repeater correspond to the transmission facilities, and transmitter terminal equipment and receiver terminal equipment correspond to the transmission terminal equipment. The transmitter transmits a given information signal by converting it into a format suitable for the transmission medium. The information signal is delivered to the receiver after several stages of repeaters and is subsequently recovered.

The transmission facility is composed of the transmission medium and the corresponding repeater. The transmission medium is the link for conveying transmitted signals, and the repeater equipment assumes the role of recovering, compensating, converting, and regenerating the transmitted signals. The transmission terminal equipment converts the transmitted information signal into a format appropriate for the transmission medium, and subsequently recovers the original signal. For this purpose, it performs such functions as information-to-electrical signal conversion, analog-to-digital conversion, modulation, multiplexing, coding, and electrical-to-optical/electromagnetic wave conversion, whereas the receiver terminal equipment performs the reverse set of functions.

1.1.3 Switching Technology

Switching refers to the function of establishing communication links by interconnecting a multiple number of subscriber service pieces of equipment. Achieving such interconnection in a cost-effective way is the rationale for switching. This is due to the fact that $n(n - 1)/$ transmission lines are required for interconnecting n nodes without switches, while only n transmission lines will do if switches are used. While the connection function is the main duty of switches, it also performs functions related to the control of communication network and subscriber services. Switching is categorized into manual, mechanical, and electronic types, depending on the switching means.

As shown in Figure 1.3, a switching system is composed of a switching network, subscriber loop interfaces, trunk interfaces, switch control, *operation and maintenance* (OAM) interface , and control network interface. The actual switching of signals is executed within the switching network, and the subscriber loop and trunk interfaces convert signals arriving from subscriber loops and trunks into a format suitable for the switching network. OAM interface performs the operation and maintenance functions of the switching system and interfaces with operation support systems. The control network interface enables accessing of the signaling network and the databases in the communication networks. The switch control is responsible for the control of all of these operations.

Switches are divided into space-division switches or time-division switches depending on the form of the switching network, into wired-logic control or *stored program control* (SPC) depending on the control scheme used, and into a concentrated control scheme or a distributed control scheme depending on the degree of distribution of the control functions. The electronic switches currently in use all use SPC. Among them, those that employ the space-division scheme are sometimes called *semielectronic* switches, and those that employ time-division multiplexing are called *digital switches* or *all-electronic switches*. Most of the current digital switches adopt the distributed control scheme.

1.1.4 Signaling Technology

Signaling refers to the process of transferring information for the control of the communication connection setup within the communication network as well

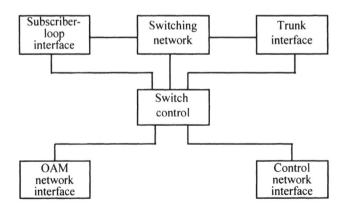

Figure 1.3 Block diagram of switching systems.

as related operations. Signaling can be classified, depending on the given function, into surveillance, addressing, and information. Surveillance implies control functions related to circuit connection, such as service request, response, alarm, and release. Address means the phone number of the destination. Information refers to the information on the progress of communication connections, such as the dial tone and ringing tone.

Signaling can be divided into subscriber network signaling and trunk signaling, depending on the applicable range. Subscriber network signaling designates the signals transferred between subscribers and the central office, whereas trunk signaling refers to the exchange of signaling information between central offices. In addition, there is special signaling for foreign exchange loops or *private branch exchange* (PBX) interoffice loops. Depending on whether signaling information is transferred via the same transmission line as that of the user information, trunk signaling is categorized into channel-associated signaling or *common-channel signaling* (CCS). These two signaling schemes are compared in Figure 1.4.

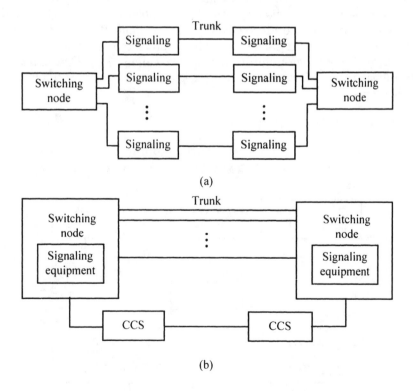

Figure 1.4 Signaling schemes: (a) channel-associated signaling and (b) common-channel signaling (CCS).

The common-channel signaling scheme is a channel-separated signaling scheme in which a collection of signals is transmitted through a separate channel that is independent of the line through which information is transferred. Here, signals are conveyed to the central office of the receiving side via *signal transfer points* (STPs), and the STPs form a signaling network. Signals are transmitted through a path different from the transmission information channel; hence, the employment of CCS shortens call setup time, allowing efficient utilization of the communication network and provision of various subscriber services in an expanded manner. On the other hand, CCS has a drawback in that partial damage of the signaling network can incur paralysis of the entire communication network.

Subscriber network signaling functions include addressing, dial tone, busy tone, ringing tone, and on-hook/off-hook indications, while interoffice-transit signaling functions include addressing, wink, and on-hook/off-hook indications.

Among the large number of signaling functions recently provided, there are many OAM-related ones. An example is the communication maintenance signaling used in adjusting the traffic load distributed in interoffice networks. Also, a database installed in the communication network is consulted to determination alternate-billing telephone or credit-card-calling services, and this process requires associated signaling functions. Also, for the central offices that are remotely monitored via OAM systems, related signaling functions are provided separately. Development of such signaling functions can bring about the establishment of *intelligent networks* (INs).

1.1.5 Packet Communication Technology

So far, the transmission, switching, and signaling technologies have been examined for the circuit-switched public communication networks. In this subsection, its counterpart, the packet-switched communication network, is examined.

Packet communication is a mode of communication in which digital data information is transmitted and switched in the packet format. Therefore, in contrast to circuit-switched networks, information in packet communication networks is not transmitted in a consecutive flow of bit units, but in intermittent packet units. If digital circuit-mode transmission is compared with packet-mode transmission from this standpoint, circuit-mode transmission is manifested as a consecutive and regulated digital bitstream, whereas in packet communication, packets in a digital bistream flow in an irregular and intermittent manner. Here, occupation of consecutive channels is not required for packet delivery, and this represents the key difference between circuit-switched networks and packet-switched networks.

Packet communication was introduced and developed for data

communication between computers and computer terminals. Consequently, packet communication has a close link with data communication, as well as with computer communication.

To realize packet communication, transmission, switching, and network operation schemes must be systematized appropriately. That is, schemes on packet format, packet delivery, and transmission path must be defined in a clear manner. Communication procedures regulated in this way for packet communication are called *protocols*. In general, packet communication is applied in a complex setting composed of numerous systems and communication networks; hence, the associated communication handling procedures are quite complicated. Therefore, if all of these communication handling functions are regulated through a single protocol, then the protocol becomes extremely complex. If, instead, information transfer procedures are appropriately separated into layers, then the development and implementation of communication handling functions become systematized and thus reduced in complexity. The reference model that was standardized in this spirit was the *open systems interconnection* (OSI) model, and here communication functions are divided into the seven-layer architecture shown in Figure 1.5.

Examples of packet communication networks include PSDN, local-area communication network, and packet wireless communication network. In PSDN, packet transmission is accomplished through packet switching at each node via a combination of interconnected nodes. Local communication networks are small-scale communication networks constructed to serve local areas, and in most cases have the character of wired data broadcast networks. That is, packets can be delivered directly to each receiver without going through intermediate nodes for switching. Local communication networks are used as

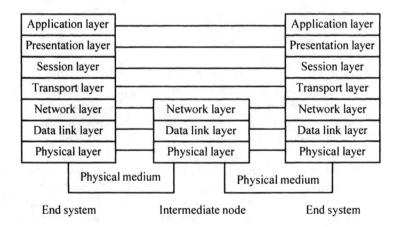

Figure 1.5   Layered architecture for packet communication.

small-scale private communication networks for local areas, and the representative examples are the *local-area network* (LAN) and the *metropolitan-area network* (MAN). A LAN is a packet communication network that provides transmission speeds up to 10 Mbps for areas within a few kilometers, whereas MAN can provide speeds up to 150 Mbps within the range of 50 km. The characteristics of regional communication networks are defined according to the transmission medium, network topology, and *medium access control* (MAC) protocol. On the other hand, the wireless version of the packet communication networks can be called *wireless packet communication networks*.

1.1.6 ISDN

ISDN is a digital communication network that provides various types of services in an integrated fashion. In other words, the ISDN is an integrated service digital communication network that not only encompasses basic voice communication, but also allows integrated provision of low-quality video and data services. Its targeted services are mainly narrowband services such as telephone, facsimile, teletex, telemetry, telewriting, and data terminals. The main feature of the ISDN is that the subscriber is interfaced with the communication network through a single access so that several services can be provided in unison, and that connection between the subscribers is accomplished digitally. That is, the subscriber network becomes digital, and various services become integrated on that foundation.

ISDN allows handling of various voice as well as nonvoice services in an integrated manner. This is possible because various information handling procedures are carried out digitally. Therefore, circuit switching and packet switching can be provided simultaneously, and transmission is done in the digital domain. The ISDN also provides a common access point for the existing circuit switching network and the packet switching network, and allows sharing of various information resources within the communication network. In other words, various databases are maintained inside the communication network that can be used in common by all the subscribers and service providers. In addition, the ISDN can provide various kinds of communication processing and information handling functions and simplifies operation and administration of the communication network.

The basic structure of the ISDN is shown in Figure 1.6. In the figure, we can see that the ISDN user terminal equipment is connected to the ISDN through the S/T reference point. Here, signaling information between the user and the network is transferred through the D channel. The network interface distributes user information to the appropriate destinations of the circuit switched network and the packet switched network, and various signals for call handling are transmitted through common-channel signaling. In the figure, *high-layer*

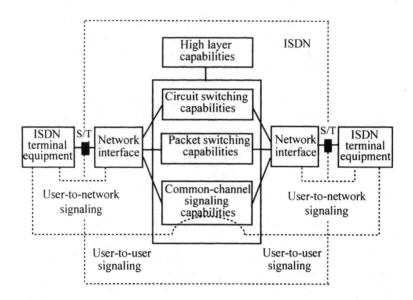

Figure 1.6   Basic architecture model of an ISDN.

capabilities represent auxiliary functions such as protocol conversion and rate conversion.

1.2 EVOLUTION OF TELECOMMUNICATION TECHNOLOGIES

Now that we have reviewed the technologies supporting the communication networks, we can examine their respective developmental processes. We review how the transmission, switching, and signaling technologies progressed in the past to establish today's circuit-switched public network. Then we consider how the packet-mode communication network became equipped with today's systematic layered protocol structures. Following this, we examine how the existing circuit-mode and packet-mode networks evolved into the ISDN, and eventually into the BISDN. Based on this overview of the developmental processes of the telecommunication technologies, we then summarize the general trends of telecommunication technology developments.

1.2.1 Evolution of Transmission Technology

The developmental process of transmission can be summarized as an evolution from analog mode to digital mode. In the early days of the telephone after its invention in 1876, a single channel was assigned per line for analog transmission; but with the development of the vacuum tube and the advances in filter theory, an analog system employing *frequency-division multiplexing* (FDM) was introduced in 1925 and a coaxial cable system was installed in 1936. Analog wireless transmission was developed simultaneously, leading to the trans-Atlantic wireless telephone system in 1927, and the microwave wireless broadcast system was proposed in 1941. The analog transmission system made great advances from the 1950s to 1970s. With the maturation of vacuum tube technology, the microwave transmission system extended into the 4-, 6-, and 11-GHz frequency regions, and the development of digital wireless transmission systems was initiated in 1969. The invention of transistors helped coaxial cable systems grow substantially in the 1960s, leading to the establishment of the L4 system consisting of 3,600 channels per coaxial cable and the L5 system of 10,800 channels.

The concept of *pulse code modulation* (PCM), which is fundamental to digital transmission, was issued a patent as early as 1926. But commercial application of PCM did not materialize until the introduction of the 24-channel T1 carrier system in 1962. With the maturation of device technology, the digital channel bank that featured toll quality was developed by the year 1972, and the 96-channel T2 carrier system was introduced as well. It was followed by testing of high-bit-rate digital systems employing *time-division multiplexing* (TDM), but they were not put to practical use because of their high costs relative to the equivalent analog carrier systems. Their economic viability began to be demonstrated when optical transmission finally became realizable. The first DS-3 optical transmission system was commercialized in 1979, and it was installed for long-distance transmission in 1983.

Satellite transmission began with the launching of the communication satellite INTELSAT I in 1965. The number of transponders used at that time was two, the frequency band was 50 MHz, and the communication capacity amounted to 480 voice channels. The numbers grew with succeeding launches, reaching a total transponder bandwidth of 2,400 MHz, and communication capacity of approximately 36,000 voice channels by the time INTELSAT VII was launched in 1995. Carrier frequency used was typically in the C-band (4/6 GHz) and Ku-band (12/14 GHz), and a Ka-band (20/30 GHz) is currently being tested for commercial deployment. The multiple-access techniques are required, because several earth stations must communicate by sharing a single communication satellite. *Frequency-division multiple access* (FDMA) and *time-division multiple access* (TDMA) are frequently employed for this purpose. Because the medium of satellite transmission is free space, it shares technical aspects with terrestrial microwave transmission. The modulation employed was analog FM in the early stages, but currently digital modulation, especially QPSK modulation, is the dominant modulation means of choice. *Low Earth orbit* (LEO) and *middle Earth orbit* (MEO) satellites are also commercially planned for personal telecommunications use.

In the early stages of optical transmission, a method of propagating light through the atmosphere as in the case of wireless transmission was attempted, but waveguide transmission using optical fiber was ultimately chosen as the transmission medium. Optical transmission began to be practical when its attenuation characteristics were lowered to 20 dB/km in the early 1970s. Optical transmission is accomplished by modulating transmitted information by a laser diode or a light-emitting diode, passing the information signal over optical fiber, and reconstructing the information by the *pin* diode or the *avalanche photo diode* (APD) device. Digital transmission based on PCM and TDM became more practical only after being supplemented by optical transmission. The 45-Mbps optical communication system was put to use in the early 1980s, followed by the development of 90-, 145-, 180-, 410-, 560-Mbps, and 1.2-, 1.7-, and 2.4-Gbps systems. These systems will eventually converge in a synchronous digital transmission at rates of 155 Mbps, 622 Mbps, 2.5 Gbps, and 10 Gbps founded on SDH/SONET.

1.2.2 Evolution Process of Switching Technology

The developmental process of switching technology can be examined in the light of advances in switching networks and control schemes. Switching networks evolved from the electromechanical space-division types to the electronic space-division types, and subsequently into those employing the SPC scheme. The initial control schemes evolved from the concentrated control to the wired-logic control and then into the SPC. Therefore, the step-by-step switches of the electromechanical direct control widely used up to the 1920s have finally evolved into today's digital switches of the electronic SPC.

The concept of automatic switching, first introduced in 1879, was not put to practical use until the advent of step-by-step switches. In step-by-step switching, a single rotation of the dial causes the point of contact to move linearly or circularly by a proportional amount, making connection in proper sequence. Step-by-step switches were in wide circulation up until the introduction of crossbar switches in the 1930s. In crossbar switching, the distributed common control scheme is used independently of the switching networks, while switching networks use coordinated crossbar switches in which horizontal and vertical lines are cross-connected. The patent application for the crossbar switching concept was first filed in 1913, and the first commercial crossbar system was installed in 1938. The No. 4 Crossbar system was installed in 1943, and the No. 5 Crossbar system in 1948.

Propelled by advances in vacuum tubes and transistors, electronic switching technology accelerated in the 1950s. Implementations of various electronic devices for the logic circuit of the interface and the switching contact of the circuit-switching network were attempted, and a wired-logic control was also proposed during this period. However, the commercial electronic switch No. 1 *electronic switching system* (ESS) was not actually installed until 1965. No. 1

ESS is a switch that employs space-division switching and SPC. The time-division digital electronic switch began to be developed in the 1970s, and No. 4 ESS and No. 5 ESS were installed in 1976 and 1982, respectively. Other large-capacity digital electronic switches such as AXE-10, DMS-100, EWSD, S1240, GTD-5, and TDX-10 were developed around the same period. However, widespread installation of such digital switching systems in communication networks was not possible until much later.

As stated earlier, time-division digital switches are more economical than the space-division analog switches. That is, switching of digital information is possible without extra conversion or multiplexing processes. Transmission systems are becoming digital due to the various advantages inherent in digital transmission, and the corresponding digitalization of switching systems can be considered a natural outcome. A digital switch that possesses both circuit-switching and packet-switching capabilities will eventually be needed, and it will be succeeded by a packet-based ATM switch (see Section 4.10).

1.2.3 Evolution of Signaling Technology

Signaling schemes should be simple and universal, and thus be efficiently used in the communication network. Consequently, subscriber network signaling has been providing a simple way of connecting subscribers to the telephone office, while trunk signaling, which is ultimately a signaling between machines, has evolved in the direction of raising the efficiency and flexibility of the communication network, even at the cost of added complexity. As a result, trunk signaling has advanced in parallel with trunk transmission facilities, influencing in turn the development of switching systems.

The early signaling schemes were based on direct current until the introduction of the in-band/out-of-band signaling scheme, which accompanied the appearance of analog transmission carrier systems. As digital transmission equipment such as the T1 carrier system began to be installed after 1962, digital bit signaling based on robbed bit or out-slot bit schemes began to be employed as well. The channel-separated signaling scheme was tried out after the establishment of SPC switching system No. 1 ESS in 1965, and was superseded by the *common-channel interoffice signaling* (CCIS) scheme, which began to be provided with the commercialization of No. 4 ESS digital switch in 1976. It subsequently became widespread, and was systematized into international reference protocols such as ITU-T CCS systems No. 6 and No. 7. When the CCS scheme was first used, it was based on 2.4-Kbps packet transmission, but it is currently based on 64-Kbps packet transmission.

In summary, the developmental process of signaling schemes can be said to be an evolution of interoffice trunk signaling from channel-associated signaling to common channel signaling. Channel-associated signaling renders relatively simple surveillance and addressing functions for communication call setup. But it has difficulty with transferring network OAM signals, in detecting calls, or collectively transmitting various different types of traffic. Furthermore, channel-associated signaling has almost nothing to contribute to flexible operation of the communication network or intelligent provision of communication services. Therefore, the evolution of channel-associated signaling to common-channel signaling is being propelled by the objective of raising communication network flexibility and intelligence. Further development and specialization of signaling has led to the formation of intelligent networks.

1.2.4 Evolution of Packet Communication Technology

The development of packet communication is closely related to that of computers. The concept of packet communication first appeared on the scene with the development of large-scale computers in the 1960s. Packet communication was required in order to link a great number of computer terminals to the large central computers. The primary example of a packet communication network that was constructed to meet this very objective was the TYMNET by Tymshare Co.

Afterwards, computer manufacturing companies contributed in numerous ways to the further advancement of the packet communication network. Communication processors for use in packet communication among computers were developed, along with the corresponding software. Such a drive ultimately led to the establishment of the concept of layered communication architecture. The *systems network architecture* (SNA), introduced by IBM in 1974, can be said to be its most representative product.

Afterwards, each computer company created its own proprietary communication topologies, and interconnection consequently became difficult. The responsibility of studying possible standard communication topologies for open systems connection was entrusted to ISO, and the outcome was the OSI reference model (see Section 5.2).

ARPANET, developed by the U.S. Department of Defense, contributed greatly to the progress of packet communication technology. The ARPANET project was initiated in the late 1960s with the objective of interconnecting large computers, but it also propelled other associated research areas such as routing and flow control algorithms. It led to advancement in packet communication technologies, the most typical example of which is TCP/IP (see Section 5.3), and also contributed to the formation of packet-switched public data networks throughout the world. Further, it influenced the standardization of the packet-switched network interface protocol X.25 by ITU-T, and gave birth to the Internet in early 1980s, which has grown tremendously in the 1990s (see Section 5.9).

LANs were originally designed to provide low-speed data transport in the local area, but LAN transmission speeds and application areas have continued to grow. This is manifested in a series of high-speed and high-density area data networks such as *fiber distributed data interface* (FDDI) LANs and *distributed queue dual bus* (DQDB) MANs. For interconnection of LANs across the public wide-area network (WAN) services such as frame relay and *switched multi-megabit data services* (SMDS) have been developed. In addition, gigabit network development has been under way as a high-speed packet network to succeed LANs and MANs (see Sections 5.4 through 5.6).

Today, various types of computers are in wide distribution, and these computers in turn support and control various forms of communication. Furthermore, demand for novel data information systems is increasing rapidly, greatly outpacing the demand for voice services. Therefore, packet communication will assume a more important role in the future, with voice continuing to maintain a superior role. As telephone networks gradually become digital and optical, it will be necessary to conceive a developmental scenario in which packet-switched and circuit-switched networks can coexist simultaneously. Such changes will be manifested in circuit-mode-oriented integrated access in the ISDN and in a packet-mode-integrated network based on ATM cells in the BISDN.

1.2.5 Evolution of ISDN

In examining the evolutionary process of the ISDN, it is necessary to note the development of the communication network that preceded it. As was observed previously, communication network has evolved from the analog mode into the digital mode. First, transmission became digital with the appearance of the T1 carrier system, and digital transmission provided a solid foundation when the optical transmission systems were developed in the 1980s. The switching mode began to become digital when switches E-10 and No. 4 ESS were installed. Digital switching systems attracted attention by being highly cost effective when used in conjunction with readily established digital transmission, and the 1980s saw the development and installation of a great number of digital switches. As a result, considerable progress was made in the digitalization of today's communication networks, especially in the case of interoffice networks. However, subscriber networks have on the whole remained analog. As transmission and switching became digital, signaling schemes also went through many changes, as discussed earlier, finally reaching today's common-channel signaling scheme via the in-band/out-of-band analog signaling and the robbed-bit/out-slot bit digital signaling schemes.

Packet communication networks began to appear independently in the 1960s with the objective of providing a link between computers and data terminals, and they advanced rapidly with the developments in the associated software and hardware, and began to be employed as packet-switched public data networks when the OSI reference model was established in the 1980s.

Packet communication networks are constructed and operated independently of circuit-switched networks, and their size is relatively small. But demand for packet communication networks has grown rapidly as data information capacity becomes larger, and study on processing real-time signals such as voice in packets has also been under way.

To absorb such data information needs, circuit-switched networks have been providing data services through modems, *digital data services* (DDS), and *circuit-switched data capability* (CSDC). This can be said to be an integration of voice and nonvoice services achieved via public telephone networks. On the other hand, study has also been under way in the packet communication field for local communication networks in order to accommodate voice services. However, although these approaches can provide partial solutions, they are not able to do so worldwide. As a means of providing a more fundamental and comprehensive solution, international standardization activities with the ISDN have been in progress since the late 1970s under the supervision of ITU-T. The standardization study carried out around ITU-T bore fruit by producing the ISDN's basic framework by 1984, and even specifics were standardized by 1988. Such a groundwork paved the way for the construction of ISDN systems and the provision of commercial services throughout the world in the mid 1980s.

The basic philosophy of the ISDN is first to digitalize subscriber networks and provide various types of communication services in an integrated manner. This is guided by the possibility of dealing with various services in a generalized manner if the communication network is digitalized in bit units. Transmission and switching in trunk networks are already being performed in digital mode, and complete digitalization of the communication network is possible if subscriber networks become digital as well. As far as signaling schemes are concerned, common-channel signaling is already possible in trunk networks; hence, the only required step is to provide a signaling scheme that is suitable for the digitalization of subscriber loops in subscriber networks. Here, voice and data information passes through the subscriber loops after being integrated in a digital mode; hence, the central office needs to possess the capability of differentiating each type of signal and applying circuit switching and packet switching, respectively, or separating each and delivering it respectively to the public switched telephone network and packet-switched public data network. Signaling information is also transmitted after being integrated with voice and data information, and hence the central office also needs to maintain functions for separating and processing signaling information and transferring them to common-channel signaling networks. Therefore, the ISDN has a structure in which various services are first integrated in digital mode at the communication network terminations inside the subscriber premises, and then passed through subscriber loops and divided and delivered appropriately to the circuit-switched network, packet-switched network, and signaling network via the central office.

1.2.6 Network Evolution Toward BISDN

In the midst of the evolutionary process of integrating services and forming the ISDN, the demand for high-speed packet communication and video communication services has increased continuously, and the capability for their provision has steadily improved as well. High-speed packet communication has come to be established in the form of FDDI LANs and DQDB MANs, and further development into gigabit communication networks is under way. In the video communication service field, the main enterprises involve making existing CATV networks optical, providing new HDTV services, and transforming ordinary telephone services into video phone and video conferencing services. Also, a great deal of research has been carried out to provide multimedia communication services associated with high-speed packet communication and video communication services.

Such high-speed packet and video communication services by nature require high-speed and broadband communication channels. Such a requirement cannot be accommodated by 64-Kbps-based ISDN, and this brought about the need for the *broadband ISDN* (BISDN), which is able to provide such broadband services in an integrated fashion. Consequently, CCITT began to proceed with BISDN standardization activities near the latter half of 1980s, and basic frameworks for the BISDN were completed near the end of the decade. With the emergence of the BISDN, the existing ISDN began to be called *narrowband ISDN* (NISDN).

Conceptually, the BISDN is no more than an extension of the ISDN to accommodate broadband services. In other words, the BISDN will function as a communication network that can provide integrated broadband services such as high-speed data services, video phone, video conferencing, high-resolution graphics transmission, and CATV services, along with such NISDN services as telephone, data, telemetry, and facsimile. Therefore, if the basic structure of the ISDN shown in Figure 1.6 is supplemented with broadband service functions, the result becomes the basic structure of the BISDN. In implementation, however, the BISDN uses a communication mode that is completely different from the NISDN because of the sheer diversity of services. That is, as a means of accommodating various characteristics and distribution properties of broadband service data, ATM is employed. It achieves packet-oriented integration of broadband services, from which arises the fundamental difference from the NISDN, whose service integration is circuit-oriented.

The BISDN is based on transmission speeds and capacities at the 155-Mbps, 622-Mbps, and 2.5-Gbps levels. The associated standards include ITU-T G Series Recommendations for SDH, SONET standards of the T1 Committee, and the ITU-T I Series Recommendations, which support the concept of ATM-based BISDN. Here, SDH is a novel system whose goal is integrated accommodation of the existing *plesiochronous digital hierarchy* (PDH), whereas ATM is an attempt to accommodate packet-mode transmission inside existing circuit-mode communication networks. Consequently, the BISDN can be said to be a highly innovative communication network in many aspects compared to existing communication networks.

Figure 1.7 shows the conceptual composition of the BISDN. It can be observed that SDH/SONET and ATM are used as transmission and switching modes, respectively, and that BISDN *network terminations* (NT) are used for broadband access. It can also be inferred from the figure that, in the BISDN, existing NISDN services and various new broadband services are accommodated collectively and that PBXs, LANs/MANs, and other networks such as PSTN, PSDN, PCN, and CATV can be interconnected.

However, actual implementation of the BISDN has a great number of associated fundamental difficulties. An example is the difficulty of broadband signal switching, and this is due not only to the high speed of broadband switching, but also to the diversity of signal speeds and service durations. On the other hand, traffic control in the BISDN emerges as an especially crucial task. In the BISDN, all information exists inside public networks in the form of packets, or ATM cells, and hence critical damage to the given service is contingent on the success or failure of proper resource management.

In spite of the difficulties, developments for realizing the BISDN have been successful, and various studies have been conducted to overcome the obstacles facing the implementation of the BISDN. In some countries BISDN testbeds have been constructed and BISDN service provisions have been tested.

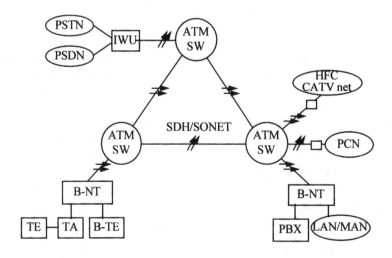

Figure 1.7 Architecture of BISDN (refer to Appendix B for acronyms).

1.2.7 Developmental Trends of Communication Networks

According to the evolutionary processes of the communication networks examined so far, the following general trends can be discerned.

First, communication networks have steadily developed in the direction of further digitalization. Digitalization in transmission technology was initiated in conjunction with PCM, followed by digitalization of switching technology, and signaling schemes also changed accordingly. Digital transmission has the inherent drawback of consuming more bandwidth per channel compared to analog transmission. For instance, whereas a 4-kHz band might suffice for analog transmission, a 64-Kbps bit rate is required for transmission in the digital mode. Nonetheless, digital transmission has continued to progress in spite of these unfavorable conditions, due to the following advantages of digital transmission. In fact, the bandwidth problem itself has been resolved by optical transmission means.

To begin with, digital transmission is free from transmission-induced noise. Consequently, no deterioration in transmission quality occurs, even in long-distance transmission involving multiple-stage multiplexing and demultiplexing processes. The main source of noise in digital transmission is quantization, but this can be reduced to imperceptible levels by increasing the number of bits.

In digital mode, information is handled in terms of time slots (that is, in terms of bits), which renders assembling and separation of several types of information simple, and this is achieved by the insertion and extraction of corresponding time slots. Consequently, it becomes possible to treat information with various different characteristics such as voice, video, and data information in an integrated manner. Signaling is also easy to perform in digital mode, surveillance and OAM of the network is easy, and the quality and stability of transmission can be raised through the employment of error detection and correction coding.

The digital mode is also simpler than the analog mode in design and manufacturing of integrated circuits, and it is also superior with respect to information storage. After the introduction of digital switches, the digital mode became particularly cost effective compared with the analog mode in interoffice transmission. Consequently, interoffice transmission links have been continuing to become digital, and subscriber transmission links will also become digital in step with the advances in optical transmission technology.

The second trend is the expansion of data services and the corresponding advancement of data communication networks. Public communication networks were originally voice oriented, and as computers became more widespread, the necessity for data communication emerged more clearly. However, initially only a simple form of packet communication was required to connect large computers and terminal equipment, but as large numbers of computers came into use and the personal computers became universal, data communication services grew dramatically. Moreover, in step with such trends as *office automation* (OA) and *factory automation* (FA), various sorts of manufacturing and handling tasks became more reliant on computers, which aided in universalizing data communication. Furthermore, the introduction of multimedia services, on-line services, and, especially, Internet services has brought about an explosive increase in the amount of data communications. Such a drastic increase in demand for data services helped regional communication networks grow rapidly, and it subsequently brought about a need for the ISDN.

The third trend is the integration of communication networks in order to cope with the diversity of the services provided. As communication network services that are generally voice oriented expand in scope into various data and video services, communication networks are becoming integrated to accommodate this change. If these diversifying services must be accommodated separately, as was the case in the communication networks of the past where telephone networks, data networks, and telex networks existed independently, then disorder or complexity is unavoidable, which is undesirable from the standpoint of economical viability and communication efficiency. Therefore, a solution has been sought that would allow each service to access existing communication networks in an integrated manner regardless of type, and ultimately lead the communication network itself to become integrated as well. The ISDN was conceived as the initial solution. Conceptually, the BISDN is no more than an expansion of the ISDN, so even packet communication, video phone, video conference, and CATV video services can be integrated and provided jointly.

The fourth trend is that communication networks are becoming more intelligent. This has been made possible by the introduction of common-channel signaling, which allows the transfer of information needed in communication connection through a separate network unrelated to the communication information itself. That is, intelligent communication networks have materialized by first constructing an independent signaling network based on common-channel signaling, and on that foundation, several types of intelligent connection services have been provided. This permits the provision of diverse types of intelligent network services to the user, such as alternate billing services and virtual private network services.

Further trends can be found in the individualization and personalization of communication services. As users desire the mobility and reliability of wireless communication that can transcend the geographical and time restrictions of wired communication, we see the emergence of *personal communication networks* (PCN), which can overcome the limitations of wireless or mobile phones. The motivation behind the personal communication network is to employ small-scale personal portable phones and the concept of signaling networks to allow the user access to the network with a personal number and also to use the network in an interworking relationship with existing public communication networks. Multimedia communication, which can also handle data and video, is sought for the future, but bandwidth restrictions inherent in wireless communication emerge as a fundamental problem.

Such trends in communication network development are also compatible with the subsequent evolutionary trends outlined in Figure 1.8. The figure also illustrates the role of the broadband subscriber network, which acts as the parent that integrates PSTN/PSDN, data networks, and video service networks, which in turn evolve into the BISDN. The development of the wireless access network via PCN and FPLMTS is also depicted.

1.3 KEY TECHNOLOGIES FOR BROADBAND TELECOMMUNICATIONS

As stated earlier, the broadband telecommunication system is in essence a communication system that provides various high-speed and broadband services in an integrated manner via the BISDN. For it to be realized in practice, maturation of several technologies is essential. First, optical transmission technology is indispensable for enabling broadband data transport, and it can be divided into the optical medium technology for efficiently using the optical physical medium and the broadband transmission technology for constructing trunk and subscriber network transmission systems over the optical medium. High-speed switching technology is required to enable broadband transport. Since the BISDN is founded on ATM, the required high-speed switching technology is no other than the ATM switching technology. Also, since the BISDN is an ATM-based public network, traffic control technology within the network is extremely critical. On the other hand, high-speed data communication technology that can effectively handle high-speed packet services, broadband video communication technology for efficient compression, transmission, and recovery of broadband video services, and the multimedia technology for combining various services and provide them as multimedia services are also important. This section reviews such key technologies required to realize broadband communication systems.[1]

1.3.1 Base Technologies for Broadband Telecommunications

Before reviewing the key technologies for the construction of broadband communication systems, it is worth examining the developmental status of the various base technologies required for realization of broadband telecommunication systems.

---

1. The contents of Sections 1.3.2 to 1.3.6 are treated in detail in Chapters 2 through 6 of this book. These chapters can therefore be used as references for more detailed explanation of the new terms and concepts in these sections.

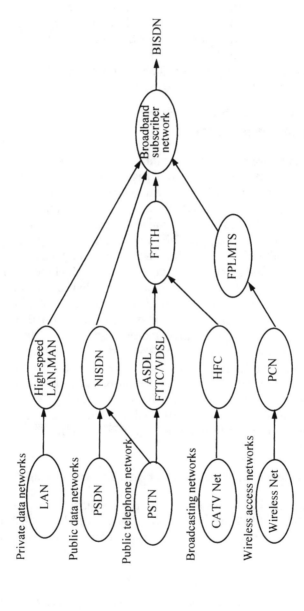

Figure 1.8 Evolution of telecommunication networks (refer to Appendix B for acronyms).

Base technologies for broadband telecommunications have matured in step with the increasing demand for broadband services. Maturation of optical communications is crucial for the realization of broadband communication. Accordingly, optical fiber loss has been reduced to below 0.5 dB / km, dispersion has also been lowered to negligible levels, and the emergence of the erbium-doped fiber amplifier has made unrepeatered, long-distance transmission possible. In the case of light-emitting devices (LEDs), the performance of laser diodes has been drastically improved, and edge-emitting LEDs have advanced to the point where they can be satisfactorily connected with optical fiber for use in subscriber networks.

Advances in integrated circuit and transistor technologies are also worthy of close attention, with bipolar *emitter-coupled logic* (ECL) or *gallium arsenide* (GaAs) transistors currently possessing high-speed processing capability from hundreds of megabits per second up to a few gigabits per second, while *complementary metal oxide semiconductor* (CMOS) technology permits processing speeds up to the 150-Mbps level. Memory devices crucial for the storage of high-speed data and image / video signals have been evolving to an unprecedented degree.

The recent advances in signal processing have made compression, conversion, and recovery of video signals simple, and if the above advances are complemented by *very-large-scale integration* (VLSI) technology, efficient terminal equipment at the user's side can be constructed. Also, developments in high-quality monitors and high-sensitivity cameras have advanced to the point that BISDN terminal equipment associated with various video services can now be made. Related TV technology has improved significantly as well, encompassing *extended-definition television* (EDTV) and HDTV.

Recent advances in software and microprocessor technologies have made high-speed control a possibility, which in conjunction with fast operating devices enables high-speed data communication, high-speed switching, and synchronous transmission.

Therefore, the base broadband technologies have matured for realization of broadband telecommunication systems, whose specific embodiment is the ATM-based BISDN. Figure 1.9 provides a pictorial representation of the above-mentioned technological environments.

1.3.2 Subscriber Network Technology

The subscriber network creates a bridge between the subscriber's communication terminals and the network provider's trunk network. Rapid increases in the amounts of information exchange and steep growth of data and multimedia services have triggered the renovation of the trunk network. Deployment of optical fibers, adaptation of SDH/SONET transmission systems, and development of ATM systems all contribute to this trunk network renovation.

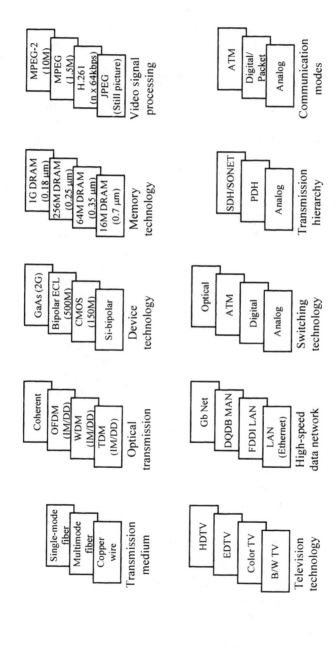

Figure 1.9　Base technologies for broadband telecommunication.

On the subscriber's premises side, comparable changes have been made in terms of services demanded and terminal capabilities. Telecommunication services have been evolving from simple POTS to broadband video telephony and video conferencing services. The video services have been evolving to include interactivity or to provide on-demand services. The data services have been ever increasing in quantity according to the diffusion of personal computers in the residential area, and increasing in bandwidth according to the popularization of multimedia and Internet services.

To bridge such subscriber's demands to the broadband trunk network, the subscriber network must also carry broadband transmission capability. This necessity triggered research and development for establishing a broadband subscriber network and consequently has produced a number of enabling technologies, which can be classified into four categories: twisted-pair-based, cable-based, fiber-based, and wireless-based technologies.

The twisted-pair-based technology takes advantage of advanced modulation techniques such as DMT and CAP, and the resulting broadband subscriber loops are DSL, HDSL, ADSL, and VDSL. The transmission rate and distance differ from one another and the application varies accordingly.

The cable-based technology refers to the HFC, in which the existing CATV networks are replaced with optical fiber up to the carrier serving area, with the remaining copper cable to the subscriber's premises left as it used to be.

The fiber-based technology can be implemented in multiple stages, such as FTTO, FTTC, and FTTH. In the FTTC, the copper wire in the feeder network is replaced by the optical fiber, with the distribution network unchanged. In the FTTO, fiber is deployed to the building and, in the FTTH, fiber is completely deployed down to the subscriber's premises.

The wireless-based technology has been also conceived and utilized in narrowband communication access and video distribution services. However, further studies are needed before wireless-based technologies can be provided.

1.3.3 Broadband Transmission Technology

The broadband transmission technology that will have a critical role in broadband telecommunication networks is no other than synchronous digital transmission. Synchronous digital transmission technology is a novel transmission mode based on SDH or SONET. It systematically accommodates existing North American and European digital tributaries, and makes possible ATM transmission of the BISDN.

SDH or SONET offers numerous innovations compared to the existing PDH. SDH employs a 125-μs frame structure, and can thus easily access DS-0 level signals and collectively accommodate all the existing digital tributaries inside the 155-Mbps STM-1 signal (STS-3c in the case of SONET). Transmission processing procedures are systematized in accordance with the layering concept, and this is reflected in the formats of overheads. A novel pointer technique is used to achieve synchronization, and on that basis it makes the establishment of a global communication network possible. Furthermore, it applies the concept of one-step multiplexing in the synchronous multiplexing procedures of forming the STM-1 signal, and thus can manage repetitive *add-drop multiplexing* (ADM) and *digital cross-connect* (DXC) efficiently.

The name *synchronous digital hierarchy* arises from the fact that SDH adopts a synchronous multiplexing structure in multiplexing tributaries into the STM-1 signal. The use of synchronous multiplexing renders multiplexing/demultiplexing simple, makes it easy to approach low-speed tributaries, can enhance the efficiency of OAM functions, and facilitates expansion into higher bit rates of the future. Here, overheads contained inside the STM-1 (or STS-1) effectively form the OAM network, and higher speed signals can be obtained by further multiplexing the base STM-1 signal [for instance, multiplexing 4, 16, and 64 of STM-1's results in STM-4 (or STS-12c) at 622.080 Mbps, STM-16 (or STS-48c) at 2,488.320 Mbps, and STM-64 (or STS-192c) at 9,953.280 Mbps, respectively.] On the other hand, STM-1 becomes an important signal component for ATM communication also, because it carries the ATM cells in its VC-4 payload space.

In broadband telecommunication systems, therefore, it is important to construct a communication network that can fully utilize the inherent benefits of SDH. ADM and DXC, especially, are constituent elements that can apply the special features of SDH to make possible the formation of point-to-multipoint network configuration through the employment of a hub, and hence assume the most important role in synchronous digital transmission. They also allow the constituent elements of the network to be connected in various ways, such as star and ring configurations, and thus allow flexible reconfigurations of the network structure. Considering that the existing public networks and subscriber networks are based solely on the star topology, we can expect that the introduction of SDH along with ring topology would add various new features to future networks.

1.3.4 ATM Communication Technology

The ultimate goal of broadband telecommunication can be said to be the BISDN, and the unique feature of the BISDN is the use of ATM. ATM was devised as a way of collectively accommodating all of the diverse characteristics possessed by broadband services. In other words, it is a means of accommodating services with opposing characteristics, such as circuit-mode and packet-mode services, low-speed and high-speed services, and *continuous bit rate* (CBR) and *variable bit rate* (VBR) services. This is made possible by first segmenting all service signals into packets of fixed size, called *ATM cells*, so that communication networks can only take part in the proper transfer of them. Here, ATM cells are multiplexed through *asynchronous time-division multiplexing* (ATDM), which is the source of the term *asynchronous* in ATM.

The ATM communication technique can be said to be a communication mode that integrates the existing circuit-mode digital communication method with the packet-mode communication method. First of all, ATM has a close connection with the packet-mode communication in that it uses ATM cells as its basic means of transport; but there is a difference in that the packet mode was created for VBR non-real-time data signals, whereas ATM can manage equally as well with real-time CBR signals. Also, packet mode is generally used for LANs, whereas ATM can be used for vast public networks and hence is accompanied by various problems inherent in any large network, such as address assignment, access and flow control, switching, and transmission. On the other hand, the fundamental difference between ATM and circuit-mode communication is that, whereas the circuit mode functions by allocating a separate service channel and transferring information signals in a continuous bit stream through it, ATM operates by segmenting the information signal so as to fit it onto the ATM cells, then it transfers them through a *virtual channel* (VC) in a *virtual path* (VP). Thus, the accompanying ATM procedures such as establishing connections, signaling, transmission, and switching present various new problems.

ATM is a novel communication technique which is introduced together with the BISDN and includes several complex new technologies. Among the most difficult are ATM traffic control technology and ATM switching technology. ATM traffic control is a critical factor that dictates the success of communication services through the ATM network, and hence can be regarded as a fundamental problem of ATM communication. Traffic control encompasses *connection admission control* (CAC) for allowing user's call setup, *usage parameter control* (UPC) for monitoring the observance of the traffic characteristics agreed on at call setup, *congestion control* (CC) for monitoring traffic and the occurrence of congestion within the ATM network, as well as the application of appropriate measures to alleviate the congestion, and *priority control* (PC) for ensuring the priority of each service. Of course, similar traffic control methods have been studied for existing packet communication networks, but they differ considerably in terms of network scale, transmission distance, transmission speeds, and traffic volume.

ATM switching also emerges as a fundamental problem to resolve. ATM switching is high-speed packet switching employing a 155.52-Mbps bit rate as the base unit. Consequently, its main associated difficulty is in the high speed, and concepts such as parallel processing, self-routing, and optical switching have been considered as possible solutions. Also, ATM switching requires a short time delay, low cell-loss rate, and high throughput, in addition to point-to-multipoint switching capability. Large-capacity ATM switches are required, especially for use in the BISDN, and must be easily implementable in VLSI. It is a challenging task indeed to realize an ATM switch that can satisfy all these various requirements.

1.3.5 High-Speed Data Communication Technology

With increasing processing speeds and distributed processing, as well as the need for wide-area connection, LANs have been expanding their speeds and geographical coverage to establish MANs and WANs and other high-speed LANs. As a consequence, the means for high-speed data transmission and wide-area networking have been the main subjects of research and development in data communication.

The related network technologies have been in progress in various areas such as FDDI, IEEE high-speed LANs, DQDB, SMDS, data service over ATM networks, high-speed and real-time communication protocols, and the Internet.

FDDI and IEEE high-speed LANs contrast in several aspects: The FDDI LAN was promoted by ANSI in the early 1980s and has been widely used, but IEEE high-speed LANs were standardized by the IEEE in the early 1990s. An FDDI LAN provides 100-Mbps transmission capacity using optical fiber, but IEEE high-speed LANs provide 100-Mbps capacity using the existing low-speed LAN equipment and facilities. The IEEE's approach can be categorized into the IEEE 802.3 Committee's conventional MAC-based approach and the IEEE 802.12 Committee's new MAC-based approaches. In either case, a data capacity comparable to that of FDDI will be offered over the copper cables.

DQDB is the IEEE 802.6 MAN standard. It has a dual-bus structure and features a distributed queue that depends on the reservation scheme. It transmits data in 53-byte slot units, which is the same as the ATM cell, and it thus has the advantage of easy interconnection with the BISDN. DQDB can provide three types of services: connectionless data services, connection-oriented data services, and isochronous services. SMDS is a packet service standard that is designed to provide connectionless packet-switched data among distant high-performance computers, LANs, and MANs over the public network. It was developed by Bellcore in 1989, and its basic framework of protocol is the same as that of the DQDB protocol.

As ATM technology becomes practical, various attempts have been made to utilize the ATM network for data services in conjunction with the existing local-area data networks. In contrast to most existing networks that provide connectionless data services, the ATM network is connection oriented. So, the bottleneck of high-speed data services in the existing network, which originates from the shared-media interconnection, can be resolved in the ATM network-based data services. Among various possible arrangements for ATM-based data services, the CLSF, IP over ATM, and LAN emulation methods have been recognized as most promising.

Newly emerging communication networks offer a wide bandwidth with low error rate and high reliability. Network services have been evolving toward real-time multimedia services such as video conferencing, distributed processing, and virtual reality, which require highly advanced communication functions such as multicast, synchronization, and cell control. Existing transport protocols such as TCP and OSI TP, however, are not suitable for this new communication environment. Therefore, various high-speed transport protocols and real-time protocols have been studied. Among the multitude of proposed protocols, the high-speed protocols of XTP and TP + + and the real-time protocols of RSVP and RTP have been comparatively well received.

The Internet is the network formed by connecting computer networks through the TCP/IP protocols. The Internet was initiated by ARPANET in the early 1980s and is coordinated by ISOC, which is technically sponsored by IETF for development of new standards. As the Internet's speed becomes faster, as the number of users explodes, and as the multimedia services become available, the existing TCP/IP has begun to face limitations, so a new generation Internet protocol, namely IPng, has been studied since early 1990s. With the standardization of IPng nearly completed, other high-level protocols that can support high-speed real-time multimedia communication services have been also studied. Typical examples are the RSVP, RTP, and RTCP.

1.3.6 Broadband Video Communication Technology

Usually, a broad bandwidth is required for the transmission of video signals; hence, the video signal compression and restoration technologies are essential for the effective use of the transmission line. So broadband video technology usually consists of the coding technology for compressing tremendous amounts of video information, the transmission technology for the efficient transmission of compressed video information, and the terminal technology integrating the first two.

Video compression technology adopts various video/image coding techniques to eliminate redundant information from an image signal within a certain range such that its characteristics do not change perceptibly. Image coding techniques can be categorized into prediction coding, transform coding, vector quantization, subband coding, and entropy coding. These coding techniques have been integrated in various video coding standards such as H.261, MPEG-1, and MPEG-2.

The MPEG-2 standard is the final product of years of lengthy standardization activities for coding of broadband video signals. It fully integrates existing image compression techniques, and is a generic coding algorithm that can be adapted to all specific video communication applications, including the HDTV signals. MPEG-2's full-featured capabilities including forward compatibility with MPEG-1, diverse coverage of bit rates, selectable resolutions and functionalities, adoption of a scalability concept, and accommodation of both interlaced and progressive scanning schemes.

For the economical provision of video communication services, which include audio and data signals, a communication network that can support versatile forms of information media and broadband capability is required. However, the existing PSTN or NISDN is not adequate for this, but an alternative can be found in ATM-based BISDN. The ATM transmission can offer various advantages over the PSTN or NISDN, including multimedia and broadband capabilities. Unfortunately, since ATM is a packet-based communication mode, the packet jitter and packet loss problems that are inherent to packet communications appear. In ATM-based packet video transmission, in particular, performance differs depending on the AAL types and the relevant forms of MPEG-2 packet streams.

Terminal technology has a close relationship with the attributes of the services to be provided. In designing terminal equipment that provides conversational video services, a number of options are available for the constituent functions such as video/audio processing functions, telematic and signaling functions, multimedia MUX/DMUX functions, and other transport functions. ITU-T's H.321 and H.310 recommend two most desirable sets of options for providing video communication services in the BISDN environment.

The HDTV, which will become the terminal used to receive high-definition video distribution services and may even evolve into high-resolution communication terminal equipment in the future, has finally settled down in digital form in the United States and Europe. In the early stages of HDTV development during the 1970s and 1980s, Japan led the development of analog-type HDTV, but in the 1990s, the United States has pioneered the development of digital HDTV technology. The digital HDTV uses MPEG-2 video coding standards and therefore renders a high-level of compatibility with the video signals for communications and digital storage.

Selected Bibliography

Acampora, A. S., *An Introduction to Broadband Networks: LANs, MANs, B-ISDN, and Optical Networks for Integrated Multimedia Telecommunications*, New York: Plenum Press, 1994.

Ambrosch, W. D., A. Maher, and B. Sasscer, *The Intelligent Network*, New York: Springer-Verlag, 1989.

Andrews, F. T., "The evolution of digital loop carrier," *IEEE Commun. Mag.*, March 1991, pp. 31–35.

Bell Telephone Laboratories, *Transmission Systems for Communications*, 5th ed., Bell Telephone Laboratories, 1982.

Bellamy, J., *Digital Telephony*, 2nd ed., New York: John Wiley and Sons, 1991.

Bellcore, *Telecommunications Transmission Engineering*, Vols. 1–3, 3rd ed., 1990.

Benedetto, S., E. Biglieri, and V. Castellani, *Digital Transmission Theory*, Englewood Cliffs, NJ: Prentice-Hall, 1987.

Berman, R. K., and J. H. Brewster, "Perspectives on the AIN architecture," *IEEE Commun. Mag.*, Vol. 30, No. 2, February 1992, pp. 27–33.

Bhargava, A., *Integrated Broadband Networks*, Norwood, MA: Norwood, MA: Artech House, 1991.

Black, U., *ATM: Foundation for Broadband Networks*, Englewood Cliffs, NJ: Prentice Hall, 1995.

Bocker, P., *ISDN*, New York: Springer-Verlag, 1988.

Bylanski, P., and D. G. W. Ingram, *Digital Transmission Systems*, London: Peter Peregrinus, 1980.

Calhoun, G., *Wireless Access and the Local Telephone Network*, Norwood, MA: Artech House, 1992.

Cheong, V. E., and R. A. Hirschheim, *Local Area Networks*, New York: John Wiley and Sons, 1985.

De Prycker, M., *Asynchronous Transfer Model: Solution for Broadband ISDN*, 2nd ed., Ellis Horwood, 1993.

Dicenet, G., *Design and Prospects for the ISDN*, Norwood, MA: Artech House, 1987.

Elbert, B. R., *Primitive Telecommunication Networks*, Norwood, MA: Artech House, 1988.

Fluckiger, F., *Understanding Networked Multimedia: Applications and Technology*, Englewood Cliffs, NJ: Prentice Hall, 1995.

Freeman, R. E., *Telecommunication Transmission Handbook*, 2nd ed., New York: John Wiley and Sons, 1981.

Goralski, W. J., *Introduction to ATM Networking*, New York: McGraw-Hill, 1995.

Handel, R., "Evolution of ISDN Towards Broadband ISDN," *IEEE Network*, Vol. 3, No. 1, January 1989, pp. 7–13.

Handel, P., and M. N. Huber, *Integrated Broadband Networks*, Reading, MA: Addison-Wesley, 1991.

Hawley, G. T., "Historical perspectives on the U.S. telephone loop," *IEEE Commun. Mag.*, March 1991, pp. 24–28.

Hui, J. Y., *Switching and Traffic Theory for Integrated Broadband Networks*, New York: Kluwer Academic Publishers, 1990.

Kreager, P. S., *Practical Aspects of Data Communications*, New York: McGraw-Hill, 1983.

Libois, L. J. (CNET), *Electronic Switching*, Amsterdam: North Holland, 1983.

McDonald, J. C., *Fundamentals of Digital Switching*, New York: Plenum Press, 1983.

McDysan, D. E., and D. L. Spohn, *ATM: Theory and Application*, New York: McGraw-Hill, 1994.

Miller, M. A., *Analyzing Broadband Networks*, M&T books, 1994.

Millman, S., *A History of Engineering and Science in the Bell System: Communication Science (1925–1980)*, AT&T Bell Laboratories, 1984.

Minoli, D., *Telecommunications Technology Handbook*, Norwood, MA: Artech House, 1991.

Murano, K., et al., "Technologies Towards Broadband ISDN," *IEEE Commun. Mag.*, April 1990, p. 66.

Noll, A. M., *Introduction to Telephone and Telephone Systems*, Norwood, MA: Artech House, 1986.

O'Neil, E. F., *A History of Engineering and Science in the Bell System: Transmission Technology (1925–1975)*, AT&T Bell Laboratories, 1985.

Owen, F. F. E., *PCM and Digital Transmission Systems*, New York: McGraw-Hill, 1982.

Partridge, C., *Gigabit Networking*, Reading, MA: Addison-Wesley, 1994.

Proakis, J. G., *Digital Communications*, New York: McGraw-Hill, 1983.

Schindller, Jr., G. E., *A History of Engineering and Science in the Bell Systems: Switching Technology (1925–1980)*, AT&T Bell Laboratories, 1984.

Schwartz, M., *Telecommunication Networks: Protocols, Modeling and Analysis*, Reading, MA: Addison Wesley, 1987.

Sklar, B., *Digital Communications*, Englewood Cliffs, NJ: Prentice Hall, 1988.

Spragins, J. D., J. L. Hammond, and K. Paulikowski, *Telecommunications: Protocols and Design*, Reading, MA: Addison Wesley, 1991.

Stallings, W., *Tutorial: Computer Communications: Architecture, Protocols, and Standards*, 2nd ed., Los Alamitos, CA: IEEE Computer Society Press, 1987.

Stallings, W., *Tutorial: Integrated Services Digital Networks (ISDN)*, 2nd ed., Los Alamitos, CA: IEEE Computer Society Press, 1988.

Stallings, W., *Tutorial: Local Network Technology*, 3rd ed., Los Alamitos, CA: IEEE Computer Society Press, 1988.

Stallings, W., *Advances in ISDN and Broadband ISDN*, Los Alamitos, CA: IEEE Computer Society Press, 1992.

Stallings, W., *Data and Computer Communications*, New York: Macmillan, 1994.

Steinmetz, R., and K. Nahrstedt, *Multimedia: Computing, Communications and Applications*, Englewood Cliffs, NJ: Prentice Hall, 1995.

Tanenbaum, A. S., *Computer Networks*, Englewood Cliffs, NJ: Prentice Hall, 1989.

Toda, I., "Migrations to broadband ISDN," *IEEE Commun. Mag.*, Vol. 28, No. 4, April 1990, pp. 55–59.

Weinstein, S., "Telecommunications in the coming decades," *IEEE Spectrum*, November 1987, p. 64.

Broadband Subscriber Network 2

The subscriber network is a domain that has been dormant until recently in the long history of the evolution of communication networks. Copper-based twisted-pair wire was deployed more than a century ago in the beginning of the telephone network evolution, and is still widely used in the midst of an abundance of advanced physical media. This continued use of twisted-pair demonstrates well how slow the growth of information generation and transmission has been until recently and how dominant the telephony services have been throughout the past century. On the other side, this, in turn, signifies how rapidly the information services have grown in recent years.

Spurred by the rapid increase of information exchange and the accelerated growth of video, data, and other multimedia services, the trunk network has undergone drastic changes during the past decade. Physical transport network has been reconstructed on the fiber-based SDH/SONET systems, and the ATM network has been getting ready for deployment over them. With the infrastructure in the trunk network getting near completion, the subscriber network has been drawing ever more attention.

In fact, the subscriber network is the bridge that connects subscribers to the service providers. The information services demand has evolved to include the conversational telecommunication services, broadband video distribution services, and interactive data services, which originally stemmed from the use of telephones, televisions, and computers, respectively. The distinctive nature of the new services is that they require large bandwidth and high interactivity. The existing subscriber network cannot accommodate such services. Therefore, the time is ripe for activating the dormant subscriber network and for applying advanced communication technologies to accommodate interactive multimedia services.

A number of technologies have been developed during the past decade that have aided the evolution of the subscriber network. They can be divided into four categories: twisted-pair-based enhancements, coaxial-cable-based improvements, optical-fiber-based renovations, and wireless-based supplements.

The twisted-pair-based enhancements, which include DSL, HDSL, ADSL, and VDSL, exhibit significant bandwidth increase by employing advanced modulation technologies. The coaxial-cable-based improvements, namely, the HFC, can provide interactivity with the existing CATV network. The optical-fiber-based renovations, being carried out in three stages, FTTO, FTTC, and FTTH, will eventually lead to an optical subscriber network. The wireless-based supplements have been developed with narrowband services in mind, but progress has been made in the broadband video distribution services too.

The evolution and technologies of the subscriber network are examined in this chapter. The four approaches mentioned are discussed one by one, and optical device technology is also introduced in support of the optical-fiber-based approach. In the final section, a full-service subscriber network that is expected to provide integrated subscriber access to telecommunication, video distribution, and interactive data services is investigated.

2.1 INTRODUCTION

The subscriber loop is the transmission and signaling facility that connects the subscriber terminal equipment to the *central office* (CO). The subscriber loop provides channels to transfer service information and signaling between the terminal equipment at the subscriber premises and the terminal equipment in the CO. In addition, the subscriber loop often feeds power to terminal equipment in the subscriber premises.

The subscriber loop is the bridge to communication services for subscribers so it is important, in this connection, to improve and fully utilize it for the benefit of improved subscriber services. This introductory section discusses the characteristics of the subscriber network and its evolution toward broadband service access, in support of the diversified discussions to follow in later sections.

2.1.1 Characteristics of Subscriber Lines

The subscriber loop must meet the requirements for the particular transmission service, and, in addition, it must accommodate signaling and control requirements related to the transmission. Therefore the subscriber loop may require special treatments of service signals depending on the physical characteristics, distance from the serving CO, and the type of communication service.

The subscriber loop forms a *subscriber network* with a star topology centered around the CO. In this sense, use of the terms *subscriber loop* and *subscriber network* may be intermixed. The subscriber network can be divided into *feeder* and *distribution* networks, with the junction points called *remote nodes* (RNs). The feeder consists of bundles of subscriber lines with each bundle containing 500 to 1,000 subscriber lines. The bundles are divided into small-size bundles consisting of about 50 pairs of lines to form the distribution network. The small-size bundles are finally split into each individual pair of lines called a *drop* to reach individual subscriber's premises. The overall subscriber network formed in this manner uses a star topology.

In case the feeder is a simple aggregate of subscriber lines to be split in bundles in the RN, the RN functions like a passive device. But if the multitude of lines in the feeder gets multiplexed into a high-speed carrier, then the feeder plant changes to a loop carrier network, with the RN turning into a loop carrier system including switching and concentration functions. The loop carrier systems may become a *digital loop carrier* (DLC) system, or may further evolve to a *fiber loop carrier* (FLC) system. The distribution area centered around the RN where the subscribers can get communication services through the carrier systems is called the *carrier serving area* (CSA). A CSA, in general, has a maximum radius of 3.6 km and a maximum number of the subscribers of 3,000.

The subscriber network distinguishes itself from the trunk network in various ways. First, the number of subscriber lines is as large as the number of the subscribers, which amounts to about 700 million worldwide. As such, the economic aspect of the subscriber network is very important. The readily invested asset is enormous, and any new deployments or changes in the network will be equally costly.

Second, the *twisted-pair* (TP) copper wire, which constitutes more than 90% of the existing subscriber network, has limited bandwidth and transmission distance. Transmission bandwidth has been limited to the voice-band signals, which is less than 4 kHz, and the transmission distance is limited to 5 to 10 km.

Third, since the subscriber loop, in most cases, has been used solely for *plain ordinary telephone services* (POTS) so far, it has room for enhancement to accommodate various multimedia services. The subscriber loop has been used mainly for voice communications, not because the subscriber loop was unable to support other services but because other services were not widely available or not technically applicable. So the emergence of multimedia services will fully exploit enhanced use of the subscriber loop.

Fourth, the 4-kHz bandwidth is what the subscriber loop environment can maintain without applying any modulation means. It has room for significant improvements. In fact, it turned out in recent year that the amount of information that can be carried by the subscriber loop can go up from the conventional rate of 64 Kbps to 6 to 8 Mbps by applying advanced modulation techniques, as discussed in Section 2.2.

2.1.2 Evolution of the Subscriber Loop

Recent advances in multimedia technology has fostered research and development of the subscriber network technology and has motivated its evolution.

The twisted-pair (TP) subscriber loop, as it is, cannot transport the high-speed data of several megabytes per second that is required by broadband video services. There are two ways to resolve the problem: One is to apply advanced modulation techniques to the TP-based transmission, and the other is to replace the TP-based subscriber loop with optical fibers.

The TP-based enhancement takes advantage of advanced modulation techniques such as QAM, CAP, DMT and DWMT. The resulting subscriber loops are in the form of HDSL, ADSL, and VDSL, but their application may differ depending on the use and transmission distance. The TP-based enhancement does not require any new deployments or changes to the loop facility other than the attachment of the corresponding converters, and therefore it is perceived as the first stage of evolution of the existing subscriber loop.

Optical-fiber-based evolution can be realized in multiple stages such as FTTO, FTTC, and FTTH. In the FTTC, copper wire in the feeder and distribution networks is replaced with optical fiber, with the final drop network left unchanged. Another comparable approach can be found in the evolution of the community antenna television (CATV) network, called the HFC, in which the existing coaxial CATV network is enhanced with bidirectional capability and expanded bandwidth for interactive digital services. Both FTTC and HFC are recognized as comparable methods in various aspects, and are regarded as the second stage of evolution toward the eventual opticalization of the subscriber network, FTTH.

Aside from the wireline-based improvements that can be made to the subscriber network, a wireless-based renovation has also been considered. Wireless access has the fundamental limitation of bandwidth and information security. However, the wireless subscriber network can be an economical alternative in certain environments and can render terminal mobility and network tractability. Therefore, the wireless subscriber network is expected to evolve to complement the advanced TP-based and fiber-based subscriber network. (Refer to Section 2.6 for a more detailed discussion on wireless subscriber loop technologies.)

2.1.3 Twisted-Pair-Based Bandwidth Expansion

There are various ways to expand the bandwidth of the TP subscriber line, each of which has different applications with different distances and bandwidth allowances. The simplest form of duplex subscriber line is called a *digital subscriber line* (DSL).

The ISDN standardization has brought about a new modulation technique called *two-binary one-quaternary* (2B1Q), which enabled TP based transmission of the 144-Kbps 2B + D *basic rate access* (BRA). The DSL can expand the TP transmission capacity almost threefold such that, for example, the POTS (1B), facsimile (1B), and data (1D) services can be provided concurrently. To make such bandwidth expansion available, however, the existing terminal equipment has to be replaced with the ISDN terminal equipment or a *terminal adapter* (TA) has to be attached.

The bandwidth of the TP-based subscriber line significantly increased with the introduction of the *high-speed digital subscriber lines* (HDSL), which enabled 784-Kbps duplex transmission over one twisted-pair. The transmission capacity was further increased to 1.5-Mbps duplex transmission over two pairs of TP. One-pair-based solution to these transmission rates will become available very soon. The HDSL has been used mainly for T1 (or E1) transmission in the feeder network, and also for service access in the CSA.

In consideration of the asymmetric character of the video services in the subscriber network [a typical example of which is the *video-on-demand* (VOD) service], *asymmetric digital subscriber line* (ADSL) has emerged in recent years. The ADSL is basically the same as the HDSL but its downstream has much larger transmission capacity than its upstream. In the early stage of ASDL evolution, 1.5-Mbps downstream and 16-Kbps upstream transmission capacities were allocated, but the downstream capacity has been upgraded to 1.5 Mbps for a 5.4-km distance and 6 Mbps for a 3.6-km distance over 0.5 mm (i.e., 24 gauge) and the upstream capacity has been upgraded to 384 Kbps. Such significant bandwidth expansions are achieved by advanced modulation techniques such as *carrierless amplitude-modulation and phase-modulation* (CAP), and *discrete multitone* (DMT).

In accordance with the spread of optical fiber deployment in the subscriber network, the territory for the TP or coaxial copper lines shrinks to the distribution network or to the CSAs. The shortened transmission distance, in this case, can help to further increase the transmission rate. Such consideration yields the short-distance version, or the large-capacity version, of the ADSL, which is called the *very-high-speed digital subscriber line* (VDSL). Transmission capacities of 13 to 52 Mbps downstream and 1.5 of 2.3 Mbps upstream are expected for transmission distances of 300 to 1,500 m for the VDSL. (See Section 2.2 for a more detailed discussion of various TP-based technologies.)

2.1.4 Optical-Fiber-Based Evolution

As discussed earlier, the subscriber network occupies the major portion of investment due to its fundamental character of point-to-point connection to individual subscribers. So the enhancement of the subscriber loop is directly tied to the investment strategy, and the "chicken-and-egg" problem gets involved in the process of investment and demand considerations. This is why phased enhancement of the subscriber network and phased deployment of fiber in the subscriber loop are inevitable.

As such, the strategies to deploy the optical subscriber network differ among countries depending on network environments. Nonetheless, a common trend is found in the FTTO-FTTC (or FTTZ) -FTTH evolution scenario.

Fiber-to-the-office (FTTO), which intends to make the optical subscriber lines serving large buildings in the business district, has progressed to the mature level in many leading countries. In most cases, optical fiber deployment is done in new buildings first, with gradual replacement occurring in existing buildings of the congested copper-based lines. For FTTO, *central office terminals* (COTs) and *remote terminals* (RTs) are installed in the CO and the buildings, respectively, and optical fibers are deployed to connect them in star or ring topologies. Recently, FTTO has been further spread toward smaller sized buildings and apartment complexes due to decreased fiber prices and increased bandwidth demand.

Fiber-to-the-curb (FTTC) is the strategy used to expand fiber deployment to the vicinity of subscribers and to small- or medium-sized buildings in the business distinct. Since the full deployment of fiber is too often costly, whereas the forecast demand for broadband services is still low, FTTC renders a compromised intermediate stage of evolution. Further, the evolution of high-performance TP-based digital subscriber lines such as VDSL supports the promotion of FTTC. For FTTC, *optical network units* (ONUs) are installed in the neighborhood of a group of subscribers, and optical fiber is placed between the CO and the ONU, with the remaining final drop left for the TP-based subscriber lines. (See Section 2.4 for a more detailed discussion on FTTC.)

Fiber-to-the-home (FTTH) may be regarded as the ultimate goal of subscriber network evolution. In fact, FTTH is necessary for virtually unlimited broadband services in the BISDN. Although the economical factor is the biggest barrier on the road toward FTTH, the FTTO-FTTC-FTTH phased approach will possibly render an economical compromise. The *hybrid fiber coax* (HFC) can also contribute to the evolution of the CATV network toward the FTTH by making the optical connection from the *head end* (HE) terminal to the last amplifier in the neighborhood of the subscriber with the remaining distribution network left for the existing coaxial cable. (See Section 2.3 for a more detailed discussion on HFC.) Therefore, FTTH can eventually render a solid foundation for a complete integration of the networks and services for telecommunications, data communications, and broadcast, which have been independently developed in the past.

2.2 TWISTED-PAIR-BASED SUBSCRIBER NETWORK

More than 90% of the existing 700 million subscriber lines worldwide are twisted-pairs, and TPs will continue to be the absolute majority even by the year 2000 when the number of subscriber lines is expected to grow to about 900 million. This illustrates well how important it is to maximize the use of the TP subscriber lines.

Some early technologies already provided the means to connect more than one telephone to a single TP wire, which may be the conceptual starting point of bandwidth expansion of the TP subscriber lines. The emergence of ISDN in the 1980s demonstrated that a TP can transport much more than one voice channel, and the resulting DSL may be regarded as the practical starting point of technology-based bandwidth expansion. The modulation technique that enabled such expansion of bandwidth was further developed to yield various advanced digital subscriber lines such as HDSL, ADSL, and VDSL. The transmission bit rates and distances of this high-speed subscriber lines are as listed in Table 2.1.

In the following subsections the technologies and characteristics of those digital subscriber lines are examined one by one, then the modulation techniques that enable them to be functional are introduced.

2.2.1 DSL

The DSL became available with the introduction of ISDN in the 1980s, which was the standardized embodiment of the concept of access integration of multiple different services. The integration was rooted on the digital domain in which every different service signal takes the same external form of digital bit streams.

In ISDN, access of the subscriber to the CO was made possible in two different ways: via *base rate access* (BRA) and *primary rate access* (PRA). BRA has the 2B + D channels at 144 Kbps, which goes up to 160 Kbps with the inclusion of a 16-Kbps maintenance channel. The PRA has 23B + D (or 30B + D) channels to be transmitted at the T1 (or E1) rate of 1.544 Mbps (or 2.048 Mbps).

Among the 2B + D channels of the BRA, the B channel is a 64-Kbps channel

Table 2.1
Comparison of Digital Subscriber Line Technologies

Technology	Bit Rates (Mbps)	Distances* (km)
DSL	0.16	5.4
HDSL	1.5–2	4
ADSL	1.5–2	5.4
	6	3.6
VDSL†	13–14	1.5
	26–28	1
	52–56	0.3

*Distance for 0.5-mm (24-gauge) TP wires.
† Estimated values.

that can be used for voice or data, and the D channel is a 16-Kbps channel that can be used for call-control messages or packet data. The BRA provides the subscriber's direct connection to the CO, and PRA is for the connections of PBXs, host computers, and LANs.

For the DSL that supports the 2B + D BRA, ISDN terminals (i.e., TE1) or a terminal adapter is necessary on the subscriber premises to generate the multiplexed 144-Kbps bit stream. The final bit rate of 160 Kbps including the maintenance bits reduces to the transmission rate of 80 Ksps (kilo symbol per second) when modulated by the 2B1Q line coding, and this DSL signal can travel over 5.4 km of distance over 0.5-mm (i.e., 24-gauge) TPs.

The modulation techniques used for the transmission of the 160-Kbps BRA signal over TP subscriber lines were 2B1Q, 4B3T, and *time compression multiplexing* (TCM). Among them 2B1Q and 4B3T are normal line coding (or modulation) techniques, but TCM is the time-shared duplex transmission technique that utilizes a single channel for transmission of two-way data streams in a "ping-pong" manner. The TCM has the advantage that it does not require an echo canceller but it requires twice the transmission bandwidth, which, in turn, reduces the transmission distance. (Refer to Section 2.2.5 for more details on 2B1Q.)

2.2.2 HDSL

HDSL is a digital subscriber line technology that can deliver symmetric T1 (or E1) rate digital channels over TP subscriber lines. HDSL is a proven technology with more than a quarter million systems in operation worldwide. It is referred to as a "plug-and-play" system because it can be installed quickly and used immediately. HDSL applies advanced electronics to existing TP wires to transform them into high-speed transmission lines. It is an outgrowth of the 2B1Q coding, which enables the 784-Kbps rate transmission of the 1.544-Mbps T1 rate data.

T1 (or E1) lines are the traditional transmission lines carrying the fundamental digital signal DS-1 (or DS-1E), which is an aggregate of 24 (or 30) DS-0 user data at 64 Kbps each and overhead channels, of the transmission rate 1.544 Mbps (or 2.048 Mbps) in North America (or Europe). T1 (or E1) was originally deployed in the interoffice trunk, and later began to be deployed in the feeder plant. Since then T1 (or E1) has grown to be the major way to feed DLC systems in the remote node, which concentrates multiple subscriber lines for connection to the CO. In the case of T1, the transmission signal is *alternate mark inversion* (AMI) coded, which takes 1.5-MHz bandwidth with a signal peak occurring at 750 kHz. Due to this inefficient frequency spectrum arrangement, a repeater is required at the first 900-m (i.e., 3,000-ft) position from the CO repeater or from the subscriber equipment, and also every 1.8 km (i.e., 6,000 ft) thereafter.

Overall transmission distance of T1 (or E1) is 5.4 km (or 4.8 km) over the repeatered 0.5-mm (i.e., 24-gauge) TP wire [see Figure 2.1(a)].

HDSL can transmit the T1 (or E1) rate signals at distances up to 4 km over 0.5-mm TP wires without using repeaters, and the transmission distance can be increased if repeaters are installed. For the T1 rate (i.e., 1.5-Mbps) transmission, two TP wires are necessary; and for the E1 rate (i.e., 2.0-Mbps) transmission, three TP wires are necessary. Each TP wire, in this case, carries a 2B1Q modulated 784-Kbps rate signal [see Figure 2.1(b)].

HDSL can transmit signals at normal power levels, and can maintain or restore signal integrity even if TP wires are not perfect. This becomes possible because the included equalizer can compensate for the TP signal distortion. This equalization function works in such a way that the transmitted signal can be restored in the receiver regardless of changes in environmental conditions. Thanks to such advanced electronics, HDSL does not require cable conditioning or bridged tap removal, and becomes immune to crosstalk and polarity reversal. Because repeaters are completely eliminated, overall system reliability and transmission performance improve.

HDSL is comparable to the T1 (or E1) line in every aspect, except that repeaters are not needed in the HDSL case [see Figures 2.1(a) and (b)]. This indicates that HDSL is appropriate for use in place of a T1 (or E1) line, or in applications where symmetric high-speed channels are necessary. Therefore, HDSL can be used as feeder lines or as business subscriber lines located within 4 km from the remote node or CO.

HDSL is beneficial to business users in public as well as in private network users. It can be used as an extension for fiber connection to the public network, or as a dedicated private line for LAN interconnection. By using HDSL instead of leasing the T1 (or E1) line in the MAN, for example, the participating organizations can get high-speed digital communication channels in support of various broadband multimedia services at reduced cost and under flexible management.

2.2.3 ADSL

ADSL is a digital subscriber loop technology that has been devised in consideration of the asymmetric nature of the broadband multimedia services to be provided to residential subscribers. It can provide high-speed asymmetric transmission channels of 1.5 to 6 Mbps (or 2 to 8 Mbps for E1) downstream and upstream transmission rates of 16 to 640 Kbps. It can transmit these rates over a single pair of TP subscriber lines at distances up to 5.4 km without repeaters. All these properties indicate that ADSL is adequate for providing broadband multimedia services including VOD or *video-dial-tone* (VDT) to residential areas.

The ADSL system consists of an ADSL unit at both ends of the TP wire

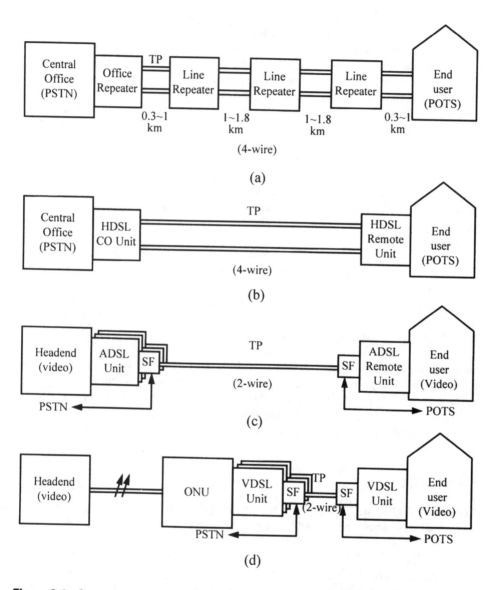

Figure 2.1 Comparison of digital subscriber loops: (a) T1/EI; (b) HDSL; (c) ADSL; and (d) VDSL.

that modulates and demodulates three information channels for transmission: a high-speed downstream channel, a medium-speed duplex channel, and a POTS channel [see Figure 2.1(c)]. The POTS channel is split off from the digital modem by using *split filters* (SFs) such that POTS can be provided uninterrupted even when the ADSL fails. The high-speed channel has bit rates of 1.5 to 6.0 Mbps, and the duplex channel has a rate of 16 to 640 Kbps.

ADSL units support data rates conforming to the North American (or European) PDHs: It provides 1.5-Mbps T1 (or 2.0-Mbps E1) downstream transmission and 16-Kbps duplex channels on the low end, as the minimum configurations, and can provide 6-Mbps T2 (or 8-Mbps E2) downstream transmission and 64-Kbps duplex channels on the other end. ADSL units supporting 8-Mbps downstream transmission and up to 64-Kbps duplex channels may also be available soon. Further, ADSL units that can accommodate ATM transport with variable bit rates may become available as market demand matures.

The data rate of the high-speed downstream channel varies depending on a number of factors, among which the length of the TP wire and the wire diameter are the most important. In general, transmission attenuation increases with wire length and signal frequency, and decreases with wire diameter. For example, the 1.5- or 2.0-Mbps channel can reach a distance of 5.4 km over a TP wire of diameter 0.5 mm, but can reach a distance of 4.6 km over 0.4-mm TP wire. Likewise, the service distance for the 6-Mbps channels is reduced from 3.6 to 2.7 km when the wire diameter decreases from 0.5 to 0.4 mm.

ADSL relies heavily on modem technology as well as signal processing technology because these technologies enable enormous amounts of data to be squeezed into the TP wire. In creating multiple channels for different uses, ADSL applies either the FDM or echo-canceling technique. With either technique ADSL can split off the 4-kHz low-frequency band for POTS. The ADSL unit attaches an error correction code to each data block, which is an aggregate of data streams created by multiplexing downstream channels, duplex channels, and maintenance channels. This code is used by the receiving ADSL unit in correcting the intervening errors in the received blocks during transmission. The modulation techniques used for ADSL are DMT and CAP.

Compared with HDSL, ADSL has properties that are desirable for residential connection: It can operate on one TP wire, whereas HDSL requires two TP wires. It reaches a distance that can encompass almost all subscriber loop areas, whereas HDSL reaches rather limited distances. Its transmission distance is longer than that of HDSL because it is free from self *near-end crosstalk* (NEXT), which is the limiting impairment of the bidirectional system, but, instead, the distance is subject to signal loss. Its asymmetric nature of information flow also better fits the asymmetric residential service environment than the business district where the symmetric data flow of HDSL is more desirable.

2.2.4 VDSL

As the telephone companies continue to deploy optical fibers in the subscriber loop near the subscriber's premises but not quite all the way (the last drop of 100 to 1,000m is still TP wires), high-speed distribution over the short-distance TP wires has become an important issue. VDSL is the short-distance high-speed digital subscriber line that has become the answer to this last mile of required connection distance. In this sense VDSL is the TP subscriber line technology that complements the FTTC network for the connection from ONU to the subscriber's premises [see Figure 2.1(d)]. In fact, VDSL can help to increase the service capability of ONU from the 8 to 32 subscribers today to 100 to 200 subscribers, thereby achieving considerable cost savings.

VDSL is still in its definition stage.[1] The upstream and downstream transmission bit rates have not yet been determined, but are expected to be asymmetric as for the case of ADSL. The bit rates will vary depending on the transmission distance: for example, 52 to 56 Mbps, up to 300 m; 26 to 28 Mbps, up to 1 km; and 13 to 14 Mbps, up to 1.5 km downstream; and 1.6 to 2.3 Mbps, up to 300m upstream. The downstream transmission rate of 52 Mbps matches the rate of SONET STS-1, or a third the rate of SDH STM-1, which indicates that ATM cell transmission can reach the subscriber premises over the TP pairs. Note that the bit rates for other distances are all submultiples of this SONET rate.

The asymmetric data channels that carry broadband multimedia services will be separated in frequency from symmetric bidirectional service channels such as POTS and ISDN channels. The frequency band can be allocated, for example, such that the ISDN signal occupies the lowest 80-kHz band (including the lowest 4-kHz band for the POTS signal), with the upstream and downstream data, respectively, occupying the 300- to 700-kHz and beyond 1-MHz bands. This arrangement enables service providers to overlay VDSL on the existing services. The POTS channel can be arranged to split off at the entrance to the VDSL unit, as was the case for ADSL [see Figure 2.1(d)].

For the connection between the VSDL unit and multiple subscriber terminal equipment, the subscriber premises network may form a star network with each *customer premises equipment* (CPE) connected to a switching or multiplexing hub that is directly connected to, or integrated with, the VDSL unit. Aside from this active mode termination, a passive termination may be also possible in which a VDSL is attached to, or integrated with, each individual CPE. In this case an appropriate MAC scheme is necessary to effectively share the common TP channel among all CPEs.

---

1. Several standards organizations or voluntary groups, such as ANSI T1E1.4, ETSI, DAVIC, ATM Forum, and ADSL Forum, study the VDSL issue.

VDSL is comparable to ADSL in various aspects. It offers asymmetric data services to subscribers. The offered data rate is significantly higher than that of ADSL, which, however, is due to the shortened transmission distance. So the advanced modulation technologies that enabled the high-speed transmission of ADSL, such as DMT and CAP, will be equally applicable to VDSL. Cost and power will be lower for VDSL because the service coverage is smaller. After all, the distinctions of the VDSL stem primarily from the shortened distance factor, and the involved technologies are essentially the same. Therefore, VDSL may be considered as a variation of the ADSL.

2.2.5 Modulation Techniques for DSLs

In the past the TP subscriber line was commonly perceived as a small-bandwidth channel that could support 4-kHz POTS only. However, the emergence of a series of DSLs (DSL, HDSL, ADSL, and VDSL) has changed this old notion. For example, the ADSL with 6-MHz downstream transmission can support services whose bandwidth is 100 times that of POTS. This tremendous increase in bandwidth has become possible owing to various communication and signal processing technologies, among which modulation techniques may be singled out as the most contributing technology.

For the many different types of modulation techniques, the performance changes depending on the application environment. For the various DSLs, the 2B1Q, DMT, and CAP modulation techniques have proved to be most useful. The 2B1Q method has been applied to DSL and HDSL, and the other two have been applied to ADSL and may be applied to VDSL also. In the following, 2B1Q, DMT, and CAP techniques are examined one by one.

2B1Q

The 2B1Q line coding converts a pair of binary data into one quaternary symbol among the values -3, -1, 1, and 3. In the conversion, the first binary data of the pair is interpreted as the sign (i.e., $-$ for "0" and $+$ for "1"), and the second binary data is interpreted as the level (i.e., 3 for "0" and 1 for "1"), as shown in Table 2.2. For example, the binary data stream "0110110010110001" is converted to the symbol stream $-1, +3, +1, -3, +3, +1, -3, -1$.

The 2B1Q line coding method has several advantages over other existing line coding techniques. First, the symbol rate reduces by half after line coding, so the 160-Kbps BRA signal can be transmitted over the 80-kHz bandwidth. Also, 2B1Q was experimentally proved to outperform the existing MMS43 or AMI line codes in enduring the NEXT problem. In addition, such multilevel line coding is comparatively simple for VLSI implementation. However, the 2B1Q coding has the drawback that it can contain a DC component.

Table 2.2
2B1Q Symbol Coding

1st Bit	2nd Bit	Quarternary Symbol
1	0	+3
1	1	+1
0	1	−1
0	0	−3

DMT

DMT is a modulation technique that belongs to a general class called *multicarrier modulation* (MCM), which divides the data stream into multiple data blocks, and then modulates each data block to different subcarriers.

The transmitter and receiver structures of DMT are shown in Figure 2.2. The input bit stream is divided in blocks of b bits each and then stored in buffer. The b bits are then divided among N subchannels with the allocated bit number differing depending on the SNR characteristics of each subchannel. The allocated bits of each subchannel are encoded to the corresponding QAM symbol on the QAM constellation. These functions are performed by the first block of Figure 2.2(a). The collection of N QAM symbols is then modulated by passing through the *inverse discrete Fourier transform* (IDFT) process in the second block. Real signals are taken from the IDFT output, parallel-to-serial converted,

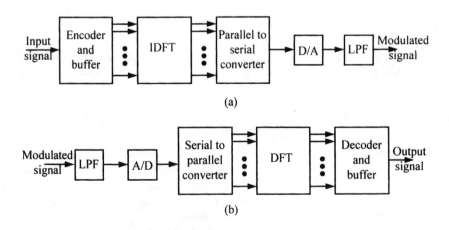

(a)

(b)

Figure 2.2   DMT tranceiver structure: (a) transmitter and (b) receiver.

D/A-converted, and finally *lowpass-filtered* (LPF) before transmission. The demodulation process of Figure 2.2(b) is the reverse process of the above.

DWMT is a variation of DMT in which the wavelet transform is used instead of the Fourier transform.

DMT has several advantages over other modulation techniques. First, frequency efficiency is good because the subcarrier frequencies can be densely aligned on the frequency spectrum. In other words, the frequency efficiency can be enhanced close to the theoretical limit. Additionally, as the number of subcarriers increases, the whole signal spectrum assumes a rectangle-like shape, and the sideband becomes very small, which contributes to decreasing the *adjacent channel interference* (ACI). Second, the overall frequency spectrum becomes flat within each subchannel and the sideband becomes very small as the number of subcarriers increases. Therefore, *intersymbol interference* (ISI) is minimized, and hence an equalizer is not necessary. But, in reality, the number of subchannels is limited to several hundred in consideration of design complexity, and thus simple equalizers are employed that are subchannel based. Third, since the symbol length per subchannel becomes longer than in the case for the single carrier system, the DMT system is robust to temporal interferences such as echo and ISI. Fourth, wideband noise such as impluse noise becomes less influential as its energy is distributed all over the subchannels. On the other hand, since data are processed in several hundred separate subchannels, the overall system becomes very complex and the timing and synchronization become critical factors.

CAP

CAP is a two-dimensional passband modulation technique that was originally derived from the closely related QAM. CAP has the same special characterictics and the same theoretical performance as QAM, but is in general less complex to implement digitally.

The transmitter and receiver structure of CAP is shown in Figure 2.3. The encoder does multidimensional encoding on the input data stream. That is, the input data stream is blocked in units of m bits and the m bits are then mapped into several separate symbol streams, which is two in the case of Figure 2.3 (a). This multidimensional encoding allows for using a fractional number of symbols. For example, in the case of 32-CAP, which has 32-point constellations with each point representing a different block of 5 bits, a two-dimensional coding yields 2.5-bit assignment for each symbol. The symbol streams then pass through independent filtering processes, for example, the in-phase and quardrature-phase passband shaping filters as shown in the figure. The filtered streams are added (or subtracted), D/A-converted, and lowpass-filtered before transmission. In the receiver, the reverse processing is performed, and the decision device [shown in Figure 2.3(b)] makes decisions on the symbol values.

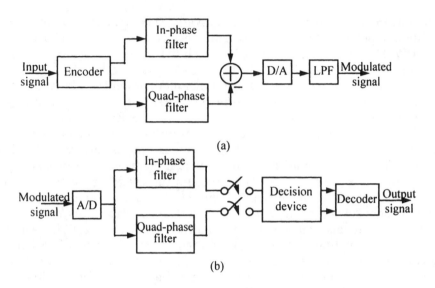

Figure 2.3   CAP tranceiver structure: (a) transmitter and (b) receiver.

CAP is particularly well suited for all-digital implementations in the applications where the bandwidth of the signal spectrum is of the same order of magnitude as the center frequency of the spectrum. CAP is comparable to QAM in spectral characteristics and theoretical performance, but is more amendable to digital implementation.

2.3 HYBRID FIBER COAXIAL CABLE

CATV networks have long been deployed solely for the distribution of television programs to the subscribers and the medium of transmission has been mainly coaxial cables. Present CATV networks are therefore optimized for unidirectional transmission of distribution services. However, as the demand for broadband and interactive services increased recently, CATV networks have been evolving to accommodate the demand. Since accommodation of broadband services was in the interest of the telephone subscriber network also, network evolution has progressed in a competitive environment.

Among several possible CATV network architectures that can provide interactive services to subscribers, a hybrid architecture that deployed optical fiber to the proximity of subscribers and utilized the existing coaxial cable networks for the remaining subscriber side appeared to be promising. Therefore, HFC was adopted as the first goal of CATV network evolution, with new interactive services overlaid on the existing CATV services.

In this section, HFC topics are discussed. The conventional CATV network is reviewed first, and then various aspects of the bidirectional HFC network are discussed. Finally, a digital CATV network is briefly considered.

2.3.1 Conventional CATV Network

The conventional CATV network basically consists of tree and branch connections of coaxial cables with the headend and subscriber's equipment positioned at both ends [see Figure 2.4(a)]. Coaxial cable in the trunks connecting the

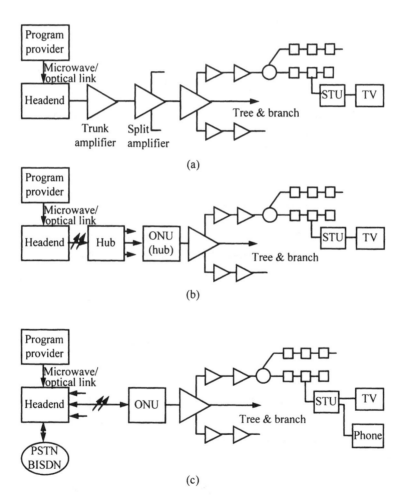

Figure 2.4 Architectures of CATV network: (a) all coaxial cable; (b) fiber and coaxial cable; and (c) hybrid fiber/coax (bidirectional).

headend to hubs can be replaced with fiber trunks, but the same analog CATV channel signals are carried over the optical fiber [see Figure 2.4(b)].

The headend receives programs from program providers through microwave links or optical fiber links, and then sends them to the primary hub through fiber, where the CATV signal is transmitted optically. Depending on how the devices are equipped, the headend can directly split the CATV signals optically and then transmit to multiple ONUs. Beyond the ONUs the CATV signals are distributed over tree and branch coaxial cables.

The ONU is the optical termination in the neighborhood of the subscriber, and is normally installed on the pole, in an on-ground hut, or in an underground manhole. It not only converts the optical signal to an electrical signal, but also feeds power to various amplifiers on the tree and branch coaxial cables. Because the amplifiers on the tree and branch coaxial cables, and even the cable itself, are in general sensitive to temperature and humidity, various self-adapting arrangements such as automatic gain control (AGC) and automatic slope control are employed.

The CATV program channels of existing CATV networks occupy the 450-MHz frequency band stretched in the 50- to 550-MHz frequency range. Each CATV program channel has a 6-MHz bandwidth, and its frequency spectrum is the same as the NTSC TV signal spectrum shown in Figure 2.5. The video signal is VSB-AM modulated, and the audio signal is FM modulated.

2.3.2 Bidirectional HFC CATV Network

The conventional CATV networks in Figures 2.4(a,b) are only able to transmit a CATV signal from the headend to televisions and very limited upstream

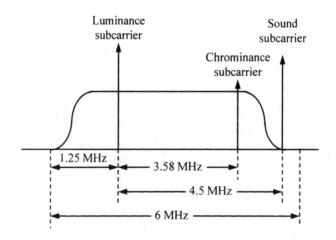

Figure 2.5 Frequency spectrum of NTSC television signal.

control signals for some cases. To accommodate bidirectional services, the architecture of the existing CATV networks has to be modified.

For the fiber and coax hybrid network in Figure 2.4(b) simple bidirectional services can be supported with a minimal hardware change. That is, upgrade or replacement of transmission equipment, amplifiers, and repeaters can help provide bidirectional services such as PSTN telephony. Figure 2.4(c) shows the CATV network architecture that has been obtained through this modification. As shown in the figure, the only change that has been made is the reversed direction of arrows and the inclusion of PSTN/BISDN.

For the Figure 2.4(c) architecture to be functional, an appropriate frequency allocation plan must be supported. Figure 2.6 shows a typical example of the HFC frequency spectrum. The frequency band for the analog CATV services remains the same, and the 25-MHz channels in the 5- to 30-MHz range are used for the upstream telephony signal as well as control signals. Another 25-MHz channel is added in the 575- to 600-MHz range for transmission of upstream and downstream telephony signals. In addition, the 375-MHz band in the 625- to 1,000-MHz region is allocated for digital interactive services such as VOD. In fact, the spectrum plan can differ among different countries and areas.

In the case of downstream digital signals, the digital information capacity that can be transported over the given bandwidth depends on the modulation techniques. For example, if QPSK is used, 8-Mbps capacity can be transported over the 6-MHz channel, and if 256-QAM or 16-VSB is used, 30 to 43 Mbps can be transported. Table 2.3 lists the available capacity for different modulation techniques.

The upstream channel at the 5- to 30-MHz region includes the channel for the return path of the CATV signals such as the set-top control. However, if additional return spectrum is provided on the top of the CATV spectrum, this frequency band may be allocated for upstream use. The 5- to 30-MHz low-frequency band could be utilized as a bidirectional channel operating completely passively, without any amplifier. The cable loss is very low in this band, and

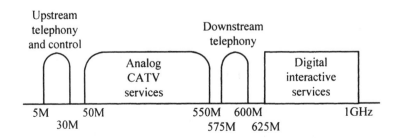

Figure 2.6 Example of HFC frequency spectrum.

Table 2.3
Modulation Technology for Digital Data Transport in HFC

Modulation Techniques	bps/Hz	Mbps in 6-MHz Channel
BPSK	1	4
QPSK	2	8
16-QAM	4	16
64-QAM	6	24–27
256-QAM/16-VSB	8	30–43

highly reliable transmission is possible even with power failure occurring in the amplifiers. These are positive implications for bidirectional services such as telephony.

This 25-MHz low-frequency band can accommodate about 24 DS-1 signals if QPSK is employed for the modulation. This number is calculated as follows: Since QPSK has the coding efficiency of 2 bits/Hz, the DS-1 rate of 1.544 Mbps can be modulated into a 0.772-MHz bandwidth. So if about 30% of guardband is allowed, a DS-1 signal can be carried in a 1-MHz channel. Therefore, the given 25-MHz channel can accommodate about 24 DS-1 signals, which is equivalent to 576 voice channels at 64 Kbps each. In fact, this number can increase further if a higher efficiency modulation technique is employed.

If the Figure 2.6 spectrum plan is used in deploying the HFC of Figure 2.4(c), then the internal architecture of the headend and the ONU can be arranged as depicted in Figure 2.7(a). Further, if the digital interactive service in the 625- to 1,000-MHz band is also taken into account, the architecture may change to the shape of Figure 2.7(b). Note that for the digital interactive service, it is assumed that the return path, or the upstream data, has comparatively small data and thus can be loaded in the upstream telephony channel.[2]

In a typical HFC network, each ONU serves a CSA of 200 to 2,000 subscribers. The HFC network can provide the services equivalent to 500 voice channels, 80 analog CATV channels, and about 300 shared digital interactive channels, each of which can provide a VOD service.

2.3.3 Digital CATV Network

Among the optical transmission technologies applied to HFC, the digital-modulation-related parts are developing faster than the analog-modulation-related

---

2. The architectures in Figure 2.7 are for illustration purposes. In practice, much more refined and diverse architectures can be devised. For example, the two-fiber arrangement can be modified to a one-fiber version by employing various multiplexing technologies such as DDM, TCM, FDM, and WDM (see Section 2.4.2).

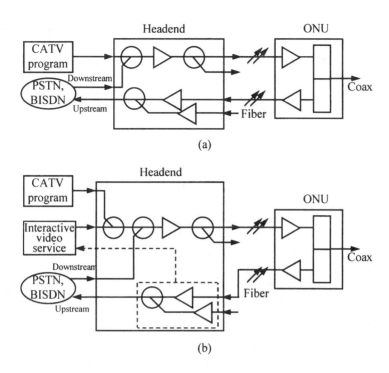

(a)

(b)

Figure 2.7 Illustration of HFC architecture: (a) for CATV and telephony services and (b) for CATV, digital, and telephony services

parts. In addition, the backbone networks being deployed in most countries are SDH/SONET-based digital optical networks. This trend provides strong motivation to digitalize the CATV network, because, if digitalized, the digital CATV network would support cost-effective operation and internetworking with the backbone networks in the future. Further, the backbone network has been standardized to support the ATM network interface, in order to be able to function as the information superhighways of the upcoming information age. Therefore, simple digitalization of CATV is not enough: "ATMization" should be considered as well.

In reality, digital CATV makes sense from the signal compression point of view. The advanced video compression technology enables video signals of NTSC broadcast quality to be compressed to 3- to 6-Mbps rates. This is possible owing to the advent of MPEG-2 technology (see Section 6.4 for MPEG-2). In fact, MPEG-2 enables us to compress the HDTV signal into the normal 6-MHz TV channel.

According to the frequency spectrum recommended by DAVIC for digital HFC, the 8- to 26.5-MHz band is allocated for upstream transmission and the

70- to 130-MHz band for downstream transmission. For modulation of the up-stream and downstream channels, 16-QAM, 64-QAM, and 256-QAM techniques are suggested. For mapping to ATM cells, MPEG-2 transport streams and ATM protocol stacks are recommended. (Refer to Section 4.4 for ATM protocols.)

It is also conceivable that the digital ATM channel could be overlaid on the conventional analog CATV channel. The two overlaid channels are trans-mitted from the CO or headend up to the ONU by fiber, and in the ONU, the two channels are split. Then the digital ATM channel is delivered to the sub-scriber over the VDSL twisted-pair for broadband conversational services, and the analog CATV channel is loaded on the coaxial cable for entertainment video services. This overlaid digital-analog network may be referred to as an HFC-FTTC combination network.

The ultimate form of digital CATV network may be the FTTH network, which can be realized by making the last mile of coaxial cables to subscribers' premises optical instead. In this stage, however, HFC will confront strong com-petition with FTTC; or the expectation of competition will influence the evo-lution strategy and construction of both HFC and FTTC.

2.4 Fiber-to-the-Curb

Whereas HFC is the CATV provider's approach toward interactive services, FTTC is the telecommunication service provider's approach toward broadband multimedia services. FTTC is on the track of the natural evolutionary process of the subscriber loop. The evolution was initiated by digitalizing, and then making optical, the feeder plant down to the remote nodes. With FTTC, opti-calization will further penetrate the subscriber loop to the curbside in the neigh-borhood of subscribers' premises. Each optical termination, ONU, will then interface with a small number of subscriber terminal units, about 20 or fewer, which is much smaller than the number of subscribers within a CSA (about 200 to 2,000) of the HFC network. For the very last fraction of the distribution net-work, VDSL technology can strongly back up FTTC with very high-speed TP-based connections. Once FTTC is accomplished, the remaining road to the final destination FTTH will be relatively easy.

FTTC belongs to a general class of subscriber loop opticalization, namely, *fiber-in-the-loop* (FITL). FITL includes all the processes including FTTO (or FTTB), FTTC (or FTTZ), and FTTH. Among them FTTO has considerably pro-gressed in the major cities, and FTTC is on the way. FTTH will become of practical interest when FTTC shows substantial progress, and, at that point, FTTH will be a rather simple extension of FTTC.

In this section various aspects of FTTC, or FITL in a broad sense, including the network architecture and optical distribution network topology are consid-ered. In addition, among the two important supporting technologies, powering and optical technologies, powering issues are considered at the end of this section, and the optical technology issue are discussed separately in the following section.

2.4.1 FTTC Network Architecture

Figure 2.8 shows the general FTTC architecture. It basically consists of headend, ONU, STU in the subscriber's premises, and the optical distribution network. Optical fiber is deployed all over the optical distribution network down to the ONU and twisted-pair wire takes over the remaining network down to the subscriber's STU. The VDSL/ADSL facility shown in the figure to connect the ONU with the end user is not a required part but a recommended option.

The headend is the host of all the services such as video and POTS provided in the corresponding subscriber network. On the other hand, it provides connection to PSTN/BISDN, program providers, management networks, and other systems. The headend may be located within, outside of, or away from a central office. Although it provides video and POTS services downstream toward ONUs on one side, it also concentrates and grooms upstream traffic and passes to the local switching system on the other side. In addition, it manages, in coordination with local switches and remote management systems, the operation and management signals coming from the ONUs and also provisions the ONUs.

The ONU does the normal optical network termination functions as for other networks. On the subcriber side of the ONU twisted-pair fills the remaining gap to the the subscriber's premises. Each ONU interfaces with about 20 or fewer subscriber's STUs. If this part is rearranged such that each ONU serves for one STU and, if the ONU is moved next to the STU accordingly, then FTTH is realized. In the interim period, which could be a considerably long time, advanced TP transmission technologies such as VDSL/ADSL will continue evolving to provide high-rate transmission support.

The optical distribtion network is an optical fiber network that provides

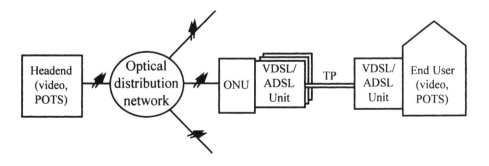

Figure 2.8 Architecture of FTTC network.

point-to-multipoint connections from the headend to the ONUs. It may or may not include active optical devices. Figure 2.9 shows two possible topologies of the optical distribution network: *passive optical network* (PON) and *active double star* (ADS) network. PON consists of optical fibers, connectors, splitters, couplers, passive WDM devices, and other passive optical devices, but does not contain active devices such as O/E converters. The ADS network can contain all active and passive devices and optical transmission systems.

The network beyond the headend can vary depending on the network and system environment. If broadband digital transmission and switching systems are fully installed, the headend may be connected to the digital switching systems via the deployed digital cross-connect and transmission facility. This digital facility can even form a ring topology that connects multiple headends on the ring, thus offering enhanced connection reliability. (Refer to Section 3.10 for information on self-healing rings.) If the available switching system is analog or is unable to support a direct interface with the headend, intermediate terminal equipment such as a *central office terminal* (COT) may be installed to aid the connection.

2.4.2 Passive Optical Network

As discussed above, PON is a special form of FTTC or FTTH that consists only of passive optical devices in the optical distribution network.[3] PON has a physical-star but logical-bus topology and uses passive optical splitters to distribute service signals from the headend to a multiple number of ONUs or subscribers [see Figure 2.9(a)]. PON is a cost-effective technology suitable for distribution services in the subscriber network. At the same time, it can also support interactive services in the following fashion.

In the downstream direction, *time-division multiplexed* (TDM) subscriber signals are transmitted to the remote node over a single-mode optical fiber. There, a passive splitter is used to broadcast the duplicated signal to all ONUs connected to the headend. Each ONU then accesses the time slots assigned to itself. As such, a large number of subcribers can share the same optical distribution network, which helps to reduce the quantity of optical fibers and the number of optical transmitter/receiver modules. Therefore PON proves itself to be an economical system. However, since all downstream information is broadcast to all ONUs, a means needs to be incorporated in the communication protocols to secure subscriber privacy.

In the upstream direction, the *time-division multiple access* (TDMA) technique is used for all ONUs to share the upstream transmission channel without

---

3. PON originated as the *telephony over PON* (TPON), which was first developed by *British Telecom Research Laboratories* (BTRL).

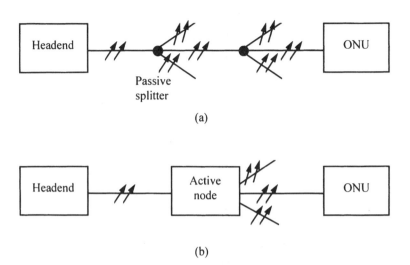

Figure 2.9 Topogies of optical distribution network: (a) passive optical network (PON) and (b) active double star (ADS).

contention. The information from the subscriber terminals is loaded onto the time slot dynamically assigned to the corresponding ONU, and is then delivered to the headend, from which the synchronization required to perform TDMA is remotely provided. However, since the distance from the headend to the ONU differs among ONUs, synchronization becomes a tricky issue.

Various modulation technologies exist that enable bidirectional optical transmission over a single optical fiber in the PON such as DDM, TCM, FDM, and WDM: *Direction-division multiplexing* (DDM) realizes bidirectional transmission over one fiber by using an optical directional coupler. *Time compression multiplexing* (TCM) provides an optical ping-pong operation, and WDM uses a different wavelength for each direction.

2.4.3 Powering of FTTC

Powering is one of the fundamental issues in the optical subscriber network, which arises because optical fibers cannot carry electric power. In this connection, powering in the optical subscriber network is a key concern in constructing FITL, including both FTTC and FTTH.

Two approaches are used for powering the ONUs in FITL: *central powering* from the CO (or headend) and *local powering* from nearby power sources (see Figure 2.10).

Central powering refers to the scheme to feed power through the electrical power line that is deployed in parallel with the optical cable. In this case power

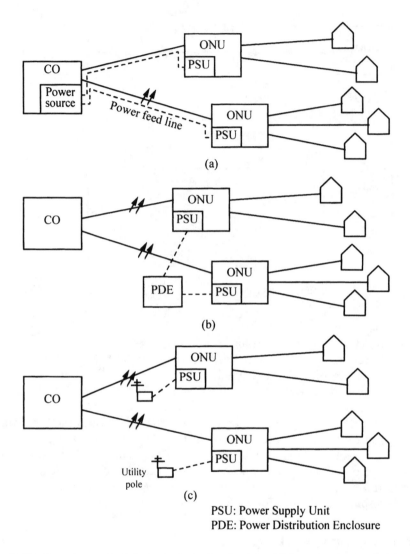

PSU: Power Supply Unit
PDE: Power Distribution Enclosure

Figure 2.10   Powering methods in the optical subscriber loop: (a) Central powering; (b) PDE based
powering; and (c) AC utility powering.

can be supplied stably, because the power generation or regulation is done in
the central office environment. However, the power line cost grows high as the
feeding distance becomes greater, and the power loss increases as well. In general, central powering is justified economically for a feeding distance of 1.8 km
(i.e., 6,000 ft) or less.

Unlike the case of central powering, local powering acquires power locally from the power supplies near by ONU. Thus, the power line deployment cost is less than the central powering case, and the power loss is minimal as well. However, maintenance is difficult because power supplies and batteries are distributed geographically. Further, the tolerance range of the power supply has to be wide because the installation environment is rather uncontrollable, with the temperature and humidity varying to a large extent.

The two methods of local powering are *AC utility powering* and *power distribution enclosure* (PDE)-based powering [see Figures 2.10(b,c)]. AC utility powering takes the AC power from the utility pole or from the subscriber's premises. PDE-based powering takes DC power from the PDE, which is installed near to ONUs and shared by multiple ONUs. The PDE powering method helps to overcome the drawbacks of central powering, such as costly power line deployment and lossy power transmission. It also eases the difficulties that arise from individual AC powering, for example, the management of the power facility and the control of the operational environment.

Table 2.4 compares the three powering methods (central powering, PDE powering, and AC utility powering) in terms of cost, efficiency, replacement, and others.

2.5 DEVICE TECHNOLOGIES FOR THE OPTICAL SUBSCRIBER NETWORK

Optical communication technology has significantly improved in many aspects since the early 1970s, and the application of optical fibers to almost the entire transmission spectrum is commonplace nowadays. Optical communication systems using optical fiber have numerous advantages over copper-wire communication systems. Low-loss and broadband characteristics of optical fibers can

Table 2.4
Comparison of Powering Methods

Category	Central Powering	PDE Powering	AC Utility Powering
Equipment first costs	High	Medium	Low
Power system operating efficiency	Low	Medium	High
Battery maintenance and replacement points	One or two points per CSA	Tens of points per CSA	Hundreds of points per CSA
System survivability	Immune to outages; aerial feed subject to environment	Catastrophic AC outages probable	Catastrophic AC outages probable

increase repeater spacing to hundreds of kilometers with channel capacity of tens of gigabits per second. Most repeater stations required under a copper-wire-based communications system can be eliminated with optical fiber systems. In addition to the inherent broadband property of optical fiber, lightwaves of different wavelengths can be simultaneously multiplexed and transmitted on a single optical fiber; thus, the number of channels or channel capacity can be increased with optical fiber installation without increasing transmission speed. In addition, due to the thinness and the light weight of optical fiber, cable space in the cable duct can be significantly spared, which can contribute to economical construction and maintenance of the cable network. Also, optical fiber's noninductive property allows the proximity of electrical cables generating electrical inductions and offers additional advantages, such as enhanced operator safety and quality assurance of the telecommunication services. Progress in the development of these key technologies will facilitate the realization of the optical subscriber network's FTTH stage, rendering introduction of new services easier.

In this section, optical fiber is examined first, then passive optical devices such as directional and star couplers and WDM are introduced, followed by a description of the optoelectronic devices and circuits, which can integrate semiconductor lasers and photodiodes into electronic ICs on a single chip. Finally, principles of the optical fiber amplifier and coherent optical communications are reviewed.

2.5.1 Optical Fiber Cable

The optical fiber used in interoffice trunk transmission can also be used in the optical subscriber network. As an aid to a more thorough understanding of the characteristics of optical communication, optical fiber's waveguide operation and its transmission characteristics are discussed first in the following.

Propagation Principles in Optical Fiber

Optical fiber, which is a slender strand of silica, is designed so that the refractive index in the center portion is slightly greater than that of the outer portion. The center portion is called the *core*, and the outer portion is called the *cladding*.

As shown in Figure 2.11, an optical fiber can be broadly categorized into three types according to its refractive index profile and the lightwave propagation characteristics. In the *step-index multimode* fiber of Figure 2.11(a), when the light reaches a boundary between two materials such as core and cladding, each possessing a different refractive index, a part of the light is transmitted through while the rest is reflected. But when the angle that the incident light from a dense medium, such as the core, forms with the surface of the less dense medium becomes less than the critical angle, then light is totally reflected

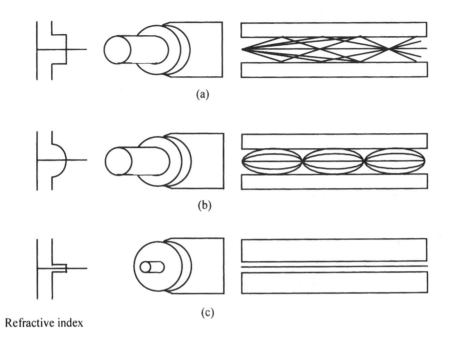

Refractive index

Figure 2.11 Light transfer types of fibers: (a) step-index multimode; (b) graded-index multimode; and (c) single mode.

internally. This phenomenon is called *total internal reflection*, and by applying this principle, a beam of light entering the core portion of the optical fiber at less than the critical angle experiences total reflection at the core-cladding boundary. If optical fiber is bent abruptly, the angle that the optical path forms with the boundary may become greater than the critical angle causing light to stray to the cladding portion, resulting in increased loss.

In these step-index multimode optical fibers, the optical path difference between a straight beam of light and one that is totally reflected with the maximum reflection angle results in a proportional amount of time offset at the optical receivers; thus, transmission capacity greater than tens of megahertz-kilometers is difficult. Such a phenomenon is called *mode dispersion*.

An optical fiber designed to reduce this time offset is the *graded-index multimode* optical fiber, shown in Figure 2.11(b). Inside this optical fiber, the index at the core is the greatest and reduces monotonically toward the cladding, resulting in a graded-index distribution. In this arrangement, the speed of the light bent along a far path increases with decreasing index value; thus, the reflected beam of light can arrive at almost the same time as the straight beam, which increases transmission capacity by up to a few gigahertz-kilometers.

ITU-T has standardized the core diameter of the multimode optical fiber to 50 μm and the cladding diameter to 125 μm.

Inside the step-index or graded-index multimode fibers, hundreds of light modes with different angles of reflection can propagate along the fiber; thus, these types of fibers are called *multimode fibers*. But, as depicted in Figure 2.11(c), if the diameter of the core is reduced even further and the index disparity between the core and the cladding is also reduced, then it can be arranged in such a way that only one mode of light is supported to propagate. Such an optical fiber is called *single-mode fiber*, and in this case no time offset is generated, since there is only one mode; thus, bandwidth greater than 100 GHz-km can be obtained. ITU-T has standardized the core diameter of the single-mode fiber to around 10 μm.

Transmission Characteristics of Optical Fiber

The transmission performance of the optical fiber is characterized by *transmission loss*, which limits repeater spacing, and *light dispersion,* which limits maximum channel capacity. Figure 2.12 shows transmission characteristics as a function of the optical wavelength, and it can be seen that the loss valley occurs

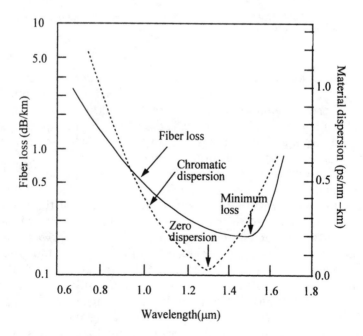

Figure 2.12 Transmission characteristics of fiber.

in the neighborhood of 1.55 μm and the minimum chromatic dispersion occurs at around 1.3 μm.

The transmission loss is caused by scattering, absorption, structural instability, and microbending. Scattering loss occurs when the light comes in contact with an object whose size is not much greater than the wavelength, which causes scattering of the light in all directions. Absorption loss is due to the phenomenon in which incident light is absorbed by the impurities and then converted to heat inside the fiber.

Dispersion refers to the phenomenon in which light pulses, when transported along the optical fiber, get distorted in shape, with their pulse width widened, as they propagate over a long distance. Therefore dispersion can cause adjacent pulses to overlap and thus limits the transmission capacity of optical fibers. Dispersion can be categorized into mode dispersion, chromatic dispersion, and structural dispersion, depending on the instigator.

Characteristics of Optical Cables

In the early field trials for the optical subscriber network, multimode optical fibers were widely used, but currently single-mode fibers are predominantly used in field tests or commercial applications. Since most subscribers are distributed within a few kilometers of the central office, multimode cables have sufficient channel capacity if just the initial services are considered. However, the almost unlimited bandwidth of single-mode cable is attractive in preparation for future expansion, since the unit price is not so different from that of the multimode cable. However, since connection or splicing is more difficult and time consuming compared to that of multimode cables, preassembled pigtails or connectors are frequently used.

Optical cable at 1.3-μm wavelength shows good cost-performance compromise in subscriber networks as well as in trunk networks, and the present technology level is considered mature. Optical cable at 1.55-μm wavelength has even smaller transmission loss, as shown in Figure 2.12, but it requires dispersion-shifted fibers that shift the zero-dispersion point to the 1.55-μm wavelength. Wavelengths up to 1.55-μm are considered useful for information services and WDM applications of the future. Within an optical subscriber network, if a central office needs to accommodate more than 1,000 subscribers, multistrand optical cables composed of more than 100 strands of optical fibers are required. In addition, it is convenient to place remote nodes and SAPs between the central office and the subscriber for convenient splicing and branching of subscriber cables. Also, while two-strand optical fibers should basically be provided to every subscriber, it is worthwhile to examine the possibility of the WDM accommodation for remote subscribers.

Structures of Optical Cables

Feeder cable is applied between central offices and remote nodes. It is mainly used as duct-type cable, but can be also used as aerial-type cable when required. Distribution cable, applied between the remote node and SAP, is used as duct-type and aerial-type cable. There is also service line cable, which is composed of one to two optical fibers. Service line cable is used to transmit signals from the SAP to the subscriber. It is connected to the transmitter/receiver module and test equipment inside the subscriber premises. In contrast to multi-fiber-optical cable, it is composed of one or two strands of doubly coated optical fibers, and it must be designed so that its long-term reliability is ensured under various working conditions inside a building.

Most optical cables possess unit configuration; hence, splicing and branching are easily accomplished. The feeder cable from the central office, especially, is usually composed of more than 100 fibers, and it is anticipated to increase to more than 1,000 fibers in certain cases. Distribution cables and access line cables contain tens or hundreds of fibers or one to two fibers. Figure 2.13 shows typical structures for optical cables grouped and the number of fibers.

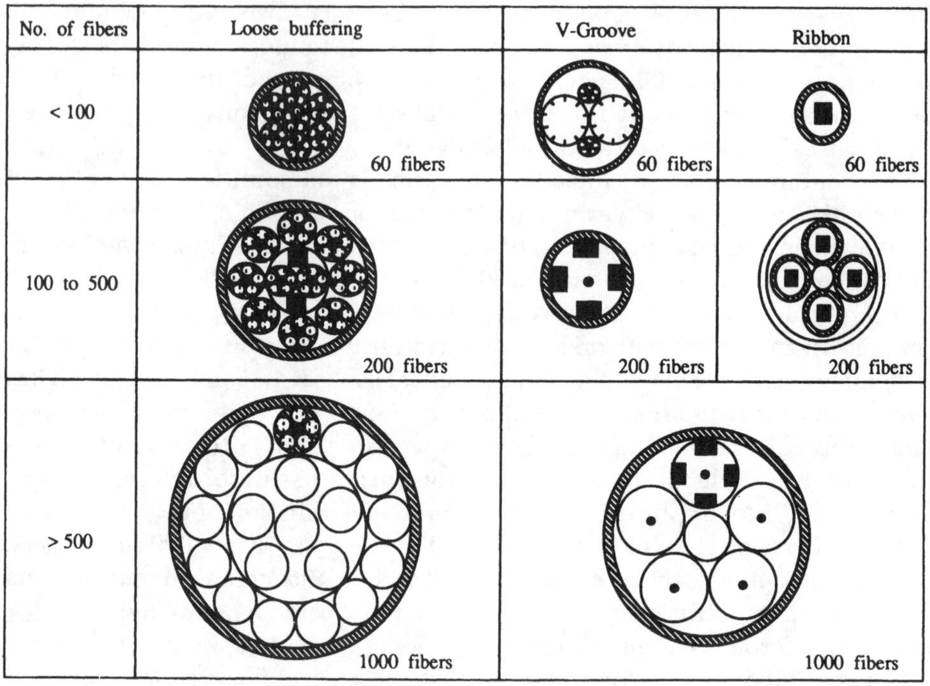

No. of fibers	Loose buffering	V-Groove	Ribbon
< 100	60 fibers	60 fibers	60 fibers
100 to 500	200 fibers	200 fibers	200 fibers
> 500	1000 fibers		1000 fibers

Figure 2.13   Examples of subscriber loop optical cables.

In contrast to the existing interoffice types, optical subscriber cables must be easy to handle, separable into units, and easy to manufacture with high fiber density, since diverse structures are required depending on the usage and location.

Optical cables used in optical subscriber networks are categorized into feeder-type and distribution-type cables according to their functionality and depending on the location of usage, and can be subcategorized into duct-type, aerial-type, service line-type, and indoor-type cables.

Splicing and Connection

The difficulties of single-mode subscriber cable splicing are apparent in two aspects: they use single-mode fibers instead of multimode fibers and the number of fibers in the subscriber cable is much more than that in the trunk cable. But the difficulties associated with single-mode subscriber cable splicing appear to have been solved through such means as multifusion splicing or mechanical splicing techniques. Currently, many telephone companies use *biconic* connectors and single-fiber fusion splicing for interoffice trunk networks. Since high density is one of the optical subscriber network's special features, quick and precise splicing are prerequisite for multiple fibers. As a way of resolving the multifiber problem, multifusion splicing and multimechanical splicing have been developed. On the other hand, it is especially desirable to use preassembled optical connectors for *optical distribution frame* (ODF) and optical transmitter/receiver modules.

2.5.2 Passive Optical Devices

Passive optical devices are being used not only in optical subscriber networks, but also in instruments, signal processing, sensors, and other optical engineering fields, and even more diverse applications are foreseen for the near future. In passive optical devices, the light signal is not converted into an electrical signal, but the light signal itself changes propagation direction depending on the wavelength, and incident light beams of different wavelengths in the same direction can be separated into different directions, or vice versa, according to the wavelengths.

The passive optical devices applicable in the optical subscriber network include the optical directional coupler, star coupler, and WDM device. The optical directional coupler and the WDM device are used between the central office and the remote office to transmit signals in both directions using just a single fiber, and the star coupler is widely used in central offices for distributing broadband signals such as video to multiple subscribers.

Optical Directional Coupler

Optical directional couplers are widely applied in bidirectional optical communication systems with single fiber. Directional couplers are also used for transmission line monitoring, instrument applications, and optoelectronic parts. The optical directional coupler can be divided into the half-mirror type and the distributed-mirror type, depending on the principle employed. The half-mirror type uses a semireflecting mirror to cause light to separate and diverge in two directions, whereas the distributed-mirror type geometrically couples the optical fiber. The latter type is widely employed in communication and instrumentation systems, as indicated in Figure 2.14.

Figure 2.14(a) is a representation of the basic configuration of bidirectional communication at the same wavelength, and here the optical directional coupler's isolation capability is important. In Figure 2.14(b), a safety device has been added to raise the reliability of the system so that even if either side malfunctions communication is still possible. In Figure 2.14(c), a directional coupler has been applied in the *optical time domain reflectometer* (OTDR), which indicates the time offset between the original light beam and that reflected at the break point, and thus can be used as a position identifier.

Star Coupler

The star coupler splits a single optical signal into several uniform signals. Figure 2.15 represents the different structures of star couplers. The mixing element type relies on internal optical multiple reflection to equalize the divided optical output.

In the tapered fiber type, a bundle of optical fibers is stretched thin under heat so that a lightwave propagating inside the core becomes radiated and split into the cladding, and resumes its course inside the adjacent fiber's core. In the spatial-division type, the graded-index lens splits the light into spatially separated optical fibers. This technique has optical splitting loss, and the maximum possible number of split outputs is limited to about four. In contrast, the number for the tapered type is around ten, and that for the mixing element type can be greater than ten. The respective optical losses are as listed in Figure 2.15, and it can be seen that the spatial-division type is superior in this respect.

Applications of star couplers can be found in the optical database system and the TV distribution systems shown in Figure 2.16, in which a star coupler is employed to enable simultaneous communication among a multiple number of terminals. The star coupler is suitable in any system that requires a division of the optical signal among an unspecified number of terminals.

Optical Wavelength Division Multiplexer

Silica optical fiber manifests low loss in the wide-wavelength range from 0.8 to 1.6 μm, and this low-loss wavelength band can be used to transmit

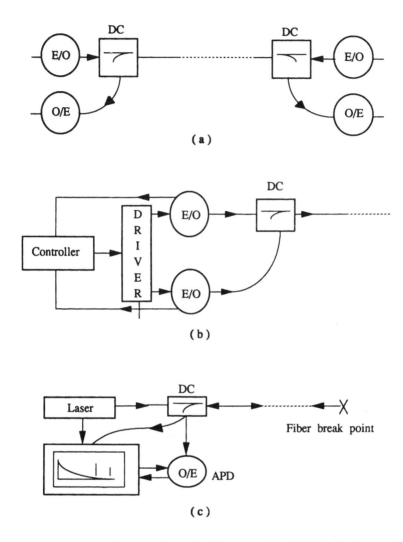

Figure 2.14   Applications of directional coupler: (a) one-wavelength bidirectional transmission; (b) optical sources protection circuit; and (c) optical fiber break detection.

simultaneously a multiple number of optical signals on different wavelengths through a single optical fiber. Two indispensable components of WDM transmission are the wavelength-division multiplexer, which multiplexes several signals with different wavelengths onto a single optical fiber, and the wavelength-division demultiplexer, which performs the reverse function.

Figure 2.17 shows three examples of possible WDM implementations. In Figure 2.17(a), a high-pass (short-wavelength) filter and a low-pass (long-wave-

Types	Structures (Examples)	Characteristics
Mixing element type	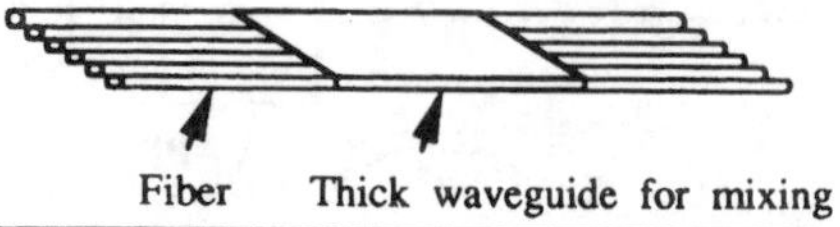 Fiber Thick waveguide for mixing	Optical loss : 3~4 dB Deviation : ± 0.5 dB
Tapered fiber type	 Fiber Tapered (Twist) region	Optical loss : 1.5~2.5 dB Deviation : ± 1~2 dB
Spatial division type	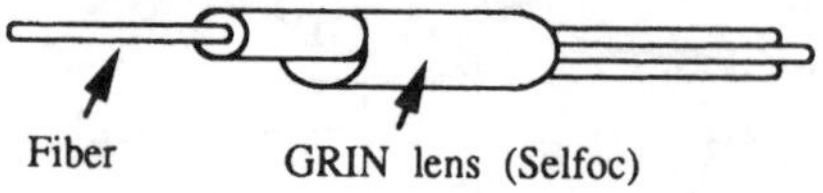 Fiber GRIN lens (Selfoc)	Optical loss : 1~2 dB Deviation : ± 0.5 dB

Figure 2.15 Structure and characteristics of star couplers.

length) filter are used to separate a light into its respective wavelength components. In Figure 2.17(b), parallel beams of light are passed through a diffraction grating, and different directions of diffracted light corresponding to each wavelength are used to separate beams. In Figure 2.17(c), as the light passes through the prism, the angle of refraction varies according to wavelength, which enables multiplexing or demultiplexing.

2.5.3 Optoelectronic Devices and Circuits

Semiconductor Laser

Ever since the continuous oscillation of a GaAlAs semiconductor laser at room temperature was first reported in 1970, the semiconductor laser has been the most widely used light source for optical communications and optical information processing applications.

Without the invention of the semiconductor laser, optical fiber communication as we know it now would not have been possible. Since the low-loss wavelength band of optical fiber is at the wavelength region of 1.2 to 1.6 μm, as shown in Figure 2.12, the development of the InGaAsP lasers at these wavelengths has replaced the GaAlAs laser diode whose wavelength is around 0.8 μm.

Compared to other types of lasers, semiconductor lasers have high electrical-to-optical power conversion efficiency, can be modulated easily to higher speeds, and are extremely convenient due to their very small size. Also, depending on the material and design, the light-emitting wavelength can be

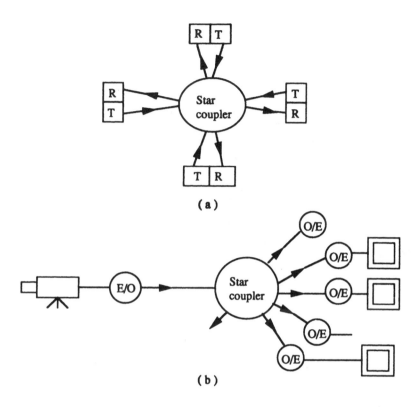

Figure 2.16 Applications of star coupler: (a) optical database system and (b) TV signal distribution system.

chosen to be anywhere from visible to far infrared regions, and more than 10 years of lifetime is possible.

GaAs and InP possess a direct transition energy band structure that is suitable for emitting light. Electrical current flows when a voltage is applied in the forward direction across the *pn* junction of these semiconductors. The electrons from the *n*-region and the holes from the *p*-region, respectively, migrate to opposite sides, and light is emitted as they recombine in the neighborhood of the *pn* junction.

In case the electrical current is lower than the threshold, the recombination of electrons and holes occurs in a random manner. Hence, there is no correlation in phase, wavelength, or direction among the radiations, and thus the stimulated emission prerequisite for laser action cannot be accomplished. But, if the pumping current is increased to a certain level called the *threshold level*, then population inversion is accomplished in the neighborhood of the *pn* junction as shown in Figure 2.18. Population inversion is a state where more

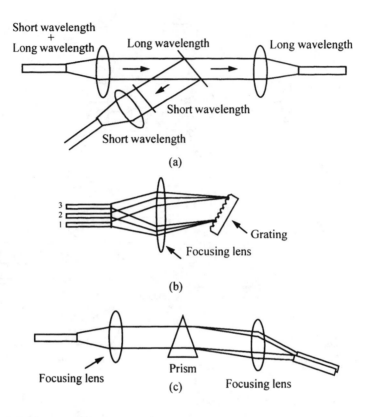

Figure 2.17 WDM device structure: (a) filter type; (b) grating type; and (c) prism type.

electrons exist at the high energy level than at the low energy level. In this state, light is released in a controlled manner proportional to the rate of recombination. Here, the region at which the controlled release of light due to recombination occurs is called the *active* or *gain region*. Laser light is emitted if the gain obtained from the stimulated emission is greater than the loss of the resonator. The threshold current is an important parameter for the evaluation of the performance and operating conditions of lasers.

The early semiconductor lasers were of the GaAlAs/GaAs family, operated at 0.85-μm wavelength. Figure 2.19(a) shows the structure of a buried *double heterostructure* (DH) semiconductor laser that belongs to this family. However, silica fibers have shown dramatically lower loss and higher bandwidth at 1.2- to 1.6-μm wavelengths, and, accordingly, semiconductor lasers of InP that operate in the 1.2- to 1.6-μm range have emerged. Besides the changes in the material used, variations in the structures of the growth layer and the electrode have led to the improvements in laser diode characteristics for optical

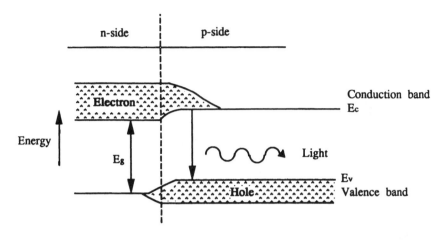

Figure 2.18 Principles of light emission in semiconductor lasers.

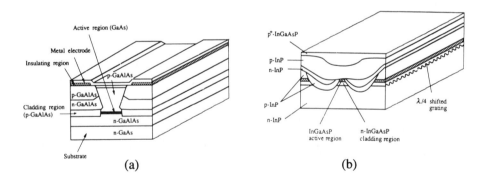

Figure 2.19 Structures of semiconductor laser: (a) buried DH and (b) DFB-LD.

communication applications. An example is the *distributed feedback laser diode* (DFB-LD) with longitudinal single-mode operation whose structure is shown in Figure 2.19(b). Its narrow-spectrum light can reduce the chromatic dispersion of the optical fiber; hence, the single-mode laser diode can be widely used in broadband optical communication systems. As shown in the figure, the cladding region of the DFB-LD is structured as a diffraction grating functioning as a distributed filter that screens only single-mode light with a narrow-spectrum width. Also, the wavelength of the output light can be precisely adjusted by varying the periodic spacing of the diffraction grating at the time of manufacture. This control over the wavelength makes the DFB-LD extremely suitable for high-density WDM and coherent optical communications.

Optoelectronic Integrated Circuit

Since optical transmission speeds of tens of gigahertz are demanded along the optical fiber, efforts have been made recently to improve the performance of optical devices and electrical circuits at such speeds. As a means of increasing device speeds by reducing the size of the devices and the influence of external conditions on its performance, as well as to reduce costs and enhance reliability, *optoelectronic integrated-circuit* (OEIC) technology, which integrates both optical devices and electrical circuits onto a single chip, has emerged. OEICs can reduce parasitic reactances and have low operation noise; thus they are useful for optical fiber submarine communication systems, which require high-speed laser diode modulation and high receiver sensitivity at the modulation speeds, and for optical subscriber systems, which require low cost and enhanced reliability.

The OEIC employed for optical communication systems can be divided into GaAs-based and InP-based OEICs, depending on the substrate used. The GaAs-based OEIC uses GaAs to produce the substrate and is used in short-wavelength (0.8- to 0.9-μm) systems, and the InP-based OEIC, which uses an InP substrate, is used in long-wavelength (1.2- to 1-μm) systems.

The GaAs-based OEIC technology has sufficiently matured by adapting existing *field-effect transistors* (FET) or *metal semiconductor FETs* (MESFET), so that, as shown in Figure 2.20(a), its integration density is higher than that of the InP OEIC. Consequently, it is employed for short-distance high-speed transmission applications such as the subscriber loop, CATV network, and optical LANs, as well as computer networks requiring low associated cost that are able to realize the full benefits of high integration density. InP-based OEICs, shown in Figure 2.20(b), are more appropriate as high-speed transmission devices because the long wavelength matches the fiber characteristics. Optical device technology using InP has already reached a stable stage, and electrical devices such as the *junction FET* (JFET), *metal-insulator semiconductor FET* (MISFET), and *heterojunction bipolar transistor* (HBT) using InP have been successfully fabricated.

2.5.4 Optical Fiber Amplifiers

As a means of coping with the limitation of repeater spacings due to losses and dispersion of optical fibers, and with the restrictions in the number of WDM

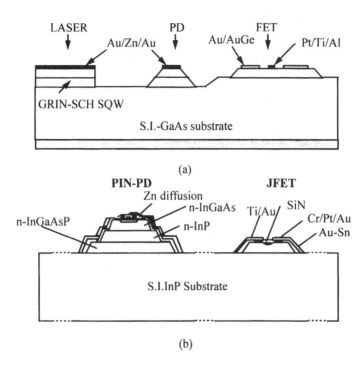

Figure 2.20 OEIC device structures: (a) GaAs-based OEIC and (b) InP-based OEIC.

channels or optical couplers, purely optical amplifiers, which amplify the optical signal without an electrical conversion, are desirable.

Depending on how the amplification is achieved, optical amplifiers can be divided into semiconductor laser amplifiers and optical fiber amplifiers. The semiconductor laser amplifier is produced by applying a nonreflective coating with a quarter-wavelength thickness to both facets of a common laser diode using such materials as SiO_2, ZrO_2, and Eu_2O_3. This coating causes the threshold current of the laser diode to increase by 1.5 to 1.6 times. If a pumping current slightly lower than the threshold current is supplied to the laser diode amplifier, then light incident only on the coated facet of the laser diode amplifier is amplified without lasing. Such semiconductor laser amplifiers are used in optical switching systems.

Optical fibers can also achieve direct amplification of a light signal without converting into its electrical equivalent by doping with a special material called *erbium*. An *erbium-doped fiber amplifier* (EDFA) can amplify a weak light signal if pumped with a laser. Since these optical fiber amplifiers can amplify light signals without any complicated equipment, they are useful for nonrepeated long-distance communications such as transoceanic undersea optical cables and for the multicasting of general optical subscriber systems or optical CATV systems. While optical fiber amplifiers have greater gain and can also amplify independently of polarized states of light compared to the semiconductor laser amplifiers, they are difficult to integrate on a single chip, and the pumping light is hard to separate.

If the light having a wavelength in the neighborhood of 0.98 or 1.48 mm is pumped into the erbium-doped fiber, electrons in the fiber get excited by absorbing the energy of the pumped light and gather around the energy level of the 1.52- to 1.56-mm wavelength. If input light at a 1.55-µm wavelength passes through the erbium-doped fiber, then the excited electrons transfer their energy to the input light, thus amplifying the input light. It takes about 10 ms for the excited electrons to return to the ground state after emitting their energy into the 1.55-µm light, so it is possible to accumulate enough energy for light amplification.

An EDFA consists of a pumping light source, wavelength selective coupler, isolators, and erbium-doped fiber, as depicted in Figure 2.21. The pumping light source provides energy for light amplification, and semiconductor lasers at the resonance wavelength of 0.98 or 1.48 µm are used as the pumping light source. The wavelength-selective coupler couples or decouples the pumping light and the input light, and the isolator isolates the reflected lights. Shown in the figure is the forward-pumping EDFA in which the pumping light travels toward the same direction as the input light, but the backward-pumping EDFA is also available in which the pumping light travels in the reverse direction.

If an EDFA is employed, it is possible to amplify the light signal by 10 dB or more, which enables the transmission distance to be increased by 20 to 50 km. The EDFA can be installed in the transmitting end, in the receiving end, or in the repeater. The EDFA installed in the transmitter works as a power

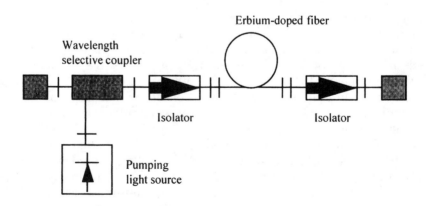

Figure 2.21   Structure of forward pumping EDFA.

amplifier, whereas the EDFA in the receiver takes the role of a prefilter, which increases the receiver sensitivity by amplifying the weak received signal. So, it is important for the EDFA installed in the receiver to be capable of suppressing the noise generated in the fiber amplifier to the minimum level. On the other hand, for the EDFA installed in the repeater, both power amplification and noise suppression are important. Transmitters equipped with the EDFA can be used for trunk transmission, subscriber loop distribution, and CATV distribution, and repeaters equipped with the EDFA can be used in submarine cables and other long-distance cables.

2.5.5 Coherent Optical Communication

In existing optical communications, the high frequency (around 200 THz) of the lightwave is used as a carrier, but the wide-frequency bandwidth is not fully utilized for information transfer. That is, existing optical communication is an intensity modulation/direct detection method that simply turns the source on and off according to the input signal, so the phase portion of the lightwave cannot be used for carrying effective information. In contrast, coherent optical communication fully exploits the lightwave's characteristics for information transfer and has the advantage of having high receiving sensitivity and frequency selectivity. Accordingly, it further increases repeater spacing and makes high-density WDM possible, and it has a tremendous transmission capacity at the terabit-per-second class and, hence, is considered the future optical communication technology.

The basic principle of coherent optical communication is depicted in Figure 2.22. The transmitter modulates input signals using *amplitude-shift keying* (ASK), *frequency-shift keying* (FSK), or *phase-shift keying* (PSK) to a carrier lightwave at frequency ω_S with a very narrow frequency linewidth. The optical signal transferred along the optical fiber is heterodyne-mixed with light received from a separate local oscillation light source of frequency ω_{LO}. It is then

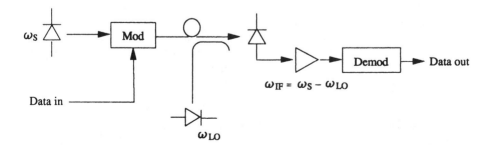

Figure 2.22 Basic principle of coherent optical communication.

converted into an electrical signal by way of an optical detector. Therefore, electrical output from the optical detector is manifested as an *intermediate frequency* (IF) band signal, which is the frequency difference component of the carrier lightwave and local lightwave or $\omega_{IF} = \omega_S - \omega_{LO}$.

In such coherent optical communication, the primary factor influencing communication quality is the phase noise of the light source. Phase noise is related to frequency linewidth, and in order to minimize it, the linewidth of the laser diode used for the light source should be as small as 0.1 to 0.0001 times the bit rate. This coherent method can increase receiving sensitivity by more than 10 to 15 dB over that of the existing intensity modulation/direct detection, and since it can accommodate the information of many channels by changing the frequency of the local oscillator, *optical FDM* (OFDM) becomes possible. Employment of OFDM enables high-volume distribution of video, as well as high-bit-rate data that can supply about 100 channels simultaneously to 1,000 subscribers per system, and facilitates channel expansion. From the subscriber's point of view, channels can be easily selected via an optical tuner, as in the case of a radio broadcast system. Bidirectional services such as video telephone are also possible. For multichannel transmission technology employing coherent optical communication, there has been a test of 400 Mbps × 16 channels by FSK modulation, and the increase in the number of channels is not a big problem in principle.

2.6 WIRELESS SUBSCRIBER NETWORK

The idea of a wireless subscriber network has been drawing attention in recent years as a complement to the existing copper-based subscriber network. If the drawbacks inherent to the nature of wireless communications could be fixed, a wireless subscriber network would possibly provide more flexible and versatile services than the copper-based network. From the network provider's point of view it has the advantage of prompt and economical introduction of new services. Such a wireless subscriber network could first be used for supporting the existing PSTN, ISDN, and digital data services. But, in addition, it could also be used for narrowband mobile services or wireless broadband distribution and on-demand services.

In this section wireless subscriber network technology is considered. To begin with, an overview of the wireless subscriber network is given. Then narrowband wireless technologies are discussed in terms of point-to-multipoint, cellular, and cordless telephone services. Finally wireless broadband distribution technologies are discussed in terms of local multipoint and multichannel multipoint distribution services.

2.6.1 Wireless Subscriber Loop Technology

The concept of the wireless subscriber network emerged in the late 1980s as a by-product of cellular wireless technology. As the demand for cellular service exploded worldwide, the cost of wireless network components decreased, while the cost for the conventional copper-based subscriber network has increased. The subscriber network, even though it appears to be a small part of the overall telecommunications network, in reality occupies as much as 25% of the overall network expenses, most of which is spent for deployment, operation, and maintenance of the subscriber lines. For this reason, the wireless subscriber network was first deployed in rural areas at the beginning where the initial cost is comparatively low. Recently, however, it has become an effective alternative to the copper-based subscriber network in urban areas too.

Aside from the advantage that the initial investment is small, the wireless subscriber network has the advantage that services can be initiated soon after the request, because all that is necessary to provide service is to install a fixed wireless terminal at the subscriber's premises. Further, a wireless subscriber network has the advantage that it can be flexibly modified in response to changes in the subscriber's needs.

The wireless subscriber network enables services to be initiated cost effectively in areas where a wireline subscriber network does not exist and allows for supplementary connections in areas where the existing wireline subscriber network is highly congested. It is preferred by the new network providers entering the market in competition with the incumbent network provider, and by the incumbent network providers who want to extend the wireline connections to other areas. It has been adopted by many developing countries as the primary means of access to PSTN.

An important issue to consider when designing wireless subscriber network systems is the wireless network architecture. Three different types of architectures are applicable to wireless subscriber networks: *microwave point-to-multipoint* (MPMP), *cellular,* and *cordless* systems. In the sparsely populated rural areas, the MPMP system is adequate because it is designed to connect subscribers by means of *line-of-sight* (LOS) microwave. For mobile users in urban and suburban areas, the cellular system is most appropriate. In highly dense areas such as inside residences, office buildings, and villages, the cordless system is a desirable choice.

If the broadband aspect of a wireless network is considered, wireless distribution of analog video services has been provided in areas with low subscriber density. A typical example of this is the *multichannel multipoint distribution service* (MMDS), which is provided over the 2-GHz frequency range. However, since the bandwidth of MMDS is rather limited, wireless systems that can possibly provide wider bandwidth services have been sought, and the most promising candidate is the *local multipoint distribution service* (LMDS).

For broadband wireless subscriber network architectures, microcellular or picocellular architectures can be employed depending on the service areas. In urban areas microcellular or picocellular architectures with diameters of a few kilometers to hundreds of meters will be adequate due to high user density. In the case of high-frequency systems such as LMDS, the high attenuation property can be utilized for frequency reuse design. In rural areas the microcellular architecture with a diameter of tens of kilometers will be most efficient due to sparse user density.

2.6.2 Narrowband Wireless Subscriber Access

If a wireless distribution system is to be competitive against existing wireline distribution networks, it must be capable of meeting diverse requirements. To begin with, it must provide inexpensive, good-quality services. Its usage must be simple and compatible with existing methods. Service provision must be transparent to the existing services. Security against eavesdropping is essential, and fast installation is important. Existing telephones, facsimiles, and modems must be still usable in the wireless environment. If these requirements are summarized from the user's point of view, the service availability, service quality, and terminal price are of main concern. Likewise, from the network provider's point of view, infrastructure cost, system flexibility, and system capacity are important.

Figure 2.23 shows a general view of the wireless subscriber network: microwave can be used in the feeder network in conjunction with coaxial cables and optical cables, and for the distribution network, the aforementioned three wireless network components—MPMP, cellular, cordless systems—can be used.

The three wireless subscriber systems were originally designed to match their own characters: MPMP for use in low-density areas, cellular for medium- to high-density areas, and cordless for high-density areas. However, variations are now available, for example, MPMP systems for rural areas or dedicated fixed cellular systems for urban and rural areas. For each of the three wireless systems, TDMA and *code-division multiple-access* (CDMA) techniques are both applicable. In the case of cellular systems, the analog *frequency-division multiple-access* (FDMA) technique also applies.

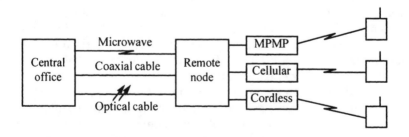

Figure 2.23 Architecture of wireless subscriber networks.

The performances of the three systems differ depending on the applied multiple-access techniques. But, in general, voice quality and service transparency are good for MPMP (for both TDMA and CDMA) and TDMA cordless systems, and the terminal cost is low for TDMA cordless and analog cellular systems. From the network provider's point of view, CDMA-based systems, in general, are robust against the fading and interference of the wireless environment but the relevant base station cost and complexity is higher than other systems. The TDMA cordless system has the advantage that its base station cost is low and frequency planning can be easily done using dynamic channel allocation, but it also has the weak point that the propagation range is short. A summary of performance comparisons among the three wireless subscriber systems is listed in Table 2.5 [Pedraja96].

Table 2.5
Performance Comparison of Wireless Subscriber Systems

Category	MPMP		Cellular			Cordless	
	TDMA	*CDMA*	*Analog*	*TDMA*	*CDMA*	*TDMA*	*CDMA*
Speech quality	Good (32 Kbps)	Good (32 Kbps)	Fair (4 kHz)	Fair (13 Kbps)	Fair (13 Kbps)	Good (32 Kbps)	Fair (13 Kbps)
Service transparency (fax, data, etc.)	Good (up to ISDN)	Good (up to ISDN)	Limited	Limited	Limited	Good (up to ISDN)	Limited
Terminal cost/ complexity	Medium	Medium/ High	Low	Medium	Medium	Low	Medium
Base station cost/ complexity	Medium	High	Low	High	High	Low	High
Modularity (channel/base station)	High	High	Low	Medium	High	Medium	High
Wireless propagation robustness	Fair	Good	Good	Fair	Good	Fair– Good	Good
Subscriber density supported	High	High	Medium	Medium	High	Very High	High
Maximum range	35 km	35 km	40 km	30 km	25 km	5 km	25 km
Available product	Propriety	Propriety	AMPS TACS	GSM	IS-95	DECT PHS	IS-95

2.6.3 Broadband Wireless Subscriber Access

As discussed earlier, broadband wireless distribution networks can be divided into two categories: MMDS and LMDS. The MMDS system operates at the low-frequency band, typically under 4 GHz, but the LMDS system operates at the millimeter-wave range. Figure 2.24 shows the network architectures of the two systems. The MMDS system is a low-frequency long-range system, whereas the LMDS system is a high-frequency short-range system in which distribution service is transmitted via optical cable to the radio transmitter's nearby subscribers.[4]

The MMDS system is usually used to distribute analog CATV services (pay-per-view) in the rural areas or in the countries where cable networks do not exist. In general, the 2.5- to 2.7-GHz frequency range is used for the MMDS in the 6- to 8-MHz channel, which can deliver video services in NTSC or PAL format. If 16-QAM modulation is used, the available net data rate for the 6-MHz band becomes 25 to 28 Mbps, which can contain four to six video programs of 6 Mbps each. With these specifications the maximum propagation distance is about 50 km.

Whereas MMDS systems are widely deployed, the LMDS system is still in its early stage. In general, the frequency range of LMDS is 27.5 to 29.5 GHz in the United States (which, in fact, is shared among geostationary and new non-geostationary satellite systems) and 40.5 to 42.5 GHz in Europe. Since precipitation attenuation is severe in this frequency range and 16-QAM is very sensitive to noise and interference, the LMDS system employs 4-PSK modulation. With these specifications the maximum propagation distance is limited to 4 to 5 km, but it can be increased to 10 to 15 km by using high-power *traveling-wave tubes* (TWT). Table 2.6 lists specification comparisons for the MMDS and LMDS systems [Cornag96, DAVIC1.0].

For rural areas the MMDS system with much longer transmission distance is more adequate, but for urban areas the LMDS system is more suitable, particularly because of its density. A return channel is expected to be provided soon in both systems in order to deliver interactive services to residential subscribers.

2.7 FULL-SERVICE SUBSCRIBER NETWORK

The evolution of the subscriber network discussed so far is based on two types of physical networks: a telecommunication network and a CATV network. The TP-based technologies such as HDSL, ADSL, VDSL, and the FITL technologies

---

4. The satellite-based broadband data distribution systems, for example, DirecTV and Spaceway, are not included in the broadband wireless subscriber network.

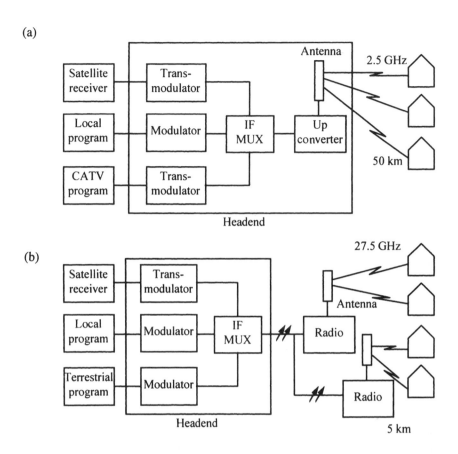

Figure 2.24 Architecture of broadband wireless subscriber distribution networks: (a) MMDS and (b) LMDS.

such as FTTO, FTTC, and FTTH are evolutions of the telecommunication network. In contrast, HFC is the first-step evolution of the CATV network. The telecommunication network and CATV network represent the first two of the three major information services: *voice communications, video distribution,* and *interactive data* services. However, they exhibit sharp contrast to each other in their fundamental characters: One is interactive but the other is broadband. Figure 2.25 portrays this relation.

The third component, interactive data service, is also indicated in the figure. It is narrowband and interactive, even if not as interactive as real-time communication services. This component usually exists within local areas or metropolitan areas in the form of LANs or MANs, and has its own evolution scenarios, which include opticalizing and fast-protocalizing (refer to Section 5.4

Table 2.6
Specification Comparison of Broadband Subscriber Distribution Systems

	MMDS		*LMDS*	
Frequency band	2.5–2.7 GHz		27.5–29.5 GHz	40.5–42.5 GHz
Channel raster	8 MHz	8 MHz	37 MHz	39 MHz
Modulation	16-QAM	64-QAM	QPSK	
Net data rate	16-QAM: 25 Mbps 32-QAM:38 Mbps		Code-rate ½: 25 Mbps Code-rate ¾: 38 Mbps	
Frame	MPEG-2TS		MPEG-2 TS	
Scrambling	$1 + X^{14} + X^{15}$		$1 + X^{14} + X^{15}$	
Outer code	RS (204, 188, 8)		RS (204, 188, 8)	
Interleaving	Convolutional ($N=204, I=12$)		Convolutional ($N=204, I=12$)	
Transmit power	1 – 100 W		600 W TWT	50 W TWT
Transmit antenna	Omnidirectional		Omnidirectional (90° sectoral)	
Receive antenna	Yagi		40 cm	20 cm
Noise figure	3–4 dB		7 dB	10 dB
Transmission range	50 km		4–5 km	

for LANs and MANs) processes. Separated local data networks can communicate with each other via the intervening telecommunication network, which is called a wide-area network or WAN. The Internet is a logical network constructed on the LANs, MANs, and WANs, which can possibly encompass all computers worldwide (refer to Section 5.9 for details on the Internet).

The three components of information services came into existence from mutually independent origins: telephones, televisions, and computers. Nevertheless, in the course of evolution each component has progressed enough to interfere, and will soon trespass, the territory of another, thus resulting in competitive relations among them. To prepare for the upcoming information age, however, some positive strategies are necessary that may turn the competitive relationship into harmonious contributions toward the construction of an information infrastructure and information superhighway (see Section 5.10).

In this context, this last section is dedicated to the discussion of a full-service subscriber network, which refers to a combined network that provides the maximum number of services to the benefit of its subscribers. Evolution of this type of service is discussed first, and then the evolution of subscriber networks is considered in association with the full-service network.

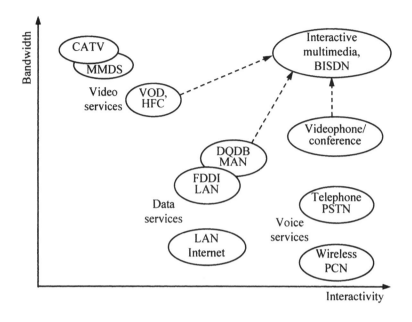

Figure 2.25 Evolution of voice, data, video services and networks.

2.7.1 Evolution of Services

There are several possible ways to classify information services depending on the classification criteria: Services can be classified into interactive and distributive services and the interactive services again subclassified into conversational, retrieval, and message services (see Sections 4.1 and 6.2). Services can also be divided into real-time services and non-real-time services; connection-oriented services and connectionless services (see Section 4.7); basic services and applied services (see Section 5.10). In the context of a full-service subscriber network, however, the network-based classification is most appropriate. This results in the three categories of services mentioned above: telecommunication services rooted in PSTN, video distribution services rooted in CATV networks, and interactive data services rooted in computers.

Telecommunication Services

The telecommunication service, which was originated by the telephone service, is not limited to voice services any longer. To the voice service, other media such as video, data, and sound have been integrated to yield broadband multimedia services. The most representative examples are videophony and video conferencing services. In the case of video conferencing service, multiple groups of people geographically separated can establish a virtual conference site assisted by real-time multimedia communications. Telecommuting and distance learning enables people to work and learn at home by overcoming the geographical distance barrier by means of multimedia communications. Videophony and video conferencing can also be extended to telemedicine to enable home-based medical diagnosis and treatment (refer to Section 6.2 for details). For all these telecommunication services to be available, however, the network must be upgraded to offer broad bandwidths to subscribers.

Video Distribution Services

Video distribution service originates from the cable-based CATV service but has been expanding to wireless distributions such as MMDS and LMDS (see Section 2.6). In either case, the main goal of its evolution is to incorporate user interaction to the distribution service. A primitive level of interaction may be realized by enabling a subscriber's channel selection, play start/stop, fast forward/backward motion, and so on. However, it needs to be further improved to provide interactive channels to each individual user.

Therefore, the next step in the evolution of video distribution will involve the provision of individual, switched, broadband channel to the subscriber with an upstream channel to convey user control information to the service control center. This enables a subscriber to access an unlimited number of servers, services, and contents. The most typical example of such interactive distribution services is the *video-on-demand* (VOD) service, which offers user program selection, instant delivery of the requested service, and full control of viewing. The wide program choice, instant availability, and interactive program control are key features of the VOD service (refer to Section 6.2 for more discussion on VOD).

Another important issue for evolution of the CATV services is the adoption of MPEG video compression as well as advanced channel coding technologies. These digital technologies enable channel capacity to be increased by a factor of 5 to 10 and improve the picture quality as well. Such technologies have been adopted by satellite and terrestrial broadcasting; in particular, the MPEG-2 video compression technology has been widely received for digital storage and transmission of video signals. By adopting those digital technologies, CATV distribution services can keep abreast of other video handling services and systems, and such compatibility can bring forth the extra benefit of cost savings (see Section 6.4 for the MPEG-2 discussion).

Interactive Data Services

The interactive data service, in a broad sense, refers to all computer-based services. Computer-based services were initially provided in the professional domain, and LAN was the major transmission means. The need for networking, however, was not limited within a LAN, and thus interconnection with remote LANs was sought across public networks. However, such LAN interconnection was limited by the small bandwidth of the subscriber network, 64 Kbps. So, applications running and exchanging data on a 10-Mbps LAN could not effectively cross over this throttling interconnection.

The need for business and professional individuals to access databases located outside their premises has fostered the growth of the Internet. The Internet has grown exponentially in recent years with about five million hosts and 40 million user stations connected to it as of early 1996. It provides access to a variety of selections, including companies, research institutes, schools, financial and market databases, software downloads, and others. The success of the Internet owes a great deal to the technical strength of the TCP/IP protocol it is founded on: The TCP/IP is an open standard that is independent of actual physical networks or computers and has a common global address.

The *personal computer* (PC) has evolved out of the business and professional domain and penetrated the residence. Such diffusion of PCs is closely linked with two key features: *multimedia* and *networking.* The multimedia PCs furnished with audio and video capabilities have attracted residential users much more than the text-only predecessors did. The enormous growth of the Internet and on-line services was made possible due, in significant portion, to those residential PCs, with the most distinctive on-line service being e-mail service. Such networking capability, when it was combined with the multimedia capability, gave momentum to the PC's proliferation in the residence.

Multimedia networking, in fact, finds a combined realization with the Internet and the *World Wide Web* (WWW), which is a network of multimedia servers constructed on the Internet. The WWW has been the fastest growing service on the Internet. However, in reality, it takes a long time to get multimedia WWW services over the Internet due to the limited bandwidth of the physical network. This results in reduced interactivity, which in turn discourages WWW users. Users desire high-speed multimedia networking, and this demand will become a strong pressure on developing the information infrastructure (refer to Section 5.9 for discussions of the Internet and WWW).

There are a number of interactive data services other than the Internet and e-mail services. For example, a transaction service is important for document handling and the telecommuting service (or computer-supported cooperative working) provides a platform for cowork among geographically separated people. Teleshopping (or home shopping) will possibly influence the distribution chain of goods. Telegame will add a new dimension to the entertainment game world. Digital library service will realize the "open library" in which no restrictions are placed on the time, place, and format of services. In fact, for industry and research sectors, high-speed solutions have been found in FDDI, DQDB, and other high-speed LANs with their wide-area connections supported by the

BISDN (see Sections 5.5 and 5.6). However, for residential users, an upgrade of the subscriber network is the most practical solution.

2.7.2 Evolution Toward Full-Service Network

The discussions thus far demonstrate how diversified the directions of service evolution have been. The service evolution, however, reflects the demand side only. To reflect the supply side also, the network evolution must be examined in connection with the service evolution.

Individual Service Evolution

For the evolution of telecommunication services, which include most bandwidth-demanding services such as videophone/video conferencing and other relevant applied services, bandwidth growth is the most important factor. Obviously the required bandwidth can be secured by developing the BISDN and the relevant ATM technologies and synchronous transport means such as SDH/SONET. (Refer to Chapter 3 for SDH/SONET and Chapter 4 for BISDN and ATM technologies.) Fibers are readily deployed or can be rather easily developed in the trunks, SDH/SONET systems are already developed for the most part, and ATM technologies are near completion. However the "ATMizing" of the subscriber network is a problem to be solved in different ways because simply making systems optical does not return the investment in this sparsely populated zone. As such, the various bandwidth-enhancing technologies that have been discussed earlier in this chapter are becoming ever more important. In this regard, several voluntary-basis standard meetings such as ATM Forum and DAVIC have begun to study ATM interface standards for the subscriber network.

For the evolution of video distribution services, the interactivity of the users is of primary interest, as indicated earlier. The upstream user message channel that allows users to control the distribution of video is indeed the minimum level of improvement without which the video services will remain "blind-sighted." However, this technology requires a significant level of architectural and technological changes because the original video distribution networks were designed solely for low-cost unidirectional video distribution only. For the CATV network to be fully functional in the interactive multimedia environment in the future, further interactivity functions need to incorporated in support of other conversational or interactive data services. Digitalizing and ATMizing the CATV service will be another drastic step in the evolution, in which environment a higher degree of interactivity can be implemented rather simply. The HFC-based CATV network can evolve to include these functions as discussed in Section 2.3.

For the evolution of interactive data services, the increase of bandwidth and the increase of interactivity both count. The degree of interactivity, in fact, very much depends on the degree of available transmission bandwidth. The long display time necessary to retrieve or receive a video display is caused primarily by the limited bandwidth. If infinite bandwidth were available, the computer's processing speed and protocol complexity would become the bottleneck. However, for a low or medium level of video quality, bandwidth is definitely the problem. The means for bandwidth expansion can be found in enhancing the LAN's and MAN's speeds and making the connections to broadband public transport networks optical. But this industry-based solution is not applicable to individual subscribers in residential areas. Therefore, the various TP-based and cable-based upgrade schemes discussed in this section are crucial for providing interactive sections also.

If Figure 2.25 is revisited with the above discussions in mind, the figure can deliver clearer messages on the orientation of the three arrows for evolution.

Full Service Network

On the other hand, for all of the telecommunication, video distribution, and interactive data services to evolve to the best of the users' satisfaction, the supporting infrastructure must be constructed in consideration of the three functional groups: service providers, network providers, and service users. For maximized efficiency in the overall infrastructure, the three different domains of functions need to be independently specialized. In particular, the network must be arranged to support open architectures such that service providers can operate on an equal basis and in a standardized manner.

Figure 2.26 illustrates how the three domains can be specialized and connected to form a full-service network: The service provider domains contains application services, content providers, and other service networks. The network provider domain includes PSTN, PSDN, ISDN, BISDN, control networks, management networks, and subscriber networks. The service user domain includes voice, video, data terminals, network termination, and subscriber terminal units (NT/STU). The service gateway on the border of the service provider and network provider domains is meant to mediate the two domains in establishing connections between the service users and service providers, and monitoring the use of network resources for subsequent charging [Alcatel95].

The subscriber network part connecting the headend node, remote node, and network termination is of the greatest concern, because it throttles the provision of interactive multimedia services. Therefore, considerable effort has been exerted for increasing the bandwidth and interactivity in the subscriber network in recent years, which has brought about the various TP-based, HFC-based, FITL-based, and wireless-based arrangements discussed in earlier

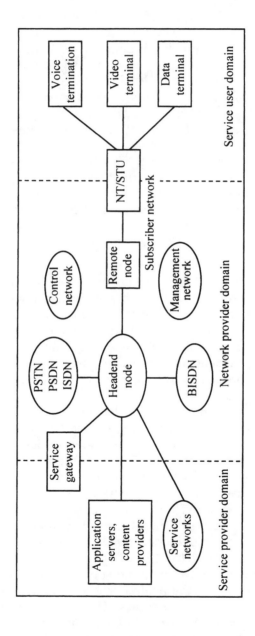

Figure 2.26   Categorization of service provider, network provider, and service user domains for full service network.

sections. All these enhancement schemes are summarized in the full-service subscriber network (or full-service access network) illustrated in Figure 2.27.

Each individual subscriber in the full-service subscriber network may need only one among ADSL, HFC, FTTC/VDSL, PON, or wireless connections. This indicates that each connection should be capable, to a certain extent, of providing the bandwidth and interactivity required for the telecommunication, video distribution, and interactive data services. In addition, it is desirable to accommodate ATM technology in the long run, so that it can become akin to the ATM-based public network to enable ATM-based end-to-end connection.

Figure 2.28 illustrates the architecture of the HFC-based full-service subscriber network. Note that it is a detailed view of the HFC architecture of Figure 2.4(c) but connections for interactive data services are also inserted. It is essentially the same as the HFC architecture in Figure 2.7(b), but interactive data service replaces the interactive video service. For the Figure 2.28 architecture, the frequency spectrum in Figure 2.6 is assumed, and the

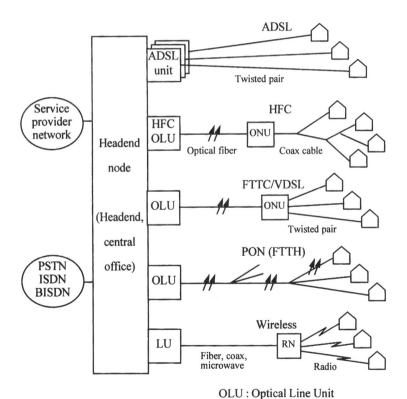

Figure 2.27 Integrated architecture of subscriber network.

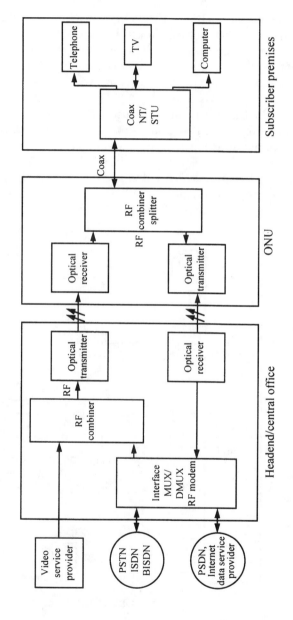

Figure 2.28   Illustration of the architecture of HFC based full service subscriber network.

interactive data service is assumed to be conveyed over the telephony channel. The full-service architecture merely illustrates how service integration can be achieved when HFC is taken as a platform. A similar arrangement should be possible for every other choice of subscriber network technology of Figure 2.27.

Selected Bibliography

ANSI T1.413, "Asymmetric Digital Subscriber Specifications."

ANSI X3T9.5 TP/PMD, Oct. 13, 1992, "Performance of 125 Mb/s 32-CAP transceiver," 1992.

DAVIC1.0 "Delivery system architectures and APIs," Part 4, Rev. 4.0, 1995.

ETSI 300–324–1 and 300–347–1, "Signalling protocol and switching: V interface at the digital local exchange and V5.2 interface for the support of access network," Part 1: V5.2 Specification, September 1994.

ETSI TM3 RG12, "Asymmetric Digital Subscriber Specifications."

ITU-T Rec. G.650, "Definition and test methods for the relevant parameters of single mode fibers," 1993.

ITU-T Rec. G.651, "Characteristics of a 50/125-μm multimode graded index optical fiber cable," 1993 (revision).

ITU-T Rec. G.652, "Characteristics of single-mode optical fiber cable," 1993 (revision).

ITU-T Rec. G.653, "Characteristics of dispersion-shifted single-mode optical fiber cable," 1993 (revision).

ITU-T Rec. G.654, "Characteristics of a 1550-nm wavelength loss-minimized single-mode optical fiber cable," 1993 (revision).

TA-NWT-001209, "Generic requirements for fiber optic branching components," Bellcore, 1991.

TA-TSV-001294, "Generic requirements for element management layer (EML) functionality and architecture," Issue 1, Bellcore, Dece,ber 1992.

TR-NWT-000909, "Generic requirements and objectives for fiber in the loop systems," Issue 1, Bellcore, December 1991.

TR-TSY-000303, "Integrated digital loop carrier system generic requirements, objectives, and interface," Bellcore, 1990.

Alcatel, "Full service access network: Join the world of networked multimedia," *Alcatel Telecom Review* (Telecom 95 issue), 1995, pp. 29–34.

Andrews, F. T., "The evolution of digital loop carrier," *IEEE Commun. Mag.*, March 1991, pp. 31–35.

Asatani, K, "Lightwave subscriber loop systems toward broadband ISDN," *IEEE J. of Lightwave Tech.*, Vol. 7, No. 11, 1989.

Balmes, M., et al., "Fiber to the home: The technology behind *hearthow*," *IEEE LCS*, August 1990, pp. 25–29.

Bernard, C. W., et al., "Bidirectional fiber amplifiers," *IEEE J. of Photonics Tech. Lett.*, Vol. 4 , No. 8, 1992, pp. 911–913.

Boinet, J. P., M. de Vecchis, and C. Verez, "Fiber in the loop," *Elec. Commun.*, Vol. 65, No. 1, 1992, pp. 44–51.

Brunet, C. J., "Hybridizing the local loop," *IEEE Spectrum*, Vol. 31, No. 6, June, 1994, pp. 28–32.

Caroll, R. L., "Optical architecture and interface lightguide unit for fiber-to-the-home feature of the AT&T SLC series 5 carrier system," *IEEE J. of Lightwave Tech.*, Vol. 7, No. 11, 1989.

Chen, W. Y. and D. L. Waring, "Applicability of ADSL to support video dial tone in the copper loop," *IEEE Commun. Mag.*, Vol. 32, No. 5, May 1995, pp. 102–109.

Chew, J. S., J. T. Tu, and J. M. Cioffi, "A discrete multitone transceiver system for HDSL applications," *IEEE JSAC*, Vol. 9, No. 6, August 1991, pp. 895–907.

Chiddix, J. A., "The use of fiber optics in cable communications networks," *IEEE J. of Lightwave Tech.*, Vol. 11, No. 1, January 1991, pp. 154–166.

Cook, A., and J. Stern, "Optical fiber access-perspectives toward the 21st century," *IEEE Commun. Mag.*, February 1994, pp. 78–86.

Cornaglia, B., and G. D'Aria, "Radio systems architectures for wireless CATV," in *Proc. ISSLS*, February 1996, pp. 128–133.

Davis, J., and C. L. Jander, "FITL spawns power concerns," *Telephony*, April 1991, pp. 24–28.

De Passoz, G., J. L. Clausse, and G. Karam, "Can digital techniques give a new boost to MMDS?," *Proc. ISSLS*, February 1996, pp. 134–139.

Eames, T. R., and G. T. Hawley, "The synchronous optical network in the loop," *IEEE LTS Mag.*, November 1991, p. 24.

Everitt, H., J. Nachef, and M. Virgin, "Full service network-operations and management," in *Proc. ISSLS*, February 1996, pp. 104–109.

Faukner, D. W., "Passive optical telephony network and broadband evolution," in *Proc. GLOBECOM'88*, Vol. 13, 1988, pp. 1579–1583.

Fukui, T., "Optical subscriber network architecture for broadband ISDN," in *Proc. ICC'88*, 1988, pp. 883–889.

Gross, R. W., "Coherent subcarrier multiplexed systems sharing transmitter and laser for video distribution in subcarrier loop," *IEEE J. of Lightwave Tech.*, Vol. 4, No. 4, 1991, pp. 524–520.

Hart, G. A., and R. Lyford, "Developing telephony services over CATV systems: Systems and architectural considerations," *NCTA Technical Papers*, 1994, pp. 206–218.

Hausken, T., and V. Brates, "Fiber-to-the-home: U.S. policy issue," *IEEE Technology and Society Mag.*, September 1991, pp. 22.

Hawley, G. T., "Historical perspectives on the U.S. telephone loop," *IEEE Commun. Mag.*, March 1991, pp. 24–28.

Hell, G., and A. Rolfe "Copper enhancement," *Ericsson Review*, No. 4, 1995, pp. 153–159.

Iguchi, Y., and S. Hashiba, "Development of a new CT/RT system," *NTT Review*, Vol. 3, No. 6, November 1991, pp. 27–33.

Im, G. H., et al., "51.84 Mbps 16-CAP ATM LAN standard," *IEEE JSAC*, Vol. 13, No. 4, May 1995, pp. 620–632.

Im, G. H., and J. J. Werner, "Bandwidth-efficiency digital transmission over unshielded twisted pair wiring," *IEEE JSAC*, Vol. 13, No. 10, December 1995.

Jager, J., H. Oesterberg, "Using DECT for radio in the local loop," *Ericsson Review*, No. 3, 1995, pp. 111–117.

Jones, J. R., "Baseband and passband transport systems for interactive video services," *IEEE Commun. Mag.*, May 1994, pp. 90–101.

Kaiser, P., "Status and future trends in terrestrial optical fiber systems in North America," *IEEE Commun. Mag.*, Vol. 25, No. 10, 1987, pp. 8–13.

Kocsis, F., "Management of HFC-based service-on-demand and full service access networks," in *Proc. ISSLS*, February 1996, pp. 98–103.

Kwok, T., "A vision for residential broadband services: ATM-to-the-home," *IEEE Network*, Vol. 9, No. 5, September 1995, pp. 14–28.

Kyees, P. J., et al. "ADSL: A new twisted-pair access to the information high-way," *IEEE Commun. Mag.*, Vol. 34, No. 4, April 1995, pp. 52–59.

Larger, D., "Creating a network for interactivity," *IEEE Spectrum*, Vol. 32, April 1995, pp. 58–63.

Latter, P. D., R. L. Fike, and G. A. Nelson, "Business and residential services for the evolving subscriber loop," *IEEE Commun. Mag.*, March 1991, pp. 109–114.

Lidoyne, O., et al., "Optical homodyne receiver using injection-locked semiconductor laser as local oscillator: Analysis," *IEEE J. of Lightwave Tech.*, Vol. 9, No. 5, 1991, pp. 659–665.

Lin, Y. K. M., "Fiber-based local access network architectures," *IEEE Commun. Mag.*, 1989, pp. 64–73.

Little, T., and A. Ghafoor, "Network considerations for distributed multimedia object composition and communication," *IEEE Network*, Vol. 4, November 1990, pp. 32–49.

Leon, V., and E. Miller, "Subscriber terminal units for video dial tone systems," *IEEE Network*, Vol. 9, No. 5, September 1995, pp. 48–57.

Midwinter, J., "Status and future trends in terrestrial optical fiber in Europe," *IEEE Commun. Mag.*, Vol. 25., No. 10, 1987, pp. 14–17.

Miki, T., and R. Komiya, "Japanese subscriber loop network and fiber optic loop development," *IEEE Commun. Mag.*, March 1991, pp. 60–67.

Miki, T., "Toward the service-rich era," *IEEE Commun. Mag.*, Vol. 32, No. 2, February 1994, pp. 34–39.

Mochida, Y., "Technologies for local-access fibering," *IEEE Commun. Mag.*, February 1994, pp. 64–73.

Morgen, D. H., "Fiber-to-the-curb power," *Telephony*, pp. 20–24, July 1991.

Ohtsuka, T., "Digital optical CATV system using hubbed distribution architecture," *IEEE J. of Lightwave Tech.*, Vol. 6, No. 11, November 1988, pp. 1728–2735.

Olson, D., et al., "Operating and powering optical fiber networks," *IEEE Commun. Mag.*, February 1994, pp. 74–77.

Paff, A., "Hybrid fiber/coax in the public telecommunications infrastructure," *IEEE Commun. Mag.*, Vol. 33, No. 4, April 1995, pp. 40–45.

Pedraja, F. G., "Wireless solution for the access network in high density urban areas: Technology analysis and experimental results," in *Proc. ISSLS*, February 1996, pp. 187–192.

Pugh W., and G. Boyer, "Broadband access: Comparing alternatives," *IEEE Commun. Mag.*, August 1995, pp. 34–46.

Reed, D. P., *Residential Fiber Optic Network*, Norwood, MA: Artech House, 1992.

Rowbotham, T. R., "Local loop developments in the U.K," *IEEE Commun. Mag.*, March 1991, pp. 50–57.

Saito, S., et al., "2223-km coherent transmission experiment at 2.5 Gbps using erbium-doped-fiber in-line amplifiers and dispersion-shifted single mode fiber," *IEEE J. of Lightwave Tech.*, Vol. 9, No. 2, 1991, pp. 161–169.

Sakaguchi, I., "Optical switching device technology," *IEEE Commun.* Mag., Vol. 25, No. 5, 1987, pp. 27–32.

Sakakibara, I., and F. Higushiyama, "Future development of optical subscriber networks," *NTT Review*, Vol 3, No. 6, November 1991, pp. 21–26.

Schaffer, B., "Synchronous and asychronous transfer mode in future broadband ISDN," in *Proc. ICC'88*, pp. 1552–1558.

Schumate, P. W., and R. K. Snelling, "Evolution of fiber in the residential loop plant," *IEEE Commun. Mag.*, March 1991, pp. 68–74.

Shimada, S., "Status and future trends in terrestrial optical fiber systems in Japan," *IEEE Commun. Mag.*, Vol. 25, No. 5, 1987, pp. 18–21.

Shimada, S., "Status and trends in fiber optic transmission systems," in *Proc. NTT International Symposium '90*, November 1990, p. 16.

Shutmate, P. W., Jr., "Optical fibers reach into homes," *IEEE Spectrum*, February 1989, pp. 43–47.

Snelling, R. K., J. Chernark, and K. W. Kaplan, "Future fiber access and systems," *IEEE Commun. Mag.*, April 1990, p. 63.

Spencer, J. L., and D. S. Kobayasahi, "Establishing reliability and availability criterion for fiber-in-the-loop systems," *IEEE Commun. Mag.*, March 1991, pp. 84–90.

Taylor, T. M., "Powering and energy in the local loop," *IEEE Commun. Mag.*, March 1991, pp. 76–82.

Tenzer, G., "The introduction of optical fiber in the subscriber loop in the telecommunication networks of DBP TELECOM," *IEEE Commun. Mag.*, March 1991, pp. 36–49.

Terada, Y., "Evolution of ISDN Towards BISDN," *NTT Review*, Vol. 3, No. 3, May 1991, p. 25.

Toda, I., "Migration to broadband ISDN," *IEEE Commun. Mag.*, April 1990, p. 55.

Toivanen, E., "High quality fixed radio access as a competitive local loop technology," in *Proc. ISSLS*, 1996, pp. 181–186.

Tsuyuki, S., K. Asano, and H. Kadoya, "CT/RT for POTS and ISDN basic services," *NTT Review*, Vol. 3, No. 6, November 1991, p. 27.

Weippert, W., "The evolution of the access network in Germany," *IEEE Commun. Mag.*, Vol. 32, No. 2, February 1994, pp. 50–55.

Yates, R. K., N. Mahe, and J. Masson, *Fiber Optics and CATV Business Strategy*, Norwood, MA: Artech House, 1990.

Yukimatsu, K., and T. Aoki, "Advanced switching technologies toward terabit communication networks," in *Proc. Int. Con. Comm. Syst. (ICCS) '90*, November 1990, p. 937.

Synchronous Digital Transmission

3

As point-to-point optical communication evolves into point-to-multipoint optical communication, the concept of synchronous digital transmission becomes an efficient means of transmission for the optical networks. The fundamental concept of synchronous digital transmission was first introduced in the early 1980s, and it matured through the decade, along with the standardization of SONET interfaces and *synchronous digital hierarchy* (SDH). Synchronous digital transmission finds its solid foundation in SDH and, it also retains compatibility with the *network node interface* (NNI) and the *user-network interface* (UNI) standards of the BISDN.

SDH is a digital transport structure that operates by appropriately managing the payloads and transporting them through (synchronous) transmission networks. Before the advent of SDH, the most common digital hierarchy in use was plesiochronous digital hierarchy, which is still widely used in the form of European and North American DS-1, DS-2, DS-3, and DS-4 signals.[1] These PDH signals are multiplexed into the *synchronous transport module* (STM) signals in SDH, *n* of which merge into an STM-*n* signal. Compared to PDH, SDH appears extremely simple in operation. However, the process of synchronous multiplexing that maps PDH tributaries into STM-*n* signals is no trivial matter.

The term *synchronous* in SDH stems from the fact that the process of multiplexing plesiochronous tributaries into STM-*n* adopts a synchronous multiplexing structure. Some of the advantages of using a synchronous multiplexing structure are:

* Simplified multiplexing/demultiplexing techniques;
* Direct access to low-rate tributaries without the need to demultiplex/multiplex all of the intermediate signals;

---

1. Strictly speaking, the existing digital hierarchies should be described as asynchronous. However, since the variance of each asynchronous tributary nominal bit rate is within the specified error tolerance, it is called *plesiochronous*. Hence, in this text, the term plesiochronous will generally be used in the place of the term *asynchronous*.

- Enhanced OAM capabilities;
- Easy transition to higher bit rates of the future in step with the evolution of the transmission technology.

Hence, one can conclude that the synchronous multiplexing structure is the very essence of SDH.

The type of communication that transmits plesiochronous digital tributaries through the baseband is called *digital transmission*, and, in the same manner, the new mode of communication that transmits synchronous digital tributaries is called *synchronous digital transmission*, or *synchronous transmission*. Accordingly, the processes that multiplex the existing DS-1 through DS-4 tributaries into an STM-*n* signal through synchronous multiplexing, reconstructs the signals via synchronous add/drop or cross-connect equipment, and finally transmits and regenerates it through a synchronous optical network, in totality, can be called synchronous digital transmission. From a synchronous transmission standpoint, the current digital transmission can be classified as asynchronous digital transmission.

The objective of this chapter is to provide an in-depth description of synchronous digital transmission. To this end, the discussions in the chapter are organized in the following manner. Following an extensive introduction, the frame structure of the STM-*n* signal and associated topics are discussed. First, we give a detailed description of the STM-*n* frame structure, followed by an examination of the synchronous multiplexing procedure and the associated mapping of asynchronous tributaries into respective containers. Description of overheads and pointers then follows. In the latter half of the chapter, synchronous multiplexers and synchronous optical transmission systems are studied, and the chapter closes with discussions of the synchronous transport network and network survivability.

3.1 INTRODUCTION

In this section, background topics of synchronous transmission are introduced in preparation for more detailed discussions in the forthcoming sections. First, SDH is compared to the existing PDH. Then, the layered approach to digital transmission, which forms the basis for synchronous transmission, is discussed, and the synchronous multiplexing structure and STM-n frame are examined. After this brief but requisite review of pertinent terms, we discuss the unique features of synchronous digital transmission and follow that up with a summary of its standardization process. Finally, we briefly compare the two representative synchronous transmission standards: the SDH and the SONET.

3.1.1 PDH Versus SDH

The existing digital hierarchies were originally prescribed by the Bell System of North America and CCITT (the predecessor of ITU-T). North American tributaries were reaffirmed as the North American standard by the T1 Committee, which emerged after the breakup of the Bell System. These existing digital hierarchies are called *PDH* to differentiate them from the nascent *SDH*.

The PDH is currently the standard digital hierarchy and can be divided into North American and European designations, as shown in Figures 3.1(a,b). North American PDH consists of DS-1 (1.544 Mbps), DS-1C (3.152 Mbps), DS-2 (6.312 Mbps), DS-3 (44.736 Mbps), and DS-4E (139.264 Mbps); and the European PDH accommodates DS-1E (2.048 Mbps), DS-2E (8.448 Mbps), DS-3E (34.368 Mbps), DS-4E (139.264 Mbps), and DS-5E (564.992 Mbps).[2] Each multiplexing step is plesiochronous, and synchronization is achieved with the use of bit stuffing and positive justification.

The SDH is composed of STM-n signals, as shown in Figure 3.1(c). Here, n is a fixed number, mainly 1, 4, 16, 64 and indicates the number of multiples of the base STM-1 with the corresponding bit rates of 155.520, 622.080, 2,488.320, and 9,953.280 Mbps, respectively. The STM-n signal is formed by synchronously multiplexing the DS-1, DS-2, DS-3, DS-4E, DS-3E, DS-2E, and DS-1E tributaries. The DS-1C and DS-5E signals are excluded in this synchronous multiplexing.[3] The STM-n signal is formed by *byte-interleaved multiplexing* (BIM) n of the STM-1 signals. In this process, the overhead composition of each STM-1 signal becomes reorganized.

Comparing Figures 3.1(a–c), it is apparent that the SDH has a much simpler structure than the PDH. In the SDH, North American and European tributaries go through only a single stage of multiplexing. In the PDH, *asynchronous multiplexing* (AM) is used when a tributary is multiplexed into a tributary of higher bit rate. In the SDH, *synchronous multiplexing* (SM) is used instead. Also, in the PDH, DS-m is a higher tributary of the DS-$(m-1)$ signal, but in the SDH, all the DS signals are equal in status.

3.1.2 Layer Concept and Overheads

Generally, a digital signal is transmitted hierarchically, via the path, the multiplexer section, the regenerator section, and the physical medium, as illustrated in Figure 3.2. Each section of the transmission procedure can be viewed as a layer, and hence we can divide the entire digital transmission procedure into the path layer, the multiplexer section layer, the regenerator section layer, and the physical medium layer (or the optical layer).[4]

---

2. The DS-4 (274.176 Mbps) signal was also among the Bell system tributaries. The T1 Committee recommended the accommodation of DS-4E in its place and additionally prescribed a multiplexing path from DS-3 to DS-4E. In indicating the digital tributaries, E denotes the European designation [*Conference of European Post and Telecommunication Administrations* (CEPT)].

3. DS-1C is included in the asynchronous multiplexing of the SONET.

4. Here, the multiplexer section layer, together with the regenerator layer, is called the *section layer*, and the section layer, together with the physical medium layer, is called the *transmission medium layer* (refer to Section 3.9).

(a)

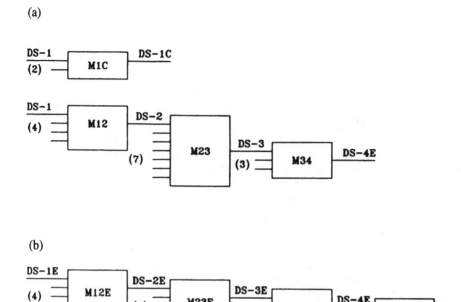

(b)

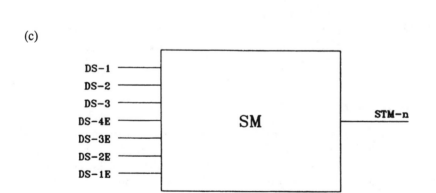

(c)

Figure 3.1   Digital hierarchy: (a) plesiochronous (North American); (b) plesiochronous (European); and (c) synchronous.

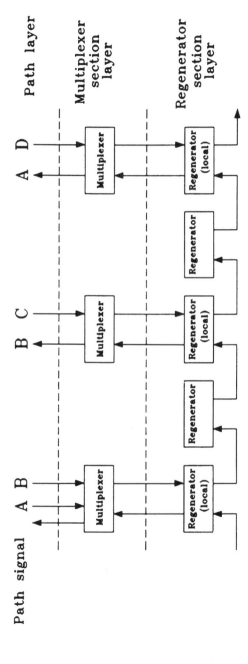

Figure 3.2 Illustration of the layering concept for digital transmission (unidirectional).

The synchronous multiplexing structure allocates the requisite bit space systematically, in accordance with the layer concept. Figure 3.3 gives an example of the application of the layer concept to organize an STM-n frame. In this figure, the STM-n frame is partitioned into five regions, four of which are providing overhead functions for different layers. The regenerator section overhead and multiplexer section overhead correspond to the regenerator section layer and the multiplexer section layer, respectively, and the path overhead corresponds to the path layer. Path overheads of lower order paths are included in the STM payload space.

The overheads used in synchronous multiplexing can be divided into the *path overhead* (POH) and the *section overhead* (SOH) in the spirit of the layer concept. The section overheads can further be categorized into the *regenerator section overheads* (RSOH) and the *multiplexer section overheads* (MSOH). The insertion of SOH bytes is the last step in the construction of an STM-n, and POH bytes are inserted every time a *virtual container* (VC) signal is constructed (see Section 3.3 for a description of VC).

The SOH bytes are added or extracted at the regenerator or multiplexer section for block framing, maintenance, performance monitoring, and other operational functions. As shown in Figure 3.3, the SOH bytes, located above and below the *pointer* (PTR), are for the regenerator and multiplexer sections, respectively. For example, B1 byte above the PTR, which contains the *bit-interleaved parity* (BIP)-8 code using even parity for regenerator section error monitoring, is checked and recalculated at every regenerator. But B2 below the PTR, which contains the BIP-24 code using even parity for multiplexer section error monitoring, is checked only at the line termination where line protection may occur.

The POH can be categorized into the higher order POH such as the VC-4 (at the level of DS-4) and the VC-3 (at the level of DS-3) and the lower order POH such as the VC-11 (at the level of DS-1), the VC-12 (at the level of DS-1E),

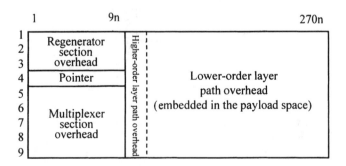

Figure 3.3   Layering concept embedded in the frame structure.

and the VC-2 (at the level of DS-2), and in both cases the POH is used for end-to-end communication between the nodes where the relevant VC is terminated.

3.1.3 Frame Structure and Synchronous Multiplexing

The STM-1 frame, the end product of the layered synchronous multiplexing procedure in the SDH, has a typical structure like the one shown in Figure 3.4, while the STM-n frame has the same basic structure but whose width is extended by n times. The STM-1 frame structure occupies 9B × 270 bit space over 125 μs, and thus acquires a 155.520-Mbps bit rate. Likewise, the STM-n frame has the structure of 9B × 270 × n and the bit rate of 155.520 × n Mbps.

If the STM-1 frame structure is observed more carefully, it can be seen that the SOH is composed of 3 × 9B and 5 × 9B partitions of space and the AU PTR is composed of 1 × 9B, while the remaining 9B × 261 is reserved for the STM-1 payload, into which one VC-4 or three VC-3s can be mapped.

As illustrated in Figure 3.1, the STM-n signal is obtained by synchronously multiplexing incoming digital tributaries. In the first stage of synchronous multiplexing, each tributary is mapped into a corresponding *container* (C). Bit stuffing with *positive/zero/negative* (P/Z/N) justification is used for synchronization. If a POH is attached to the container, it becomes a VC, and if a pointer is added as well, it becomes a *tributary unit* (TU). When a VC is mapped into an STM-1 without going through other VCs, then the corresponding TU becomes an AU.

Assembling a number of TUs creates a *tributary unit group* (TUG). The TUG can then be multiplexed into the next higher level of VC. If a PTR is

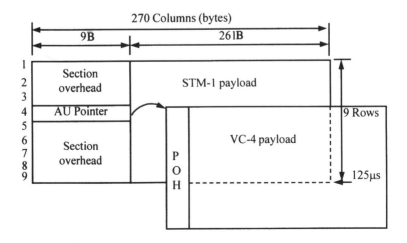

Figure 3.4 STM-1 frame structure.

added, a corresponding TU or AU is obtained. In this process, the number ($m = 1, 2, 3, 4$) attached to all signal units indicates that the corresponding signal unit is equivalent to the DS-m tributary in terms of rates. When $m = 1$, an additional number is attached to differentiate the DS-1(11) and DS-1E(12) tributaries.

A notable feature in this synchronous multiplexing is that a tributary DS-m directly enters the SM processing via VC-m and TU-m without being multiplexed into the next higher tributary DS-$m + 1$. For example, DS-1 gets into the SM procedure via VC-11, TU-11, TUG, etc., without passing through the DS-2, DS-3, and DS-4 multiplexing paths, finally reaching the destination of STM-n.

The pointer employed in the synchronous multiplexing process provides a means to synchronize a VC to a TU (or AU) that is made out of a clock with frequency difference. For this synchronization, P/Z/N byte justification is used.

3.1.4　Features of Synchronous Digital Transmission

As can be inferred from previous sections, synchronous transmission has some distinct characteristics, which are discussed in detail in this subsection.

125-μs Frame

The first distinct feature of synchronous digital transmission is the fact that its frame structure consists of 125-μs time intervals. This is a special feature that cannot be found in the existing plesiochronous digital hierarchies. Its advantage is that low-level signals, especially DS-0, can be accessed directly from high-level signals, and that all data manipulations can be carried on at the byte-unit level. But it has a drawback in that it introduces a new problem, namely, waiting time jitter. This is an unavoidable result of P/Z/N justification, which is an inherent feature of synchronous multiplexing, but viable solutions have been introduced.

Unification of Digital Hierarchy

The synchronous multiplexing structure accommodates both North American and European signals. That is, given any signal among North American DS-1, 2, 3 and European DS-1E, 2E, 3E, 4E, a generic STM-n signal can be constructed through the synchronous multiplexing procedure. Moreover, North American signals can be combined with European signals during this procedure and vice versa. This type of cross-continental unification has never been attempted previously. Of course, not all the proposed multiplexing paths are practical, but imposing the possibility paves the way toward future global network integration.

Layered Structure

One of the unique features of synchronous digital transmission is the fact that the frame structure is organized according to the layer concept. The division of overhead into SOH and POH is an instance of the application of this concept. Under this layered arrangement, SOH can perform its function independently of POH. SOH can be further divided into sublayers; for example, SOH located in the upper part of PTR can be devoted to the regenerator section, and the one in the lower part to the multiplexer section. The advantage of the layer concept is the "division of labor" of the transmission process, so that each required step can be performed independently of others.

Systematic Use of Overheads

As can be seen from the STM-1 frame, $9 \times 9B$ of space is allocated for SOH and PTR. If the POH and PTR accumulated from multiple multiplexing steps are taken into account, the size of actual overhead increases significantly. For example, when a DS-4E with a bit rate of 139.264 Mbps is mapped to STM-1, the overhead can take more than 10% of the frame space. This was a luxury in existing plesiochronous digital hierarchies, but it can be accommodated without adding additional difficulties in the optical networks. Overhead, which is systematically divided into POH and SOH, can be fully utilized to enhance the communication operation and maintenance capability of the transmission network. By reserving some of the overhead space for future use, a means of future transition and growth is also provided.

Synchronization via Pointers

The synchronization of the overall transmission network is achieved by the repeated insertion of pointers during the synchronous multiplexing process. That is, the frequency offset between the system clock and the received signal is effectively nullified by the use of pointers and P/Z/N justifications. Such a pointer-based synchronization method enables us to cope with the plesiochronous environment with a small elastic store and thus global synchronization is possible. Synchronization through the use of the pointers corresponds to a P/Z/N bit stuffing if viewed from the standpoint of existing asynchronous multiplexing. Hence, pointer synchronization generates similar low-frequency, high-amplitude jitters.

One-Step Multiplexing

In the synchronous multiplexing process, the tributary DS-1 can be multiplexed directly to STM-*n* via intermediate VC-11, TU-11, TUG-2, and VC-4 without passing through the next higher tributary DS-m + 1. This one-step multiplexing eliminates the need for intermediate steps. It is a novel scheme that is not found in existing asynchronous multiplexing systems. For communication networks that transmit high-rate signals through numerous multiplexing stages, it renders add/drop and cross-connect simple and economical. It should also be mentioned that one-step multiplexing is possible because of the container concept.

Network Concept

Another trait of synchronous digital transmission is that it is based on the network concept. The existing transmission system is inefficient because its structure has a point-to-point configuration; hence, add/drop and cross-connect at intermediate nodes are cumbersome, multistep processes. But since optical communication is becoming an increasingly common and popular means of digital transmission, a standardized transmission system based on the network concept is within the realm of possibility. In making it a reality, one-step multiplexing plays an important role. Efficient OAM through the use of the aforementioned layer concept should also help.

Global Transmission Network

Synchronous transmission can be said to encompass global communication. Global synchronization through the repeated use of pointers is what makes such a notion possible. Unification of the North American and European hierarchies is a significant step toward global unification. On that basis, and propelled by such preparatory measures as overhead allocation and a multiplexing structure based on the network concept, a global communication network should become a reality in the near future.

3.1.5 Standardization Process of Synchronous Digital Transmission

For synchronous transmission to attain its unique features, Metrobus and SONET played an important role. *Metrobus* is an internally synchronous optical communication system developed by AT&T Bell Laboratories, and *SONET* is a North American optical communication interface standard subsequently proposed by Bellcore and approved by the T1 Committee. Metrobus challenged the classical optical communication concepts of existing systems and founded the concept of an internally synchronous optical communication system by exploiting such innovative ideas as one-step multiplexing, use of containers, versatile utilization of overheads, and the use of a 150-Mbps signal as the internal standard. SONET followed and contributed the layer concept with 50-Mbps signal as the base, the synchronization method via the use of pointers, and the systematization of overhead assignment. The present SDH inherited the best of

Metrobus and SONET and makes standardized global communication a reality by employing 155.52 Mbps as the base signal and encompassing European digital hierarchy as well as the North American.

Metrobus

Metrobus is an optical communication system first proposed by J. D. Spalink of AT&T Bell Laboratories in 1982. It was developed on a full scale from early 1984 and entered the *first office application* (FOA) stage by early 1987. The prime objective of Metrobus was to develop an optimal communication system that takes full advantage of optical communications and that is most appropriate with respect to communication network evolution, device technology, and service growth. The name *Metrobus* arose from the fact that it was originally developed to accommodate all services in metropolitan areas. During the research and development stage of Metrobus, many new concepts were introduced. Among the most representative were the point-to-multipoint optical network, an internally synchronous system, the visibility of DS-0, one-step multiplexing, simultaneous accommodation of tributaries by controlling the number of containers, establishment of 150 Mbps as the internal signal standard, and maximum utilization of overhead.

The idea of a point-to-multipoint network was considered revolutionary when it was first introduced because, at that time, all existing optical systems employed point-to-point communication. The concepts listed above played an indispensable role in making it a reality. Given the seemingly infinite bandwidth of optical communications, sufficient space could be allocated for overhead such that a devoted overhead channel could be constructed to employ applications that involve the whole optical network. The bit rate assigned for overhead, however, was more than 4.5% of the total transmission capacity, an innovative concept difficult to accept at the time.

Choosing 150 Mbps (146.432 Mbps, to be precise) as the internal signal standard of the communication network was the key notion. First, 150 Mbps is the signal rate that can accommodate all signals from DS-1 (1.544 Mbps) to DS-4E (139.264 Mbps). From the service application standpoint, the existing voice, data, and video signals (including the compressed HDTV signals) can all be accommodated at 150 Mbps, and it is expected that CMOS technology can be easily used up to this rate. From the subscriber loop viewpoint, *light-emitting diode* (LED) and *pin* diodes could be used instead of laser diodes and *avalanche photodiodes* (APDs), and optical fiber coupling efficiency could be increased by employing graded-index multimode fibers instead of single-mode fibers.

One-step multiplexing was also a revolutionary concept at the time. Direct multiplexing of the DS-1 signal to the 150-Mbps internal standard signal without the involvement of intermediate DS-2 and DS-3 could not be found in the existing asynchronous multiplexers, and it acts as a cornerstone for efficient add/drop and cross-connect, which occur frequently in optical communication networks. One of the tools that makes one-step multiplexing possible is the idea of accommodating multiplexed tributaries by controlling the number of containers. In other words, by defining the size of a container, DS-1, DS-1C, DS-2, and DS-3 can be simultaneously loaded onto 1, 2, 4, and 28 containers, respectively. Since all the signals are manipulated on a container unit basis, add/drop and cross-connect can be executed efficiently.[5]

The internal standard signal, composed of 13W (words) × 88, is shown in Figure 3.5, and has a bit rate of 146.432 Mbps. During the 125-μs/13 time interval, 88 words get loaded onto 88 containers; four of these are used as overheads, and each DS-*n* signal occupies a corresponding number of containers.

The introduction of the internal synchronization concept has signaled the beginning of the synchronous communication network. The internally standardized communication network is designed to serve a specific metropolitan area, and is therefore adequately equipped to establish communication with plesiochronous neighbors within the area. Here, the clock signal can be chosen from among the *basic synchronization reference frequency* (BSRF), the local oscillator frequency, and the frequency derived from the received signal.

The visibility of the DS-0 can be said to be a direct consequence of constructing a 125-μs-based frame structure. In other words, if the construction of the frame structure and the mapping of the tributaries into respective containers are all performed on a 125-μs-unit basis, the DS-0 signals obtained from the 8-Kbps sampling rate are transparent or can be directly accessed from higher-order signals. Hence, a DS-0 signal from the 150-Mbps internal standard signal can be extracted efficiently.

The organization of the Metrobus system is shown in Figure 3.6. The line labeled "Synchronous internal bus" corresponds to the 146.432-Mbps internal standard signal. This signal is produced from DS-1 through DS-3 signals after they go through the *programmable multiplex bank* (PMB). As is shown in the figure, the direct optical communication at 146 Mbps, 878 Mbps, or 1.7G bps is possible either by passing the signal straight through *lightwave transmission equipment* (LTE) or by *word-interleaved multiplexing* (WIM) of 6 or 12 of the signals first, then transmitting the group through LTE. Also, at *programmable cross-connect* (PCC), the containers built out of 146-Mbps signals to take the DS-1 bit rate are used to perform the cross-connect function.

SONET

SONET is an abbreviation for Synchronous Optical Network, and was first conceived by R. J. Boehm and Y. C. Ching of Bellcore. It was proposed as an optical

---

5. The term *container* was first used when CCITT prescribed the STM-*n* signal.

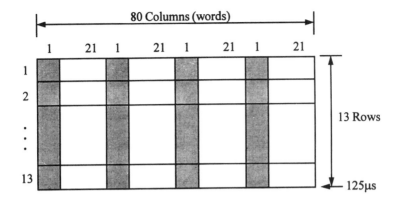

Figure 3.5 Metrobus frame structure (shaded region for overhead, blank space for payload).

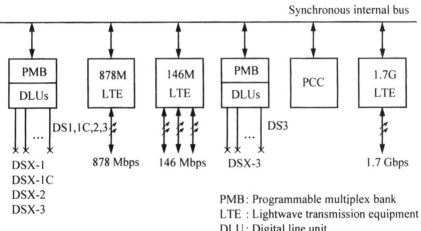

Figure 3.6 Metrobus system architecture.

communication interface standard to the T1 Committee (which functions as the North American standardization organization) at the end of 1984. As indicated in Figure 3.7(a), the frame proposed at that time had a $3 \times 8 \times 33B$ format, with a bit rate of 50.688 Mbps ($= 3 \times 8 \times 33 \times 8 \times 8$ Kbps). This base signal was named *synchronous transport signal* (STS)-1, and it was determined that

DS-3 or SYNTRAN DS-3, which carry basic tributaries, were to be byte-interleaved multiplexed to higher order signals through STS-1.[6]

The initial SONET was proposed for the purpose of "midspan meet"; hence, there was skepticism as to its realizability and little advancement achieved toward its standardization during the first year after its inception. But the standardization process was rejuvenated by the announcement of Metrobus in September 1985, and innovative ideas such as the concepts of layered system structure and pointer-based synchronization were proposed by T1 Committee participants to provide added momentum.[7] The original developers of SONET proposed the systematic refinement of the original frame structure according to the formula $(28 + L)(24 + M)(8 + N)$. This formula signifies the allocation of overhead spaces at the DS-3, DS-1, and DS-0 levels, which are, respectively, L of DS-1 size, M of DS-0 size, and N of bit size. A frame structure reorganization based on this formula, with $L = 2, M = 2, N = 0$, is the midstage SONET frame shown in Figure 3.7(b), which is a 26B × 30 structure with a 49.92-Mbps rate. AT&T Bell Laboratories then proposed that the Metrobus internal standard signal with a 26B × 88 frame structure and a 146.432-Mbps rate be adopted as a provisional SONET standard signal. This signal can be represented by a modified Bellcore formula, $\{J + K(28 + L)\}(24 + M)(8 + N)$, with $J = 1, K = 3, L = 1, M = 2$, and $N = 0$. Here, L, M, and N are the same as before, K indicates the number of DS-3-level signals, and J indicates the size of the overhead in terms of the equivalent number of DS-1 level signals.

A heated debate followed in the T1 Committee between the two alternate signal standard candidates. Even detailed functional comparisons could not clearly evince the superiority of 50 Mbps over 150 Mbps, or vice versa. Ultimately, the T1 Committee selected 49.92 Mbps as the STS-1 standard in the early part of 1986. At the time, CCITT was engaged in broadband optical channel standardization, and T1 proposed that 149.976 Mbps, which is three times the rate of the adopted SONET standard, be used as the North American standard. This signified that the T1 Committee, who selected 50 Mbps over 150 Mbps, ironically admitted the adequacy of 150 Mbps from the technical standpoint.

Afterwards, the SONET standardization process, steered by the T1X1 Working Group, sailed smoothly on, and by early 1987 even the specific standards were approved. But a full-scale mediation with CCITT ensued regarding the BISDN NNI interface standard. The 13B × 60 frame structure and 49.92 Mbps (or 149.976 Mbps in groups of three) were designed for North

---

6. SYNTRAN (Synchronous Transmission) is a synchronous DS-3 signal and has the same rate and frame structure as the DS-3, but its effective payload is restructured to be transparent to DS-0. SYNTRAN DS-3 was first proposed in 1983 by G. R. Ritchie and after two years of discussion was established as the North American signal standard by the T1 Committee.
7. The concept of a pointer was introduced to the T1 Committee by J. Ellson of Bell Northern Research in 1986.

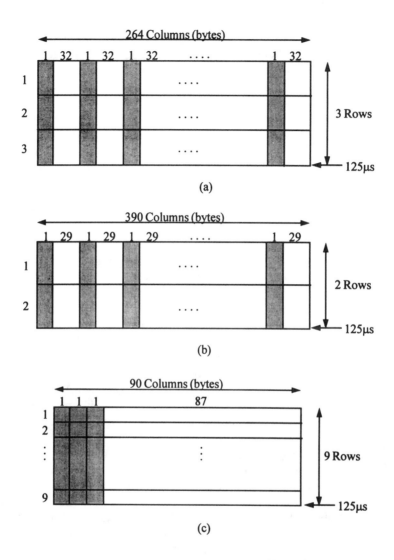

Figure 3.7   SONET frame structure (shaded region for overhead, blank space for payload): (a) initial stage structure; (b) mid-stage structure; and (c) final stage structure.

American digital signals and thus were limited in their ability to accommodate European digital signals. Hence, the SONET frame structure was once again altered, as shown in Figure 3.7(c), to 9B × 90 with a new bit rate of 51.84 Mbps (or 155.52 Mbps in groups of three). This modification was approved at a Phoenix meeting in February 1988, and after a vote at a Seoul meeting in April 1988 the SONET interface standardization was completed. The final SONET interface standard reflects various contributions from Metrobus, such as the mapping of tributaries. SDH, as established by CCITT Recommendations G.707 to 709, is based on a signal that is three times the SONET standard, and thus has a 9B × 270 structure and a 155.520-Mbps bit rate.

Synchronous Digital Hierarchy

One of the results of ISDN standardization by CCITT in the early part of 1980 was the selection of H1, H2, H3, and H4 as the user high-speed channels. Among the four, the H1 channel was standardized in two different forms: as the H11 channel with a 1.536-Mbps rate based on the North American DS-1 signal, and the H12 channel with a 1.920-Mbps rate based on the European DS-1 signal. The H2, H3, and H4 channels had a general outline appropriate to the existing digital hierarchy; but from 1985 on, they formed the basis for the ensuing broadband channel standardization. At first, 30- to 40-, 45-, or 60- to 70-Mbps bit rates were considered for the standardization, as well as the SONET standard 149.976 Mbps, which was proposed by the T1 Committee.

In July 1986, CCITT, with SG XVIII playing the central role, began the process of standardizing SDH, to be used for NNI independently of UNI. This was the instigating step toward full-scale SDH standardization, and for that purpose the T1 Committee and CCITT maintained a close relationship. For the standardization of the STM-1 signal, which is the essence of the BISDN's NNI, North America, at a meeting in Brasilia in February 1987, formally proposed the use of 50 Mbps based on STS-1 of SONET, while CEPT insisted on the necessity of using the 150-Mbps standard, which can accommodate both North American and European digital hierarchies. In the Hamburg meeting in July of the same year, the United States put forth a modified signal that was based on STS-3, with a 149.976-Mbps bit rate and a frame structure of 13B × 180, while CEPT responded by proposing a 9B × 270, 155.520-Mbps signal. The two organizations debated these alternatives heatedly for a while, with the problem of accommodating 8.448-Mbps DS-2E and 34.368-Mbps signals being the crucial point. The 9B × 270 structure won out and was finally chosen as the standard at the Seoul meeting in February 1988. This NNI standard was made official in CCITT Recommendations G.707 through 709 in June of that year, and the SDH based on the STM-1 signal with the 9B × 270 frame structure and 155.520 Mbps came into being.

Even after the 1988 CCITT recommendations were finalized, active study and research on SDH continued. When the initial SDH standard was determined in 1988, the synchronous multiplexing system already had a fixed structure like the one shown in Figure 3.8(a). But during the two years of study that followed, in which additional SDH standardization was established through CCITT Recommendations G.781–784 and G.957–958, the synchronous multiplexing system was simplified, resulting in the one shown in Figure 3.8(b). We can see from the figure that the multiplexing paths of European tributaries converge

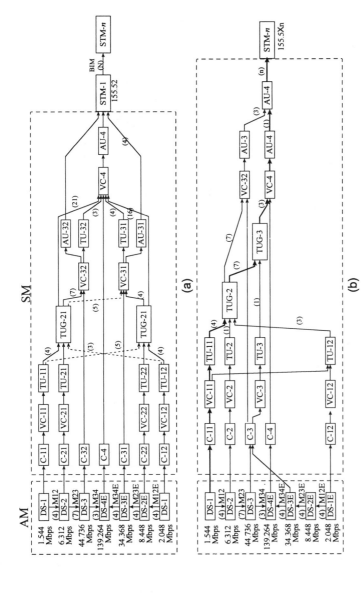

Figure 3.8 Synchronous multiplexing structure (rates in megabits per second): (a) initial structure and (b) final structure.

across the board with those of the North American tributaries, and that the concepts of AUG and TUG-3 have been applied.

As can be inferred from the preceding discussion, the contributions of Metrobus on SONET, and SONET on the standardization of SDH, cannot be overemphasized. The unique features of SDH, such as the concepts of an optical communication network, an internally synchronous system, the visibility of DS-0 through a 125-µs time unit, one-step multiplexing, the accommodation of multirate signals by controlling the number of containers, the establishment of 150 Mbps as the internal signal standard, and the enhancement of adaptability and reliability through versatile use of overheads, are all derived from Metrobus. Also, the layered system structure, systematic overhead organization, synchronization via pointers, and the possibility of establishing a global network are the by-products of SONET interface standardization. Upon such a foundation, SDH standardization, by accommodating both the North American and European hierarchies, makes the realization of a global communications network possible.

On the other hand, SDH, whose initial goal was the standardization of the NNI of the BISDN, also had a decisive influence on the BISDN UNI standard. First of all, prescribing the bit rate of BISDN UNI as 155.520 Mbps, and further prescribing the bit rate of the associated payload to be under 149.760 Mbps, is a direct result of the BISDN NNI standard. Also, the mapping of the ATM cells into VC-4 payloads for the purposes of SDH-based transmission at the BISDN UNI is another of the consequences. As described, SDH played a crucial role in introducing a new mode of transmission and finalizing the format of the BISDN.

3.1.6 SDH Versus SONET

As discussed earlier, in relation to the standardization process of synchronous digital transmission, SDH and SONET maintain an extremely close relationship. SONET standardization activities were what made SDH standards possible, while SDH was instrumental in expanding the scope of SONET so as to make it applicable for a global communications network. Therefore, a description of SDH inevitably includes a treatment of SONET.

However, minute differences do exist between SDH and SONET. For example, the base rate of SDH is around 150 Mbps, while that of SONET is around 50 Mbps. This implies that SDH employs DS-4E as the highest level tributary to accommodate lower order signals, whereas SONET employs DS-3 as the highest level tributary. Of course, this is not a crucial difference, due to the concept of concatenation. In other words, concatenation of three SONET's base transmission signal STS-1 or STS-3c is equivalent to the 155.520-Mbps STM-1 of SDH.

SDH and SONET also differ slightly in the respective diversity of transmission rates (see Table 3.1). SDH is based on STM-1 (155.520 Mbps), with

Table 3.1
Transmission Speed of SDH and SONET

SDH		SONET	
N	*STM-N*	*N*	*STS-N*
		1	51.840 Mbps
1	155.520 Mbps	3	155.520 Mbps
		9	466.560 Mbps
4	622.080 Mbps	12	622.080 Mbps
		18	933.120 Mbps
		24	1,244.160 Mbps
		36	1,866.240 Mbps
16	2,488.320 Mbps	48	2,488.320 Mbps
		.	.
		.	.
		.	.
64	9,953.280 Mbps	192	9,953.280 Mbps

STM-4 (622.082 Mbps) and STM-16 (2,488.320 Mbps) being the signals of primary interest. In contrast, STS-1 (51.840 Mbps) is used as the base signal unit in the case of SONET, with STS-3 (155.520 Mbps), STS-9, STS-12 (622.080 Mbps), STS-18, STS-24, STS-36, and STS-48 (2,488.320 Mbps) being of primary interest. Here, the STM-n signal generally has the same transmission rate as STS-3n. To be more precise, STM-n is the equivalent of the concatenated signal STS-3nc. The optical counterpart of STS-m is called the *optical carrier level m* (OC-m).

With respect to the frame format, SONET is equivalent to the reduction of SDH by a factor of one-third. SDH has a 9 × 270B frame structure in the case of STM-1, while STS-1 of SONET has a 9 × 90B structure. Similar to the SOH of STM-1, which is located at the head of the STM-1 frame with the 9 × 9B format, the SOH of STS-1 is located at the head of the STS-1 frame with the 9 × 3B format. In both cases, the fourth row is allocated to the pointer. Examined in more detail, if the first, fourth, and seventh columns of the STM-1 SOHs are taken, then what results is the same as the SOH of STS-1. The functions of the components of the respective overheads are identical.

As was alluded to earlier, SDH and SONET employ different base signal units, namely, the 150-Mbps rate STM-1 and the 50-Mbps rate STS-1. This means that in the case of SDH the construction of STM-1 requires systematic multiplexing of all the tributaries from DS-1 to DS-4, while STS-1 needs to multiplex just five of the tributaries, which are DS-1, DS-1E, DS-1C (3.152 Mbps), DS-2, and DS-3. Consequently, the creation of STM-1 necessitates the establishment of such intermediate signal units as C, VC, TU, TUG, AU,

AUG, as explained in Section 3.1.3, and also the systematic multiplexing procedures depicted in Figure 3.8(b). In contrast, STS-1 requires only one type of intermediate signal, called *virtual tributary* (VT). The VT is equivalent to the VC of SDH, and the VTs that correspond to VC-11, VC-12, and VC-2 are VT1.5, VT2, and VT6, respectively. Also, VT3 is the VT for the DS-1C signal.

SDH and SONET have different sets of intermediate signal units, and consequently the multiplexing structure of one is different from that of the other. SDH requires a systematic multiplexing structure that can link C, VC, TU, TUG, AU, AUG, and STM-n as shown in Figure 3.8(b), while the multiplexing structure of SONET only has to link DS-m, VT, and STS-1, and is therefore much simpler. Here, the method of mapping tributaries into VT1.5, VT2, and VT6 is equivalent to mapping into VC-11, VC-12, and VC-2, and the mapping of DS-1C into VT-3 is similarly based on P/Z/N justification. The mapping of these VTs into the payload space of STS-1 [i.e., the *synchronous payload envelope* (SPE)] is analogous to the mapping of related VCs into VC-3 via TUG-2. Lastly, the mapping of DS-3 into SPE is the same as the mapping of DS-3 into VC-3. However, the mapping of SYNTRAN DS-3 is additionally provided in SONET.

If the interrelationship between SONET and SDH is examined from the terminology point of view, VT1.5, VT2, and VT6, respectively, correspond to VC-11, VC-12, and VC-2; STS-1 SPE corresponds to VC-3; and STS-3c to STM-1. Also, with respect to the terms related to layering, the physical medium, regenerator section, multiplexer section, and path layer of SDH are respectively called photonic, section, line, and path layer in SONET. The respective terms relating to mapping, multiplexing, overhead, and synchronization are almost identical.

Like SDH, SONET is based on the layering concept and hence uses the 125-µs frame, systematically utilizes overheads, and has the same base transmission rate. Also, it collectively accommodates all of the North American digital tributaries as well as the European DS-1E, features one-step multiplexing, and employs the same synchronization technique based on pointers as the SDH and consequently can encompass all of North America under a single synchronous digital network.

3.2 FRAME STRUCTURE AND SYNCHRONOUS MULTIPLEXING

The STM-n frame structure is the essence of the SDH and synchronous digital transmission because all the distinct characteristics of the SDH such as the transmission rate, the visibility of DS-0, the layer concept, and synchronization via pointers are incorporated in the STM-n frame. This frame structure is formed out of the digital tributaries through the synchronous multiplexing process. In this section, the organization of the STM-n frame structure is described first, and then the various related fundamentals of synchronous multiplexing are examined.

3.2.1 Structure of STM-*n* Frame

The STM-*n* frame structure is an *n*-fold extension of the STM-1 frame structure shown in Figure 3.4. It occupies 9B × 270 × *n* bit space over 125 μs, which translates into a bit rate of 155.520 × *n* Mbps (9 × 270 × *n* × 8 × 8 Kbps). The 9B × 261 × *n* partition of the frame is used for carrying the STM-*n* payload; the 3B × 9 × *n* and 5B × 9 × *n* partitions are dedicated for the overhead; and the remaining 1B × 9 × *n* is used for the AU PTR (see the bottom half of Figure 3.9).

Organization of STM-n Frame Structure

Typically, *n* of VC-4 can be mapped into the STM-*n* payload space. An AU-4 signal is produced by adding an AU-4 PTR to the VC-4 signal, which is equivalent to the AUG. On the other hand, byte-interleaved multiplexing (BIMing) of three AU-3s results in the AUG, as shown in Figure 3.10. Additional BIMing of *n* AUGs, together with an SOH, produces the STM-*n* signal, as shown in Figure 3.9.

Hence, to study the STM-*n* frame, it is enough to examine the AUG signal, and to study SOH as well, it is sufficient to investigate the STM-1 signal. Therefore, unless a special need arises, our discussion focuses on STM-1. So BIMing *n* AUGs makes up the signal, which is equivalent to the STM-*n* payload space affixed by the AU PTR. Attaching SOH space yields the STM-*n* frame shown in Figure 3.9.

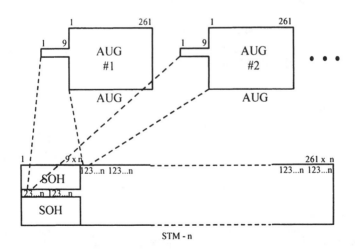

Figure 3.9 Organization of STM-n.

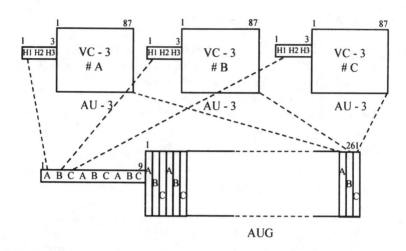

Figure 3.10   Organization of AUG.

Payload Space

The payload space of STM-1 is used to carry one VC-4 or three VC-3 signals. Loading of one VC-4 (or three VC-3) onto the STM-1 payload is done in a floating mode, with the pointer indicating the location of its first byte in the STM-1 payload. Since the STM-1 payload consists of 2,349B (= 9B × 261), if an address is assigned to each of 3B units of the payload, 783 addresses are required. The address is assigned in the row direction from 0 to 782 immediately after the AU PTR, as shown in Figure 3.11. Such an addressing scheme is useful not only for the AU-4 but also for the AU-3. In the case of AU-3, three sets of 783 addresses are required to assign a unique address to three sets of the AU-3. Therefore, at least 10 bits are required for addressing whole STM-1 payload bytes. The address thus assigned also indicates the degree of offset of each address location from the pointer location.[8]

Section Overhead Space

The SOH space is split into the upper and lower parts by the AU PTR, which is located between. The upper part is the RSOH and is used to improve the transmission reliability between regenerators. Each regenerator looks at only this part of the overhead and ignores the information carried in the rest of the

---

8. This implies that the address to be indicated by the AU-4 PTR or AU-3 PTR total 783 in number. The number of addresses to be represented by the TU-3 PTR, TU-2 PTR, and TU-11 PTR are 765, 428, 140, and 104, respectively. Hence, 10 bits are sufficient for these purposes (see Section 3.6.2).

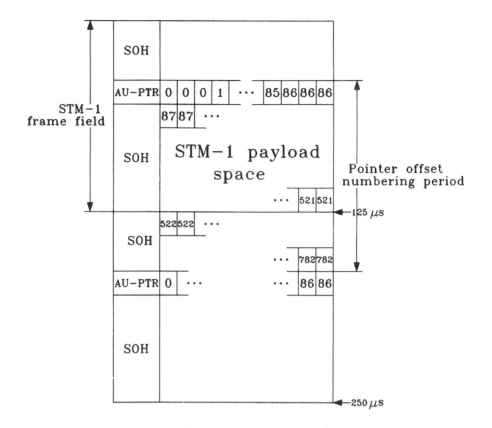

Figure 3.11 Addressing (or pointer-offset numbering) of the STM-1 payload.

frame. The lower part corresponds to the MSOH and is used to carry the information necessary to monitor the multiplexer section. When an STM-1 signal flows into a multiplexer, the multiplexer checks and examines only this part of the overhead. More details are given in Section 3.5.

AU Pointer Space

The nine bytes of AU PTR on the fourth row of the STM-1 overhead consist of three triplets of H1, H2, and H3, as shown in Figure 3.10. They are employed to keep track of the shifting location of the first byte of the VC-4 or the VC-3. If the STM-1's payload is carrying a VC-4 (i.e., in the case of AU-4), only the first triplet of H1, H2, and H3 is used. On the other hand, in the case of AU-3, each of the H1, H2, and H3 triplets independently keeps track of the address of each AU-3. Out of the 24 bits that correspond to the three bytes of H1, H2, and H3, only 10 bits are needed to indicate the addresses from 0 to 782, and hence the remaining 14 bits are used for other purposes. A more detailed discussion of this matter is given in Section 3.6.

Transmission of STM-n

The transmission of STM-n in its natural form may result in long sequences of 0's and 1's. To prevent this, the *frame synchronous scrambler* (FSS) whose generating polynomial is $1 + x^6 + x^7$ is used. This scrambler operates at the transmission rate and applies to the entire STM-n frame except for the first row of SOH, which is left unscrambled because it contains frame alignment words A1 and A2 (see Section 3.5.2). The scrambler is reset to 1111111 after the last byte of the first row of the SOH, and then commences its frame scrambling function.

3.2.2 Tributaries for Synchronous Multiplexing

The tributaries that form the SM structure can be either North American or European. In Figure 3.8(b), the right rectangle represents the SM structure, and the left rectangle represents the existing AM structure. Hence, it can be seen that all of the tributaries are mapped into the STM-n either by going through both the AM and the SM, or directly through the SM.

To elaborate, we take American tributaries as an example. As can be seen from the figure, each of the DS-1, DS-2, and DS-3 signals can be directly mapped into the corresponding container in the SM, or can be asynchronously (or plesiochronously) multiplexed into a higher tributary, and then mapped to a larger container in the SM. Hence, the DS-1 can be either loaded directly onto the C-1 or multiplexed to the DS-2 first, and then get mapped into the C-2. The same scenario applies to DS-2 and DS-3.

The European SM structure is somewhat less complete than its North American equivalent. The original SDH multiplexing structure [as shown in Figure 3.8(a)] could accommodate direct mapping of all North American and European plesiochronous tributaries to the SM structure by supplying a suitable container to each tributary. However, for practical reasons, the structure was simplified, resulting in the one shown in Figure 3.8(b). As can be inferred by comparing the two figures, the direct mapping of DS-2E into the SM structure is not possible in the simplified version. DS-2E has to be multiplexed to the higher tributary DS-3E first and then mapped into C-3 before it can enter the SM structure.

3.2.3 Signal Elements of Synchronous Multiplexing

The signal elements that form the SM structure include the container, the VC, the TU, the TUG, the AU, the AUG, and the STM.

Container (C)

The container is the most elemental unit of the SM structure in the sense that all of the North American and European PDH tributaries have to be mapped into the respective containers before they can proceed with the SM process and emerge as a part of the STM-n.

Containers of the SM structure are categorized into classes C-1, C-2, C-3, and C-4, with the number denoting the corresponding digital hierarchical levels. C-1 can further be categorized into C-11 and C-12, with C-11 accommodating the North American DS-1, and C-12 accommodating the European DS-1E. C-4 can carry either a DS-4E from the PDH, or the ATM cells of the BISDN. Particulars of the respective tributary mappings are described in Section 3.4.

Virtual Containers (VCs)

The VC's function is to support the connections between the path section layers in synchronous transmission. The VC consists of the payload, which carries the information data, and the POH. The payload portion corresponds to a container, and the whole VC frame is repeated every 125 or 500 μs. The four classes of VC, namely, VC-1, VC-2, VC-3, and VC-4, correspond to C-1, C-2, C-3, and C-4, respectively. Similar to C-1, VC-1 can be further categorized into VC-11 and VC-12. VC-1 and VC-2 are called the *lower order VCs* and VC-3 and VC-4 are called the *higher order VCs*. The POH for the lower order VCs is called *V5* and the POH for the higher order VCs is called *VC-3 POH* or *VC-4 POH*.

Tributary Unit (TU)

The TU was designed to provide an adaptability between higher and lower order path layers. For instance, lower order VCs can be mapped into higher order VCs through a TU or a TUG. A TU is created by attaching a TU PTR to a lower order VC, and here the pointer is used to indicate the degree of offset of the lower order VC relative to the starting position of the higher order VC's frame. The TU is categorized into TU-1, TU-2, and TU-3. TU-1 is further categorized into TU-11 and TU-12, depending on the type of VC it contains.[9]

Tributary Unit Group (TUG)

The role of the TUG is to collect one or more TUs and load them onto a fixed location on the payload of a higher order VC. No overhead is added when

---

9. The *lower order* VCs refer to VC-1 and VC-2 and *higher order* VCs refer to VC-3 and VC-4. In contrast, the lower VC and upper VC are relative concepts that indicate the relative relation between VCs; for example, VC-3 is a lower VC of VC-4.

a TUG is formed from the TUs. There are two classes of TUG: TUG-2 and TUG-3. TUG-2 is formed by assembling a homogeneous group of TU-1s or by a direct mapping of a single TU-2. Similarly, TUG-3 could be an assembly of TU-2s or a single TU-3.

Administrative Unit (AU)

The AU functions as an adapter between the higher order path layer and the multiplexer section layer. As before, AU consists of the payload and the AU PTR. The payload carries a higher order VC, and the AU PTR indicates the relative offset between the starting positions of the AU payload and the frame of the multiplexer section layer. In other words, the two types of AU, namely, AU-3 and AU-4, carry VC-3 and VC-4, respectively, and the AU PTR indicates the degree of offset of VC-3 or VC-4 with respect to the STM-n frame.

Administrative Unit Group (AUG)

One or more AUs occupying a fixed location on an STM payload is called an AUG. An AUG can consist of three AU-3s or a single AU-4.

Synchronous Transport Module (STM)

The STM is the final product of the SM structure and is the signal that is actually transmitted over the synchronous transmission networks. STM-n is formed by byte-interleaving n AUGs and the addition of SOH to the beginning of its frame. Here, n of the numbers 1, 4, 16, and 64 are of primary interest.

3.2.4 Synchronous Multiplexing Processes

Figure 3.8(b) depicts the synchronous multiplexing process in terms of the signal elements. If the figure is modified to reflect the various required procedures of SM, namely, mapping, aligning, pointer processing, and so on, then Figure 3.12 is the result.[10] In the figure, the solid arrow represents mapping, the dotted arrow represents aligning, and the boldfaced arrow represents multiplexing.

---

10. The bottom part of the figure shows the mapping of ATM cells into the VC-4. A detailed discussion of this procedure is given in Section 3.4.7.

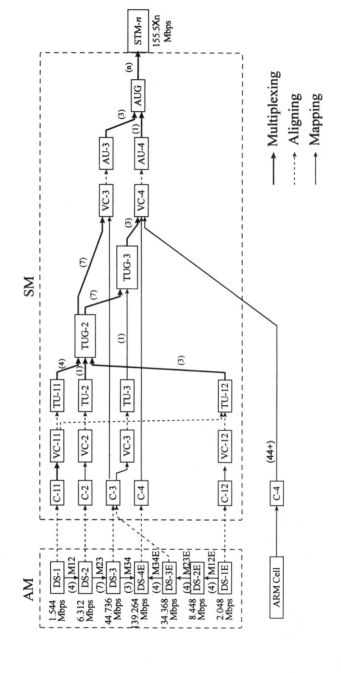

Figure 3.12 Classification of synchronous multiplexing procedure (including the path for ATM cell).

Mapping

Mapping is the appropriate transformation of tributaries into the corresponding containers or VCs across the SDH network border.[11] Since the tributaries are sent from an asynchronous environment, P/Z/N justification is required before they can be mapped into the synchronous containers or VCs.

Aligning

Aligning refers to the process of loading a VC onto a TU or an AU, along with the "frame offset" information. Here, the frame offset is due to the clock discrepancy between the VC and the corresponding TU or AU. The VC is aligned on a 1B or 3B unit basis, and the alignment status is indicated by the TU or AU PTR.

Pointer Processing

Pointer processing is employed when the frame offset occurs due to the differences in the clock frequencies between a VC and the corresponding TU or AU. Pointer processing involves the indication of the starting position (and its alteration information) of the VC on the payload space of the TU or AU, and the associated P/Z/N justification information.

Multiplexing

The process by which multiple lower order path layer signals are adapted into a higher order path layer signal, or the appropriate transformation of multiple higher order path layer signals into a signal element of SM or AM, is called *multiplexing*.[12] Figure 3.12 shows characteristic multiplexing processes, such as the adaptation of TUs through a TUG so that they can be fitted onto a higher order VC, or the adaptation of AUs through an AUG so they can be loaded onto the STM-*n*. Here, when the TU or the AU is multiplexed to the TUG or the AUG, no additional overhead is employed, while when the TUG is multiplexed to a higher order VC, a POH is added, as was the case when a container was mapped into a VC. The SOH is added when AUG is multiplexed to the STM-*n*.

---

11. A VC is nothing more than a container with a POH attached to it, and thus carries the same payload as the container. Hence, when it is said a tributary is mapped into a container, there can also be the implication of a mapping into a VC because the two mapping procedures are equivalent.
12. This is multiplexing in the narrow sense of the word. Multiplexing in the broad sense encompasses mapping, aligning, pointer processing, and concatenation.

Concatenation

Concatenating multiple VCs, then reducing the total load to the capacity of a single VC while maintaining the integrity of the bit sequences is called *concatenation*. For example, if there exists a payload that needs an x number of containers for mapping, since it is difficult to preserve the correct bit sequence when they are divided into x number of VCs and then transmitted, the x VCs are concatenated to form a new virtual container VC-2-xc and are then transmitted. This is equivalent to the formation of a VC-4-xc by concatenating x VC-4s. Here, x denotes the number of concatenated VCs, and the letter c stands for concatenation. TU-2-xc and AU-4-xc are formed in a similar way.

3.2.5 Synchronous Multiplexing Structure

If Figure 3.8(b) or Figure 3.12 is studied, we see that the synchronous multiplexing structure can be broadly categorized into the lower order structure, which involves C-1 and C-2, and the higher order structure, which involves C-3 and C-4. In Figure 3.13, these two structures are drawn separately.

Lower order multiplexing entails the transformation of the low-level tributaries DS-1, DS-1E, and DS-2 into TUG-2 after they go through multiple stages of mapping and multiplexing via corresponding containers, VCs, and TUs. Here, the peculiarity is that the DS-2E tributary is not a part of lower order multiplexing because it has to be multiplexed to DS-3E and then get loaded onto the C-3. Also, it is worthwhile noting that the DS-1 can be mapped into C-11 and VC-11, and then multiplexed to TUG-2 via TU-11 or TU-12. Creation of TUG-2 marks the end of the lower order multiplexing stage, and from then on TUG-2 is on an equal level with C-3 and C-4 and is hence a part of the higher order multiplexing.

Higher order multiplexing involves the multiplexing of high-level tributaries into the AUG and the STM-n via VC-3 or VC-4, and subsequently via AU-3 or AU-4. The associated signals include the tributaries DS-3, DS-3E, DS-4E, and TUG-2. Here, both DS-3 and DS-3E are mapped into C-3 and then VC-3. VC-3 can be aligned directly within AU-3 or multiplexed to VC-4 via TU-3, and then be aligned within AU-4.

Therefore, every multiplexing path is bound to go through the higher order multiplexing stage, and the low-level tributaries must, in addition, go through lower order multiplexing. As shown in Figure 3.13, there are four different multiplexing paths from the lower order tributaries to TUG-2. Also, two independent multiplexing paths exist from TUG-2 to AUG, and, hence, a lower order tributary can be multiplexed to AUG in eight different ways. On the other hand, higher order multiplexing paths are three in number. Hence, in sum, 11 independent multiplexing paths exist within the SM structure.

Among the 11 different paths, the path that emerges as the most

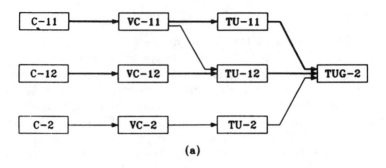

(a)

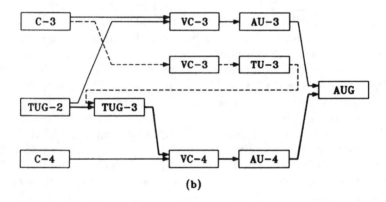

(b)

Figure 3.13 Synchronous multiplexing structure: (a) lower-order multiplexing and (b) higher-order multiplexing.

economical is the C-1\VC-1\TU-1\TUG-2\TUG-3\VC-4\AU-4\AUG path, which is highlighted in Figures 3.8(b) and 3.13. Since, from the point of view of a DS-1/DS-1E signal or the 64-Kbps DS-0 signal, TUG-2 or TUG-3 is simply an assembly of TU-1s, DS-1/D-1E is multiplexed directly to VC-4 via VC-1. Since this path evades the need for extra asynchronous multiplexing stages, it is the shortest path from DS-1/DS-1E to STM-n.

On the other hand, the path that is considered to be the most inefficient is the C-3\VC-3\TU-3\TUG-3\VC-4\AU-4\AUG path, which is indicated by a dotted line in Figure 3.13(b). Of course, in this path TUG-3 has no significance. However, the DS-3\VC-3\TU-3\VC-4\AU-4\AUG path is less efficient than the DS-3\VC-3\AU-3\AUG path. In addition, from the point of view of a DS-1/DS-1E, two stages of synchronous multiplexing are required for the signal to reach

DS-3. Therefore, the path with the dotted line is the longest and is, hence, the least efficient of the 11 STM-*n* paths.

3.3 SYNCHRONOUS MULTIPLEXING PROCEDURES

In the preceding section, an overview of synchronous multiplexing was given in terms of its signal elements, the various procedures it entails, and some characteristic multiplexing paths. With that section as the foundation, we now focus on a detailed description of synchronous multiplexing procedures. First, the multiplexing of lower order VCs to TUG-2 will be studied, then the multiplexing of higher order VCs or TUG-2s to the AUG. The respective mapping of each tributary to a corresponding container or a VC is examined in the section that follows.

3.3.1 Multiplexing of Lower Order VCs to TUG-2

As shown in Figure 3.14, the lower order virtual containers VC-11, VC-12, and VC-2 are composed of 104B, 140B, and 428B, respectively. The first of the bytes,

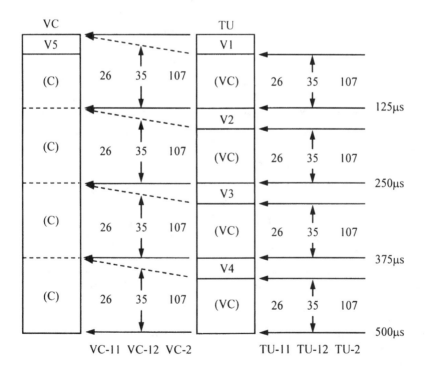

Figure 3.14 The composition of low-order VCs and TUs.

designated as V5, functions as the POH for each VC, and the remainder of the bytes are allotted for carrying the containers C-11, C-12, and C-2. The frame content of each lower order VC is a concatenation of four container frames, each 26B, 35B, or 107B over 125 μs long, resulting in a 500-μs multiframe.

As can be seen in Figure 3.14, if four pointer bytes designated V1, V2, V3, and V4 are attached to the beginning of each of the frames of a VC, a TU is created. Hence, the multiframe of a TU occupies 27B × 4, 36B × 4, or 108B × 4 over 500-μs, with the 26B × 4, 35B × 4, or 107B × 4 portion corresponding to the four sets of payloads, and four bytes of pointers reflecting the relative positions of the payloads within the TU multiframe.

In terms of the base 125-μs frame unit, the sizes of TU-11, TU-12, and TU-2 are 9B × 3, 9B × 4, and 9B × 12, respectively, consisting of 3, 4, and 129B columns. Hence, if four TU-11s or three TU-12s are joined through BIM, 12 columns result, equivalent to a single TU-2.[13] These 12 columns can be used to create a single TUG-2, and such a formation is depicted in Figure 3.15. When TUG-2 is formed, the TUs are made to have the same phase; that is, the starting points of each multiframe are identical, and the phase status is indicated by the H4 byte in the POH of the higher order VC.

As can be seen in Figure 3.13(a), VC-11 can be aligned within TU-12 and then multiplexed to TUG-2. This is made possible by transforming the shape of VC-11 to resemble VC-12 first, and subsequently performing the alignment. Since VC-11 occupies 26B over 125 μs and VC-12 takes up 35B per 125 μs, 9B of fixed stuff bytes are needed to pad the VC-11. Figure 3.16 depicts such a transformation of VC-11 to VC-12. The figure displays a multiframe consisting of four 125-μs frames. The shaded region corresponds to the fixed stuff bytes. The dummy bytes are selected to have even parity. The resulting TU-12 is displayed to the right.

3.3.2 Mapping to VC-3

The VC-3 consists of 85 9B columns. The first column is used for the POH, and the remaining 84 columns correspond to the payload, which can carry a C-3 or a TUG-2. The C-3 is mapped into the VC-3 as shown in Figure 3.17(a). A description of its internal structure is given in Section 3.4. As for TUG-2, seven can be mapped to the VC payload, as shown in Figure 3.17(b). Since each TUG-2 contains 12 9B columns, 84 columns of the VC payload can be filled to its maximum capacity. Here, all the TUGs have the same phase within the VC-3, and the phase information is conveyed by the H4 byte in the POH of the VC-3.

---

13. The fact that TU-2 and TUG-2, created by multiplexing four TU-11s, have an identical size, has the same significance as obtaining tributary DS-2 by multiplexing four DS-1s.

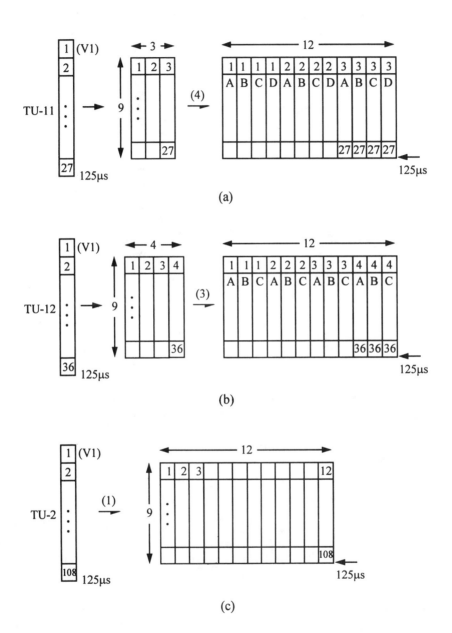

Figure 3.15 The formation of TUG-2: (a) based on TU-11; (b) based on TU-12; and (c) based on TU-2.

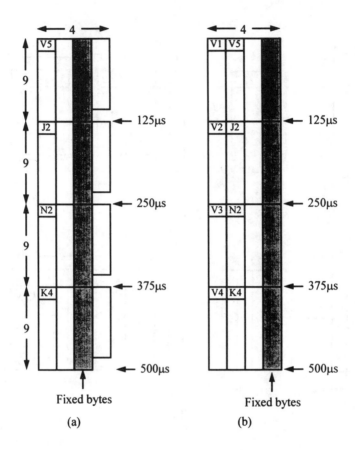

Figure 3.16 The alignment of VC-11 in TU-12 (Shaded region for fixed stuff bytes): (a) VC and (b) TU-12.

3.3.3 Multiplexing to TUG-3

TUG-3 consists of 86 9B columns, allowing the mapping of one VC-3 or TU-3, or seven TUG-2s into its payload. As shown in Figure 3.18(a), the addition of three pointer bytes H1, H2, and H3 to the VC-3 results in the TU-3, and the further addition of six fixed stuff bytes in the same column results in the TUG-3. Here, the starting position of the VC within the TU-3 (hence, within the TUG-3) can vary, and this information is indicated accordingly by the pointer bytes. Note that in mapping VC-3s into TUG-3, only the VCs mapped from the C-3s are allowed, not those mapped from the TUG-2s.

The TUG-2 is mapped into the TUG-3 in a similar fashion, but in this case a POH is not needed. This is because TUG-3s are multiplexed to the VC-4 as a single unit, and the POH can be added at that instant. Figure 3.18(b) depicts the

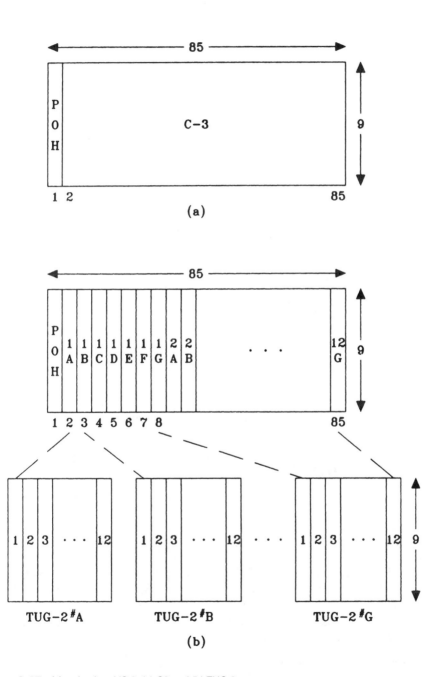

Figure 3.17 Mapping into VC-3: (a) C3 and (b) TUG-2.

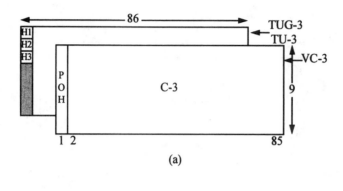

(a)

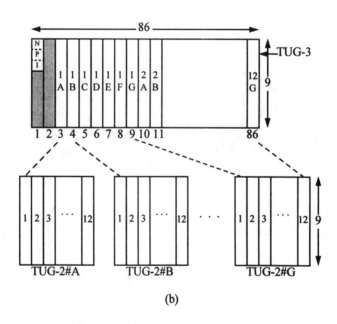

(b)

Figure 3.18 Multiplexing to TUG-3 (Shaded region for fixed stuff bytes): (a) VC-3/TU-3 and (b) TUG-2.

multiplexing of TUG-2s to the TUG-3, and here pointer bytes such as the H1, H2, and H3 from Figure 3.18(a) are lacking. The absence of a pointer is indicated by the *null pointer indication* (NPI), and this is achieved by filling the two bytes that correspond to H1H2 with the value 1001ss1111100000. (See Section 3.6.2 for a description of pointer functions.)

3.3.4 Mapping Into VC-4

The VC-4 is composed of 261 columns, with each column being 9B in size. The first column is used for the POH, and the remaining 260 columns are employed to carry the payload, to which can be mapped a C-4 or TUG-3. The multiplexing of the C-4 to the VC-4 is shown in Figure 3.19(a). The internal structure of C-4 is discussed in detail in Section 3.4.

Three TUG-3s can be byte-interleaved and mapped into the payload of VC-4, as shown in Figure 3.19(b). Since each TUG-3 is 9B × 86 in size, loading three of them to the VC-4 leaves it with two spare columns, which can be filled with fixed stuff bytes. All the TUGs are arranged to have the same phase within

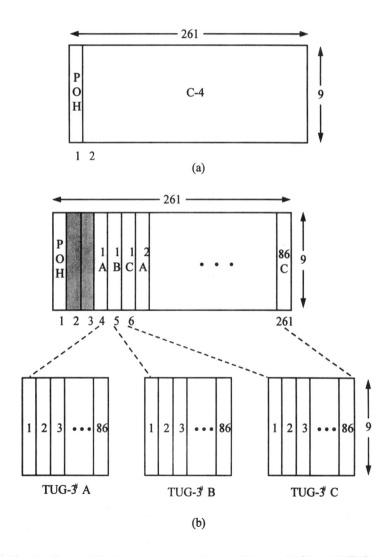

Figure 3.19 Mapping into VC-4 (shaded region for fixed stuff bytes): (a) C4 and (b) TUG-3.

the VC-4, and the phase information is carried by the H4 byte, which is a part of the VC-4's POH.

3.3.5 Multiplexing to AUG

An AUG is composed of 261 9B columns plus 9 1B columns, which are placed at the beginning of the fourth row, as shown in Figure 3.9. The addition of an SOH of 3B × 9 and 5B × 9 in size transforms the AUG into the STM-1. Also, byte-interleaving n AUGs and the addition of a corresponding SOH yields the STM-n. The AUG can be formed either from three AU-3s or from a single AU-4.

An AU-4 is produced by attaching a 9B pointer to the VC-4. The pointer bytes are H1, Y, Y, H2, 1*, 1*, H3, H3, H3, with Y and 1* representing 1001ss11 and 11111111, respectively.[14] Three H3s are due to the fact that, in the case of VC-4, P/Z/N justification is done on a 3-byte basis. A single AU-4 is directly mapped into the AUG.

An AU-3 is formed by adding two 9B columns of fixed stuff bytes to the VC-3 so that it acquires 87 columns in all, and then adding a 3B pointer. VC-3 can float within the AU-3, and its starting point is indicated by the AU-3 pointer. The fixed stuff bytes are placed in the 30th and 59th columns, and the three associated pointer bytes are H1, H2, and H3. The AUG is produced by byte-interleaving three AU-3s, and the process is depicted in Figure 3.20.

3.3.6 Concatenation of AU-4

If the payload to be transmitted is greater than the transmission capacity of a single VC-4, AU-4-xc, which is a concatenation of x AU-4s, can be used instead. Since externally AU-4-xc looks identical to the AUG-x, a pointer is used to indicate the concatenation status. The concatenation pointer's format is equivalent to x byte-interleaved AU-4 PTRs. Here, the first of the set maintains its normal pointer function, and the remaining $x - 1$ pointer elements denote the concatenation status, which is achieved by setting H1 and H2 bytes to Y and 1*.[15]

The concatenated administrative unit's payload consists of 261x columns. When $x = 1$ (e.g., in the case of AU-4), the payload is partitioned into 1 column devoted to the POH and 260 columns for carrying the C-4, which translates into a payload capacity of 149.76 Mbps (= 260 × 9 × 64 Kbps). When $x > 1$, only one of the x first columns is used for SOH, and the remaining $x - 1$ columns are stuffed with fixed overheads. Hence, the payload capacity becomes

---

14. Asterisks are used in this chapter to denote a byte consisting of identical bits. For example, 1* denotes 11111111 and, likewise, R* denotes RRRRRRRR.
15. Note that the union of Y and 1* can also function as a concatenation indicator. The term ss in Y does not signify anything (see Section 3.6.2).

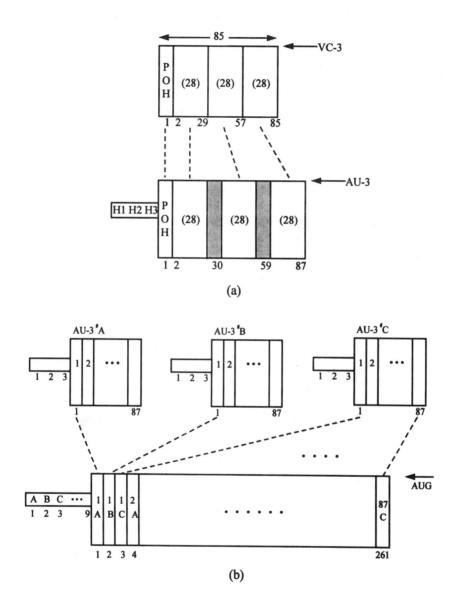

Figure 3.20 Multiplexing to AUG (shaded region for fixed stuff bytes): (a) alignment of VC-3 and (b) multiplexing of AU-3.

149.736x Mbps, and consequently, in the case of $x = 4$, the capacity becomes 599.04 Mbps.

3.3.7 Example of SM Procedure

Figure 3.21 depicts the synchronous multiplexing process; that is, the DS-1\C-11\VC-11\TU-11\TUG-2\TUG-3\VC-4\AU-4\AUG\STM-n path marked with bold lines in Figure 3.8(b). The DS-1 signal is first mapped into C-11, then C-11 becomes VC-11 with the addition of a VC-11 POH. Attaching a TU-11 PTR to VC-11 and multiplexing four of them results in a TUG-2.

Multiplexing seven TUG-2s and attaching FOH yields TUG-3, and multiplexing three TUG-3s and attaching FOH and VC-4 POH produces VC-4. Therefore, the VC-4 signal is equivalent to the signal constructed by multiplexing 21 TUG-2s and attaching VC-4 POH and FOH to the head of its frame.

As a result of this highly coordinated multiplexing procedure, each of the

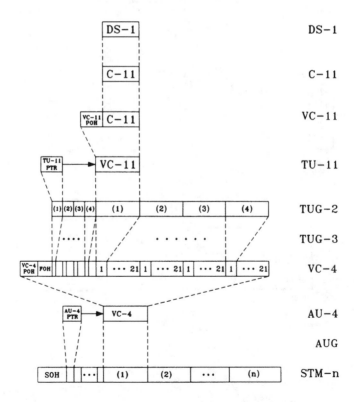

Figure 3.21   Example of multiplexing procedure for the multiplexing path DS-1\C-11\VC-11\TU-11\TUG-2\TUG-3\VC-4\AU-4\AUG\STM-n.

constituent 84 TU-11 signals can be accessed separately on a VC-4. In this case, the FOH is just an overhead to match the size of a VC-4 frame to that of the VC-3. A VC-4 together with the AU-4 PTR produces an AU-4, which is equivalent to an AUG. Multiplexing n AUGs and adding an SOH produces an STM-n.

3.4 TRIBUTARY MAPPING FOR SYNCHRONOUS MULTIPLEXING

In the preceding two sections, the synchronous multiplexing structure and the multiplexing procedures within it were examined. Hence, the multiplexing of a container or a virtual container to the STM-n via AUG and other intermediate signal elements was studied. In this section, our attention shifts to the mapping process of tributaries into the respective containers and virtual containers. This is a significant step that links the asynchronous multiplexing structure of the existing digital hierarchy to the nascent synchronous multiplexing structure. First, asynchronous and synchronous mappings are examined, then the floating mode and locked mode are studied, and finally the mapping method for each respective tributary is discussed.

3.4.1 Asynchronous and Synchronous Mapping

The mapping of a tributary into a virtual container can be classified as asynchronous, bit synchronous, or byte synchronous. Since a tributary is generally formed by using a clock that is independent of the clock used to make the corresponding virtual container, asynchronous mapping can be said to be the most common. However, those tributaries that are created as links to the synchronous multiplexers can be manipulated to use the same clock as the synchronous multiplexers, and in this case it is much simpler to use synchronous mapping. Such synchronous mapping is useful in permitting the direct entrance of DS-0 into the SM structure via DS-1 or DS-1E. Depending on whether the boundary of the DS-0 byte is exposed or not, synchronous mapping can be subdivided into bit-synchronous and byte-synchronous modes.

Asynchronous Mapping

Asynchronous mapping is applicable when the tributary clock is independent of the container clock or the virtual container clock. In practice, the respective clocks have a plesiochronous relationship. Asynchronous mapping can be applied to all the tributaries (i.e., DS-1, DS-1E, DS-2, DS-3, DS-3E, and DS-4E).

The most basic procedure in asynchronous mapping is the synchronizing process that resolves the respective clock offsets. In the case of DS-1, DS-1E, DS-2, and DS-3E, P/Z/N justification via bit stuffing is used, and in the case of DS-3 and DS-4E, only the positive justification with bit stuffing is used.

P/Z/N justification is used when the nominal transmission capacity of the signal after the synchronization procedure is identical to its nominal bit rate before the procedure; in other words, when, as in the case of mapping of tributaries, the bit rate of a VC payload is the same as the nominal bit rate of the corresponding tributary. In this case, *zero justification* is maintained in normal situations, but when the tributary bit rate becomes lower than the bit rate of the VC payload, when the accumulated offset becomes one bit long, a null bit (or a garbage bit) is sent in place of effective information data (*positive justification*). Conversely, if the tributary's bit rate becomes high with respect to the VC bit rate, a spare bit in the payload is used to absorb the offset (*negative justification*).

The P/ZshN justification function is performed through the justification opportunity bits J_1, J_2, and the execution status is indicated through the use of the justification control bits C_1 and C_2. J_1 is used as an overhead bit in normal operation, but acts as a spare bit to carry the extra data in the case of negative justification; similarly, J_2 is normally a part of the effective payload, but is used to carry the null bit in the case of positive justification.

DS-1, DS-1E, and DS-2 each have three sets of C_1 and C_2 control bits, and DS-3E has five. If the C_1's are all 0, this signifies that J_1 is a regular information bit, and if the C_1's are all 1, it implies that J_1 is carrying a null bit. Similarly, C_2 indicates whether or not J_2 is carrying a null bit. If, due to an error in the transmission process, there appears to be a lack of unanimity among the C_1 and C_2 bits, the receiver extracts the correct justification status according to the majority vote rule.

Positive justification is the synchronization procedure that is applied when the nominal transmission capacity of the signal after the synchronization becomes higher than its original nominal bit rate. In this case, since no negative justification occurs, only the opportunity bit and control bit for positive justification are needed. These bits are denoted by J and C, respectively. For every DS-3 or DS-4E, five C-bits are assigned to every J-bit. If the C's are all 0, this signifies that J is an information bit, and if the C's are all 1, this means that J is a justification bit, and the majority vote rule applies in case of bit errors.

Bit-Synchronous mapping

Bit-synchronous mapping is a mapping procedure that is used when the tributary's clock is synchronized with the container clock or the virtual container clock. Hence, synchronous mapping does not require synchronization and consequently no justification occurs. Bit-synchronous mapping is the general case of synchronous mapping and is performed on a bit basis regardless of the internal composition of the tributary involved. In other words, with the exception that no synchronization procedure is required, bit-synchronous mapping is equivalent to asynchronous mapping. Bit-synchronous mapping applies to tributaries DS-1, DS-1E, and DS-2.

Byte-Synchronous Mapping

Byte-synchronous mapping is a special case of synchronous mapping, which is used when DS-0 is mapped into the VC such that its boundary is exposed. Byte-synchronous mapping can be applied to DS-1 and DS-1E. When one of these tributaries is to be byte-synchronously mapped into a VC, its frame identity must be confirmed first, and then each DS-0 byte must be mapped into an assigned location. At this point in the operation, the frame bits and signaling bits of the tributary must be mapped also.

In the case of DS-1, the frame bits and the signaling bits are byte-synchronously mapped directly to preassigned locations. There are four signaling bits for each frame, and, hence, to represent signaling bits for 24 DS-0's, six frames in all are required. Consequently, 24 frames are required to represent the ABCD signal, which would occupy 3 ms ($= 125$ μs $\times$ 24) of time. To be able to distinguish between these frames, overheads P_0, P_1 are used, and, especially for operation in the locked mode, H4's (from the higher order VC POH) 3-ms mode is used (see Section 3.4.2).[16]

In the case of DS-1E, 30 DS-0 channels are visibly mapped into the VC–12. If DS-1E is operating in *channel-associated signaling* (CAS) mode, multiframe bits as well as signaling bits are mapped into one DS-0-sized channel in the VC-12. If it is operating in the *common channel signaling* (CCS) mode, one additional DS-0 channel is added instead, for 31 DS-0 channels in all (see Table 3.3 in a following subsection). In both cases, it is not essential to map the frame bits, but sufficient space is reserved for them just in case. For indicating the multiframe and CAS for 30 channels, 16 frames (2 ms) are required, and to be able to identify each of the frames, overheads P_0 and P_1 are used. In the locked mode, H4's 2-ms mode is used.

3.4.2 Floating and Locked Modes of Operation

In general, after the lower order tributaries are mapped into VC-n (n = 11, 12, 2), VC-n floats freely in the payload of corresponding TU-n, and its address is indicated by the TU-n PTR. This is called the *floating mode* of operation. However, when VC-n is synchronized with VC-m (m = 3, 4), VC-n maintains a fixed location within the TU-n. Consequently, the starting position of the VC is fixed to the starting position of the TU-n payload and, hence, is called the *locked mode* of operation.

In locked mode, the TU-n PTR is not separately needed. Also, the fact that VC-n and VC-m are synchronized implies that they were created at the same place, which in turn implies that they will follow the same transmission paths.

---

16. H4 is used to indicate the multiframe status and can give 500-μs, 2-ms, and 3-ms indications (see Sec. 3.5.4).

Hence, in this case, POH V5 for VC-n is not separately needed. In other words, in locked mode, V1, V2, V3, V4, and V5 bytes are not required, and, consequently, there is no need to construct a 500-μs multiframe.

Therefore, the mapping technique for the floating mode of operation is defined separately on a 125-μs frame unit basis. In the case of VC-11 and VC-12, a type of mapping that combines byte-synchronous and locked modes can be defined. Here, as a mark for the CAS phase, the H4 byte in the POH of higher order VC-m is used. To be more specific, a 3-ms mode using 5 bits from H4, namely, P_1, P_0, SI_1, SI_0, and T, is employed for VC-11, and a 2-ms mode using CI_3, CI_2, CI_1, and T-bits is used for VC-12 (see Sec. 3.6.4 for a detailed description of the H4 byte).

3.4.3 Mapping of DS-1 and DS-1E

DS-1 and DS-1E are mapped into VC-11 and VC-12 in the floating mode, and the mapping format is depicted in Figure 3.22. Such a format is common to both asynchronous and synchronous modes. In the figure, I* represents the tributary information byte (hence, I* = IIIIIIII), R* is a byte that is composed of fixed R-bits (or R* = RRRRRRRR), and the bytes W_1 to W_4 and Y_1 to Y_4 can vary depending on whether or not the mapping is performed in asynchronous mode.

In asynchronous mode, W_1 = RRRRRRIR, W_2 = W_3 = $C_1C_2OOOOIR$, W_4 = $C_1C_2RRRJ_1J_2R$, and Y_1 = R*, Y_2 = Y_3 = $C_1C_2OOOORR$, Y_4 = $C_1C_2RRRRRJ_1$.

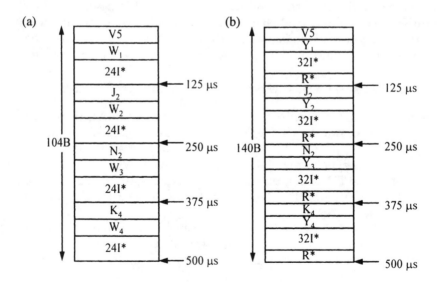

Figure 3.22　Mapping of DS-1 and DS-1E (floating mode): (a) DS-1 and (b) DS-1E (see Table 3.2 for W_1–W_4 and Y_1–Y_3).

Here I represents the effective information bit, O represents the overhead bit, and three C_1 bits and three C_2 bits are the justification control bits corresponding to the justification opportunity bits J_1 and J_2, respectively. In the case of VC-12, the J_2 bit is the first of the bits in 32I* space, which immediately follows Y_4.

In bit-synchronous mode, $W_1 = W_4 = 10RRRRIR$, $W_2 = W_3 = 10OOOOOIR$, and $Y_1 = Y_4 = 10RRRRRR$, $Y_2 = Y_3 = 10OOOOORR$. Since there is no justification required in bit-synchronous mode, J_1 and J_2 have been altered to R and I, respectively. Also, to reflect this change, the values of C_1 and C_2 have been fixed to all 1's and 0's, respectively.

In the byte-synchronous case, $W_1 = W_2 = W_3 = W_4 = P_1P_0S_1S_2S_3S_4FR$, and $Y_1 = Y_2 = Y_3 = Y_4 = P_1P_0RRRRRR$. Here F represents the DS-1's frame bit, and S_1 to S_4 represent the signaling bits. Also, P_1P_0 are used to adjust the phase of the signal frame, with the initial frame acquiring $P_1P_0 = 00$. In the case of DS-1E, the first byte from the 32I* space is filled with the R* or the first byte from DS-1E, the 17th byte is used for a multiframe indication or to contain signaling bits, and the remaining 30I* are used as information channels. If CCS is employed, the 17th byte is used to carry an additional channel of information, since signaling bits are unnecessary within VC-12 in this case.

In locked mode, POH V5 is not separately needed, and V5 always equals R*. In bit-synchronous mapping, also, $W_1 = W_2 = W_3 = W_4 = W = 10RRRRIR$ and $Y_1 = Y_2 = Y_3 = Y_4 = Y = 10RRRRRR$. Also, for byte-synchronous mapping, $W_1 = W_2 = W_3 = W_4 = W = RRS_1S_2S_3S_4FR$, and $Y_1 = Y_2 = Y_3 = Y_4 = R*$. If this is the case, the mapping can be performed on a 125-μs-unit basis, and the mapping format of Figure 3.22 simplifies to that of Figure 3.23. The format of other information bytes is identical to that of the floating-mode case.

Table 3.2 gives a summary of the contents of W_1 to W_4 and Y_1 to Y_4 bytes for the locked-mode and floating-mode operations.

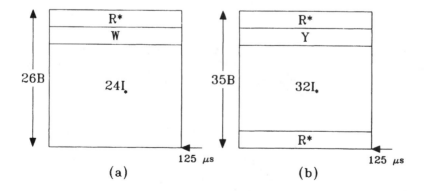

Figure 3.23 Mapping of DS-1 and DS-1E (locked mode): (a) DS-1 and (b) DS-1E.

Table 3.2
The Contents of W_1–W_4 and Y_1–Y_4
(a) Floating Mode

		Asynchronous	*Bit Synchronous*	*Byte Synchronous*
DS-1 to VC-11 mapping	W_1	RRRRRRIR	10RRRRIR	$P_1P_2S_1S_2S_3S_4FR$
	W_2	C_1C_2OOOOIR	1OOOOOIR	$P_1P_2S_1S_2S_3S_4FR$
	W_3	C_1C_2OOOOIR	1OOOOOIR	$P_1P_2S_1S_2S_3S_4FR$
	W_4	C_1C_2RRRJ$_1$J$_2$R	10RRRRIR	$P_1P_2S_1S_2S_3S_4FR$
DS-1E to VC-12 mapping	Y_1	RRRRRRRR	10RRRRRR	P_1P_0RRRRRR
	Y_2	C_1C_2OOOOIR	1OOOOORR	P_1P_0RRRRRR
	Y_3	C_1C_2OOOOIR	1OOOOORR	P_1P_0RRRRRR
	Y_4	C_1C_2RRRRRJ$_1$*	10RRRRRR	P_1P_0RRRRRR

* +J_2 is the bit right behind J_1

(b) Locked Mode

		Bit Synchronous	*Byte Synchronous*
DS-1 to VC-11 mapping	W_1	10RRRRIR	RRS$_1$S$_2$S$_3$S$_4$FR
	W_2	10RRRRIR	RRS$_1$S$_2$S$_3$S$_4$FR
	W_3	10RRRRIR	RRS$_1$S$_2$S$_3$S$_4$FR
	W_4	10RRRRIR	RRS$_1$S$_2$S$_3$S$_4$FR
DS-1E to VC-12 mapping	Y_1	10RRRRRR	RRRRRRRR
	Y_2	10RRRRRR	RRRRRRRR
	Y_3	10RRRRRR	RRRRRRRR
	Y_4	10RRRRRR	RRRRRRRR

When the locked mode is employed, the phase relationship between the signal bits is indicated by the POH H4 in the higher order VC. In the case of DS-1, since the ABCD signal marks for 24 DS-0 channels require 24 frames, this relationship is represented by H4's 3-ms mode bits $P_1P_2SI_2SI_1T$. In the case of DS-1E, each frame can convey two signaling channels, which means that for 30 signaling channels and an additional multiframe indication, 16 frames are required, and this relationship is indicated by the H4's 2-ms mode bits $CI_1CI_2CI_1T$. Table 3.3 shows a comparison of the locked mode and the floating mode in terms of the signaling phase alignment relationship. In the table, the subscripts 1 to 24 or 1 to 30 on the ABCD signaling marks indicate the corresponding signal's channel numbers.

Table 3.3
Signaling and Phase Indication
(a) DS-1

Signaling				Floating Mode		Locked Mode				
S_1	S_2	S_3	S_4	P_1	P_0 (VC-11)	P_1	P_0	SI_2	SI_1	T(H4)
A_1	A_2	A_3	A_4	0	0	0	0	0	0	0
A_5	A_6	A_7	A_8	0	0	0	0	0	0	1
A_9	A_{10}	A_{11}	A_{12}	0	0	0	0	0	1	0
A_{13}	A_{14}	A_{15}	A_{16}	0	0	0	0	0	1	1
A_{17}	A_{18}	A_{19}	A_{20}	0	0	0	0	1	0	0
A_{21}	A_{22}	A_{23}	A_{24}	0	0	0	0	1	0	1
B_1	B_2	B_3	B_4	0	1	0	1	0	0	0
B_5	B_6	B_7	B_8	0	1	0	1	0	0	1
B_9	B_{10}	B_{11}	B_{12}	0	1	0	1	0	1	0
B_{13}	B_{14}	B_{15}	B_{16}	0	1	0	1	0	1	1
B_{17}	B_{18}	B_{19}	B_{20}	0	1	0	1	1	0	0
B_{21}	B_{22}	B_{23}	B_{24}	0	1	0	1	1	0	1
C_1	C_2	C_3	C_4	1	0	1	0	0	0	0
C_5	C_6	C_7	C_8	1	0	1	0	0	0	1
C_9	C_{10}	C_{11}	C_{12}	1	0	1	0	0	1	0
C_{13}	C_{14}	C_{15}	C_{16}	1	0	1	0	0	1	1
C_{17}	C_{18}	C_{19}	C_{20}	1	0	1	0	1	0	0
C_{21}	C_{22}	C_{23}	C_{24}	1	0	1	0	1	0	1
D_1	D_2	D_3	D_4	1	1	1	1	0	0	0
D_5	D_6	D_7	D_8	1	1	1	1	0	0	1
D_9	D_{10}	D_{11}	D_{12}	1	1	1	1	0	1	0
D_{13}	D_{14}	D_{15}	D_{16}	1	1	1	1	0	1	1
D_{17}	D_{18}	D_{19}	D_{20}	1	1	1	1	1	0	0
D_{21}	D_{22}	D_{23}	D_{24}	1	1	1	1	1	0	1

3.4.4 Mapping of DS-2

The procedure for mapping the DS-2 into the VC-2 is depicted in Figure 3.24. The picture displays only a single 125-µs frame, and the 500-µs multiframe is just four replications of such a frame structure, with the exception that the first byte, denoted by V5/R*, becomes V5 in the first frame and R* in the remaining three frames.

In asynchronous mapping, the byte contents of W_1 to W_4 become W_1 = IIIIIIIR, W_2 = W_3 = C_1C_2OOOOIR, W_4 = C_1C_2IIIJ$_1$J$_2$R, and for the synchronous case, W_1 = IIIIIIIR, W_2 = W_3 = 1OOOOOIR, W_4 = 1OIIIRIR. With regard to operation of the justification bits C_1, C_2, J_1, and J_2 note that, for the

Table 3.3 (Continued)
(b) DS-1E

Signaling								Floating Mode		Locked Mode			
S_1	S_2	S_3	S_4	S_1	S_2	S_3	S_4	P_1	P_0 (VC-12)	CI_3	CI_2	CI_1	T(H4)
0	0	0	0	x	y	x	x	0	0	0	0	0	0
A_1	B_1	C_1	D_1	A_{16}	B_{16}	C_{16}	D_{16}	0	0	0	0	0	1
A_2	B_2	C_2	D_2	A_{17}	B_{17}	C_{17}	D_{17}	0	0	0	0	1	0
A_3	B_3	C_3	D_3	A_{18}	B_{18}	C_{18}	D_{18}	0	0	0	0	1	1
A_4	B_4	C_4	D_4	A_{19}	B_{19}	C_{19}	D_{19}	0	1	0	1	0	0
A_5	B_5	C_5	D_5	A_{20}	B_{20}	C_{20}	D_{20}	0	1	0	1	0	1
A_6	B_6	C_6	D_6	A_{21}	B_{21}	C_{21}	D_{21}	0	1	0	1	1	0
A_7	B_7	C_7	D_7	A_{22}	B_{22}	C_{22}	D_{22}	0	1	0	1	1	1
A_8	B_8	C_8	D_8	A_{23}	B_{23}	C_{23}	D_{23}	1	0	1	0	0	0
A_9	B_9	C_9	D_9	A_{24}	B_{24}	C_{24}	D_{24}	1	0	1	0	0	1
A_{10}	B_{10}	C_{10}	D_{10}	A_{25}	B_{25}	C_{25}	D_{25}	1	0	1	0	1	0
A_{11}	B_{11}	C_{11}	D_{11}	A_{26}	B_{26}	C_{26}	D_{26}	1	0	1	0	1	1
A_{12}	B_{12}	C_{12}	D_{12}	A_{27}	B_{27}	C_{27}	D_{27}	1	1	1	1	0	0
A_{13}	B_{13}	C_{13}	D_{13}	A_{28}	B_{28}	C_{28}	D_{28}	1	1	1	1	0	1
A_{14}	B_{14}	C_{14}	D_{14}	A_{29}	B_{29}	C_{29}	D_{29}	1	1	1	1	1	0
A_{15}	B_{15}	C_{15}	D_{15}	A_{30}	B_{30}	C_{30}	D_{30}	1	1	1	1	1	1

DS-1 or the DS-1E, justification occurs every 500 µs, whereas it occurs every 125 µs for the DS-2 signal.

3.4.5 Mapping of DS-3 or DS-3E

The mapping procedure of the tributary DS-3 into the VC-3 is shown in Figure 3.25(a). Inside the 125-µs frame, the 125/9 µs format subframe in the shaded region is repeated nine times. The subframe's organization is shown in detail in Figure 3.25(b). In the picture, W_1 = RRCIIIII, W_2 = CCRRRRRR, and W_3 = CCRROORJ, respectively. Hence, for the synchronization of DS-3, positive justification is used every 125/9 µs, and 5 C-bits are used for the justification control.

The tributary DS-3E is mapped into the VC-3 as shown in Figure 3.26(a). Within the 125-µs frame, the 125/3-µs subframe is repeated three times, and the subframe is divided into three smaller subframes of 125/9 µs each. The 125/9-µs subframes all have a format like the one shown in Figure 3.26(b), except for the Q1 to Q4 bytes. These Q-bytes become Y_1, Y_1, I*, and I* in the first two 125/9-µs subframes, and Y_1, R*, Y_2, and Y_3 in the third of the 125/9-µs subframes, where Y_1 = RRRRRRC$_1$C$_2$, Y_2 = RRRRRRRJ$_1$, and Y_3 = J$_2$IIIIIII. This relationship is listed in Table 3.4. Consequently, it can be inferred that for synchronization of DS-3E, P/Z/N justification is employed every 125/3 µs, and for justification control, five pairs of C_1 and C_2 bits are used. Also, since one

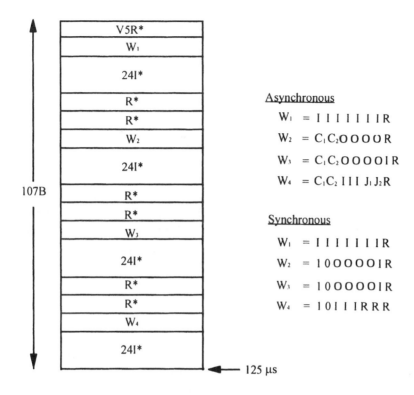

Figure 3.24 Mapping of DS-2.

fixed stuff byte is inserted for every three effective information bytes, it can be seen that about a quarter of the VC-3 payload is wasted.

3.4.6 Mapping of DS-4E

The mapping of DS-4E into the VC-4 is depicted in Figure 3.27(a). The 125/9-μs-format subframe in the shaded region is repeated nine times within the 125-μs frame, and the subframe's organization is as displayed in Figure 3.27(b). In the figure, $Y_1 = Y_2 = Y_3 = Y_4 = Y_5 = CRRRRROO$, and $Y_6 = IIIIIIJR$. For the synchronization of DS-4E, positive justification is employed, and for justification control, five C-bits are used.

3.4.7 Mapping of the ATM Cell

The ATM cell is composed of 5 bytes of header and 48 bytes of information data. The fifth of the ATM header bytes corresponds to the *header error control* (HEC) byte and is produced by applying *cyclic redundancy checking* (CRC) with generating polynomial $g(x) = x^8 + x^2 + x + 1$ to the preceding 4 bytes and

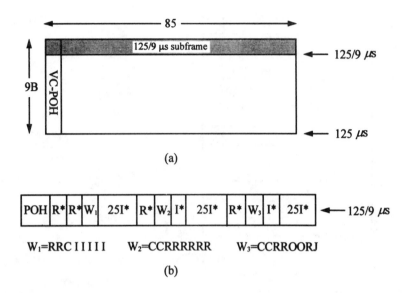

W_1=RRC I I I I I W_2=CCRRRRRR W_3=CCRROORJ

(b)

Figure 3.25   Mapping of DS-3 into VC-3: (a) 125 µs frame and (b) 125/9 µs subframe.

then adding 01010101 to the resulting CRC. Such a tracking scheme enables the detection of the boundary of an ATM cell in the ATM cell bit stream. Before any mapping or transmission, the 48 bytes from the ATM information space go through a *self-synchronous scrambling* (SSS) process based on the characteristic polynomial $x^{43} + 1$. So a treated ATM cell can be mapped into the VC-n or VC-n-xc and then go through synchronous multiplexing.[17]

In the case of the mapping of an ATM cell into the VC-4, the size of an ATM cell is 53B, and the payload of VC-4 can accommodate 2340B, which is not a multiple of 53B. Consequently, ATM crosses the VC boundary, and the starting location of an ATM cell varies from one VC-4 to another. In this case, the VC's POH H4 byte can be used for easy detection of the ATM cell boundary.[18] That is, the distance from the H4 byte to the initial location of the ATM cell is counted, and this information is stored in the H4 byte. Because the distance can vary from 0 to 52, 6 bits of H4 are allotted for this purpose. The mapping of an ATM cell into the VC-4 and the associated H4 indicator are shown in Figure 3.28.

---

17. If an ATM cell stream is mapped into the VC-4-xc, it is mapped into the container space of 260x columns, and $x - 1$ columns out of the x columns for POH are filled with fixed stuff bytes. The composition of ATM cells is described in detail in Chapter 4. When an ATM cell is mapped into the VC-4, VC-4's POH C2 byte is recorded as 00010011 to indicate this status (see Section 3.5.4.).

18. H4 is usually employed to furnish the multiframe phase information when the lower order tributaries are mapped into the VC. However, in the case of ATM cell mapping, such a function is not required and H4 is free to use for other purposes.

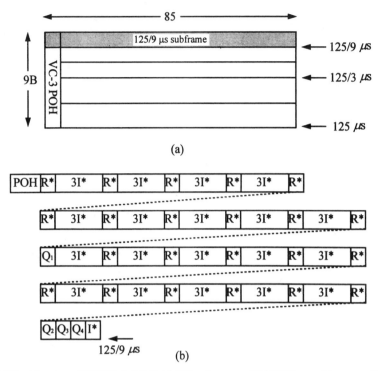

Figure 3.26 Mapping of DS-3E into VC-3: (a) 125 μs frame and (b) 125/9 μs subframe and (see Table 3.4 for Q1 to Q4).

Table 3.4
The Structure of Q_1 to Q_4 Bytes

Location	Q_1	Q_2	Q_3	Q_4
First 125/9-μs subframe	Y_1	Y_2	I*	I*
Second 125/9-μs subframe	Y_1	Y_2	I*	I*
Third 125/9-μs subframe	Y_1	R*	Y_2	Y_3

Note: $Y_1 = RRRRRRC_1C_2$, $Y_2 = RRRRRRJ_1$, $Y_3 = J_2$ IIIIIII.

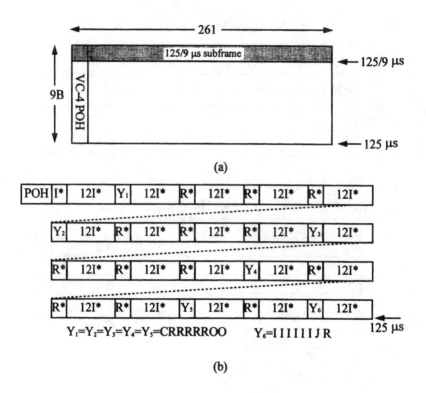

Figure 3.27   Mapping of DS-4E into VC-4: (a) 125 μs frame and (b) 125/9 μs and subframe.

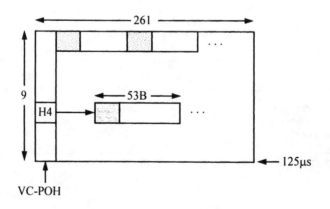

Figure 3.28   Mapping of ATM cells.

The receiver can detect the ATM cell boundary in two possible ways. First, after searching through the STM-n frame, it locates the VC-4 starting location through the AU-4 PTR, and then it can find the ATM cell boundary by reading the H4. The second method is by first ascertaining the starting position of AU-4, then picking out the ATM cells, and subsequently searching for the ATM cell boundary using the HEC byte (see Section 4.6 for a more detailed description of HEC).

3.5 SECTION AND PATH OVERHEADS

Synchronous transmission is based on the layer concept, which divides the digital transmission process into the path layer, the multiplexer section layer (sometimes called the *digital section*), the regenerator section layer, and the physical medium layer, as illustrated earlier in Figure 3.2. Such an organization is reflected in the various types of overhead. Overhead consists of the SOH, which is applicable to the multiplexer section and the regenerator section, and the POH, which applies to the transmission paths for various virtual containers. Hence, the SOH can further be divided into the RSOH and the MSOH. Similarly, the POH can also be divided into the higher order POH associated with the higher order VCs and the lower order POH, which is used for the lower order VCs. Such relationships are shown in Figure 3.3 in an earlier section.

In this section, the functions and compositions of various types of overheads are studied in detail. First, the overall organization of the SOH is examined, followed by a study of the RSOH and the MSOH. Afterwards, POH is discussed in terms of the higher order POH and the lower order POH, separately.

3.5.1 Composition of Section Overhead

The SOH applies to the regenerator and multiplexer sections, and it was designed for the reliable transmission of STM-n. After n (n = 1, 4, 16, 64) AUGs are byte-interleaved together, an SOH is added to form the STM-n signal. Consequently, for STM-n, the format of the SOH is well regulated, and the respective SOHs for n = 1, 4, and 16 are shown in Figure 3.29. In the figure, the portion above the AU PTR corresponds to the RSOH and the portion below it corresponds to the MSOH for all three cases.

As can be deduced from Figure 3.29, the SOHs for STM-1, STM-4, STM-16, and STM-64 have a well-organized relationship with respect to each other. Inside the overhead, the size of A1, A2, B2, C1, X, and Z0 can be increased by a factor of n, and the remaining parts are fixed at one byte each.[19] This is because

---

19. As can be seen in Figure 3.29, the position for C1 bytes is also used for J0 or Z0 in the case of STM-4, 16, or 64. It is also used for J0 in the case of STM-1.

- To maintain a short frame-alignment time as *n* increases, the A1 and A2 bytes have to be lengthened accordingly.
- To preserve the accuracy of the bit error confirmation function, more B2 bytes are required as the signal unit becomes bigger.
- As the size of STM increases, the STM identifier C1 has to be increased as well.

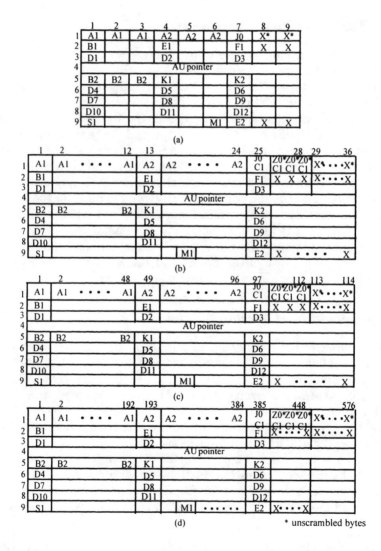

(a)

(b)

(c)

(d)   * unscrambled bytes

Figure 3.29   Comparison of section overheads: (a) STM-1; (b) STM-4; (c) STM-16; and (d) STM-64.

- With increasing n, it is natural to have more bytes for international (Z0) and national (X) uses.

On the other hand, as STM-64 is reduced to STM-16, STM-4, and STM-1, the corresponding restructuring of the overhead is even more regulated. When STM-64 is reduced to STM-16, if the $4m + 1$'th ($m = 0, 1, \ldots, 143$) columns are taken from the 576 columns of the STM-64's SOH, the end product is identical to the $9B \times 144$ SOH of the STM-16. The same relation holds for reducing STM-16 to STM-4 (in this case, $m = 0, 1, \ldots, 35$) and for reducing STM-4 to STM-1 (in this case, $m = 0, 1, \ldots, 8$). There is one exception in this regular process, which deals with the position of M1: The M1 byte is always placed at the fixed position $3n + 3$ ($n = 1, 4, 16, 64$) of the 9th row for each STM-n frame.

3.5.2 Regenerator Section Overhead

RSOH is the overhead that is used by the regenerators. It is located right above the pointer bytes within the STM-n frame and consists of the bytes A1, A2, B1, C1 (or J0, Z0), D1, D2, D3, E1, and F1, as shown in Figure 3.29.[20] A1 and A2 bytes function as alignment marks for differentiating the STM-n boundaries and are fixed at A1 = 11110110, and A2 = 00101000.

B1 is a BIP byte of period 8 used for the error monitoring function. B1's i'th ($i = 1, 2, \ldots, 8$) bit performs even parity check on only the i'th bits of the STM-n bytes. BIP-8 is computed over all bits of the previous STM-n frame after scrambling and is reordered onto B1 of the following STM-n frame before scrambling.

C1 functions as an STM identification number. That is, within the STM-n, each STM-1 is granted its unique identification number and this number is used for aligning the STM-n frame or for extracting a single STM-1 frame.

The C1/J0 byte is used as follows: It becomes C1 when used for identification of STM signals, and becomes J0 when used for tracing transmission signals in regenerator sections. The regenerator section trace refers to the continuity checking of connections within regenerator sections, and this function is carried out by repeatedly transmitting an *access point identifier* (APId) of a predefined pattern. For this APId, a number in the 0~255 range, or the 16-byte frame format in Figure 3.30 recommended by ITU-T Rec. G. 831, is used for national or single operator domain use; and the G. 831 format is used for international or different operator domain use.

D1, D2, and D3 are *data communication channels* (DCCs) used by the regenerator sections. The three bytes are used as one single channel of 192-Kbps (3×64 Kbps) capacity.

---

20. The parts of the overhead that have been left blank in Figure 3.29 currently have no designated functions.

bit 1							bit 8
1	1	C1	C2	C3	C4	C5	C6	C7 *
2	0							
3	0							
⋮	⋮			⋮				
16	0							

* C1~ C7 are for CRC-7 calculated for the previous frame, and blank space is for don't care bits.

Figure 3.30   16-byte frame format for APId.

E1 is an orderwire channel used by the regenerator section for voice communication.

F1 is a user channel reserved for communication network operators. The X and Z0 bytes are assigned respectively for national and international uses.[21]

3.5.3 Multiplexer Section Overhead

The MSOH is checked only by the multiplexers and is passed transparently through the regenerators. Here, the multiplexer is used to assemble or disassemble an AUG. The MSOH is located below the pointer in the SOH portion of the STM-n frame and consists of the bytes B2, D4 to D12, E2, K1, K2, M1, and S1.

The B2's are BIP check bytes used for the multiplexer section error monitoring function. The STM-n's B2 consists of $3n$ bytes; hence, the parity check is achieved via the BIP-$24n$ format. That is, the i'th ($i = 1, 2, \ldots, 24n$) bit of B2 performs a parity check on every $24n$'th data bit, starting with the i'th bit, and modifies itself to 1 or 0 so that the total number of 1's is even. Here, BIP-$24n$ is performed on the previous STM-n frame except for the RSOH and is recorded onto the following B2 bytes before scrambling.

The D4 to D12 bytes form an MSOH DCC, which is used as one channel with a 576-Kbps capacity.

E2 is the multiplexer section orderwire, which can be used for voice communication.

K1 and K2 form the APS channel, which carries the APS signaling to switch the operating channel to a protection channel in case the operating channel fails or its performance degrades. K2 also carries the *alarm indication signal* (AIS) or the *remote defect indication* (RDI) signal. If bits 6~8 of the received K2

---

21. In Figure 3.29, the X-bytes with asterisks (*) are the bytes that are left unscrambled. This is due to the fact that the first row of the SOH does not go through the scrambling treatment.

are 111, it means that K2 is representing the AIS, and if the bits are 110, K2 is representing the RDI, which means that the transmitting end has received AIS.

M1 is the *remote error indication* (REI) byte, which notifies the far end of the number of error BIPs in the B2 byte.

S1 is used to indicate the level of synchronization, or the quality of synchronization, of the multiplexer section.

3.5.4 High-Order Path Overhead

The high-order POH is dedicated to the high-order virtual containers such as VC-3 and VC-4. The higher order POH is placed at the first byte of VC-3/VC-4's first column and executes various functions required for the reliable VC path connection. High-order POH is made up of the J1, B3, C2, F2, G1, H4, F3, K3, and N1 bytes, as shown in Figure 3.31(a).

J1 is a path trace byte that checks the continuity of connection between path terminations. The transmitting terminal transmits the higher order path APId over the J1 byte repeatedly, and the receiving terminal confirms the APId. For the APId, an arbitrary 64-byte format or the ITU-T Recommendation G.831 frame format in Figure 3.30 is used for national or single operator domain; and the G.831 format is used for international or different operator domain.

B3 is a BIP check byte used for the path error-monitoring function. B3 executes the parity check through the BIP-8 format, which is calculated for the previous VC-3/VC-4 and is inserted to the corresponding B3 before scrambling.

C2 is a signal label byte for indicating the composition of the VC-3/VC-4.

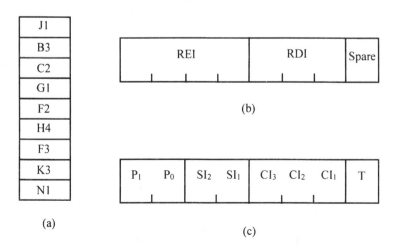

Figure 3.31 Higher-order path overheads: (a) path overhead; (b) G1; and (c) H4.

Among its 256 possible binary values, 00000000 indicates "VC-3/VC-4 un-equipped(nonspecific payloads." Other label contents are as listed in Table 3.5.

The F2 byte is allocated for path user communication purposes between equipment.

G1 is a channel used by the receiver to convey back to a VC-3/VC-4 transmitter the path condition and performance. As illustrated in Figure 3.31(b), the first 4 bits of G1 are for REI. The count of errors detected in the B3 byte's BIP code is written on this REI space. Bits 5~7 of G1 are for RDI, and the relevant indications are as listed in Table 3.6. G1's last bit is not used.

Table 3.5
C2 (Path Signal Label) Code

Code		
Binary	*Hexadecimal*	*Indication*
00000000	00	VC-3/4 path unequipped
00000001	01	Equipped nonspecific
00000010	02	TUG structure
00000011	03	Locked TU
00000100	04	Asynchronous mapping of DS-3 or DS-3E into C-3
00010010	12	Asynchronous mapping of DS-4E into C-4
00010011	13	ATM
00010100	14	MAN (DQDB)
00010101	15	FDDI
11111110	FE	Test signal defined in ITU-T Rec. O.181
11111111	FF	VC-AIS in a tandem connection link

Table 3.6
RDI Code

Code	*Indication*	*Remarks*
000	Remote in normal state	
001	Remote in normal state	
010	Remote payload in failure	Loss of cell delineation (LCD)
011	Remote in normal state	
100	Remote in failure	Alarm indication signal (AIS), Loss of pointer (LOP), Unequipped (UNEQ), Signal label mismatch (SLM)
101	Remote server failed	AIS, LOP
110	Remote connection failed	Trace Identification Mismatch (TIM), UNEQ
111	Remote in failure	AIS, LOP, TIM, UNEQ, SLM

The H4 byte is a position indicator for specific payloads. The use of H4 is mandatory for floating-mode TUs; H4's 8 bits are named P1, P0, SI2, SI1, CI3, CI2, CI1, and T, respectively. Among them, P (or P1, P0) is used for counting up to four, SI (or SI2, SI1) for counting up to three, CI (or CI3, CI2, CI1) for counting up to eight, and T for binary count. Since a single STM frame is 125 μs in duration, to indicate the 500-μs mode, CI1 and T can be used; for 2-ms mode, CI and T; and for 3-ms mode, P, SI, and T can be used (see Table 3.3 for an illustration). Here, the 500-μs mode is employed to indicate the multiframe status of the lower order TUs that have been multiplexed to VC-3/VC-4. The 2-ms and 3-ms modes indicate the signaling phases of DS-1E and DS-1, respectively. On the other hand, a VC-4 that carries an ATM-cell does not require a multiframe indicator; hence, in this case, H4 is used to denote the starting position of the ATM cell (see Sec. 3.4.7).

K3 is an APS byte for path protection of the higher order path. Bits 1~4 of K3 are for APS itself, and bits 5~8 are reserved for future use.

N1 is used for tandem connection monitoring.

3.5.5 Lower Order Path Overhead

Lower order path overhead is dedicated to a lower order VC, that is, VC-11, VC-12, or VC-2. Lower order POH bytes are V5, J2, N2, and K4, and are positioned at the beginning of each 125-μs interval within a 500-μs multiframe as shown in Figure 3.32(a). execute the various necessary functions for the reliable path connection of the lower order VC payload.

As illustrated in Figure 3.32(b), V5 includes bits for BIP check (BIP-2), REI, *remote failure indication* (RFI), signal labels (L1, L2, L3), and RDI. The RDI bit notifies the far end if a path AIS or signal defect is received. RFI notifies to the far end if, after a path failure, recovery is not made until the maximum allowed time for protection switching. A detailed indication on the RDI-related defect can be made by sending the RDI codes listed in Table 3.6 over bits 5~7 of K4 byte.

The J2, N2, and K4 (bits 1~4) are used for the same purpose as J1, N1, and K3 (bits 1~4), respectively.

A summary of various functions of SOH and POH is listed in Table 3.7.

3.6 POINTERS AND POINTER PROCESSING

In synchronous transmission, the synchronization required in the synchronous multiplexing process is achieved through the use of pointers. Such a synchronization process is needed because, in general, a VC is created using a different clock from the one associated with the AU or the TU. When a VC is aligned within an AU or TU, a pointer conveys the information regarding its starting

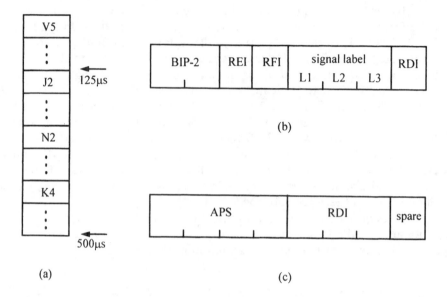

Figure 3.32   Lower-order path overhead: (a) path overhead; (b) V5; and (c) K4.

location; when the address changes, the pointer is also altered to keep track of the shifted location.

In this section, pointers in general and synchronization through the use of pointers are discussed. First, the composition and various functions of pointers are examined, and then the synchronization process via pointers are described.

3.6.1 Composition of Pointers

Pointers can be grouped into higher order pointers, such as AU-4 PTR, AU-3 PTR, and TU-3 PTR, and lower order pointers, such as TU-11 PTR, TU-12 PTR, and TU-2 PTR.

The higher order pointer is contained in bytes H1, H2, and H3. In the case of AU-4/AU-3, the higher order pointer is located in the left part of the fourth row of the AU-4/AU-3 frame, as shown in Figure 3.9 and Figure 3.20. As for TU-3, its pointer is positioned in the top portion of the frame's first column, as shown in Figure 3.18(a). Among the three bytes, H1 and H2 are used as the address indicator for the starting location (or its variance) of the corresponding VC, and H3 is used for the execution of negative justification. One byte each of H1 and H2 is assigned for every 125-µs frame; but three H3 bytes are assigned to AU-4 and only a single H3 byte is used in the case of AU-3 or TU-3.

Table 3.7
Functions of Section and Path Overheads
(a) Section Overheads

Overhead	Function	Note
A1, A2	Frame alignment	"11110110", "00101000"
B1	Regenerator section error monitoring	BIP-8
B2	Multiplexer section error monitoring	BIP-24
C1	STM-1 identifier	
D1–D3	Regenerator section data communication	
D4–D12	Multiplex section data communication	
E1, E2	Order wires	
F1	User channel	Network operator
H1, H2	AU-4 PTR/path AIS	/"11111111"
H3	Pointer action	Negative justification
J0	Regenerator section path trace	
K1, K2	Automatic protection switching	
K2 (bit 6–8)	Section AIS/section RDI	"111"/"110"
M1	Section REI	B2 error count
S1	Synchronization state indication	
X	Reserved byte for national use	
Z0	Reserved byte for international use	

(b) Path Overheads

Overhead	Function	Note
B3	Path error monitoring	BIP-8
C2	Path signal label	(see Table 3.5)
F2, F3	User channel	Path user
G1 (bits 1–4)	Path REI	B3 error count
G1 (bits 5–7)	Path RDI	(see Table 3.6)
H4	Multiframe indication	ATM cell offset (bits 3–8)
J1, J2	Path trace	
K3, K4	Path APS, RDI	(see Table 3.6)
N1, N2	Tandem path monitoring	
V5	Lower-order POH	BIP-2, REI, RFI, Signal label, RDI

The lower order pointer is contained in the bytes V1, V2, and V3. As illustrated in Figure 3.14, these bytes correspond to the first bytes of the four 125-μs partitions of a 500-μs TU frame. Hence, V1, V2, and V3 each appear once in every 500-μs frame. If the lower order TU is restructured in 9B × n (n = 3, 4, 12) format, as shown in Figure 3.15, V1, V2, and V3 are always positioned in the first row and first column. The functions of V1, V2, and V3 are identical to those of H1, H2, and H3; this relationship is depicted in Figure 3.33.

3.6.2 Functions of Pointers

The 16 bits of H1 (or V1) and H2 (or V2) can be divided into three sections, as shown in Figure 3.33, with each serving an independent function. The first 4 bits carry the NDF; the next 2 ss bits indicate the signal type of the frame; and the last 10 bits contain 5 increment bits and 5 decrement bits in an interleaved format.

The NDF is fixed to 0110 in the regular employment of the pointer. However, in case a need arises for an abrupt alteration of the pointer value or the exclusive use of the pointer for a specific purpose, the NDF is inverted to 1001.

The signal type indicator bits ss are assigned the value 10 for every higher order pointer; but in the case of lower order pointers, ss becomes 00 for TU-2, 10 for TU-12, and 11 for TU-11 (the ss bit values are not defined in the case of specialized indications such as concatenation or null pointer). This is listed in Table 3.8.

The address bits indicate the starting location of the VC in the regular mode of operation; however, when the need arises for an alteration of the pointer value, the change is reflected through the inversion of I- or D-bits. If the VC has been shifted up (farther from the beginning of the host frame), only the I-bits are inverted; in the reverse situation, only the D-bits are inverted.

As shown in Figures 3.11 and 3.34(a), the address for the AU-4/AU-3 PTR indication begins immediately after the H3 byte. The same situation applies for TU-3 as shown in Figure 3.34(b). However, in the case of TU-2, TU-12, and TU-11, the address assignment begins right after the last bit of the V2 byte, which is illustrated in Figures 3.34(c)–(e), respectively. Therefore, each address indicator denotes the degree of offset from the pointer H3, as in the case of higher order pointers, or from the V2, as in the case of lower-order pointers.

The range of addresses indicated by the pointer address bits is different for each signal element. For instance, the address ranges from 0 to 782 in the case of AU-4 or AU-3[22]; the range is 0 to 764 for TU-3. Table 3.8 lists a summary of different address ranges for various signal elements.

---

22. Even though the same address assignment method is used for AU-3 and AU-4, the number of bytes denoted by each respective address is different, with one byte of AU-3 corresponding to three bytes of AU-4. This relationship also holds for the size of H3 bytes for each respective signal.

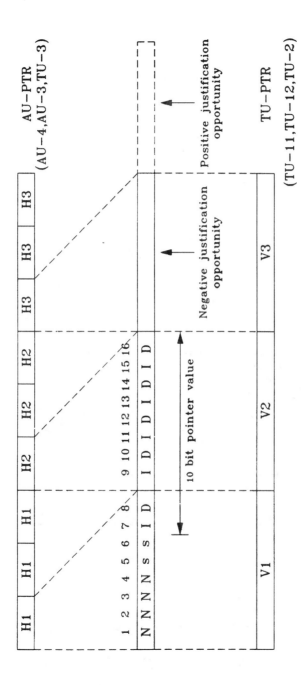

Figure 3.33 The pointer structure and function.

Table 3.8
The Range of Pointer Address

Pointer	Signal Type (ss)	Range of Address
AU-4	10	0–782
AU-3	10	0–782
TU-3	10	0–764
TU-2	00	0–427
TU-12	10	0–139
TU-11	11	0–103

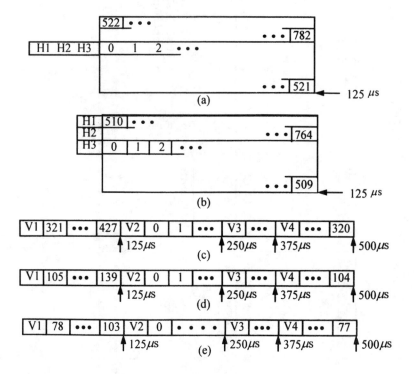

Figure 3.34   Address assignment methods: (a) AU-4/AU-3; (b) TU-3; (c) TU-2; (d) TU-12; and (e) TU-11.

H1 (or V1) and H2 (or V2) bytes are also used as a concatenation indicator. If AU-4-xc is taken as an example, the AU-4 PTR for the first of x AU-4s operates in the regular pointer mode, while the remaining $x - 1$ AU-4 PTRs designate the concatenation status. Here, the concatenation indication code is 1001ss11 11111111.

When a lower order TU is multiplexed to TUG-3 via TUG-2, a separate pointer for TUG-3 is not necessary. On the other hand, a TU-3 that has VC-3s aligned within it requires the TU-3 PTR. However, TU-3 and TUG-3 have an identical 9B × 86 structure, as illustrated in Figure 3.18. Therefore, in the case of TUG-3, the site that corresponds to the TU-3 PTR is occupied by the NPI code 1001ss11 11100000.

3.6.3 Synchronization via Pointers

In general, the VCs that are aligned within the AU or TU are created from a different source, which has a different clock from the one that produced the AU/TU. Of course these two clocks have a plesiochronous relationship, and the degree of discrepancy between the two is small and within the specified tolerance. However, since they are not perfectly synchronized either, a synchronization procedure is always needed. This can be achieved by using pointers.

The general principle of the synchronization procedure via pointers is as follows. First of all, VC does not get "locked" into the AU/TU, but is allowed to "float" within the frame's payload, and the address of the VC's first byte is recorded in the pointer. If the bit rate of the VC is lower compared to that of the AU/TU, and the accumulated data offset becomes one byte long (three null bytes if AU-4), then one null byte (three null bytes for AU-4)[23] is inserted into the payload, the starting address of the VC is shifted up by one, and the altered status is again recorded in the pointer. Hence, the clock discrepancy between the VC and AU/TU has been effectively resolved. Such a process is called *positive justification*. If the VC's bit rate is too high with respect to that of the AU/TU, an opposite treatment called *negative justification* is used to resolve the clock offset. To elaborate, when the data offset becomes one byte long (three bytes for AU-4), one spare byte is used to convey the extra byte of data and the address of the VC is shifted down by one, and, as before, the pointer value is altered to reflect the change. Here, H3 (or V3) is the spare byte that conveys the extra data.

Figure 3.35 illustrates the pointer-based synchronization techniques. In the figure, STM-1$_A$, STM-1$_B$, and STM-1$_C$ are the synchronous transport modules that have been produced at three different network nodes, A, B, and C, using

---

23. To be more accurate, it should be written, "when the accumulated offset data becomes one address location long." This length is equivalent to the size of H3, which is three bytes in the case of AU-4, and one byte for the others.

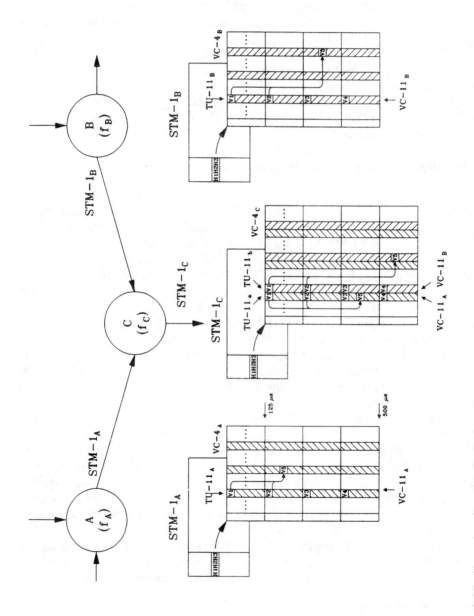

Figure 3.35   Illustration of synchronization by pointers.

three different system clocks, f_A, f_B, and f_C. Inside STM-1$_A$, VC-4$_A$ floats freely, with its starting location indicated by the pointer bytes H1 and H2. If the multiplexing path VC-11\TU-11\TUG-2\TUG-3\VC-4 is considered, the inside of the VC-4 is occupied by 84 TU-11s. If a designated TU-11 is picked out and is named TU-11$_A$, it consists of three columns, as shown in the picture. It also consists of four 125-μs frames, which together form a 500-μs multiframe. Within the first bytes from the first rows of each 125-μs frame reside the bytes V1, V2, V3, and V4. TU-11$_A$ and VC-4$_A$ are generated using the same clock, and VC-11$_A$ is allowed to float inside the TU-11$_A$ frame. VC-11$_A$'s first byte is designated V5, and its address is indicated by the pointer bytes V1 and V2. Such an internal organization also applies to STM-1$_B$, VC-4$_B$, TU-11$_B$, and VC-11$_B$.

Suppose that VC-11$_A$ from STM-1$_A$ and VC-11$_B$ from STM-1$_B$ are transported to network node C and multiplexed together to form STM-1$_C$ after an add/drop procedure. This implies that VC-11$_A$ and VC-11$_B$ each get realigned by the VC-4's clock, which is the same clock that was used to produce the TU-11$_A$ and TU-11$_B$, the TUs newly generated within STM-1$_C$ to accommodate VC-11$_A$ and VC-11$_B$. The VC-11's realignment relationship is indicated by the pointer's V1 and V2. If there is an accumulated offset due to any discrepancy between the VC-11's clock and the TU-11's clock, then the TU-11 PTR is employed to remedy the situation. If the add/drop operates at the VC-4 level and a similar problem arises, then the AU-4 PTR is employed to resolve it.

3.6.4 Execution of Justification

Now we investigate, using an illustration, the execution of the justification process, which is an inseparable part of the pointer-based synchronization procedures. For this purpose, AU and TU from Figure 3.34 have been conceptually integrated and restructured as shown in Figure 3.36. The pointer portion in the shaded region corresponds to the H1, H2, and H3 (or V1, V2, and V3) from Figure 3.34. In Figure 3.36, the address ranges from 0 to $N - 1$, with $N - 1$

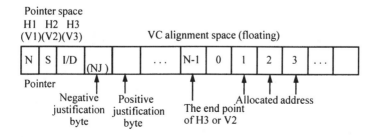

Figure 3.36 Conceptually integrated AU/TU structure.

being 782, 764, 427, 139, and 103, corresponding to the signals AU-4/AU-3, TU-3, TU-2, TU-12, and TU-11, respectively.

In the figure, the address location 0 is positioned in the middle of the frame, and, in the case of AU-4/AU-3, this corresponds to the location right after the H3 byte or after the V2 byte, as in the case of TU-2, TU-12, and TU-11. NJ and PJ represent the negative and positive justification opportunity bytes, respectively. PJ's actual position immediately follows H3 (or V3). N, *ss*, and I/D are for NDF, signal type, and increment/decrement indications, as explained in Section 3.6.2. During the justification execution process, N is always fixed to 0110, and *ss* acquires one of the values listed in Table 3.8 depending on whether the associated signal is an AU or a TU.

First, we study the execution procedure of positive justification with Figure 3.37 as a reference. In the figure, the number $N - 1$ from Figure 3.36 is specified to be 8; t denotes the t'th frame, and $t + 1$ represents the next frame. Just before the positive justification procedure is initiated, the VC's starting address is recorded as 2 by the I/D-bits, and the entire VC alignment space is filled with the VC data (Figure 3.37(a)). Once the positive justification starts, all five I-bits from I/D are inverted, the PJ execution byte is loaded with one null

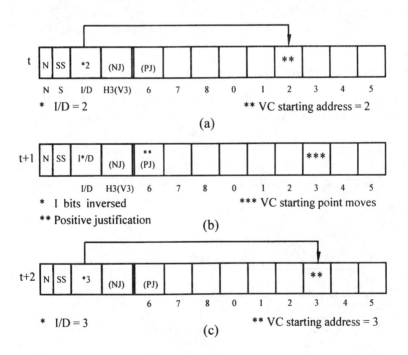

Figure 3.37   Illustration of positive justification: (a) before justification; (b) during justification; and (c) after justification.

byte (or left as a blank space), and the VC data get loaded only onto the remaining VC alignment space. In that case, the VC's starting address gets incremented by one (Figure 3.37(b)). After the termination of the PJ procedure, the VC's new starting location 3 is recorded onto the I/-bits, and the space for VC gets filled with the effective VC data (Figure 3.37(c)).

The negative justification procedure's basic technique is equivalent to that of positive justification except for the direction of execution. We examine its operation using Figure 3.38, with the condition before the execution being identical to the positive justification case (Figure 3.38(a)). When the negative justification is executed, the five D-bits are inverted, the NJ byte acquires effective VC data, and the entire VC space also gets filled with the VC data. In that case, the VC starting point address gets decreased by one (Figure 3.38(b)). Immediately following the termination of negative justification, the new VC starting point address 1 is recorded onto I/D, and the entire VC is now filled with the effective VC data (Figure 3.38(c)).

The pointer I/D address or the AU/TU size indicator can be altered without going through the normal justification procedure. The new pointer address or the new signal type indicator can be sent while the NDF is in the inverted

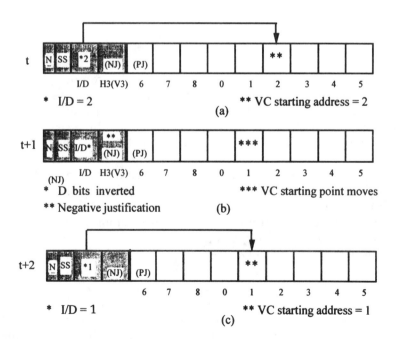

Figure 3.38 Illustration of negative justification: (a) before justification; (b) during justification; and (c) after justification.

condition 1001. (Here, the NDF can be said to be acting as an "interrupt" command.)

In case an error in transmission causes damage to the I-, D-, or N- (data flag) bits, the majority vote rule applies. That is, if three or more of the five I/D-bits are inverted, this is interpreted as denoting justification execution, and if three or more of the four N-bits are inverted, this is considered an NDF.

3.7 STRUCTURE OF SYNCHRONOUS MULTIPLEXERS

In Section 3.2, the synchronous multiplexing structure was examined, followed by an overview of the synchronous multiplexing procedure in Section 3.3. Hence, the synchronous multiplexing procedure was studied in terms of the lower order paths, higher order paths, and the synchronous transport module. It was also observed that the processes associated with the lower order and higher order paths include mapping, the insertion of POH, aligning, multiplexing, and so on. With these two sections as a basis, the present section will give a systematic and detailed examination of the synchronous multiplexer structure.

3.7.1 Functional Structure of Synchronous Multiplexer

The synchronous multiplexer's functional structure in its normal mode of operation is shown in Figure 3.39. In the picture, G.703 represents an existing plesiochronous tributary,[24] and STM-m and STM-n represent synchronous transport modules. The top path that starts from STM-m takes the STM-m and demultiplexes it to extract a higher order VC, which is subsequently multiplexed again to form the STM-n. Also, the middle path involves the multiplexing of a lower order tributary G.703 into the STM-n via both the lower order procedure and higher order procedure. Finally, the lower path represents the multiplexing procedure of a higher order tributary G.703. Hence, this diagram can be said to encompass all the basic multiplexing functions that the synchronous multiplexing structure supports, including the multiplexer for the existing tributaries, synchronous multiplexers for synchronous tributaries, add/drop multiplexers, and interworking multiplexers.

Comparing Figure 3.39 with Figure 3.12, we examine the function of each block. In Figure 3.39, the *physical interface* (PI) represents the function for interfacing with the tributary G.703; the *lower order path adaptation* (LPA) represents the function for mapping the tributary DS-n (n = 11, 12, 2, 3, 4) into the container C-n; *lower order path termination* (LPT) is the function for creating and adding the VC-1/2 POH to the container C-1/2. Lower order path

---

24. Note that sometimes it is called G.702. See, for example, Chapter 4.

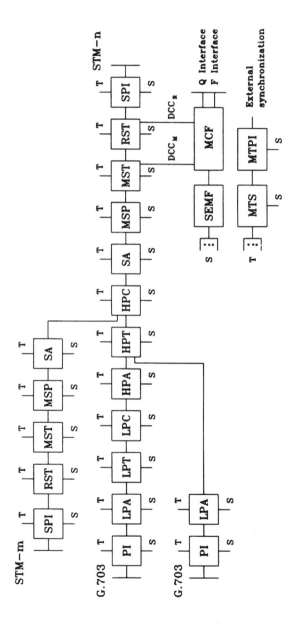

Figure 3.39 Generalized multiplexer functional block diagram.

connection (LPC) is a type of cross-connect function for the elastic storage of lower order VCs inside the higher order VCs. It is used on an optional basis depending on the type of multiplexer.

Higher order path adaptation (HPA) is the function for mapping TU-1/2 into VC-3/4 via a pointer, and *higher order path termination* (HPT) is the function for creating and attaching a higher order POH to TU-1/2. Also, *higher order path connection* (HPC) is a cross-connect function for the flexible storage of VC-3/4 inside the STM-*n*.

Section adaptation (SA) is the function for multiplexing AU-3/4 to the STM-*n* payload through the use of a pointer, *multiplexer section protection* (MSP) is the function for switching signals to another line for protection purposes, *multiplexer section termination* (MST) is the function for the creation and insertion of the MSOH (SOH rows 5 to 9), and *regenerator section termination* (RST) is the function for the creation and insertion of the RSOH (SOH rows 1 to 3). Finally, the *SDH physical interface* (SPI) represents the function for interfacing an STM-*n* signal to a physical medium.

Other functions include the *message communication function* (MCF), which is the function for utilizing the DCC, the *synchronous equipment management function* (SEMF), the *multiplexer management function* (MMF), the *multiplexer timing source* (MTS) for network synchronization, and the *multiplexer timing physical interface* (MTPI).

In the figure, *T* represents the timing signal, *S* represents the supervisory (monitor, alarm, control) signal, and *Q* and *F* are the message interfaces to MCF, with *Q* being the MCF access point to the *telecommunication management network* (TMN). Among the functions, only the PI and the LPT vary with each payload, and the remaining functions are payload independent.

3.7.2 Functions of Constituent Blocks

As shown in Figure 3.39, the synchronous multiplexer's constituent functions can be categorized into lower order path function, higher order path function, and transport termination function. The lower order path function consists of G.703 PI, LPA, LPT, and LPC, while the higher order path termination function is composed of HPA, HPT, and HPC. The transport termination function is composed of SA, MSP, MST, RST, and SPI.

Lower Order Path Function

The G.703 interface functions as the interface between the multiplexer and the physical medium. In the G.703 receiving direction, it extracts data and timing information from the G.703 signal and sends them off to an LPA block, and in the transmission direction it creates a tributary from the data and timing signals.

When the incoming G.703 signal is in the *loss of signal* (LOS) state, the PI generates an "all 1" signal and transmits it instead of the data.

LPA signifies the capacity to receive data and clock signals from the PI and create a container C-n (n = 11, 12, 2, 3, 4) out of the input information. Hence, the specific LPA function varies from tributary to tributary, and, as a way of differentiation, the nomination scheme LPA-n (n = 11, 12, 2, 3, 4) is used.

Referring to Section 3.4, we can infer that LPA-11, LPA-12, or LPA-2 has asynchronous and synchronous adaptation capability, while LPA-3 or LPA-4 only has the asynchronous adaptation capability.

LPT represents the capability for producing the VC-n by creating and attaching a POH to a C-n or, inversely, the ability to separate the POH from VC-n and process it. Here, POH is categorized into V5, which is used for the lower order paths, and VC-3 POH, which is used for the higher order paths. The associated functions of POH are path trace, signaling, path status and performance indication, path error monitoring, and so on.

The function of LPC is the flexible connection of VC-n (n = 11, 12, 2, 3) to the VC-n slots in the higher order paths. This is a function that was not elucidated in the multiplexing structure illustrated in Figure 3.12. LPC is not equipped in the multiplexer types I, II, IIa, and IV, and, in this case, VC-n is maintained in a fixed connection state in the higher order path (see Sec. 3.7.3). The type III multiplexer with the add/drop and cross-connect capabilities is also equipped with LPC. A connection matrix is employed to record the lower order path's connection status.

Higher Order Path Functions

The HPA function represents the TU-n (n = 11, 12, 2, 3, 4) pointer processing capability. The associated functions include pointer generation, pointer translation, and justification. The ability to assemble VC-n (n = 11, 12, 2) and adapt them to the VC-4 payload or, conversely, to dismantle VC-4 into the VC-n is called HPA-n/4, and the ability to process VC-3 in the same manner is called HPA-n/3.

The HPT function represents the capacity to generate and insert the higher order VC-n POHs or to separate them from the VC-n. Here, the information regarding path trace, signal identification, path status and performance, and path error monitoring is directed to SEMF. If multiframe alignment is required, the POH's H4 byte manipulation is also connected to SEMF.

HPC's function is the flexible connection of VC-n (n = 3, 4) to the VC-n slots inside the multiplexer section. The multiplexer types I, II, and IV are not equipped with HPC. As with LPC, a connection matrix is used to record the higher order path's connection status.

Transport Termination Function

SA's function is to adapt the higher order path to AU, to assemble or disassemble the AUG, to perform bit-interleaved multiplexing or demultiplexing, and to generate and translate pointers.

The MSP's function is to protect the STM-n signal from the breakdown of the multiplexer section (or RSP and SPI functions and the physical medium). The MSP function at both ends involves the continuous monitoring of the STM-n signal and the exchange of relevant information through channels K1 and K2. The protection switching is conducted on the entire multiplexer section, and the protection structure is 1 + 1 or 1:n.

The SPI functions as an interface between the regenerator termination and the SDH physical medium. Here, the SDH physical medium is usually electrical or optical. In the SDH receive direction, the STM-n data and the clock information are extracted, and, in the transmission mode, a reverse function is executed.

3.7.3 Classification of Synchronous Multiplexers

The multiplexer can be categorized into types I, II, III, and IV according to its execution capabilities and utilization purpose. It can be further categorized into a and b subtypes, depending on whether it is equipped with the LPC and HPC functions or not. Hence, the four types can be subclassified into I, Ia, II, IIa, IIIa, IIIb, and IV. Figure 3.40 is an illustration of these types from the viewpoint of interfacing signals. The specialized role of each type can be deduced from the picture, with type I/Ia being a basic multiplexer, II/IIa being a synchronous multiplexer, IIIa being a basic add/drop multiplexer, IIIb being a synchronous add/drop multiplexer, and IV being an interworking multiplexer.

Basic Multiplexer (Types I, Ia)

The basic multiplexer takes G.703 tributaries and multiplexes them together to create the STM-n. Hence, if the left upper branch from Figure 3.39 (i.e., the demultiplexing portion for the STM-n interface signals) is eliminated, then the resulting structure is equivalent to the type I multiplexer. The basic multiplexer equipped with LPC and HPC functions is classified as type Ia, and the one without is classified as type I. Therefore, type Ia differs from type I in that it can support the flexible connection of VC-1/2s or VC-3/4s.

Synchronous Multiplexer (Types II, IIa)

The synchronous multiplexer's special function is to assemble multiple STM-m signals to form the STM-n signal ($n > m$). To illustrate, if the middle and lower branches are removed from Figure 3.39 and replaced with the paths

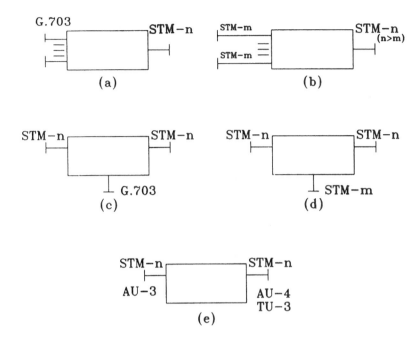

Figure 3.40   Classification of multiplexers: (a) types I and Ia (basic); (b) type II and IIa (synchronous multiplex); (c) types IIIa (basic add/drp); (d) type IIIb (synchronous add/drop); and (e) type IV (interworking).

interfaced with the STM-m, then the synchronous multiplexer results. The synchronous multiplexer, as described, is classified as type IIa; if the HPC function is lacking, then it is classified as type II. As before, type IIa differs from type II in its capacity to provide a flexible connection for the VC-3/4 signals. Without the HPC function, the type IIa multiplexer is directly connected to the SA function, and therefore two SAs can be combined to form a single SA.

Add/Drop Multiplexers (Types IIIa, IIIb)

The add/drop multiplexer takes the G.703 tributary or the STM-n, performs an add/drop procedure on the signal, and transmits it as part of another STM-n signal. The type that operates on the G.703 tributary is classified as the basic add/drop multiplexer (IIIa), while the kind that operates on the STM-m is the synchronization add/drop multiplexer (IIIb). These two add/drop multiplexers are illustrated in Figures 3.41(a) and (b). Comparing Figures 3.39 and 3.41, we can infer that the type IIIa multiplexer is equivalent to the structure that results if the two branches connected to the HPT (which has the G.703 signal interface) in Figure 3.39 are displaced to the bottom of the HPC. Here, the HPC enables

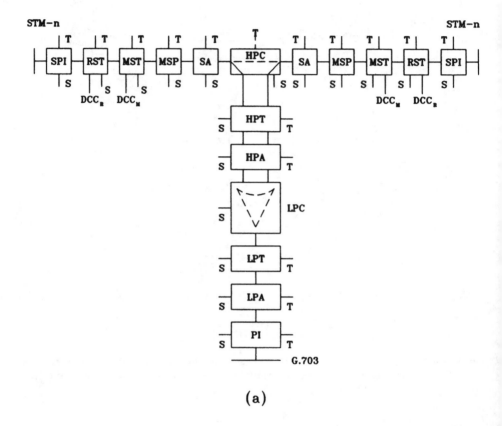

Figure 3.41 Functional structure of add/drop multiplexers: (a) basic add/drop type (type IIIa) and (b) synchronous add/drop type (type IIIb).

the add/drop of VC-1/2 within the VC-3/4. The type IIIb multiplexer has basically the same function as the type IIIa multiplexer, except that its add/drop signal is STM-m rather than G.703. Hence, type IIIb is equivalent to type IIIa if the LPC's lower path (LPT\LPA\PI\G.703), which is a part of type IIIa's function, is replaced by the LPC's upper path (HPA\HPT\SA\MSP\RST\SPI\STM-m). Here, since a separate HPC function is not required for the add/drop procedure on STM-n, it is not a part of type III's function.

Interworking Multiplexer (Type IV)

The interworking multiplexer's function is to integrate inside the STM-n those payloads that conform to the AUG-3 structure with the payloads that are

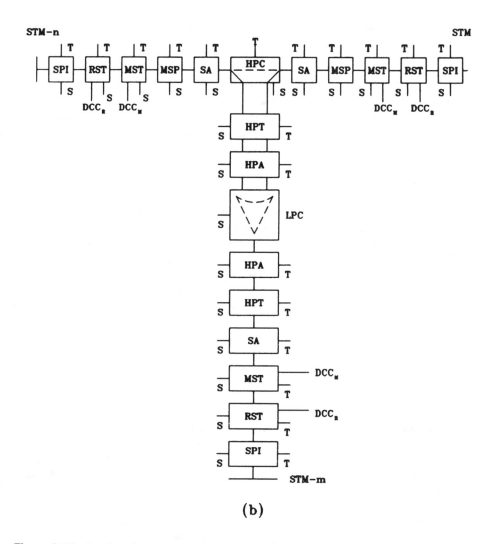

(b)

Figure 3.41 *(continued)*

organized according to the AU-4/TU-3 structure. Hence, type IV's function comprises the STM-*n*\SPI\RST\MST\MSP\SA\HPT path, which disassembles the AU-3, and the HPT\SA\MSP\MST\RST\SPI\STM-*n*, which assembles the payloads in the AU-4/VC-4 format, and vice versa. Since performing interworking is type IV's sole objective, HPC and LPC functions are not required.

3.7.4 Operation and Maintenance Functions

Use of OAM Channels

The OAM channels for synchronous transmission exist inside the POH and SOH. Inside the POH reside the channels for indicating the path AIS, *loss of pointer* (LOP), RDI, REI, RFI, and error checking. Inside the SOH there are also channels for indicating the section AIS, LOS, *loss of frame* (LOF), RDI, REI, RFI, and error checking. In addition, SOH supports the *network operator maintenance channel* (NOMC), which includes orderwire (E1, E2), user channel (F1), and data channels (D1 to D12). Also, for management purposes, additional TMN can be employed.

SEMF represents the function for accepting the OAM-related signals through regenerators or multiplexers to be sent to the MCF, or its reverse. MCF receives and transmits the message through the DCC, or delivers it to the TMN, and it can also function in the reverse direction. Figure 3.42 depicts this relationship in terms of multiplexers and regenerators. In the figure, regenerator-related DCC represents the D1, D2, and D3 channels within the RSOH (DCC$_R$), and the multiplexer-related DCCs represent the D4 through D12 channels with the MSOH (DCC$_M$).

Application of OAM Signals

In the synchronous transmission network, OAM refers to such functions as detection of failure, confirmation of failure and maintenance, location of failure, and isolation and recovery of failure. The OAM signals indicating the state of failure include an *alarm indication signal* (AIS), which notifies of failure in the forward direction; and *remote defect indication* (RDI), which notifies of failure in the backward direction. AIS is generated by setting some particular bits (such as bits in K2 in STM-n, H1 and H2 in AU, V1 and V2 in TU) to the 1 state or by replacing all lost frame with "all 1" data. The RDI message is delivered through some particular bits (such as K2 in STM-n, G1 in VC-3/4, V5 in VC-1/2) in the normal frame traveling backward to the remote end.

There are two ways to monitor performances and alarms in the synchronous transmission network: in-service-monitoring and out-of-service monitoring. The former is used when monitoring performance and alarms while providing services continuously and the latter is applied when locating failures through loopback or test access.

In the synchronous network, maintenance is done systematically in a layered approach, based on regenerator section, multiplexer section, higher order path and lower order path, as illustrated in Figure 3.43. If a failure occurs in the regenerator section and, consequently, the eastward signal in the STM-n gets lost, then the *regenerator section termination* (RST) detects this failure and

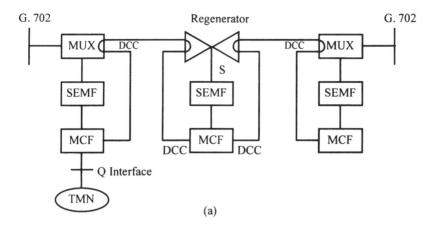

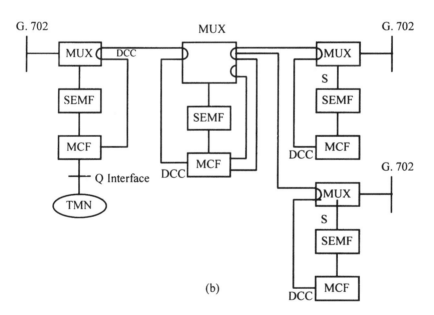

Figure 3.42 OAM-related functions and channels: (a) regenerator section and (b) multiplexer section.

sends out the "all 1" signal (or MS-AIS) in the forward direction. Detecting this MS-AIS, the *multiplexer section termination* (MST) in the east issues the AIS in the forward direction by replacing the AU-3/4 signal in the failed STM-*n* with "all 1" data and, at the same time, sends out MS-RDI backward to the MST in the west. The higher order and the lower order *path termination equipment*

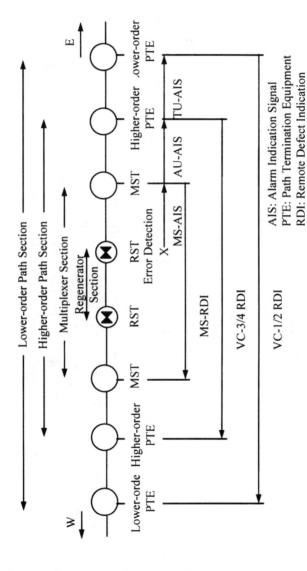

Figure 3.43   Propagation of OAM signals in synchronous transmission systems

(PTE) in the east detects the "all 1" AIS, and then transmits the RDI to the PTE in the west. As such, AIS enables the detection of the upward link failure and RDI the downward link failure in the remote end in a systematic manner. Once a failure is detected, each equipment delivers alarms to the relevant SEMF and the recovery procedure is activated.

Failure is divided into *hard failure* and *soft failure*: Hard failure causes service failure due to malfunctioning of the hardware and/or software of the system, and soft failure causes performance degradation. In general, a failure is regarded as a hard failure if it yields a bit error rate of 10^{-3} or higher; and is regarded as a soft failure otherwise. In case a hard failure occurs, protection switching is operated and alarm signal is generated, whereas in the case of a soft failure, performance is monitored more cautiously with optional protection switching.

Failure detection of the line or signal path is done mainly by checking the state of the received signal. Failure indication signals include transmitter failure, receiver failure, *loss of signal* (LOS), *loss of frame* (LOF), *signal label mismatch* (SLM), *signal unequipped* (UNEQ), *trace identification mismatch* (TIM), and failure of protection switching.

The generation of administration-related signals and the corresponding countermeasures are shown in Figure 3.44. In the figure, relevant actions are categorized into section, higher order path, and lower order path layers, with the generation and the detection of the signals indicated by the filled circles and the countermeasures by blank circles. In the section layer, in the event of the detection of LOF, LOS, section AIS in the received signal, or excessive bit error (B2), the far end is notified of the condition (RDI) and a VC-3/4 path AIS is sent to the higher order path. In the higher order path, if higher order path AIS or higher order path LOP is detected, it is reported to the far end (higher order path RDI), and lower order path AIS is sent to the lower order path. Also, if any bit error (B3) is detected, it is confirmed and notified to the far end (REI). In the lower order path layer, the far end is notified if the lower order path AIS or the lower order path LOP is detected, and AIS is sent to the associated containers. Also, any bit error is confirmed and reported to the far end (REI/V5).

The Roman numerals II, III, and I, which are displayed in the right half of Figure 3.44, represent the three classes of multiplexers and the coverage of OAM signals. That is, the class II multiplexer can detect the OAM-related signals and apply relevant countermeasures for the SOH, and the class III multiplexer can do the same not only for the SOH but also for higher order path overhead. The class I encompasses all three layers.

Protection Switching

If a hard failure is detected in the line system (the line system is part of the network's operation and maintenance function), protection switching is

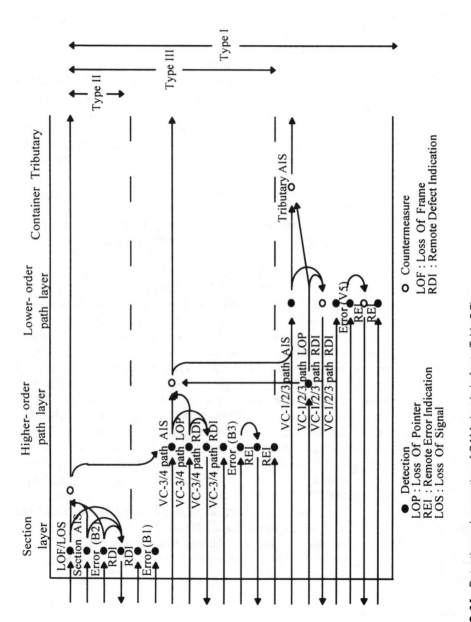

Figure 3.44   Detection and generation of OAM signals (refer to Table 3.7).

triggered. Here, the line system designates the path that starts at the multiplexer termination, goes through the regenerator, and ends at the starting point of the far-end multiplexer.[25] Protection switching is executed to protect all functions and equipment between a pair of MSPs including MST, RST, and SPI. Protection switching can be performed in the 1 + 1 or 1 : n (n = 1 to 14) format, and the employment of the protection action can be unidirectional or bidirectional and can also be revertive or nonrevertive. The protection switching architectures for the 1 + 1 and 1 : n cases are illustrated in Figure 3.45.

Automatic protection switching enhances the transmission reliability of the network. In the synchronous transmission network, reliability can be further enhanced, because a ring-type or mesh-type architecture is also available in place of the conventional star-type architecture. That is, the flexible add/drop and dynamic cross-connecting capability of the synchronous transmission network effectively renders a ring-type or mesh-type architecture with the add/drop multiplexers and digital cross-connects forming its constituent network node elements. In the ring and mesh networks, it is possible to reconfigure the network connection whenever link failure occurs such that the connection can detour the failed links, thus further improving the transmission reliability (see Section 3.10).

3.7.5 Synchronization Functions and Jitter

Timing Methods

In the functional structuring of the multiplexer in Figure 3.39, the functional block associated with timing and synchronization is the MTPI. The MTPI provides the function for physically interfacing the external synchronization signal to the internal timing source. The role of the MTS is that of providing a timing standard to each functional element of the multiplexer.

There are several timing methods depending on the timing source. They are *external timing, line timing, loop timing, through timing,* and *internal timing,* which are illustrated in Figure 3.46. External timing synchronizes the system clock to the externally received clock; line timing takes the system clock out of the received STM-n signal and uses it for transmission in both directions; loop timing uses the received clock for transmission in the backward direction: through timing uses the received clock for transmission in the forward direction; and internal timing takes the clock from the locally installed oscillator and uses it for independent synchronization or for timing replacement in the case of system clock failure.

---

25. To be more accurate, the line system starts from RST, goes through SPI and the regenerator, and terminates at the far-end SPI and RST. Here, SPI and RST functions should be considered to be part of the regenerator function.

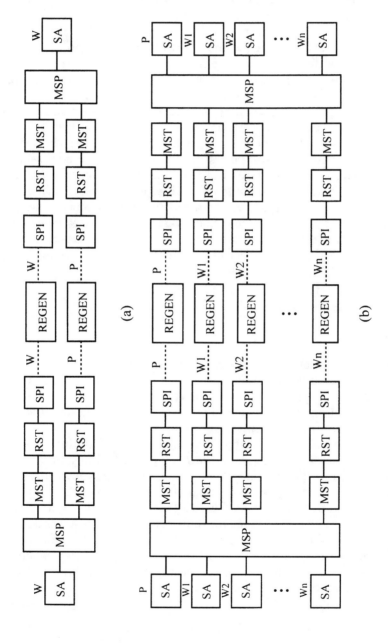

Figure 3.45   Protection switching architectures: (a) 1 + 1 and (b) 1 : n.

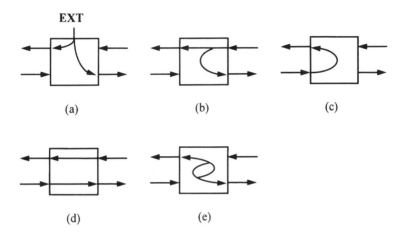

Figure 3.46 Timing techniques in the synchronous transmission network: (a) external timing; (b) line timing; (c) loop timing; (d) through timing; and (e) internal timing.

Clock references can be extracted from three different sources: The first is the G.702 external synchronization interface, the second is the G.702 tributary interface, and the third is the STM-n interface. The 2.048-kHz external timing signal is furnished by the G.702 synchronization interface; and the timing signal for line timing, loop timing, or through timing is taken from the signal received through the G.702 interface or the STM-n interface.

Synchronous transmission equipment exchanges synchronization state messages to achieve network synchronization in an effective and consistent manner. This message is delivered through bits 5 through 8 of the S1 byte in the multiplexer section overhead (MSOH) of the STM-n signal. The synchronization state message specifies how to choose a synchronization source for each section of the network and helps to make synchronization strategy compatible among all timing methods. It also helps to maintain the best possible synchronization quality by indicating the quality of each timing source, and further protects from clock looping, which can occur when each end system takes the timing source from the received signal.

Pointer Adjustment Jitter

Even though the synchronous digital transmission was originally conceived to operate within a synchronized communication network, it can also be accommodated in plesiochronous networks through pointer manipulations. Consequently, the synchronous transmission network's jitter and wander performance is determined by the internal and external clock performances of the synchronous transmission network, the output wander at the synchronous network interface, and the jitter and wander of the synchronous line system. Also, the jitter and wander of the G.702 tributary output, which is dependent on the performance of pointer manipulation, is determined by the jitter and wander performance of the synchronous transmission network and the jitter and wander handling capability of the synchronous multiplexer/demultiplexer at the boundary of the synchronous network.

Consequently, jitter and wander are important performance indicators for the synchronization of the synchronous digital network. The tolerance range of jitter and wander in the synchronous digital network is regulated for the G.702 interface and the STM-n signal, and is categorized into multiplexer-related jitter and line-system-related jitter. If optical transmission is employed, the jitter related to the transmission line system is of no concern. However, in the synchronous transmission network, waiting time jitter emerges as a particularly important problem to resolve because the pointer technique is used as a means of achieving synchronization.

The pointer technique is one of the most distinct features of synchronous digital transmission. It can be regarded as an influence of computer and software technologies on communication technology. The pointer-based synchronization method can achieve synchronization without repetitive frame search procedures, and it can also cope with a plesiochronous environment with small elastic store, making synchronization possible over a wide area. However, because the pointer technique is linked with the 125-μs duration frame, it generates low-frequency and high-amplitude jitters. This is because the justification ratio corresponds to 0.5 and the justification is performed on byte sizes of one or three. From the standpoint of waiting time jitter, the justification ratio of 0.5 is quite an undesirable value, and has therefore been avoided by existing plesiochronous transmission systems.

To reduce the jitter induced by pointer adjustment, various jitter treatment techniques have been devised, including the direct reduction scheme, which utilizes narrowband phase-locked loops (PLLs) for desynchronization, and the bit leaking control scheme, which splits the processing interval of byte size into bit size or a fraction of a bit.[26]

3.7.6 Synchronous Network Management

The synchronous transmission system is equipped with overhead channels for various maintenance functions and data communication channels (DCCs) for centralized network management. The 192-Kbps channel formed by channels D1, D2, and D3 of the RSOH and the 576-Kbps channel formed by D4 through

---

26. Refer to ITU-T Recommendations G.823, G.824, G.743, G.752, G.783, G.958 for the requirement on jitter and wander in the SDH network.

D12 of the MSOH provide the DCCs for transmission of OAM information. Such DCCs render the physical link to operate the *embedded control channel* (ECC), which enables the construction of an *SDH management network* (SMN) among synchronous transmission *network elements* (NE).

The SMN is a network that manages NEs in the synchronous transmission network, and can be regarded as a part of the *telecommunication management network* (TMN). The SMN is furnished with a management application function, network element function, and *message communication function* (MCF), which contribute to the generation, termination, collection, and transfer of the TMN messages. The MCF is carried out by operating the ECC through the DCC. The operation of the ECC provides access to the *data communication network* (DCN) and the *local communication network* (LCN), which are required in the TMN. Consequently, the LCN of the TMN can be easily accessed if synchronous transmission equipment is installed in every central office and, further, the DCN can be constructed through the SMN if the *operating system* (OS) for network management is positioned at some central offices.

Therefore, the network management function of the synchronous transmission network can be considered in the context of the TMN access function. The most important objective of network management through the TMN is the survivability of network services. For network survivability, or for an uninterrupted service provision, a real-time management capability is necessary that can react to network environment changes actively. For example, it is necessary to be able to identify degraded network elements and set up detouring paths, to be capable of evaluating the performance quality of a network element by measuring the failure occurrences and then taking appropriate reactive actions, and to be able to distribute properly the network facility in response to service traffic changes. In the existing plesiochronous transmission environment, the network management system has to be newly defined whenever a new system is developed, and the OAM information gathered from different transmission equipment has to be transformed to the data formats adequate for an integrated transmission network management. To improve such inefficiency and equipment dependency of network management, an integrated management information system is necessary, and the SMN satisfies this necessity in the synchronous transmission network environment.

Figure 3.47 depicts the process by which the SEMF, or the management host of synchronous transmission equipment, converts the OAM information into management data. It first collects OAM information via the S interfaces of each functional block shown in Figure 3.39, then processes and delivers the information to the *managed objects* (MOs). Each MO processes and stores the information in a predetermined manner. The agent converts this information into *common management information service element* (CMISE) messages and sends them to the manager. It also receives appropriate management action messages from the manager and passes them to MOs. In this process, the OSI

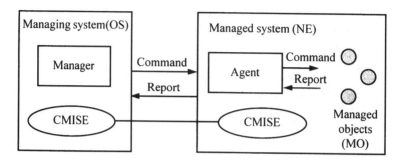

Figure 3.47 Network management system.

standard network management, the object-oriented design, the network management protocol, and the agent function are employed for conversion and processing of the OAM information and the management data.[27]

The OSI standard network management technique models the network facility in the network management aspect. That is, it represents, based on the object-oriented modeling scheme, the physical management elements of the network facility in the form of objective MOs, and uses the CMISE network management protocol as a means to exchange the management information issued by the MOs. Such a modeling and network management protocol enables the provision of commonly recognizable access references. In other words, the management information can take a format that is recognizable by the manager and the agent without restriction in MOs, and the network management protocol can operate through commonly recognizable procedures.

Therefore, network management can be referred to as an interaction among the manager, the agent, and the MOs. The manager monitors the management information issued by the MOs and commands appropriate management actions, and these functions are all handled by the OS. The agent delivers the manager's commands to MOs and reports MO management information to the manager. The CMISE is the network management protocol that is devised to make management behaviors recognizable by each other in this process. The CMISE protocol transfers various types of management information in predetermined formats and specifies various management behaviors adequately in accordance with their attributes.[28]

---

27. Refer to ITU-T Recommendations G.773, G.774, G.784, and M.3010.
28. For a more detailed description on CMISE, refer to ITU-T Recommendations X.710 and X.711, and ISO Standards 9595 and 9596.

3.8 SYNCHRONOUS TRANSMISSION SYSTEM

In the previous section the functional configuration of synchronous multiplexers was discussed, and the discussion revealed that synchronous multiplexers consist of the functional blocks shown in Figure 3.39. Based on this functional configuration, physical configurations of synchronous transmission system are considered in this section. The synchronous transmission system includes *add/drop multiplexers* (ADM), a terminal-type multiplexer, and a broadband *digital cross-connect system* (DCS). If viewed from the transmission rate aspect, synchronous transmission systems include a 155-Mbps system, 622-Mbps system, 2.5-Gbps system, and 10-Gbps system. The module configuration of these transmission systems is examined in this section.

3.8.1 General Module Configurations

If the functional configuration in Figure 3.39 is reconfigured in order to realize various synchronous transmission systems, the module configuration in Figure 3.48 is obtained. In the figure, the VC-1, VC-2, C-3, and C-4 modules map the PDH tributaries to virtual containers (VC) or containers (C); the VC-3/4 adaptation module adapts lower order VCs to higher order Cs to form administrative unit groups (AUGs) (see Figure 3.12). AU and TU connection modules reconfigure the AU and TU level connections, respectively; and STM-1, STM-4, STM-16, and STM-64 modules multiplex 1, 4, 16, and 64 AUGs, respectively, affix SOHs, and then transmit the resulting STM-n signals. On the other hand, the SEMF module is for the maintenance of the synchronous equipment; the MCF module is for OAM data processing; and the STG module is for system timing generation. Other additional modules can be attached for other additional functions at the operator's convenience.

PDH Tributary Interface Modules

The VC-1 module maps the DS-1 (i.e., DS-1/1E) tributary to C-1 (i.e., C-11/12) and attaches POH to form VC-1 (i.e., VC-11/12). Then it adapts VC-1 to TU-1 (i.e., TU-11/12) and multiplexes four TU-11s or three TU-12s to form TUG-2. Likewise, the VC-2 module maps the DS-2 tributary to C-2, attaches POH to form VC-2, and then adapts VC-2 to TU-2, which is TUG-2. In a similar manner, the VC-3 module(T) takes the DS-3 tributary and generates the intermediate signals C-3, VC-3, and TU-3, finally issuing TUG-3. On the other hand, the C-3/4 module maps DS-3/4 to form C-3/4, and the VC-3/4 module(A) does the additional functions of generating VC-3/4, AU-3/4 and finally AUG (refer to Figure 3.12). Each PDH tributary interface module also performs the reverse process of those discussed above.

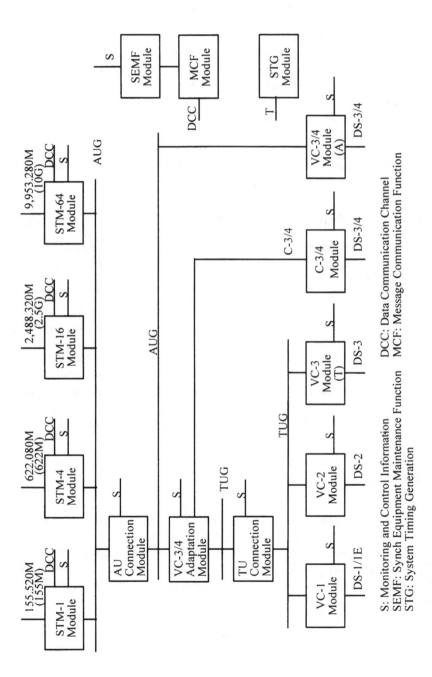

Figure 3.48 Module configuration of synchronous transmission systems.

S: Monitoring and Control Information   DCC: Data Communication Channel
SEMF: Synch Equipment Maintenance Function   MCF: Message Communication Function
STG: System Timing Generation

Higher Order VC Adaptation Modules

The VC-3/4 adaptation module first multiplexes seven TUG-2s to form VC-3, or multiplexes three TUG-3s to form VC-4. It can also form VC-3 or VC-4 by attaching VC-3/4 POH to the C-3/4 signal generated by the C-3/4 module. It then adapts VC-3/4 to AU-3/4 and finally generates AUG. It also performs the reverse processing of these functions.

SDH Signal Interface Modules

The STM-n (n = 1, 4, 16, 64) module multiplexes n AUGs, then attaches SOH to generate the STM-n signal, and does the reverse processing also. The SOHs to be attached in this process are those shown in Figures 3.29(a–d). The STM-n signal gets *frame-synchronous scrambled* (FSS), then goes through electrical-to-optical conversion, and is optically transmitted in the manner specified in ITU-T Recommendation G.958. The resulting transmission rates are 155.520, 622.080, 2,488.320, and 9,953.280 Mbps, respectively, for STM-1, STM-4, STM-16, and STM-64, and in this context the corresponding optical transmission systems are called 155-Mbps, 622-Mbps, 2.5-Gbps, and 10-Gbps transmission systems, respectively.

TU/AU Connection Modules

The TU connection module performs the lower order path connection (LPC) function in the TU level among the TUG signals, and also does the lower order path monitoring function. Similarly, the AU connection module performs the higher order path connection (HPC) function in the AU-3 or AU-4 level among the AUG signals and does the higher order path monitoring function. The connection function performed by the TU and AU modules is bidirectional and symmetrical, and the interface signals are the standard TUG and AUG signals, respectively.

Synchronous Equipment Maintenance Function Module

The SEMF module processes the OAM messages collected through the S interface and controls the OAM-related functions of each module. It also generates MOs by applying object-oriented modeling process to the OAM messages. The OAM functions for the VC path and the STM-n section are applied in common, independent of the STM-n transmission systems, ADMs, and DCSs. The OAM functions for the equipment-dependent modules are also applied in common at the level of modules, but signal capacity or processing amount may differ depending on the type of equipment.

Message Communication Function Module

The MCF module processes the data in the DCC of the STM-*n* signals and provides an interface to the OS of the external TMN. It relays, terminates, generates,
and processes the OSI seven-layer protocol to provide an interface between the
ECC protocol of the DCC and the OS of the TMN. Also, it can optionally provide
an interface with the *local communication network* (LCN). Higher layer processes are common among ECC, TMN, and LCN, but lower order layer processes
differ depending on the interface type. The MCF module is used commonly
among all synchronous transmission equipment, but the amount of processing
differs depending on the STM-*n* signal capacity.

System Timing Generation Module

The STG module generates timing clocks based on dependent, external, and
independent timing modes and distributes them to each module in the synchronous transmission system. The dependent timing mode extracts timing
clocks from the 8-kHz signals received from the PDH and the SDH interface
modules, the external timing mode takes the externally provided timing signal,
and the independent timing mode is used when the dependent and the external
timing modes both fail. The STG module is required to generate the timing
clocks within the allowed tolerance for a certain duration of time even after all
synchronization sources are lost. The STG module is optional for the ADM in
the through timing and the loop timing modes.

3.8.2 155-Mbps Transmission System

The 155-Mbps transmission system is the synchronous transmission equipment
that multiplexes various PDH tributaries into the 155.520-Mbps STM-1 signal
and then transmits it. There are two different types of 155-Mbps systems: the
terminal type and the ADM type. The module configurations of these two types
are shown in Figures 3.49(a,b), respectively.

The terminal type 155-Mbps system does a simple function: It accommodates the PDH tributaries and multiplexes them to form an STM-1 signal and
then transmits it optically. The terminal type system consists of the PDH interface modules (i.e., VC-1 module, VC-2 module, and C-3/4 module), the SDH
interface module (i.e., the STM-1 module), and the VC-3/4 adaptation module.
Among the modules, the PDH interface modules can be protected by board
switching, and the VC-3/4 adaptation module and the STM-1 module can be
protected by 1 : n or 1 + 1 protection switching.

The ADM-type 155-Mbps system is formed when TU and AU connection
modules are added to the terminal-type system. That is, TU exchange function
is added to the VC-3/4 adaptation module, and the AU connection module is

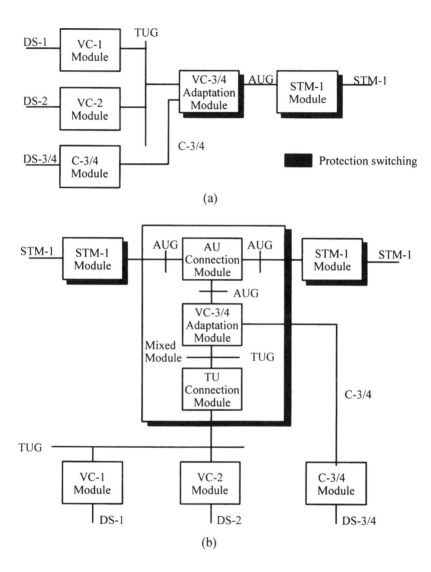

Figure 3.49 Module configuration of 155 Mbps transmission system: (a) terminal type and (b) ADM type.

additionally attached. In the ADM system, a part of the VCs in the original STM-1 signal are extracted while passing through the AU and TU connection modules and then get dropped as the corresponding PDH tributaries. Instead, some other PDH tributaries are inserted through a reversed process to fill the blank positions in the STM-1. Among the modules in the ADM system, the mixed module for the AU connection, VC-3/4 adaptation, and TU connection functions, as well as the STM-1 module, can be protected by 1 + 1 protection switching.

3.8.3 622-Mbps, 2.5-Gbps, and 10-Gbps Transmission Systems

These high-speed synchronous transmission systems multiplex and add/drop higher order PDH tributaries(i.e., DS-3 and DS-4) and lower speed SDH signals (i.e., STM-1, STM-4, STM-16), and transmit high-speed STM-n (n = 4, 16, 64) signals optically. It is also possible to accommodate lower order PDH tributaries (i.e., DS-1 and DS-2), but in this case it is desirable to employ another 155-Mbps ADM system to process them separately, aside from the high-speed transmission system. The module configuration of the high-speed transmission system is as shown in Figure 3.50. The interfaces between modules in the figure can be arranged to be identical to those of the 155-Mbps system. The STM-n modules can be protected by 1 : n or 1 + 1 automatic protection, but the AU connection module can be better protected by adopting 1 + 1 automatic protection.

The high-speed transmission system in Figure 3.50 exhibits a joint system that contains the basic add/drop and the synchronous add/drop multiplexers in one system (refer to Sec. 3.7.3). If it is reconfigured to include one STM-n module and several STM-m modules only, it then forms an SDH terminal-type transmission system.

In the case of the 10-Gbps (i.e., STM-64) optical transmission system, it is also possible to consider *wavelength-division-multiplexing* (WDM) or *optical frequency-division multiplexing* (OFDM) four of the 2.5-Gbps (i.e., STM-16) signals. Such an approach can be equally applied in constructing very-high-speed optical transmission systems such as 40-Gbps (i.e., STM-256), 100-Gbps (i.e., STM-640), and 1-Tbps (i.e., STM-6400) systems.

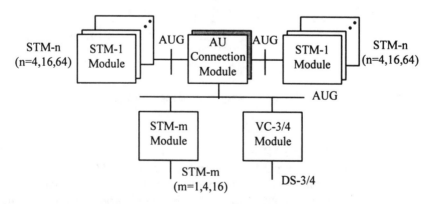

Figure 3.50 Module configuration of 622 Mbps/2.5 Gbps/10 Gbps transmission systems.

3.8.4 Broadband Digital Cross-Connect Systems

Broadband DCS provides cross-connection in the lower order path and the higher order path levels among multiple STM-n (n = 4, 16, 64) signals, and also adds/drops STM-1 and PDH tributaries. The module configuration of the DCS is shown in Figure 3.51.

The broadband DCS is usually located at a concentrating point of the synchronous transmission network, so its network management capability is very important. For this reason, internationally standardized network management capability is installed in the DCS, along with the performance and failure monitoring capability for the tandem connection links, and the automatic failure recovery capability for the mesh networks and ring networks.

The TU/AU connection module in the DCS enables, through VC-n (n = 1, 2, 3, 4) level switching, to reconfigure dynamically the connections for distribution, add/drop, path switching, test access, and broadcasting. This capability helps make the *digital system cross-connect* (DSX) functions of the *main distribution frame* (MDF) more automatic and electronic (the functions

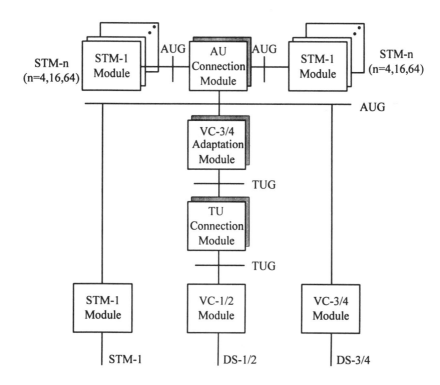

Figure 3.51 Module configuration of broadband DCS.

used to rely on manual operation in the past). It also helps to improve the transmission service quality by consolidating various types of transmission facility into one DCS.

3.8.5 Synchronous Transmission Network

Figure 3.52 illustrates an example of the synchronous transmission network consisting of the synchronous transmission systems considered so far. Note that

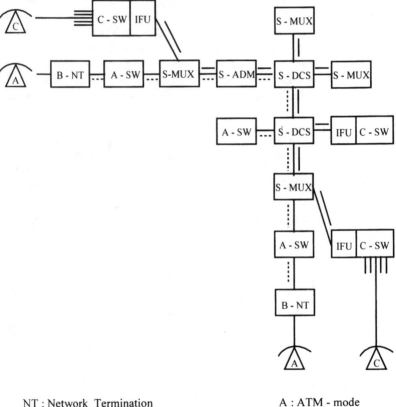

NT : Network Termination
SW : Switch
MUX : Multiplexer
IFU : Interface Unit
ADM : Add - Drop Multiplexer
DCS : Digital Cross-Connect System

A : ATM - mode
B : Broadband
C : Circuit - mode
S : SDH
—— STM/SDH
······ STM/ATM

Figure 3.52 An example of synchronous transmission network.

the synchronous multiplexer (MUX), ADM, and DCS are the main systems of the synchronous network. *BISDN network termination* (B-NT) and *ATM switch* (A-SW) are also included together with the *circuit switch* (C-SW). This signifies that the BISDN will be constructed on the basis of the synchronous transmission network, which in itself is built on the foundation of the existing circuit-mode transmission network. The conventional circuit switches will continue to be an important network element in the synchronous transmission network for a considerably long period of time, and thus STM-n *interface units* (IFU) are necessary to effectively connect them to the synchronous network. On the other hand the BISDN signals coming into the synchronous network via the B-NTs will take the forms of ATM cells and thus will be switched by ATM switches. In the figure, "A-" represents ATM mode; "B-" BISDN; "C-" circuit mode; and "S-" SDH/SONET.

Three different types of signals exist in the synchronous transmission network. The first is the STM-n signal obtained by synchronously multiplexing the PDH tributaries (namely, "STM/SDH"); the second is the STM-n signal obtained by mapping ATM cells into VC-4 payload space in the STM-n format (namely, "STM/ATM"); and the third is the ATM signal formed by a stream of the ATM cell itself (namely, "cell/ATM"). The first two among the three signals appear in Figure 3.52 as solid and broken lines, respectively. The second signal is an STM-n signal, which carries ATM cells internally, so it may be named "STM/ATM." On the other side, the signal is transferred in terms of ATM cells but takes the SDH mapping instead of a pure cell stream, so it may be named "ATM/SDH." If this type of interpretation is applied to the third signal, it yields the representation "ATM/cell."

The above three types of signals all have 155.520-Mbps transmission rates. The STM/SDH signal and the STM/ATM signal are compatible on the STM-n signal transmission layer; and the STM/ATM signal and the cell/ATM signal are compatible on the ATM layer. However, the STM/SDH signal and the cell/ATM signal do not share anything compatible even though their external transmission rates are the same. In this respect, it is desirable to limit the signal type in the synchronous transmission network only to the STM/SDH and the STM/ATM signals.

The notion of mapping ATM cells into the STM-n signal to form the STM/ATM signal can be equally extended to map packets into the STM-n signal. In this case the naming strategy for the STM/ATM yields the name "STM/packet." The STM/packet signal thus obtained is also compatible with the STM/SDH in the transmission layer, and therefore can coexist with the STM/SDH and STM/ATM signals in the synchronous transmission network.

3.9 TRANSPORT NETWORK FOR SYNCHRONOUS TRANSMISSION

The term *transmission* denotes the actual physical process of transmitting an information signal over a physical medium, whereas the term *transport* applies at a more conceptual or logical level, signifying the functional procedure of delivering information from one point to another. Therefore, the conceptually systematized communication network is called a *transport network*, whereas the physical network consisting of the actual equipment such as the multiplexer, the switching system, the cross-connect system, optical transmission system, and so on, is termed the *transmission network*.

The transport network is a large, complex set with various components. Therefore, for the design, operation, and management of a communication network, an accurate network model with well-defined functional entities is essential. In other words, it is necessary to approach a complex transport network by dividing its operation into specific functional blocks of interest. For this purpose, the concepts of vertical division (*layering*) and horizontal division (*partitioning*) are useful. The layering concept has already been examined in detail as being the very foundation of synchronous digital transmission. The partitioning concept can be viewed as being a naturally occurring phenomenon in step with the evolution of the communication network.

In this section fundamental aspects of the transport network, such as layering and partitioning concepts, functional elements, and layer networks, are discussed.

3.9.1 Layering Concept and Layer Network

As observed in Sections 3.1.2 and 3.5, the layering concept forms the basic foundation for synchronous digital transmission. The transmission process becomes layered, which means that it is divided into the path layer and the section layer. This is reflected in the STM-n structure in the form of POH and SOH. The path layer is further divided into the lower order path layer associated with the VC-1/2 and the higher order path layer associated with the VC-3/4.

Table 3.9 is an illustration of the layering concept in its idealized form. In

Table 3.9
Layering Concept

Coarse Layering	Medium Layering	Fine Layering
Circuit layer	Circuit layer	Circuit layer
Path layer	Path layer	Lower-order path layer
		Higher-order path layer
Transmission medium layer	Section layer	Multiplexer section layer
		Regenerator section layer
	Physical medium layer	Physical medium layer

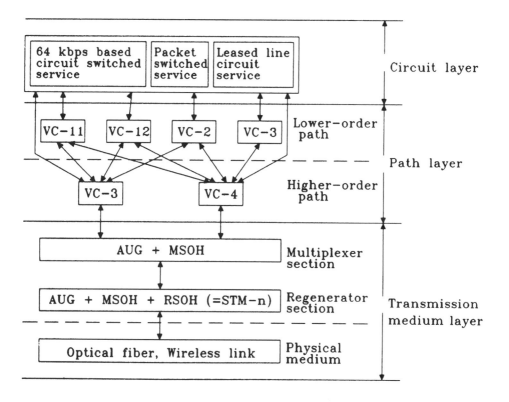

Figure 3.53 Illustration of layering concept.

the table, the circuit layer is placed on top of the path layer, with the physical medium layer put below the section layer. The circuit layer represents various kinds of services that are transported through a common path layer. The physical medium layer forms the lowest part of the transmission process and represents optical or radio media.[29] The physical medium layer and section layer can be combined and labeled the *transmission medium layer.*

If the layering concept is illustrated in conjunction with the synchronous transmission processes, then Figure 3.53 is the result. We can see from the figure that the circuit layer consists of the 64-Kbps circuit-switched service, packet-switched service, leased-line circuit service, and so on. Also, the figure

---

29. The basic physical medium for synchronous transmission is optical fiber. However, transmission based on coaxial cable or radio should not be excluded from the discussion. In this case, a slight variation might result due to the special characteristic of each type of transmission medium.

illustrates that inside the path layer the lower order path layer consists of VC-11, VC-12, VC-2, and VC-3, while the higher order path layer consists of VC-3 and VC-4.[30] AUG and MSOH are a part of the multiplexer section layer, and, finally, the RSOH is included in the regenerator section layer.

If the layering concept is applied as shown in Table 3.9, the communication network can be divided into the circuit layer network, path layer network, transmission medium layer network, and so on. Here, the adjacent layer networks maintain a server/client relationship, and each layer network possesses its own OAM capability.

The circuit layer network provides communication service to users through the circuit layer access points. Its target services include circuit-switched service, packet-switched service, and leased-line service. The configuration of the circuit layer network varies depending on the kinds of services a specific network can provide. The circuit layer network operates independently of the path layer network.

The path layer network delivers information to the path layer access points in support of the circuit layer network. The path layer network functions as a lower order layer network that can be shared by different sets of services. The path layer network can be divided into the lower order path layer and the higher order path layer, and is independent of the transmission medium network.

The transmission medium layer network supports the path layer by transporting information from one path layer access point to another. The transmission medium layer network is dependent on the actual physical medium used, such as optical fiber and radio link. The transmission network's internal layers consist of the multiplexer section layer, the regenerator/section layer, and the physical medium layer.

The circuit layer network, path layer network, and the transmission medium network are illustrated in terms of an actual physical network in Figure 3.54(a–c). In the figure, the circuit layer is the network connecting service transport termination points, while the path layer network can be seen as its sublayer which connects path layer access points. Here, the respective circuit from each circuit layer network follows one unique path in the path layer network. The transmission medium layer network becomes a physical layer that is established in accordance with the path layer network.

3.9.2 Partitioning Concept and Subnetwork

Whereas the layering divides the transport network in the vertical direction, the partitioning corresponds to a division along the horizontal direction of the

---

30. For convenience, the layers associated with the VC-1/2 and VC-3/4 are called the *lower order layer* and *higher order layer*, respectively, while the *lower* and *upper layers* are used as relative concepts. If there is no conceptual ambiguity, they will be used interchangeably.

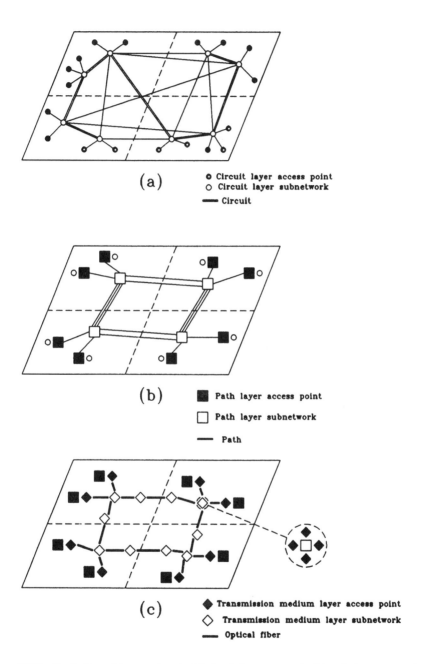

Figure 3.54 Illustration of layer network: (a) circuit layer network; (b) path layer network; and (c) transmission medium layer network.

transport network. The partitioning applies to each layer network; hence, the transport network can be divided into three classes according to the layering concept, and each layer network can be subdivided into subnetworks according to the partitioning concept. Figure 3.55 is a graphical comparison of the layering concept with the partitioning concept.

A partitioned layer network consists of subnetworks and the corresponding link connections. Subnetworks can further be partitioned into smaller subnetworks and link connections. Examples of subnetworks include the international network, national network, transit network, and access network.

Figure 3.56 illustrates the partitioning of a layer network into subnetworks and link connections. Figure 3.56(a) demonstrates the partitioning concept through an actual layer network, and Figure 3.56(b) is a conceptual demonstration of partitioning. Here, the second figure can be considered as illustrating a magnified version of the subnetwork in the first figure along the dotted line connecting trail termination points A and B.

3.9.3 Functional Elements of Transport Network

To establish functional reference modeling of the transport network, it is necessary to examine the constituent elements, the transport entities, the transport

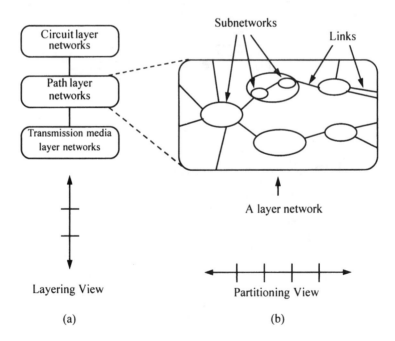

Figure 3.55 Comparison of of the layering concept and the partitioning concept: (a) layering and (b) partitioning.

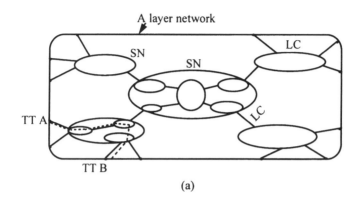

(a)

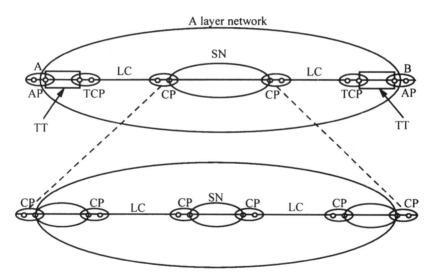

LC : Link Connection   AP : Access point
SN : Subnetwork   TCP : Termination connection point
TT : Trail Termination   CP : Connection point

(b)

Figure 3.56 Illustration of partitioning: (a) physical configuration and (b) conceptual configuration.

handling functions, and the reference points. Among them, the network constituent elements refer to the previously examined layer network, subnetwork, and link connections, which can be said to be a topological representation of the transport network. In the following section, the transport entities, transport handling function, and reference points are examined with the illustration in Figure 3.57 as a basis.

Transport Entities

Transport entities refer to the connection that provides transparent transport of information between the layer network's access points. In other words, between any pair of network access points linked by a transport entity, no change or degradation is made to the transported information. Transport entities include network connection, subnetwork connection, link connection, and trail.

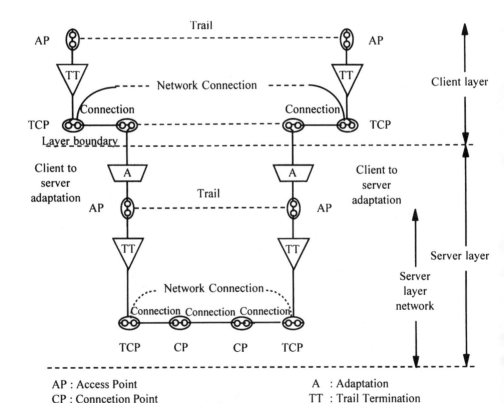

AP : Access Point
CP : Conncetion Point
TCP :Termination Connection Point

A : Adaptation
TT : Trail Termination
NC : Network Connection

Figure 3.57   Illustration of network constituent elements.

Network connection (NC) provides transparent information transfer across a network at a given layer. It is a concatenation of subnetworks and link connections and is delimited by the network *connection points* (CP) located at the network boundary and hence by the *termination connection points* (TCP).

Subnetwork (SN) connection is a network connection across a subnetwork that is delimited by the connection point on the subnetwork boundary.

Link connection (LC) is the capability for the transparent transport of information between end points connecting two subnetworks. Link connection is bounded by the connection points on the boundary of link and subnetwork, and can be formed by concatenating smaller link connections.

Trail refers to the capability for transporting characteristic information between two *access points* (AP). Characteristic information is the information that is transferred within subnetworks. It is a signal of specified bit rates and format that goes through an adaptation process at the server network boundary. An instance of characteristic information would be the VC-n ($n = 11, 12, 2, 3, 4$). The trail is created by adding the *trail termination* (TT) function between an access point and a termination connection point. Examples of the trail include transport between terminal equipment, between multiplexers, and between line terminations.

Transport Handling Function

The transport handling function encompasses adaptation and trail termination. The *adaptation* function is the process of suitably adapting characteristic information so that it can be transported from one layer to another. Examples of interlayer adaptation include multiplexing, channel coding, bit rate conversion, frame alignment, and justification.

The *trail termination* function appropriately handles information so as to ensure the integrity of information transfer within a trail. In general, the trail termination function is embodied in the form of adding spare information at a trail termination source point and monitoring the information at the trail termination sink point.

Reference Points

The reference point is the point that forms the boundary for the handling function or transport entities, which include CP, TCP, and AP. The CP acts as the boundary for the network connection, subnetwork connection, and link connection. The TCP is the connection point at which trail termination and the link connection meet, and is located at the network boundary. The AP forms the boundary between the adaptation function and the trail termination function and is the point of access between the server network and the client network.

3.9.4 Functional Elements and Layer Model

Figure 3.58 illustrates interlayer association within a transport network including the transport entities, transport handling functions, and access points. In the figure, the circle represents the termination function and subnetworks associated with the circuit layer, the square represents the termination function and subnetworks associated with the path layer, and the rhombus represents the termination function and subnetworks associated with the physical medium layer. The darkened objects correspond to the termination function and the blank ones correspond to the subnetworks. The figure also illustrates various kinds of trails such as the circuit layer trail, lower order path layer trail, higher order path layer trail, and transmission medium layer trail. In the figure, TE, NT, SW, DCS, MUX, REG, and LT stand for terminal equipment, network terminations, switch, digital cross-connect, multiplexer, regenerator, and line termination, respectively.

If the layer model and the constituent functional elements of the synchronous transport network of Figure 3.58 are redrawn in contrast to the synchronous transmission functional blocks of Figure 3.39, the layer model shown in Figure 3.59 is the result. From the figure, we see that PPI belongs to the circuit layer; LPA, LPT, LPC functions to the lower order path layer; HPA, HPT, and HPC functions to the higher order path layer; and SA, MST/RST, and SPI to the transmission medium layer. In each layer network, there exist NC, LC, and SN: SN forms the *DS-n path subnetwork* (DPSN), the *lower order path subnetwork* (LPSN), the *higher order path subnetwork* (HPSN), and the *STM-n section subnetwork* (SSSN) depending on the layer. In the functional aspect, LPA, HPA, and SA provide the trail adaptation function and LPT, HPT, and MST/RST provide the trail termination function.

3.10 SYNCHRONOUS NETWORK SURVIVABILITY

The synchronous transmission network is a large-capacity transmission system with transmission rates ranging from several hundred megabytes per second to tens of gigabytes per second, so any failure in a transmission line or equipment can lead to catastrophic service failures. In this context, it is very important to arrange the network in such a way that services can be provided without interruption even when a failure occurs in the network. Such network survivability can be considered at the transmission network level (i.e., transmission path and section layers), at the ATM network level (i.e., virtual channel and virtual path layers), or at the optical technology level (i.e., physical medium layer). Among them network survivability in the SDH/SONET network level is considered in this section.

Network survivability in the synchronous transmission network can be acquired in different ways depending on the network architecture. In the case

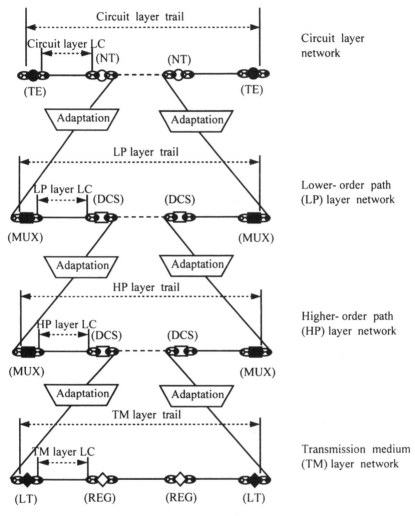

TE : Terminal Equipment   NT : Network Termination
SW : Switching   MUX : Multiplexing
DCS : Digital Cross-Connect   LT : Line Termination
REG : Regenerator   LC : Link Connection

Figure 3.58 Interlayer association in the transport network.

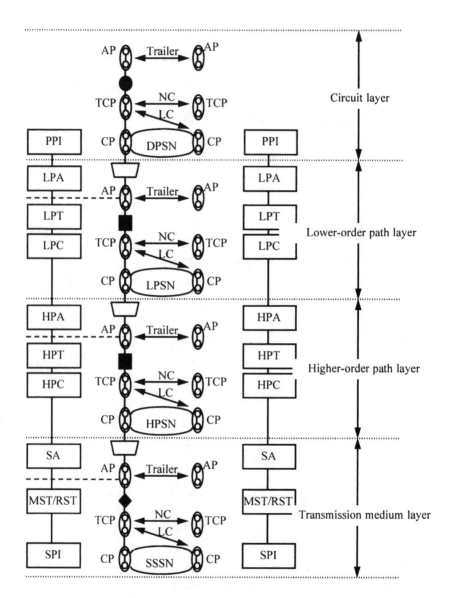

LPSN: Lower-order Path Sub-Network
SSSN: STM-n Section Sub-Network
NC: Network Connection

HPSN: Higher-order Path Sub-Network
DPSN: DS-n Path Sub-Network
LC: Link Connection

Figure 3.59 Relation of SDH functional blocks and transport network functional elements.

of the point-to-point network, automatic protection switching (APS) can be applied; in the case of the ring network consisting of add/drop multiplexers (ADM), a *self-healing ring* (SHR) arrangement can be employed; and in the case of the mesh network including broadband digital cross-connect systems (DCS), detouring paths can be established within the network.

In this section network survivability is discussed for the three approaches above. In discussing each approach network survivability is considered from the aspect of the trail of the transport network introduced in the previous section. Basic concepts for network survivability are explained first, and then network survivability in the point-to-point network, ring network, and mesh network is detailed. Finally, survivability of the ATM network is discussed briefly.

3.10.1 Basic Concepts and Terminology

To secure network survivability, various ways of protection and restoration schemes may be applied : *Protection* refers to the arrangement that switches an operation channel to some protection channel when the operation channel fails, whereas *restoration* refers to the arrangement to utilize spare channel capacity to establish detouring paths when failure occurs in the operating network. In the following, various basic concepts and terminology are introduced in relation to the network protection and network restoration.

Point-to-Point, Ring, and Mesh Networks

In the point-to-point network, protection channels are prepared separately from operation channels, and when an operation channel fails it is automatically switched to the protection channels. Since the span between two nodes is switched to each other in this case, automatic protection switching (APS) is also referred to as *span switching*. The protection channels for APS may lie in parallel with the operation channels along the same line as shown in Figure 3.45, or may be separately placed to take a diverse path as shown in Figure 3.60(a).

In the case of the ring network including ADMs, a protection ring is deployed in addition to the operation ring, as shown in Figure 3.60(b). If failure occurs in a span within the operation ring, the neighboring nodes take APS procedures to activate the protection ring. If a ring network has the capability to "heal" the failure itself in this manner, it is called a *self-healing ring* (SHR). The operation channel in the SHR is protected 1 : 1, and thus has a fast failure-recovery capability as in the case of the APS-based point-to-point network.

In contrast to the APS-based point-to-point network or the SHR network, separate protection channels are not prepared to protect against failure if the mesh network includes DCS. Instead, the mesh network is restored after failure by utilizing spare capacity in the network. For example, if the link connecting

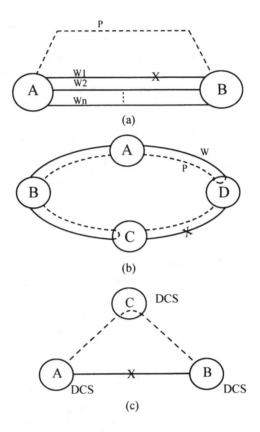

Figure 3.60 Survivability of point-to-point, ring, and mesh networks: (a) automatic protection switching point-to-point network; (b) self-healing ring network; and (c) path detouring mesh network.

nodes A and B fails in the mesh network in Figure 3.60(c), then the DCSs in nodes A, B, and C are activated to establish a detouring path via node C using spare capacity in the pertaining links.

Uniform Routing and Diverse Routing

There are two types of routing: uniform routing and diverse routing. *Uniform routing* applies when operation channels and protection channels share the same physical transmission line [see Figure 3.61(a)], whereas *diverse routing* applies when they lie in geographically separate lines [see Figure 3.61(b)]. In the figures, the traffic is transmitted over both the operation and the protection paths at the same time, but the receiver selects only one path. If the operation path in the west (W) fails, the receiver selects the traffic coming in through the protection path in the east (E) for protection switching. Consequently, two-way traffic shares the same links and equipment even after protection switching in the case of uniform routing, whereas they pass through geographically separate links and equipment in the case of diverse routing. Figure 3.60(b) illustrates the uniform routing ring network. For the ring network, however, diverse routing arrangement is also possible [see Figure 3.63(b)].

Single-Ended Switching and Dual-Ended Switching

Single-ended switching refers to the protection switching scheme to switch the failed channel in one direction only, while *dual-ended switching* refers to the scheme to switch both channels in both directions even if failure occurs in only one channel. These two schemes are illustrated in Figures 3.62 and 3.63, respectively, for the point-to-point network with 1 + 1 span switching and the ring network with diverse paths.

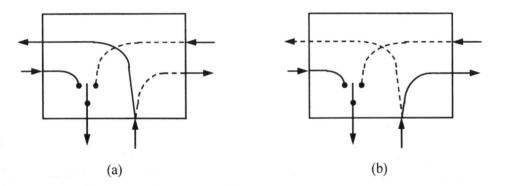

(a)   (b)

Figure 3.61   (a) Uniform routing and (b) diverse routing.

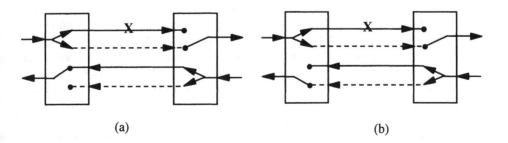

(a)   (b)

Figure 3.62   1 + 1 span switching point-to-point network: (a) single-ended switching and (b) dual-ended switching.

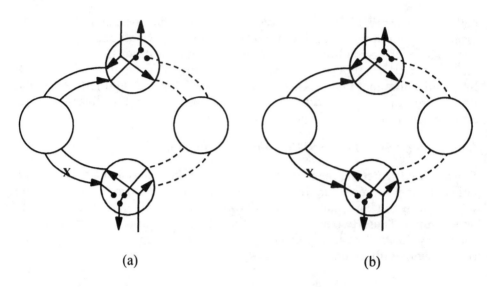

(a)   (b)

Figure 3.63   Diverse path ring network: (a) single-ended switching and (b) dual-ended switching.

Because single-ended switching does not require a protection switching protocol, it is simple to implement and yields fast switching. So when failure occurs frequently, the traffic can be recovered with a higher probability if a single-ended switching scheme is employed. In contrast, in the case of a dual-ended switching scheme, the traffic in both directions is switched to the same protection links and equipment, and hence it is easy to fix the failure. In this case the transmission delay is kept identical in both directions even after protection switching, which is advantageous in case the lengths of the operation link and the protection link are significantly different from each other.

Dedicated Protection and Shared Protection

Dedicated protection refers to the protection scheme that assigns a dedicated channel to protect total traffic capacity in the operation channel, whereas *shared protection* refers to the scheme to make multiple operation channels share one or multiple protection channels. The 1 + 1 APS and the 1 : n APS for the point-to-point network are typical examples of these protection schemes (see Figure 3.45). On the other side, protection switching can be operated in the *revertive* switching mode in which the operation channel returns to the operational stage after the failure recovery or in the *nonrevertive* switching mode in which the protection channel stays in operation even after failure recovery. Dedicated protection may be operated in either mode, but shared protection must be done in the revertive switching mode.

Operation of APS

In the synchronous transmission network, link protection is done by the multiplexer section protection (MSP) functional block for the pertinent multiplexer section, and data communication related to the protection switching is done by bit-oriented protocols over the K1, K2 bytes in the protection channel. For trail protection and subnetwork connection protection, bit-oriented protocols are also applied as in the MSP case over the K3 byte (in the case of the higher order path) and the K4 byte (in the case of the lower order path).

3.10.2 Point-to-Point Network

In the point-to-point network, an additional sublayer is inserted between the access point (AP) and *trail termination* (TT) for trail protection, which consists of TTp (TT Protection), TCPp (TCP Protection), MC (*Matrix Connection*), CPp (CP Protection), Ap (Adaptation Protection), and APp (AP Protection) (see Figure 3.64). Among the added functions, the Ap provides access to the APS channels (i.e., K1, K2, K3, K4); the TTp provides the operation status of the protected channel; and the MC provides protection switching between the operation and the protection channels. Trail protection switching is performed based on the operation status information reported by TT and the protection switching request delivered by the APS channel.

Trail protection in the point-to-point network can be divided into *multiplexer section* (MS) trail protection and *higher order path* (HP) trail protection depending on the adaptation points to protect. Also, it can be divided into 1 + 1 trail protection and 1 : n trail protection depending on the protection configuration.

1 + 1 Trail Protection

In the case of the 1 + 1 trail protection, the operation trail is protected by a dedicated protection trail as shown in Figure 3.64. The MC provides connection to the operation trail (W) in a startup or normal operation state but switches to the protection trail (P) if either failure occurs or the APS channel requests such action [see Figures 3.64(b,c)]. After protection switching, the protection trail is used as the operation trail. That is, the 1 + 1 trail protection is nonrevertive.

In the case of 1 + 1 MS trail protection, an MS trail protection sublayer is formed with Ap replaced by MSPA (MS protection adaptation), MC replaced by the MS protection switch, and TTp replaced by MSPT (MS protection termination), and trail protection switching is performed based on the failure status information detected by MST and MSPA. In the case of the 1 + 1 HP trail protection, a HP trail protection sublayer is formed in a similar manner

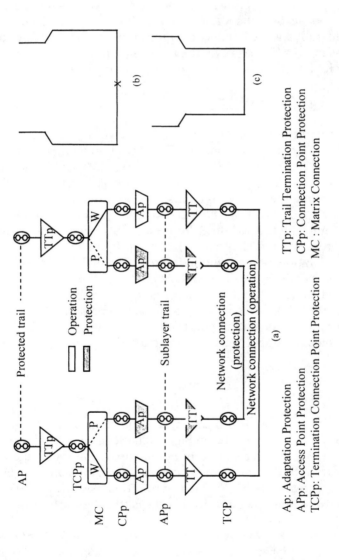

Figure 3.64   Point-to-point network trail protection (1 + 1 protection switching); (a) functional configuration; (b) connection in normal operation; and (c) connection in protection state.

and protection switching is carried out based on the failure status detected by HPT and HPPA (HP protection adaptation) (refer to Figure 3.66 for acronyms).

1 : n Trail Protection

In the case of the 1 : n trail protection n operation trails share one protection trail for protection. Figure 3.65 illustrates the 1 : n trail protection for $n = 2$. The MC in this case provides connections to each individual operation trail in normal operation, but, if failure occurs in a trail, it switches the corresponding connection to the protection trail [see Figures 3.65(b) and (c)]. When the failed trail recovers its normal operation state, the MC returns the connection to the original operation trail. That is, the 1 : n trail protection is revertive. Furthermore, the 1 : n trail protection switching is dual ended so that the connection of both directions can be switched at the same time.

3.10.3 Self-Healing Ring

There are two types of SHRs—dedicated and shared—and the shared SHR itself can be divided into two categories—*two-fiber SHR* (SHR/2F) and a *four-fiber SHR* (SHR/4F). These three types of SHRs are depicted in Figures 3.66 to 3.68. For both dedicated and shared SHRs, bit-oriented ring APS protocols and algorithms are applied, and the ring protocol can support up to a maximum of 16 nodes.

As can be seen in Figures 3.66 and 3.67, there is not much difference between the MS dedicated SHR and the MS shared SHR in protection switching link configurations, but the traffic characteristics of the channels in the protection ring are very different from one another. Since the dedicated SHR provides bidirectional traffic using a unidirectional ring, a particular time slot in each MS passing through the dedicated protection ring occupies under failure a particular channel allotted from the overall ring network level. In contrast, since the shared SHR provides bidirectional traffic between two adjacent nodes using two opposite-directed rings, the channel occupied by each time slot can differ depending on the MS. Therefore the traffic passing through the protection ring can differ under failure depending on the location of the failed MS. In other words, since the shared SHR provides detouring paths through ring switching when MS channel fails, all the operation channels in the overall ring network share the protection ring.

Dedicated SHR

A dedicated SHR is equipped with a dedicated protection ring that can protect the whole traffic capacity. It consists of two opposite-directed rings with one carrying live traffic in operation and the other reserved for traffic protection in

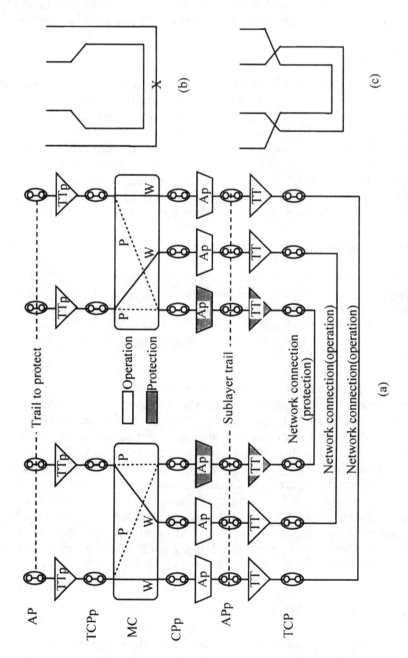

Figure 3.65 Point-to-point network trail protection (1:2 span switching): (a) functional configuration; (b) connection in normal operation; and (c) connection in protection state.

the ring failure state. The maximum traffic capacity of a dedicated SHR is the same as the capacity of a span in the ring. Failure recovery of the dedicated SHR can be accomplished through the MS dedicated self-healing process or by HP trail protection of the dedicated SHR.

Figure 3.66 depicts the functional configuration of a node in an MS dedicated SHR, where the link incoming from the west (W) and outgoing to the east (E) is for the operation ring and the link of the reverse direction is for the protection ring. As for the case of trail protection in the point-to-point network, a trail protection sublayer exists for MS trail protection between the MS trail termination (i.e., MST) and the MS adaptation (i.e., SA). The trail protection sublayer consists of MS protection adaptation (MSPA), MS protection termination (MSPT), and MS matrix connection (MS MC). The HP matrix connection (HP MC) in the figure controls the connection of the higher order paths.

In normal operation state, the MS MC selects the MS operation trail incoming from the west and the MS operation trail outgoing to the east. If a failure occurs on the operation trail incoming from the west or if a protection switching request is received via the APS channel accessible by the MSPA, the MS MC selects the protection trail incoming from the east instead of the operation trail incoming from the west and it selects the protection trail outgoing to the west instead of the operation trail outgoing to the east. This switches protection from the operation ring to the protection ring [see Figures 3.66(b,c)]. Both revertive and nonrevertive protection switching modes are applicable to MS dedicated SHR, but the revertive mode is advantageous over the other for a simpler operation and maintenance.

In the dedicated SHR, end-to-end operation trail and protection trail are configured on the HP layer for trail protection. The HP is protected through protection switching when a failure occurs in the higher layer or performance degrades in the HP layer. For HP trail protection in the dedicated SHR, both dual-ended switching and single-ended switching schemes are applicable. The APS channel (i.e., K3 byte) in the HP can be used for dual-ended switching but not necessarily for single-ended switching.

Shared SHR/2F

The shared SHR/2F consists of two opposite-directed fiber rings, with a half of the total capacity in each ring allotted for the operation channel and the other half for the protection channel. When a failure or performance degradation occurs in a ring, the operation channel is switched to the protection channel in the other ring. Figure 3.67 depicts the functional configuration of a node in the shared SHR/2F, in which multiple MS operation trails share one MS protection trail. In this case half of the MS trail capacity is allocated for the operation trail, with the other half allocated for the protection trail. In normal operation, the MS MC connects the operation trail of the MSPA to the operation path of the

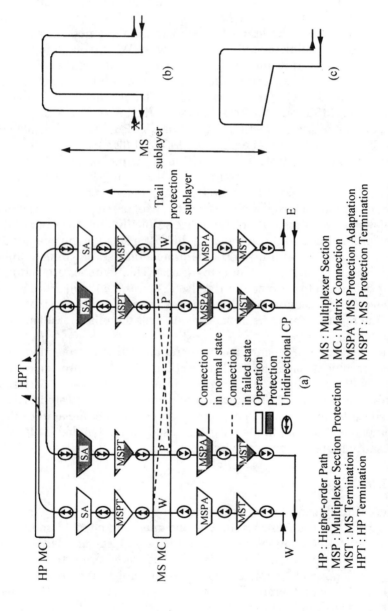

Figure 3.66   MS dedicated SHR (MS SHR): (a) functional configuration; (b) connection in normal operation; and (c) connection in protection state.

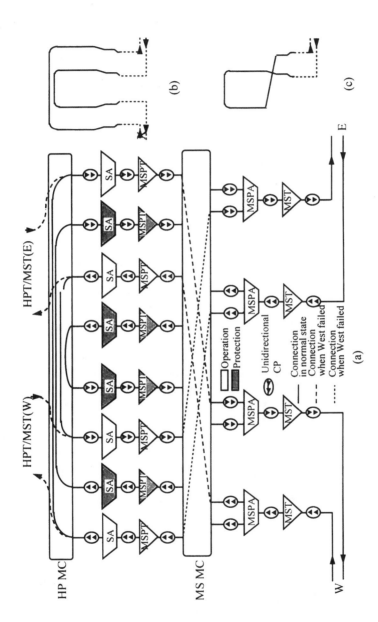

Figure 3.67 MS shared SHR / 2F: (a) functional configuration; (b) connection in normal operation; and (c) connection in protection state.

MSPT and connects the protection trail to the protection path. When a failure occurs, it switches the connection such that the traffic on the protection trail, which flows to the reverse direction of the failed traffic, can be taken for operation [see Figures 3.67(b,c)]. The MS MC performs protection switching of MS trails in units of HPs, and the HP MC controls the connection of the HPs to be added/dropped as in the case of the MS dedicated SHR.

The shared SHR/2F is operated in revertive protection switching mode, and the dual-ended switching scheme is applied based on the ring APS protocol over the APS channel. The shared SHR/2F is more complicated to operate than other SHRs since it utilizes only half the MS trail capacity as the operation channel. In case the fiber link carries the STM-n signal, the shared SHR/2F can provide a maximum traffic capacity of (155.520 Mbps $\times$ $n/2$) $\times$ $(m-1)$, where m denotes the number of nodes in the ring.

Shared SHR/4F

The shared SHR/4F consists of two sets of opposite-directed fiber rings, with one set of fiber rings used for operation channels and the other set for the protection channels. If a failure occurs in the operation ring, it protects the trail by switching the overall MS trail to the corresponding protection ring.

Figure 3.68 depicts the functional configuration of a node in the shared SHR/4F. In normal operation, the MS MC provides connections to the operation trails in both directions by connecting the operation MSPA to the corresponding operation MSPT, and it does the same for the protection trails. When a failure occurs in a fiber, it switches the connection such that the MSPT in the operation trail is connected to the MSPA in the oppositely directed protection trail [see Figures 3.68(b,c)]. The HP MC controls the connection of the HPs to be added/dropped as in the shared SHR/2F case.

The shared SHR/4F employs the revertive protection switching mode and the dual-ended switching scheme as in the shared SHR/2F case. However, in contrast to the shared SHR/2F case, it supports both ring protection switching and the span protection. If the fiber link carries the STM-n signal, the shared SHR/4F can provide a maximum traffic capacity of (155.520 Mbps $\times$ n) $\times$ $(m-1)$ for the ring having m nodes.

3.10.4 Mesh Network

For the mesh network, which is a collection of network nodes linked with multiple transmission paths, it is more efficient to restore failures by utilizing the spare capacities than to arrange dedicated protection trails or protection subnetworks. Such a restoration scheme is an economical way to survive a network failure because it does not require the network to exclusively secure a large amount of channel capacity for protection. However, it requires protocols to

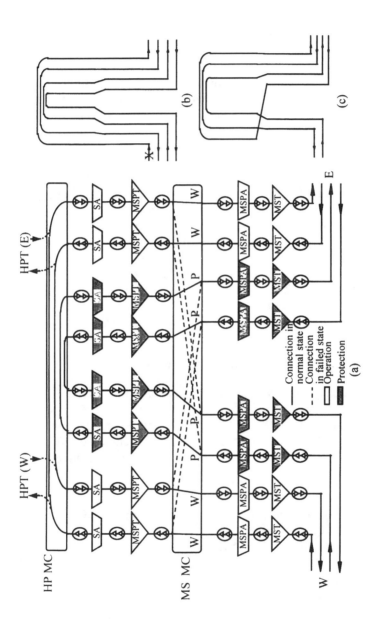

Figure 3.68 MS shared SHR/4F: (a) functional configuration; (b) connection in normal operation; and (c) connection in protection state.

coordinate failure restoration among related nodes in order to configure the optimal detour paths using spare capacity. Therefore the restoration time becomes comparatively long and the restoration processing becomes complicated.

The mesh network can be restored either in the form of line restoration or in path restoration, both of which are illustrated in Figure 3.69. In the case of *line restoration*, detouring paths are established with respect to the two nodes where the failed line terminates, by utilizing the spare channels in all the links that connect the two nodes. In contrast, in the case of *path restoration*, detouring paths are established with respect to the virtual channels between the two relevant access nodes (i.e., trail termination or access points) independently of the failed line. In Figure 3.69, S and C indicate, respectively, the sender and the chooser of the failed traffic.

There are two ways to control the mesh network restoration: centralized control and distributed control. In *centralized control*, the operating system that manages overall network elements directly controls the network elements to configure detour paths for restoration. In contrast, in *distributed control*, the nodes that are adjacent to the failed links or equipment (i.e., the sender and the chooser) initiate collaborative works among the relevant nodes to establish detour paths. For this, the sender that detects the failure notifies adjacent nodes of the failure state, and the adjacent nodes do the same to their adjacent nodes, and the process is repeated until the failure message gets propagated to the chooser and the chooser finally returns the message to the sender.

3.10.5 Survivability in SDH and ATM Networks

In the case in which the synchronous transmission network (i.e., SDH/SONET network) forms the infrastructure for the BISDN (i.e., ATM network), another dimension can be added to the network survivability because the main interest of the ATM network lies in the recovery of virtual channels (VCs) and virtual paths (VPs). In this case network survivability can be achieved not only by

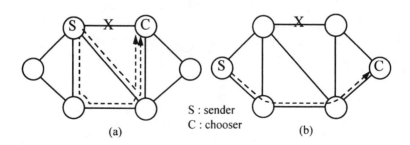

Figure 3.69 Restoration in mesh network: (a) line restoration and (b) path restoration.

recovering the relevant paths in the SDH/SONET layer but also by restoring detour paths in the ATM layer.

Figure 3.70 illustrates the relation of the SDH/SONET layer and ATM layer with respect to failure occurrence. We can see that a failure that occurred on a link in the SDH/SONET layer can lead to the failure of multiple VPs in the ATM layer. In this situation, the recovery of the failed link in the SDH/SONET layer renders a shortcut to recover the multiple VP failures at the same time.

Figure 3.71 illustrates two ways to restore the VP in the ATM layer when

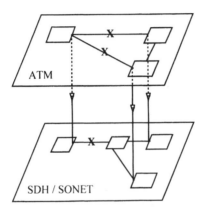

Figure 3.70   The Relation of the SDH/SONET layer and ATM layer with respect to failure occurrence.

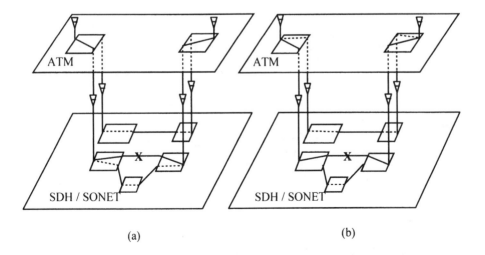

(a)   (b)

Figure 3.71   Restoration of the VP in the ATM layer when a link in the SDH/SONET layer fails: (a) in the SDH/SONET layer and (b) in the ATM layer.

a link in the SDH/SONET layer fails. Figure 3.71(a) shows how to restore the failure by setting a detour path in the SDH/SONET layer and Figure 3.71(b) illustrates how to restore the failure by reconfiguring the VP in the ATM layer. This kind of interlayer collaborative restoration, in which the paths or virtual paths are restored through collaborative working among different layers, is called *escalation*. Escalation enables numerous ways to restore the network failure and thus helps to enhance network survivability by adopting the most economical and effective means from among those available.

Selected Bibliography

ANSI T1.102, "Digital hierarchy interfaces."

ANSI T1.105–1988, "American national standard for telecommunications-digital hierarchy-optical interface rates and formats specification," 1988.

ANSI T1.105–1991, "Digital hierarchy-optical interface rates and formats specifications (SONET)," 1991.

ANSI T1.105a-1991, "Supplement to T1.105," 1991.

ANSI T1.106–1988, "American national standards for telecommunications-digital hierarchy-optical interface specifications (single mode)," 1988.

ANSI T1.117, "Digital hierarchy-optical interface specifications (SONET) (single mode-short reach)," 1991.

ITU-T Rec. G.702, "Digital hierarchy bit rates," 1988.

ITU-T Rec. G.703, "Physical/electrical characteristics of hierarchical digital interfaces," 1991.

ITU-T Rec. G.707, "Synchronous digital hierarchy bitrates," 1993 (revision).

ITU-T Rec. G.708, "Network mode interface for the synchronous digital hierarchy," 1993 (revision).

ITU-T Rec. G.709, "Synchronous multiplexing structure," 1993 (revision.).

ITU-T Rec. G.774, "Synchronous digital hierarchy (SDH) management information model for the network element view," 1992.

ITU-T Rec. G.781, "Structure of recommendations on equipment for the synchronous digital hierarchy (SDH)," 1994 (revision).

ITU-T Rec. G.782, "Types and general characteristics of synchronous digital hierarchy (SDH) equipment," 1994 (revision).

ITU-T Rec. G.783, "Characteristics of synchronous digital hierarchy (SDH) equipment functional blocks," 1994 (revision).

ITU-T Rec. G.784, "Synchronous digiatl hierarchy (SDH) management," 1994 (revision).

ITU-T Rec. G.803, "Architecture of transport networks based on the synchronous digital hierarchy (SDH)," 1993 (revision).

ITU-T Rec. G.804, "ATM cell mapping into plesiochronous digital hierarchy (PDH)," 1994.

ITU-T Rec. G.825, "The control of jitter and wander within digital networks which are based on the synchronous digital hierarchy (SDH)," 1993.

ITU-T Rec. G.831, "Management capabilities of transport networks based on the synchronous digital hierarchy (SDH)," 1993.

ITU-T Rec. G.957, "Optical interfaces for equipments and systems relating to the synchronous digital hierarchy," 1995 (revision).

ITU-T Rec. G.958, "Digital line systems based on the synchronous digital hierarchy for use on optical fiber cables," 1994 (revision).

ITU-T Rec. G.SHR-1 (Draft), "SDH protection: Rings and other architectures," May 1994.

SR-NWT-001756, "Automatic protection switching for SONET," Issue 1, Bellcore, 1990.

SR-NWT-002224, "SONET synchronization planning guidelines," Issue 1, Bellcore, 1992.

TA-NWT-001042, "Generic requirements for operations interfaces using OSI tools: SONET path switched ring information model," Issue 3, Bellcore, 1992.

TA-NWT-001250, "Generic requirements for synchronous optical network (SONET) file transfer," Issue 2, Bellcore, 1992.

TN-NWT-001042, "Generic requirements for operations interfaces using OSI tools: Synchronous optical network (SONET) transport information model," Issue 1, Bellcore, 1992.

TR-NWT-000253, "Synchronous optical network (SONET) transport systems: Common generic," Issue 2, Bellcore, 1991.

TR-NWT-001230, "SONET bidirectional line switched ring equipment generic criteria," Issue 2, Bellcore, 1992.

TR-TSP-000496, "SONET add/drop multiplex equipment (SONET ADM) generic criteria," Issue 3, Bellcore, 1992.

TR-TSY-00023, "Wideband and broadband digital cross-connect generic requirements and objectives," Issue 2, Bellcore, 1989.

TR-TSY-000303, "Integrated digital loop carrier system generic requirements, objectives, and interface," Issue 1, revision 3, Bellcore, 1990.

Aprille, T. J., "Introducing SONET into the local exchange carrier network," *IEEE Commun. Mag.*, Vol. 28, No. 8, August 1990, pp. 34–38.

Asatani, K., K. R. Harrison, and R. Ballart, "CCITT standardization of network node interface of synchronous digital hierarchy," *IEEE Commun. Mag.*, Vol. 28, No. 8, August 1990, pp. 15–20.

Balcer, R., et al., "An overview of emerging CCITT recommendations for the synchronous digital hierarchy: multiplexers, line systems, management, and network aspects," *IEEE Commun. Mag.*, Vol. 28, No. 8, August 1990, pp. 21–25.

Ballart, R., and Y. C. Ching, "SONET: Now it's the standard optical network," *IEEE Commun. Mag.*, Vol. 27, No. 3, March 1989, pp 1–15.

Bars, G., J. Legras, and X. Maitre, "Introduction of new technologies in the French transmission networks," *IEEE Commun. Mag.*, Vol. 28, No. 8, August 1990, pp. 29–43.

Bates, R. J., *Introduction to T1/T3 Networking*, Norwood, MA: Artech House, 1992.

Bellamy, J. C., "Digital network synchronization," *IEEE Commun. Mag.*, Vol. 33, No. 4, April 1995, pp. 70–83.

Boehm, R. J., "Progress in standardization of SONET," *IEEE LCS Mag.*, Vol. 1, No. 2, May 1990, pp. 8–16.

Boyer, G. R., "A perspective on fiber in the loop systems," *IEEE LCS Mag.*, Vol. 1, No. 3, August 1990, pp. 6–11.

Ching, Y. C., and G. W. Cyboron, "Where is SONET?" *IEEE LTS Mag.*, Vol. 3, No. 4, pp. 44–51, November 1991.

Ching, Y. C., and H. S. Say, "SONET implementation," *IEEE Commun. Mag.*, Vol. 31, No. 9, September 1993, pp. 34–41.

Day, C. N., and C. H. Lin, "SONET and OSI: Making a connection," *IEEE LTS Mag.*, Vol. 3, No. 4, November 1991, pp. 52–59.

Doverspike, R. D., et al., "Network design sensitivity studies for use of digital cross-connect systems in survivable network architectures," *IEEE JSAC*, Vol. 12, No. 1, January 1994, pp. 69–78.

Eames, T. R., and G. T. Hawley, "The synchronous optical network and fiber-in-the-loop," *IEEE LTS Mag.*, Vol. 3, No. 4, November 1991, pp. 24–29.

Haque, I., W. Kremer, and K. Raychaudhuri, "Self-healing rings in a synchronous environment," *IEEE LTS Mag.*, Vol. 3, No. 4, November 1991, pp. 30–37.

Hasegawa, S., et al., "Control algorithms of SONET integrated self-healing networks," *IEEE JSAC*, Vol. 12, No. 1, January 1994, pp. 110–119.

Hibino, M., and F. Kaplan, "User interface design for SONET networks," *IEEE Commun. Mag.*, Vol. 30, No. 8, August 1992, pp. 24–27.

Holter, R., "SONET: a network management viewpoint," *IEEE LCS Mag.*, Vol. 1, No. 4, November 1990, pp. 4–13.

Holter, R., "Managing SONET equipment," *IEEE Network Mag.*, Vol. 5, No. 4, January 1992, pp. 36–41.

Hugles, J. P., and W. R. Franta, "Geographic extension of HIPPI channel via high speed SONET," *IEEE Network Mag.*, Vol. 8, No. 3, May/June 1994, pp. 42–53.

Jakubson, J., "Managing SONET network," *IEEE LTS Mag.*, Vol. 3, No. 4, November 1991, pp. 5–31.

Johnston, C. A., "Architecture and performance of HIPPI-ATM-SONET terminal adapters," *IEEE Commun. Mag.*, Vol. 33, No. 4, April 1995, pp. 46–51.

Kasai, H. T., Murase, and H. Ueda, "Synchronous digital transmission systems based on CCITT SDH standard," *IEEE Commun. Mag.*, Vol. 28, No. 8, August 1990, pp. 50–59.

Kawamura, R., and I. Tokizawa, "Self-healing virtual path architecture in ATM networks," *IEEE Commun. Mag.*, Vol. 33, No. 9, September 1995, pp. 72–79.

Klein, M. J., and R. Urbansky, "Network synchronization—a challenge for SDH/SONET?," *IEEE Commun. Mag.*, September 1993, pp. 42–50.

Kunieda, T., S. Suginoto, and N. Sasaki, "A synchronous digital hierarchy network management system" *IEEE Commun. Mag.*, Vol. 31, No. 11, November 1993, pp. 84–91.

Mazzei, U., et al., "Evolution of the Italian telecommunication network towards SDH," *IEEE Commun. Mag.*, Vol. 28, No. 8, August 1990, pp. 44–50.

Miura, H., K. Maki, and K. Nishihata, "SDH network eVolution in Japan," *IEEE Commun. Mag.*, Vol. 33, No. 2, February 1995, pp. 86–92.

Nederlof, L., et al., "End-to-end survivable broadband network," *IEEE Commun. Mag.,* Vol. 33, No. 9, September 1995, pp. 63–71.

Omidyar, C. G., and A. Aldridge, "Introduction to SDH/SONET," *IEEE Commun. Mag.*, September 1993, pp. 30–33.

Passeri, P., et al., "Introducing SDH systems in Europe," *IEEE LTS Mag.*, Vol. 3, No. 4, November 1991, pp. 38–43.

Pelegrini, G., and P. H. K. Wery, "Synchronous digital hierarchy," *Telecom. J.*, Vol. 58, 1991, pp. 815–824.

Sandesara, N. B., G. R. Ritchie, and B. Engel-Smith, "Plans and considerations for SONET development," *IEEE Commun. Mag.*, Vol. 28, No. 8, August 1990, pp. 26–33.

Sexton, M. J., and A. B. D. Reid, *Transmission Networking: SONET and the Synchronous Digital Hierarchy*, Norwood, MA: Artech House, 1992.

Sharifi, M., and B. Mortimer, "The evolution of SDH: A view from telecom New Zealand," *IEEE Commun. Mag.*, Vol. 28, No. 8, August 1990, pp. 60–66.

Shirakawa, H., K. Maki, and H. Miura, "Japan's network evolution relies on SDH-based systems," *IEEE LTS Mag.*, Vol. 3, No. 4, November 1991, pp. 14–18.

Sosnosky, J., and T. H. Wu, "SONET ring applications for survivable fiber loop networks," *IEEE Commun. Mag.*, Vol. 29, No. 6, June 1991, pp. 51–58.

Sosnosky, J., "Service applications for SONET DCS distributed restoration," *IEEE JSAC*, Vol. 12, No. 1, January 1994, pp. 59–68.

Spears, D. R., "Broadband ISDN switching capabilities from a services perspective," *IEEE JSAC*, Vol. SAC-5, No. 8, October 1987.

To, M., and J. MacEachern, "Planning and deploying a SONET-based metro network," *IEEE LTS Mag.*, Vol. 3, No. 4, November 1991, pp. 19–23.

Wasem, O. J., et al., "Survivable SONET networks—Design methodology," *IEEE JSAC*, Vol. 12, No. 1, January 1994, pp. 205–214.

Wu, T.-H., and M. E. Burrowes, "Feasibility study of a high-speed SONET self-healing architecture in future inter-office networks," *IEEE Commun. Mag.*, Vol. 28, No. 11, November 1991(a), pp. 33–51.

Wu, T.-H., "SONET ring applications for survivable fiber loop networks," *IEEE Commun. Mag.*, Vol. 29, No. 6, June 1991(b), pp. 51–58.

Wu, T.-H., *Fiber Network Service Survivability*, Norwood, MA: Artech House, 1992(a).

Wu, T.-H., et al., "An economic feasibility study for a broadband virtual path SONET/ATM self-healing ring architecture," *IEEE JSAC*, Vol. 10, No. 9, Dec. 1992(b), pp. 1459–1473.

Wu, T.-H., "Cost-effective network evolution," *IEEE Commun. Mag.*, Vol. 31, No. 9, September 1993, pp. 64–73.

Wu, T.-H., et al., "The impact of SONET digital cross-connect system architecture on distributed restoration," *IEEE JSAC*, Vol. 12, No. 1, January 1994, pp. 79–87.

Wu, T.-H., "Emerging technologies for fiber network survivability," *IEEE Commun. Mag.*, Vol. 33, No. 2, February 1995, pp. 58–74.

Yamagishi, K., N. Saski, and K. Morino, "An implementation of a TMN-based SDH management system in Japan," *IEEE Commun. Mag.*, March 1995, pp 80–85.

BISDN and ATM Technology

4

Conceived under the influence of the standardization process of the SDH, the BISDN is an expanded version of the ISDN, which has the capability to accommodate various types of broadband signals while retaining the basic intent of ISDN. The BISDN's fundamental objective is to achieve complete integration of services ranging from low-bit-rate bursty signals up to broadband continuous real-time signals, including voiceband services such as telemetry, data terminal, telephone, and facsimile, and broadband services such as video telephony, video conference, HDTV transmission, high-speed data transmission, and video signal transmission. Consequently, an efficient technique for dealing with such a diverse set of services in a generalized manner was desired, and ATM is the technique that was proposed as the solution.

In the background surrounding the emergence of the BISDN concept, there is an increasing demand for various types of broadband services, including video services. To accommodate all such broadband signals, the capability to integrate interactive services such as video telephone with distributive services like CATV is needed, as well as the capability to provide both the circuit-mode services and packet-mode services in a generalized manner. In addition, a technique that enables the joint accommodation of signals over a wide range of bit rates, including low-rate telemetry signals (a few bits per second), midrange voice-speed signals (tens of kilobits per second), and high-rate video signals (hundreds of megabits per second) is also required. A possible solution for meeting these requirements is the scheme in which various service signals are first made to have a common external shape and then are piled up one by one and multiplexed together. Here, the cell (or a fixed-size packet) is the standardized external form, and the method for multiplexing a collection of cells is *asynchronous time-division* multiplexing (ATDM). The communication mode based on cells and the ATDM is ATM.

The ATM communication technique can be said to be a transfer mode that integrates the existing circuit-mode digital transfer method with the packet mode transfer method. First of all, ATM has a close connection with the packet-mode transfer method in that it uses ATM cells as its basic means of transport; but there is a difference in that the packet mode was created for variable-rate, non-real-time data signals, whereas ATM can manage real-time fixed-rate signals as well. Also, the packet mode is generally used for LANs, whereas ATM can be used for a vast public network and is hence accompanied by various problems inherent in any large network, such as routing, access and flow control, switching, and transmission. On the other hand, the fundamental difference between ATM and the circuit-mode transfer method is that whereas the circuit mode functions by allocating a separate service channel and transferring through it information signals in a continuous bit stream, ATM operates by segmenting the information signal so as to fit it onto the ATM cell, then transferring it through a virtual channel. Thus, the accompanying ATM procedures such as connection setup, data processing, transmission, and switching raise various new problems.

The purpose of this chapter is to provide a detailed examination of the BISDN and ATM. Because ATM is a new transfer technique that was conceived as a means of realizing the BISDN, *BISDN* will at times be used interchangeably with *ATM network.*

The structure of this chapter is as follows. First we provide a comprehensive introduction to the BISDN and ATM. Then the functional architecture of the BISDN and basic principles of ATM are discussed. Next, we examine the protocol reference model of the BISDN, followed by a discussion of the physical layer, ATM layer, and the *ATM adaptation layer* (AAL) of the BISDN. Then, network control, network management, and traffic control of the BISDN are investigated, after which we end with an examination of ATM switching.

4.1 INTRODUCTION

This section provides the background for BISDN and ATM technology. First, BISDN services and service characteristics are examined, followed by discussions of the technological background and standardization process. Then the basic underlying concept of BISDN is considered, and the relationship between the ISDN and BISDN is discussed.

4.1.1 BISDN Services

The services that can be provided by the communication networks of the future include not only narrowband services such as telephone, data, facsimile, telewriting, telemetry, teletex, videotex, and electronic mail, but also broadband services such as video telephone, high-speed data, color facsimile, CATV, HDTV distribution, high-fidelity sound, video mail, video surveillance, high-resolution image, video conferencing, broadband videotex, and so on. These services can be classified into interactive services and distributive services, depending on the direction of information transfer. Interactive services can be further categorized into conversational services, message services, and retrieval services. Distributive services can be subcategorized into controllable services and noncontrollable services (see Table 4.1).

Conversational Service

In general, a conversational service is a type of service that provides real-time, bidirectional communication between users at two ends. On the average, service information flows symmetrically in both directions, but not strictly so. Examples of conversational services include video telephony, video conferencing, and interactive data transmission.

Message Service

A message service provides communication between users at two ends by way of storage devices. Here, the storage device can store, transfer, and process messages; that is, it has the capacity to compile, process, and transform information. Examples of message services include mailing services for film, high-resolution images, voice information, and a *message handling service* (MHS).

Retrieval Service

A retrieval service involves the retrieval of public-sector information stored in a central information center. On request, the desired information is transferred to the user on an individual basis. Here, the starting time of the information sequence can be controlled by the user. The targeted areas of retrieval service include film, high-resolution image, voice information, and recorded information.

Table 4.1
Classification of BISDN Services

Main Categories	*Subcategories*	*Examples of Services*
Interactive services	Conventional services	Video telephony, video conference
	Messaging services	MHS, video mail service
	Retrieval services	Retrieval of film, document, etc.
Distributive services	Noncontrollable distributive services	Television, sound
	Controllable distributive services	Video graphy, near VOD

Uncontrollable Distributive Service

An uncontrollable distributive service refers to any general service that is broadcast. Typically, a continuous stream of information is transmitted by a central service provider to an unlimited number of users within the network. Here, the users have access to the information flow, but have no control over the starting time or the program of the information flow. A typical example of an uncontrollable service would be television and radio broadcasting.

Controllable Distributive Service

Controllable distributive services are similar to uncontrollable services in the sense that a central service provider furnishes a continuous flow of information to a multiple number of users. But unlike the uncontrollable distributive services, the information flow consists of information packets that are periodically repeated. Consequently the user can selectively choose among the periodically distributed information packets, and thus can control starting time and program content of the packets. In other words, due to the periodic nature of the information content, the user can always receive the chosen information packets from the beginning. An example of a controllable service is the *near video-on-demand* (NVOD) service.

4.1.2 Characteristics of BISDN Services

The goal of the BISDN is to encompass all of the services that may come into being in the communication network of the future. Accordingly, all of the aforementioned narrowband services and broadband services fall within the scope of BISDN services. Consequently, various types of services with differing characteristics coexist within the BISDN. These services include multimedia services that have wide or narrow bandwidths, those that are circuit mode or packet mode, and real-time or non-real-time. The manifestation of such a diverse set of characteristics is the most distinct feature of BISDN services.

Provision of Multimedia Services

The first characteristic of BISDN services is that they are multimedia in nature. For example, video telephone is a multimedia service that simultaneously involves three different media: audio, video, and data. Depending on the particular service provided, text or graphics can also be included. Also, dissimilar terminal equipment can be interconnected within the BISDN. For instance, to set up a video conferencing service, video terminals are usually connected, but a user with just a telephone terminal can be connected as well.

Coexistence of Interactive Services and Distributive Services

As studied in the previous section, the integrated provision of both interactive and distributive services is one of the BISDN's unique features. In existing communications networks, a separate network has to be constructed for each type; consequently, it is impossible to provide interactive services such as telephony through a distributive services network such as CATV. Within the BISDN, however, CATV service and video telephone service can be provided simultaneously.

Widely Ranging Bandwidth and Service-Time Distributions

One of the notable characteristics of BISDN services is that the associated bandwidth and service-time distributions are extremely wide. In the case of narrowband ISDN, the basic component signals are distributed around the 64-Kbps-rate signal. But the BISDN also encompasses all kinds of digital tributaries, various video signals, as well as high-speed data signals. Consequently, from the bit rate standpoint, BISDN signals evince a widely ranging distribution, from the few bits per second used for the telemetry signal to the hundreds of megabits per second required for video signals as illustrated in Figure 4.1. Service time can also range from a few seconds of low-speed data to the hours required for video services.

Therefore, *broadband* in the narrow sense means that the BISDN has the ability to provide broadband services whose speeds go up to hundreds of megabits per second, but its meaning in the broad sense is that its bandwidth and service-time distributions are spread over the broadband.

Coexistence of Continuous-Type and Bursty-Type Services

Another feature of the BISDN is that continuous signals such as sound and images can coexist with bursty signals such as terminal data. Depending on how they are digitized, voice and video signals can be converted into a signal with a constant bit rate or into a signal whose bit rate varies slightly. But in the case of data signals, the bit rate always varies widely. Also, although voice or image services require real-time processing capability, data services do not. Consequently, in the BISDN, services with fixed bit rates can coexist with services with variable bit rates; similarly, real-time services can coexist with non-real-time services. On the other hand, in existing communications networks, voice services are provided by way of circuit switching, whereas data services are furnished by way of packet switching. Hence, circuit-mode services and packet-mode services coexist within the BISDN.

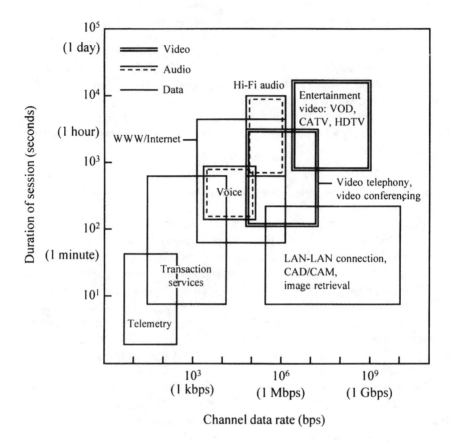

Figure 4.1 BISDN service distribution.

4.1.3 Conception of BISDN

As social and business activities have become more diversified, the demand for various multimedia and broadband services has increased rapidly. This is manifested in the sudden proliferation of data terminals and personal computers, the ubiquity of fax machines, the increased installation of video conferencing systems, and the success of the CATV industry and the increase in the number of its subscribers. The services in demand are BISDN services, which include both interactive services and distributive services.

There is a chance that each type of service sporadically forms its own communication network. Since the circuit-mode services and packet-mode services are also mixed together, each could develop a separate communication network of its own. However, the construction of an independent service-oriented network imposes a heavy financial burden and can bring about an obstruction of communication information transfer and disorder in the management of the communication network. Therefore, it is desirable to integrate various networks into one universal communication network so that all the services can be provided in an integrated manner. The BISDN concept is designed to achieve such an integrated service network by expanding the already standardized narrowband ISDN.

Therefore, the BISDN can be described as a digital communication network that utilizes broadband transmission and switching technologies to interconnect concentrated or distributed subscribers and service providers, and to support integrated services with wide-bandwidth distribution that ranges from a few bits per second to hundreds of megabits per second.

From the service standpoint, the BISDN integrates narrowband services with broadband services and, consequently, the BISDN can be said to be a communication network that takes the concept of the existing ISDN and extends it to enable the provision of various kinds of broadband services.

As such, the BISDN is a concept that was proposed as a means of satisfying the increasing demand for broadband multimedia services. The BISDN's basic purpose is to provide integrated interactive and distributive services, and to furnish narrowband services and broadband services simultaneously by employing mature high-speed transmission, switching, signal processing, computer, software, and device technologies to construct a digital network.

To meet these objectives, a BISDN must be capable of supporting point-to-point/point-to-multipoint connections and on-demand/reserved/semipermanent connections. A BISDN must be intelligent enough to allow future growth and improvement in service characteristics, and it must be equipped with network operation, maintenance, control, and management functions. However, the specifications and regulations set to achieve these requirements should be carefully arranged so as not to hinder the advances in future technology or the evolution of the implementation means.

As a way to realize the BISDN's basic objective, the ATM concept is employed. ATM is a communication means that segments various service signals, maps them onto ATM cells of a fixed size, and subsequently transports them using ATDM. ATM is a connection-oriented method that establishes virtual paths and virtual channels for the transfer of ATM cells. Thus, the use of ATM allows highly flexible network access, real-time service provision, and variable assignment of bandwidth. Because ATM is defined independently of the particular transport means of the physical layer, information transfer can be accomplished through diverse types of physical media and transport networks.

4.1.4 Technological Background of BISDN

Since the services furnished by the BISDN have a diverse set of characteristics, as was examined in the preceding section, the maturation of several basic technologies is extremely crucial for the realization of the BISDN ideal. To begin with, since high-speed and broadband service signals will form the main axis of BISDN, high-speed processing and device technologies are indispensable, as are broadband transmission and switching technologies. Also, since various video services will become the BISDN's primary services of interest, the maturation of image processing and implementation technologies is needed. Because low-speed and high-speed services are intermingled, circuit-mode and packet-mode services coexist, and the corresponding communication network technology is essential.

These basic technologies have steadily grown into maturity, in step with the increasing demand for broadband services. To begin with, communication network technology has already matured; optical fiber attenuation has been reduced to below 0.5 dB / km, and the prices of light-emitting and light-detecting devices have fallen drastically. Advances in integrated circuit and transistor technologies are also worthy of close attention, with silicon bipolar or GaAs transistors currently possessing high-speed processing capability from hundreds of Mbps up to tens of Gbps, while the CMOS technology can operate in the 150-Mbps range. Recent advances in software and microprocessor technologies have made high-speed control a possibility, which, along with fast operating devices, renders high-speed switching possible.

Due to the advances in signal processing techniques, the compression, conversion, and regeneration of various service signals have become an easy task, and the advances in computer technology have also made the collecting, handling, and processing of service signals a simple matter. In addition, if the above advances are complemented by VLSI technology, efficient terminal equipment at the user premises can be constructed. Also, developments in high-quality monitors and high-sensitivity cameras have advanced to the point that practical BISDN terminal devices associated with various video services can now be realized.

On the other hand, the BISDN standardization process, which has been progressing on a grand scale since the mid 1980s with CCITT as its center of operation, inspired a great deal of research on the possible integration of various types of broadband service signals and the digitalization of communication networks, which also contributed significantly to advances in communication network technology. Such groundwork bore fruit in the form of ATM, which can enable the harmonious coexistence of various BISDN services possessing a disparate set of properties, and this signifies the maturation of the BISDN's service integration technology.

4.1.5 BISDN Standardization Background

During the ISDN standardization process in the early 1980s, CCITT established the H1, H2, H3, and H4 channels as the high-speed channels. Among them, the

H1 channel became the ISDN's primary access in the form of the 1.536-Mbps H11 channel and the 1.920-Mbps H12 channel; thus, together with the 2B+D basic access of 144 Kbps, came to form the very foundation of the ISDN. For the remaining H2, H3, and H4 channels, only rough outlines were given that correspond to the broadband channels of the existing digital hierarchy.

As interest gradually shifted toward the broadband channels, starting around 1985, the bit rates of 30 to 40 Mbps, 45 Mbps, and 60 to 70 Mbps were examined as the target standard for the H2, H3, and H4 channels. On the other hand, the T1 Committee once proposed the SONET-based 149.760-Mbps broadband channel as the candidate rate. Then, in July 1987, CCITT went ahead with the NNI standardization process separately from the UNI standardization.

The NNI standardization, which unfolded with the SG XVIII's working party (WP) 7 as the center of operations, experienced numerous hardships and finally settled on the STM-1 signal with a 9B × 270 structure and 155.52-Mbps bit rate as the standard at the Seoul meeting in February 1988 (see Section 3.1.5). This is the very SDH standard that was subsequently confirmed in Recommendations G.707 to G.709. On the other hand, the BISDN's UNI standardization operation, which was propelled by the SG XVIII's *Broadband Task Group* (BBTG) completed Recommendation I.121 in 1988. This is the first document that established BISDN's basic outline, and it offered regulations on such issues as whether BISDN should be based on ATM, whether to divide the BISDN into interactive and distributive services, and whether the BISDN's functional configuration and basic structure should be the same as the ISDN's. Also, the protocol model for ATM was produced, and the size of the ATM cell was prescribed to be around 32 to 120 bytes. In addition, the broadband channels H21, H22, and H4 were assigned the unofficial bit rates of 32.768 Mbps, 43 to 45 Mbps, and 132 to 138.240 Mbps, respectively, and UNI was unofficially separated into the 150-Mbps class and the 600-Mbps class.

Afterwards, BBTG became WP8, and the UNI standardization operation suffered through another difficult time with questions regarding the ATM cell size, the bit rate at the interface, and the frame structure. Europe's proposal of a 4 + 32 byte size (cell header + payload) for the ATM cell was in conflict with the U.S.'s proposal of a 5 + 64 byte size, and ultimately the 5 + 48 byte size was agreed on as a compromise.[1] The bit rate to be used at the interface was selected as 155.520 Mbps under the strong influence of the previously completed SDH standardization as discussed in Chapter 3. Also, the physical medium's frame structure was decided either to have the STM-1 frame structure or to consist purely of a flow of ATM cells. The mapping of other G.702 signals

---

1. Some of the candidate ATM cell sizes included the 6 + 66 size proposed by Japan and the 5 + 60 size proposed by Korea.

was approved as well.[2] At the Matsuyama conference in November 1990, WP8 finalized the matters that had been approved up to that point on 13 I Series Recommendations. In this manner, various basic frameworks relating to the BISDN were completed.

In 1993 ITU reorganized internally. In that reorganization, CCITT was transformed into ITU-T within the WTSC (World Telecommunication Standardization Conferences). Since then the BISDN-related standardization work has been succeeded by SG13 of the ITU-T. The ITU-T recommendations on BISDN issued to date are as listed in Appendix A.

On the other hand, the ATM Forum was formed in October 1991 among computer vendors and telecommunication service providers, whose principal interest was in accelerating ATM product development and ATM service expansion. Although not an actual standards body, the ATM Forum has been generating ATM-related specifications under close collaboration with other standards organizations. The ATM Forum's specifications issued to date include ATM UNI, ATM DXI (Data eXchange Interface), and ATM B-ICI (Broadband Inter-Carrier Interface) specifications (refer to Appendix A). In addition, the IETF (Internal Engineering Task Force) also has been generating IP (Internet Protocol) suite-related ATM standards.

4.1.6 BISDN Versus ISDN

The concept of BISDN started out as an extension of ISDN standardization. Consequently, the BISDN has many similarities with the ISDN at the conceptual level. However, from the practical implementation standpoint, the BISDN and ISDN share no common traits.

To begin with, from the viewpoint that the basic motivation is to achieve integration of services, the BISDN's role is the same as that of the ISDN; however, in the case of the BISDN, the services that are targeted for integration include broadband signals as well. Consequently, the BISDN's basic structural model is the same as the ISDN's, except that the ISDN's 64-Kbps base narrowband capacity has been supplemented by the BISDN's broadband capability.

In terms of the functional structure or the basic configurations, the BISDN is identical to the ISDN. The fact that the functional groups consist of TE1, TE2, NT1, NT2, and TA and that the reference points are R, S, and T, is applicable to both networks (see Figure 4.5 in Section 4.2). The two are equivalent in basic physical configurations at the *user-network interface* (UNI) (see Figure 4.6 in Section 4.2).

However, they are equivalent only in concept and are not compatible in practice. That is, a BISDN cannot be created simply by augmenting the ISDN

---

2. The G.702 signal refers to the existing plesiochronous digital hierarchy signals specified in Recommendation G.702.

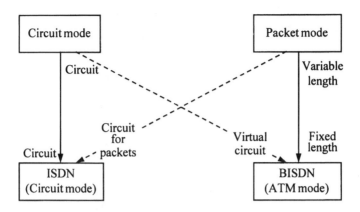

Figure 4.2 BISDN versus ISDN.

with broadband service equipment; nor can the ISDN's TE be connected directly to the BISDN's NT. Consequently, the ISDN's functional groups or reference points and BISDN-related elements are equivalent only at the conceptual level and are entirely different in practice.[3] But it is possible to interface the ISDN's TE1 or TA through the BISDN's NT2.

From the practical communication standpoint, the BISDN and ISDN are two fundamentally different types of networks as illustrated in Figure 4.2. Whereas the ISDN's communication method is equivalent to including packet-mode communication as a part of the existing digital circuit mode communication, the BISDN opts for ATM, which is an entirely different technique. In other words, if the ISDN can be said to be a circuit-mode-oriented network that can accommodate the packet mode, the BISDN is a packet-mode-oriented network that can accommodate circuit mode. Consequently, the BISDN differs from the ISDN in all aspects, including transmission, switching, signaling and network administration. Therefore, it is reasonable to treat the BISDN and the ISDN as two entirely different entities except for the conceptual similarities.

4.2 FUNCTIONAL ARCHITECTURE AND UNI CONFIGURATIONS OF BISDN

The BISDN's functional architecture is fundamentally equivalent to that of the ISDN. That is, in terms of such concepts as reference configuration, functional

---

3. From the signaling point of view, the long-term goals and objectives of BISDN signaling diverge from the ISDN. However, the current Q.93B and B-ISUP draft recommendations represent just minor changes to their ISDN counterparts. The ATM Forum is currently defining a modified version of Q.93B, which supports limited point-to-point connections.

group, and reference point, the BISDN is nothing more than an extension of the ISDN, except that the internal functions of reference configuration, internal composition of the functional groups, and interface signal of the reference points are slightly different in each case.[4]

The UNI physical configuration of BISDN is equivalent to that of ISDN's basic configuration, but the configuration for the shared medium is totally different. In fact, BISDN's UNI accommodates the physical configurations of ISDN and, in addition, all physical configurations related to shared medium structures existing in LANs or MANs. This implies that the BISDN-based public network can achieve a direct connection with the innermost parts of the subscriber premises networks.

This section focuses on the BISDN's basic architectural model, reference configuration, functional groups, UNI physical configurations, and reference points.

4.2.1 Basic Architectural Model

The BISDN's basic architectural model, as shown in Figure 4.3, is composed of lower order layer capability and higher order layer capability. Lower order layer capability consists of broadband capability, narrowband ISDN capability, and interexchange signaling capability. Broadband capability refers to the ATM-based information transfer capability provided at the BISDN's UNI, as well as at the internal exchange entities of the communications network. Narrowband ISDN capability implies the circuit-switching and packet-switching capability based on the 64-Kbps bit rate. Interexchange signaling capability is provided for performing signaling between offices, and here signaling information is also transferred by way of ATM. Lower order layer capability also includes switching and transmission capability provided by the narrowband ISDN.

Higher order layer capability in general is a function related to terminal equipment. However, depending on the type of service, the associated higher order layer capability can be provided through a special node inside the BISDN. This node can be a part of a public communications network, or belong to some center operated by another system, which is linked to the BISDN through the UNI or the NNI.

The *local function capability* (LFC) in Figure 4.3 corresponds to the switching and transmission function provided for local switching equipment, digital cross-connect equipment, and multiplexing equipment.

As can be seen in Figure 4.3, user-to-network signaling applies between

---

4. The functional architectural model of the ISDN is described in ITU-T Recommendation I.324, and that of the ISDN is described in Recommendation I.327.

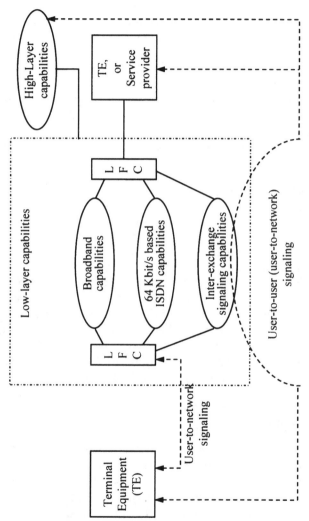

Figure 4.3 Basic architecture model of BISDN.

the BISDN's terminal equipment and LFC, and user-to-user signaling applies between the terminal equipment and the higher order layer capability of the service provider.

4.2.2 Reference Configuration

Fundamentally, the BISDN can be divided into user equipment and the public BISDN. User equipment includes terminal equipment and user networks, and here user networks are either the *broadband integrated services private branch exchange* (BISPBX) or private BISDNs. The reference point that links terminal equipment and user networks is called S_B, and the reference point that links user networks and public networks is called T_B. In case the user network has the same connection format as the public BISDN, the connection for the entire BISDN terminates at the S_B reference points, and S_B coincides with T_B if the user network is not present. Figure 4.4 is a pictorial representation of such a BISDN configuration.

4.2.3 UNI Functional Group

In the BISDN, the functional groups are designated in the same manner as in the ISDN, and to indicate broadband, the prefix B is attached to each designation. That is, terminal equipment is called B-TE, and private or public BISDN network terminations are called B-NT. Among the B-TEs, BISDN terminal equipment is designated B-TE1, while the non-BISDN terminal equipment is designated B-TE2. Also, among the B-NTs, the private BISDN network terminations are indicated by B-NT2, and public BISDN network terminations are indicated by B-NT1. Accordingly, T_B becomes the reference point between B-NT1 and B-NT2, and S_B is the reference point between B-NT2 and B-TE1. On the other hand, B-TA acts as an adaptor for linking B-TE2 to BISDN, and R becomes the reference point between them. Figure 4.5 illustrates these relationships.

Network Termination 1 (B-NT1)

In general, B-NT1 encompasses functions that correspond to the physical layer of the OSI protocol reference model. Functions of the B-NT1 include line transmission termination, transmission interface management, and OAM, and functions that relate to the transmission system itself can also be included. For a case in which B-NT1 terminates a cell-based OAM information flow, a separate cell delineation function is required.

Network Termination 2 (B-NT2)

B-NT2 encompasses functions corresponding to the physical and upper layers, which include the following functions: medium adaptation (the function for

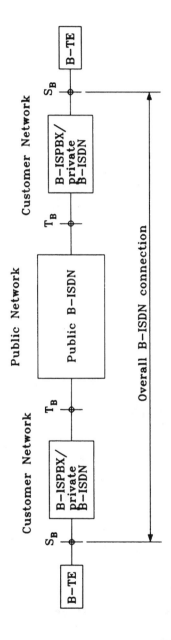

Figure 4.4 Overall BISDN configurations.

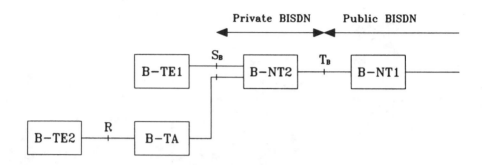

Figure 4.5   BISDN reference configuration.

adapting media and network structures that are different from one another), cell delineation, concentration, buffering, multiplexing/demultiplexing, resource allocation, usage parameter control, adaptation layer for signaling information about internal traffic, T_B and S_B interface management, OAM, signaling protocol management, and switching for internal connections. In special situations, B-NT2 can consist solely of physical layer functions. In case B-NT2 is absent, reference point S_B coincides with T_B (this coincidence is formed separately depending on whether the system in question is SDH-based or cell-based). B-NT2 can be implemented in the concentrated or distributed mode.

Terminal Equipment 1/2 (B-TE1/2)

Terminal equipment also includes functions that correspond to the physical and upper layers. Functions of B-TE include user-to-user and user-to-machine interface and protocol functions, the interface termination function and other physical layer functions, as well as the protocol management function for signaling information, the management function for connection with other equipment, and OAM functions. B-TE is divided into the B-TE1 type, which possesses the interface function that conforms to the BISDN's interface standard, and the B-TE2 type, which does not.

Terminal Adaptor (B-TA)

The terminal adaptor includes the functions corresponding to the physical and upper layers, and assumes the role of adapting B-TE2 or TE-2 to the BISDN interface.

4.2.4 Physical Configurations of User–Network Interface

Figure 4.6 is an illustration of various forms of basic physical configurations at UNI in terms of the previously studied functional groups B-NT1, B-NT2, B-TE1,

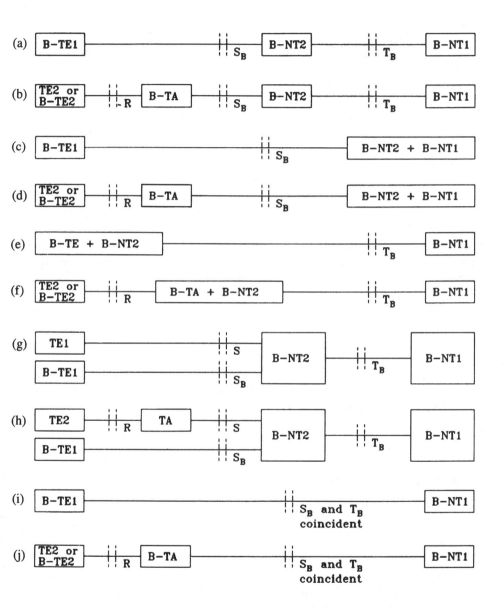

Figure 4.6 Physical configurations for UNI: (a,b) physical interfaces occur at S_B and T_B; (c,d) physical interfaces occur at S_B only; (e,f) physical interfaces occur at T_B only; (g,h) physical interfaces occur at S, S_B, and T_B; and (i,j) physical interfaces occur at a location where S_B and T_B coincide.

B-TE2, and B-TA, and the reference points R, S_B, and T_B.[5] Figures 4.6(a,b) represent the case in which the BISDN's physical interface occurs at the two reference points S_B and T_B; in Figures 4.6(c,d) it occurs only at S_B; and in Figures 4.6 (e,f), it occurs at T_B only. Figures 4.6(g,h) represent the case in which the physical interface occurs simultaneously at the BISDN reference points S_B and T_B and at the ISDN's reference point S; and in (i) and (j), the reference points S_B and T_B coincide due to the absence of B-NT2. On the other hand, Figures 4.6(b,d,f,h,j) can be viewed as representing instances in which the interface occurs only at reference point R. It can be inferred that these basic physical configurations of the BISDN are equivalent to those of the ISDN except for the cases of Figures 4.6(g,h).

The ability to accommodate various UNI physical configuration topologies for shared-medium connection is one of the BISDN's most notable features, allowing star, ring, bus, and starred-bus topologies. In addition, the configuration of B-NT2s can be provided in both the centralized or distributed types.

Figure 4.7 illustrates the instances of physical configurations employed for shared-medium access. In the figure, (a) illustrates the centralized B-NT2 configuration, and (b) illustrates a generic distributed B-NT2 configuration. Also, (a) represents the star structure, (c) the distributed ring structure, (d) the starred-bus structure, (e) the bus structure, and (f) the multiaccess ring structure. MA is a medium adaptor for accommodating distributed B-NT2 with a special network structure, and W is the interface between MAs. Interface W is a nonstandard interface, but can be made to be the same as the interface at the S_B reference point. B-TE* is a B-TE1 that provides shared-medium access capability, and SS_B is the interface between B-TE*s and is equivalent to the interface at the S_B reference point.

4.2.5 Basic Properties at the Reference Points S_B and T_B

The physical layer interface at the reference points S_B and T_B can be chosen to be either SDH-based or cell-based, and the ATM layer is common to both cases. At the T_B reference point, just a single interface exists per B-NT1, and the physical medium is also a point-to-point connection with just a pair of transreceivers.

More than one interface can exist per B-NT2 for the S_B reference point, and the interface at S_B can provide point-to-point connection for the physical layer and point-to-multipoint connection for the upper layers.

Since mutual compatibility exists between the interfaces at S_B and T_B, the

---

5. The specifications on the ISDN UNI and the BISDN UNI physical configurations are given in CCITT Recommendations I.411 and I.413, respectively.

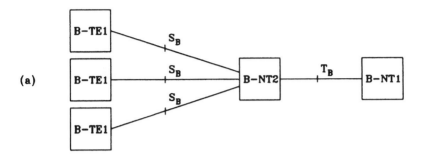

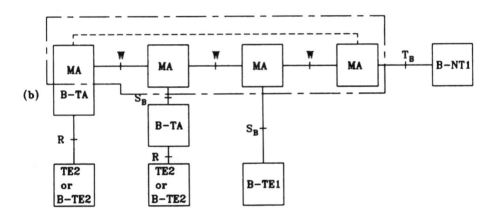

Figure 4.7 Physical configurations for multipoint applications: (a) centralized B-NT2 configuration (star configuration); (b) distributed B-NT2 generic configuration; (c) distributed ring configuration; (d) starred bus configuration; (e) bus configuration; and (f) multiaccess configuration.

two points coincide when the functional group B-NT2 is absent. Of course, here the cell-based S_B coincides with the cell-based T_B, and the SDH-based S_B with the SDH-based T_B (for a discussion of differences between cell-based and SDH-based systems, see Section 4.5).

At the UNI, user information is transferred in the ATM cell format. Information regarding connection-related functions is also transferred by way of ATM cells. Physical layer-related OAM information is transferred via ATM cells in the case of cell-based transmission, and STM frame overhead in the case of

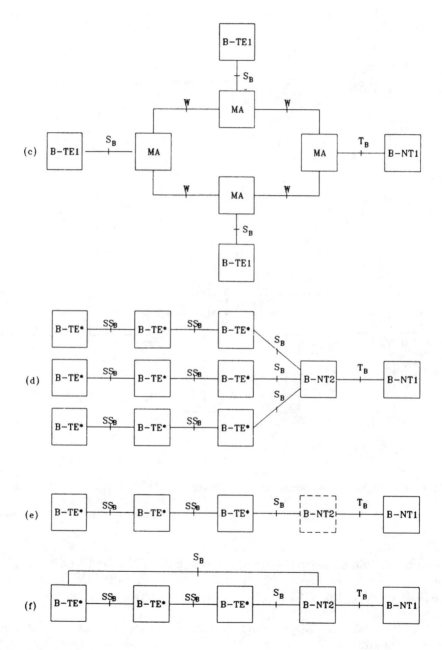

Figure 4.7   (continued)

SDH-based transmission. But ATM layer-related OAM information is always transferred by way of ATM cells.[6]

At the UNI, the timing information is delivered at the same instant as the ATM cell or the STM frame. The physical medium or the associated transmission system at the interface must be able to provide dependence on the bit sequence integrity.

4.3 ASYNCHRONOUS TRANSFER MODE

ATM was conceived as a means to realize the basic goal of the BISDN. The BISDN's fundamental goal is the integrated provision of various types of services with a great number of disparate characteristics (see Section 4.1.2), and ATM is a means of accommodating this requirement of versatility. In the present section, the topics associated with the ATM communication mode itself, i.e., the asynchronous transfer mode, as well as asynchronous time-division multiplexing and the ATM cell, are investigated.

4.3.1 Asynchronous Transfer Mode

ATM is a packet-mode transfer technique with a special format that employs *asynchronous time-division multiplexing* (ATDM). In the BISDN, service information is transferred by way of a continuous flow of packets of a fixed size, and these packets are called ATM cells. Accordingly, service information is first partitioned down to a fixed size and subsequently mapped into an ATM cell, and this cell is asynchronously time-division-multiplexed with other ATM cells to form the basic units of BISDN transmission. ATDM is a type of statistical multiplexing technique that time-division-multiplexes mutually asynchronous ATM cells coming from several different channels.

When ATM is employed, the capacity of a service channel is measured by the number of ATM cells. Consequently, the amount of transmitted information is reflected in the corresponding number of ATM cells, and the burstiness of the service information is indicated by the degree of ATM cell crowding. Here, the transmission capacity is assigned at the user's request at the time of call setup, and this scheme endows a versatile transfer capability on all the services, including connectionless (CL) services.

ATM is a connection-oriented (CO) method that transfers service information through the establishment of virtual channels (VCs). A connection identifier is assigned whenever a VC is established, and when the connection is released, the identifier is removed. The order of ATM cells inside a VC is

---

6. The UNI-related OAM information includes maintenance/administration signals, performance monitoring signals, and communications control provision signals. Details regarding OAM are given in Section 4.8.

maintained by an ATM layer function. Signaling information for setting up a call is delivered via dedicated ATM cells.

Hence, it can be inferred that the use of ATM allows the integration of various BISDN services possessing many different characteristics. That is, broadband and narrowband services can coexist within the same communications network by using the ATM cells of the same format, differing only in the number of cells that each type requires. Constant bit rate (CBR) services are made up of ATM cells with a uniform distribution; variable bit rate (VBR) services, which are widely distributed, are made up of the same ATM cells. Also, the delay problem associated with real-time services is solved through the use of VCs, thus making their provision a possibility.

The ATM communications technique can be said to be an integration of the existing circuit-mode digital communications technique with the packet-mode communications technique. First, in the sense that the ATM communications system uses ATM cells as its basic unit of transmission, it has a close connection with packet-mode communication. But there is a significant difference in that the packet mode was developed to support non-real-time VBR data signals, whereas ATM can manage real-time, CBR signals as well. Also, packet-mode communication is designed for use in regional LANs, whereas ATM is to be used for public networks; thus, differences arise in terms of address assignment, access and flow control, switching, and transmission. On the other hand, the circuit mode has a fundamental difference from ATM in that, in the former, the information signal is transmitted in continuous bit streams by allocating a separate channel for this purpose, whereas, in the latter, segmented service information is fitted onto ATM cells and transmitted through a VC. Consequently, a new set of problems is introduced in the accompanying connection setup, synchronization, signal processing, transmission, and switching.

For systematic and flexible information transfer, ATM prescribes a three-layer protocol reference model. The associated layers include the physical layer, ATM layer, and *ATM adaptation layer* (AAL). The AAL performs the function of mapping service signals into ATM cell's payload space, and the ATM layer executes the ATM cell header-related functions for the transparent delivery of the ATM payload space. The physical layer's function is to transfer ATM cells by converting them into transmission bit streams (refer to Section 4.4). If the three-layer protocol reference model is depicted in the form of the seven-layer architecture in Figure 1.5, Figure 4.8 is obtained.

Based on the preceding discussion, a comparison can be made among the circuit mode, packet mode, and ATM as listed in Table 4.2. As is evident from the table, the ATM can be stated in one phrase as "a packet mode that can provide real-time multimedia services by adopting fixed-size packets (i.e., cell) and ATDM, together with a simplified header function (i.e., without flow control and error control) and connection-oriented virtual circuit." In fact, this is very nature that enables simple and fast ATM switching.

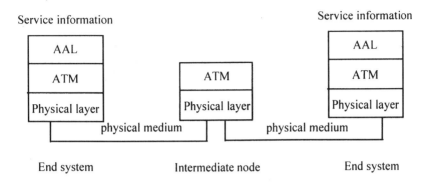

Figure 4.8 Layered architecture for ATM communication.

Table 4.2
Comparison of Communication Modes

	Circuit Mode	*ATM Mode*	*Packet Mode*
Service type	Real-time CBR	Real/non-real-time CBR/VBR	Non-real-time VBR
Connection type	Circuit (physical connection)	Virtual circuit (CO)	Virtual circuit/datagram (CO/CL)
Multiplexing	TDM (STM)	ATDM (ATM)	Statistical packet multiplexing
Transmission	Continuous bitstream (no delay)	Continuous cell stream (cell processing delay)	Intermittent packet stream (large delay)
Switching	Circuit switching	ATM switching	Packet switching
Network	Public network	Public/local network	Local network
Other			
• Protocol structure	—	3 layer architecture	7 layer architecture
• Packet type	—	Fixed length	Variable length
• Flow control	—	Not included	Included
• Error Control	—	Not included	Included

4.3.2 Asynchronous Time-Division Multiplexing

The TDM, which is widely used for multiplexing existing plesiochronous digital tributaries, is essentially a *synchronous* multiplexing technique as far as the system clock is concerned. This is because the TDM signals are constructed through the repetition of multiplexed frames created using the multiplexer

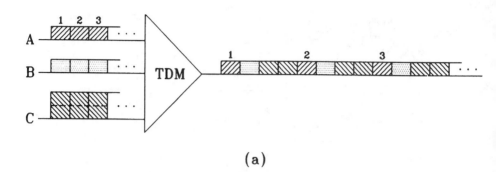

(a)

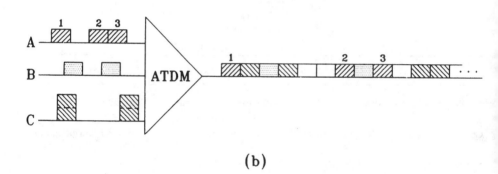

(b)

Figure 4.9 Comparison of TDM and ATDM: (a) TDM and (b) ATDM.

system clock. This results in the appearance of low-speed signals at fixed locations inside the frames, as depicted in Figure 4.9(a). In other words, low-speed signals always exist at locations that are synchronized with the system clock.[7]

ATDM is a type of multiplexing technique that stores each of the incoming low-speed signals inside a buffer, then retrieves and inserts the stored signals one by one into a multiplexing slot according to a priority-scheduling principle. The simplest example of the priority scheduling principle would be *first-in/first-out* (FIFO), and here the input signals are ATM cells in the case of an ATM communications system. Therefore, as shown in Figure 4.9(b), the low-speed input signals do not occupy locations inside the ATDM signal in a

---

7. Due to this synchronous property the TDM-based transfer mode is often called *synchronous transfer mode* (STM). That is, TDM-based transfer mode is the STM and, in contrast, the ATDM-based one is the ATM.

well-regulated manner, and thus behave *asynchronously* compared to their TDM equivalent.

ATDM is superior to TDM in that it has a higher channel utilization factor. TDM assigns an exclusive channel to each of the incoming service signals; thus, even when a given channel is in a vacant state containing no effective information, it is not possible to pass other service information through it. But since there is no exclusive channel allocation in ATDM, a blank channel can be taken by any incoming signal, resulting in a higher channel utilization factor.

Such channel utilization relationships are illustrated in Figure 4.10. In the figure, the length in the vertical direction denotes the channel capacity of the multiplexed signals, while the horizontal length corresponds to the time duration. Also, the parts in slanted lines or those that are darkened indicate the presence of effective information corresponding to the size of an ATM cell. In the case of TDM, since the multiplexed signal is no more than a combination of several independent channels, it can be seen that any vacant space in each channel is maintained as it is. But in ATDM the multiplexed signal consists of just a single channel; hence, any information vacancy can be collected and used for providing a new service, consequently increasing the channel utilization factor.

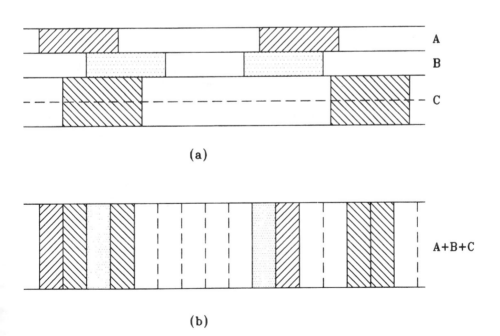

Figure 4.10 Comparison of channel use: (a) TDM (dedicated) and (b) ATDM (shared).

4.3.3 ATM Cell Structure

The ATM cell acts as the basic unit of information transfer in the ATM communication. As shown in Figure 4.11(a), the ATM cell is composed of 53 bytes. The first 5 bytes are for the cell-header field, and the remaining 48 bytes form the user information field. The cell-header field is divided into *generic flow control* (GFC), *virtual path identifier* (VPI), *virtual channel identifier* (VCI), *payload type* (PT), *cell loss priority* (CLP), and *header error control* (HEC) fields. The associated bit sizes differ slightly at the NNI and the UNI. The bit sizes for the two interfaces are as shown in Table 4.3 and Figures 4.11(b,c).

Although the main function of the GFC header is the physical access

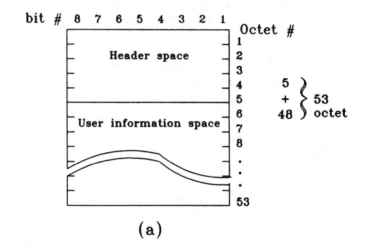

(a)

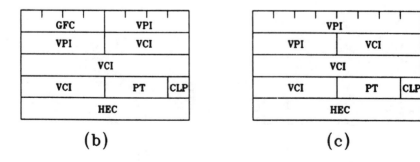

(b)   (c)

Figure 4.11   ATM cell structure: (a) cell structure; (b) header structure at UNI; and (c) header structure at NNI.

Table 4.3
Bit Allocation of Cell Header

	Bit Allocation	
Function	UNI	NNI
GFC	4	0
VPI	8	12
VCI	16	16
PT	3	3
CLP	1	1
HEC	8	8

control, it can also be used for reduction of cell jitter for CBR services, fair capacity allocation for VBR services, and traffic control for VBR flows. Such a function requires the capability to control any UNI structure, whether it be a ring, a star, a bus configuration, or any combination of them.

The role of the VPI/VCI field is to indicate VC or VP identification numbers in order to distinguish cells belonging to the same connection. A set of separate, fixed VPI/VCI identifiers is preassigned for indicating unassigned cells, physical layer OAM cells, metasignaling channel, and generic broadcast signaling channel. (For a detailed discussion of VPs, VCs, and preassigned identifiers, see Section 4.6.)

PT is used for indicating the presence of user information and for indicating whether the given ATM cell suffered from traffic congestion. CLP is a bit used for indicating whether the corresponding cell may be discarded during the time of network congestion. HEC is a CRC byte for the cell-header field and is used for detecting and correcting cell errors and delineating the cell header (see Section 4.5 for details about the use of HEC).

As shown in Table 4.4, ATM cells can be classified according to the associated layers and functions. (Details regarding the ATM layer and the physical layer are given in Sections 4.5 and 4.6.) For instance, the *ATM layer cell* is a cell that is formed at the ATM layer, and the *physical layer cell* is similarly formed at the physical layer. The ATM cells are divided into assigned cells and unassigned cells, and physical layer cells are divided into idle cells and physical layer OAM cells. *Assigned cells* refer to those cells that are allocated to ATM layer services, and the *unassigned cells* refer to the remaining types. *Idle cells* are created in order to fill the vacant space that results when there are no cells to be transmitted, and *OAM cells* are used for the transfer of OAM information of the physical layer. On the other hand, the ATM cells can also be divided into valid cells and invalid cells from the physical layer viewpoint. Here, *valid cells* designate those that have no errors or whose errors have been

Table 4.4
Classification of ATM Cell

Layer-Based	Function-Based	Function
ATM layer	Assigned cell	Services related to upper layer
	Unassigned cell	Services inherent to ATM layer
Physical layer	Idle cell	Stuffing blank space
	Physical layer OAM cell	OAM cell

corrected, and *invalid cells* designate the other cells. The invalid cells are discarded at the physical layer.

4.4 BISDN PROTOCOL REFERENCE MODEL

In the BISDN, a protocol reference model (PRM) is adopted in order to perform all of the various funcions required by the network. The PRM involves separating the overall communications functions into several layers and defining the appropriate set of functions for each layer. Here, a lower layer has the role of providing a prescribed set of services to the adjacent higher layer, and transparent connection is achieved between two ends of the same layer. Conceptually, the PRM of the BISDN is an extension of the ISDN PRM, and is rooted in the OSI concept (see Figures 1.5 and 4.8).[8]

4.4.1 Protocol Reference Model

The protocol reference model for the BISDN has a composition like the one shown in Figure 4.12, consisting of the user, control, and management planes. The management plane can be further separated into layer management and plane management. The user plane provides user-information-related functions; the control plane performs various control functions for service provision; and the management plane provides communications network management functions.

The user plane and the control plane each consist of the physical, ATM, AAL, and upper layers. The physical layer provides the physical medium and transmission functions, the ATM layer provides call transfer function for all of

---

8. The PRM for the ISDN is described in CCITT Recommendation I.326, and the one for the BISDN is described in Recommendation I.321. The OSI-based layered communications concept is described in Recommendation X.200.

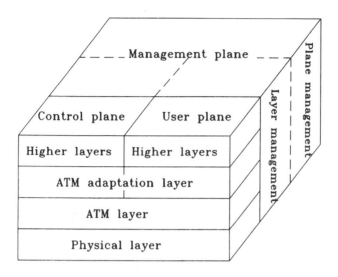

Figure 4.12 BISDN protocol reference model.

the BISDN services, and the AAL provides service-related functions for its upper layers. The upper layer of the user plane provides service information management functions, and the higher layer of the control plane provides functions associated with call control and connection control. Table 4.5 gives a list of the respective functions of each layer.

4.4.2 Functions of Each Plane

The *user plane* provides functions for transferring the flow of user information, as well as associated control functions such as flow control and error correction. Here, the term *user information* implies various BISDN service information, such as voice, image, data, text, and graphics. User information can be passed transparently through the BISDN, or can be transferred after an appropriate processing procedure.

The *control plane* provides call connection and connection control functions. That is, the control plane provides the function associated with call establishment, call monitoring, and call release. It can also provide a control function for changing the characteristics of a readily established service.

The *management plane* provides the communications network monitoring function associated with user information and control information transfers. The management plane is divided into plane management function and layer management function. The plane management function performs management of the entire network through its role as an interplane arbitrator. Layer

Table 4.5
Functions of Each Layer of BISDN PRM

Layer	Sublayer	Functions
ATM adaptation layer (AAL)	Convergence sublayer (CS)	Convergence functions
	Segmentation and reassembly (SAR)	Segmentation and reassembly functions
Asynchronous transfer mode (ATM) layer		Generic flow control Cell-header generation/extraction Cell VPI/VCI translation Cell multiplex and demultiplex
Physical layer (PL)	Transmission convergence (TC)	Cell rate decoupling HEC header sequence generation/verification Cell delineation Transmission frame adaptation Transmission frame generation/recovery
	Physical medium (PM)	Bit timing Physical medium

management function refers to the management function associated with the parameters and resources within each protocol entity (i.e., TE, network terminations, and so on) and manages the OAM information flows associated with each layer.[9]

4.4.3 Physical Layer

The physical layer is divided into the physical medium sublayer and the transmission convergence sublayer. The functions of each sublayer are listed in Table 4.5. The physical medium sublayer provides the physical medium and bit timing-related functions, and the TC sublayer provides functions for converting ATM cell flow into data bit/symbol streams and the reverse functions.

Physical Medium Function

The physical medium function is associated with the transmission medium itself. In the case of optical transmission, for example, it is a function related to

---

9. Details on management plane functions are given in CCITT Recommendation Q.940.

optical fibers, light-emitting devices, light-detecting devices, optical connectors, and so on.

Bit Timing Information Function

This function involves the conversion of data bit flow into a waveform adapted to a particular physical medium or the reverse conversion process, the insertion or the extraction of timing information, and line coding or decoding. Consequently, the information transferred from the physical medium sublayer to the transmission convergence sublayer consists mainly of data bit/symbol stream and the corresponding timing information.

Transmission Frame Generation and Extraction Function

This function involves the generation and extraction of the transmission frame. This function does not apply in cell-based transmission, since a separate transmission frame is not needed in this case. However, STM-n frames are required in SDH-based transmission, and DS-3 signal frames are required in G.702-based transmission (see Section 4.5).

Transmission Frame Adaptation Function

The transmission frame adaptation function involves the mapping of ATM cell flow into the payload of the transmission frame, or, conversely, the extraction of ATM cell flow from the transmission frame. This function is required in an SDH-based network or in a G.702-based network.

HEC Signal Generation and Confirmation Function

The associated duties include generating and confirming the HEC signal of the ATM cell header. In the transmitting direction, it generates the HEC signal for the first four bytes of the ATM cell header and inserts it into the fifth byte of the header. Conversely, it applies an identical procedure to the received signal to inspect whether the HEC signal is correct, and in the event an uncorrectable error is detected, the cell in question is discarded (see Section 4.5.5 for further discussion of HEC signal generation and errored-cell treatment).

Cell Delineation Function

This is a function for identifying ATM cell boundaries in the ATM cell flow. In the transmitting direction, it performs the ATM cell scrambling function, and in the receiving direction performs cell delineation using the HEC, confirmation, and descrambling functions. (Details regarding ATM cell delineation and scrambling are given in Section 4.5.7.)

Cell Rate Decoupling Function

The cell rate decoupling function augments the ATM cells that are carrying valid information using idle cells in order to match the overall cell rate to that of the corresponding payload capacity, or, conversely, extracts cells with effective information by removing idle cells.

4.4.4 ATM Layer

The ATM layer is independent of the physical layer, and its functions are listed in Table 4.5.

Cell Multiplexing and Demultiplexing Function

This function provides the capability of multiplexing ATM cells from different VPs and VCs to form a composite cell flow, or the opposite demultiplexing capability. Here, the multiplexed cell flow does not have to be continuous.

Cell VPI/VCI Translation Function

This function is required at the ATM switch or the ATM cross-connect node, and its role is to map the values stored in the VPI/VCI field of each ATM cell header into a new set of values. (Refer to Section 4.10 for a detailed description of ATM switching.)

Cell-Header Generation and Extraction Function

This function applies at the ATM layer's terminations, and involves the generation or extraction of the first four bytes in the ATM cell header. For the generation of the cell header, the associated information received from the upper layer is mapped into the corresponding field, and the opposite is executed for the cell-header extraction process. This function also encompasses the translation of the *service access point identifier* (SAPI) into the VPI and the VCI.

Generic Flow Control Function

The GFC function is the function for controlling access and information flow at the UNI. Here, the information is transferred via assigned cells or unassigned cells.

4.4.5 ATM Adaptation Layer

The AAL is divided into the *convergence sublayer* (CS) and *segmentation and reassembly* (SAR) sublayers. At the CS, the function for converting the user service information coming from the upper layer into a protocol data unit, or the opposite process, is performed. At the SAR sublayer, the function for segmenting the PDU to form the user information field of the ATM cell, or the opposite process, is performed. The ATM adaptation function varies depending on the type of upper layer service. The AAL is described in detail in Section 4.7.

4.4.6 Interlayer Information Transfer

For communication between two adjacent layers of the BISDN PRM, the information transfer in both directions must be specified in advance.

Between Physical Layer and ATM Layer

The ATM layer requests to the physical layer the transport of service data units, and the physical layer indicates that the SDUs sent by its counterpart physical layer entity are ready. The information transferred from the physical layer to the ATM layer includes valid cells (excluding idle cells and physical layer OAM cells) and the associated timing information. The information sent from the ATM layer to the physical layer consists of assigned cells and unassigned cells and the relevant timing information (timing information refers to the clock and the indication of the presence of the transferred data).

Between ATM Layer and AAL

The AAL requests to the ATM layer the transfer of ATM SDUs, and the ATM layer indicates that the ATM SDUs from its counterpart AAL entity are ready. The information exchanged between the ATM layer and the AAL includes ATM cell payload, SAPI, and the relevant timing information.

Others

OAM-related information is exchanged between the physical layer and the management plane. The information transferred from the physical layer to the management plane can include indications of the loss of input signal, received errors, and degradation in error performance. The information exchanged between the physical medium sublayer and the transmission convergence sublayer of the physical layer consists of logic symbol flow, or bit streams, and the associated timing information.

4.5 PHYSICAL LAYER OF BISDN

The main function of the BISDN's physical layer is to collect and organize ATM cells sent down from the ATM layer, transport them to the physical medium, and also perform the reverse of the process. The physical layer must also be able to deal with various obstacles that can occur during transmission. To accomplish the above objectives, the physical layer handles various procedures, which can be grouped into the transmission convergence sublayer and the physical medium sublayer, as shown in Table 4.5.

4.5.1 Interface Characteristics of the Physical Layer

The BISDN maintains a 155.520-Mbps or 622.080-Mbps transmission rate at the T_B reference point. Its structure is symmetric, but it can also be asymmetric in the case of 622.080 Mbps. The physical medium can be optical cable or coaxial cable, and the extension capability of 0 to 100m is required in the case of an electrical interface, and 0 to 800m in the case of an optical interface.[10]

Both cell-based and SDH-based physical layers are allowed at the T_B as well as at the S_B reference point (G.702-based case is allowed as well). The cell-based signal consists solely of ATM cell flows, and the SDH-based signal is formed by filling VC-4 payload space of an STM-frame with ATM cells. In the case of cell-based signals, OAM signals such as AIS, RDI, REI and RFI are transported in the form of OAM cells. But in the case of SDH-based signals, the OAM signals are transported via STM's SOH or POH.

If the transmission speed of the physical medium is 155.520 Mbps inside the BISDN, the actual transmission capacity of the cells created at the ATM layer amounts to 149.760 Mbps. This capacity encompasses user information cells, signaling cells, as well as OAM cells, and corresponds to the capacity of the VC-4 payload space of the SDH-based signal. The remaining capacity of 5.760 Mbps gets filled with physical layer OAM cells and idle cells in the case of the cell-based signal, and STM frame overheads (SOH, POH, pointer) in the case of the SDH-based signal. As a method of identifying cell boundaries, the HEC technique is used for the cell-based signal, and HEC or an SDH overhead can be used for the SDH-based signal.

4.5.2 Cell-Based Physical Layer

If the physical layer is cell based, a flow of scrambled ATM cells is transmitted as is, without the use of an external frame. In other words, the relevant HEC field is calculated and added to the ATM cells descended from the ATM layer,

---

10. If possible, extension capability of up to 200m is recommended in the case of electrical interface, and up to 2000m in the case of optical interface. Refer to CCITT Recommendation I.433.

which are then transmitted directly after a scrambling process. Consequently, a function for generating transmission frames is not required. Here, a *distributed sample scrambling* (DSS) with the characteristic polynomial $x^{31} + x^{28} + 1$ is used for scrambling (see Sec. 4.5.8 for a detailed explanation of distributed sample scrambling). ATM cell flow is a continuous flow of 155.520 Mbps, and the receiving end can extract timing information from it. Identification of ATM cell boundaries from a cell flow relies on the HEC technique (see Section 4.5.5 for a detailed explanation of HEC).

Of the cell flow capacity of 155.520 Mbps, 5.760 Mbps or more always gets filled with idle cells, physical layer OAM cells, and other physical layer reserved cells. This capacity corresponds to the section and path overheads of the SDH-based signal and is equivalent to 1/27 of a cell. Accordingly, the OAM cell portion provides various maintenance and performance monitoring functions, which the section and path overheads provide. Here, representative examples of a maintenance signal are the AIS and the RDI (see Section 4.8 for more details on OAM functions).

Idle cells, physical layer OAM cells, and other reserved cells are identified by assigning a special bit pattern to the header of each type of cell. Various header bit patterns for the physical layer OAM cells are listed in Table 4.6. These physical layer cells are transported only as far as the physical layer of the receiving side and are not returned to the ATM layer. Hence, it is meaningless to assign bit patterns in terms of fields. In the table, P represents a bit that is usable at the physical layer. The first four bits of the ATM cell overhead, which are not listed in the table, are set to "all 0" in all cases other than the physical layer reserved cell in the UNI, for which they are set to "all P."

4.5.3 SDH-Based Physical Layer

If the physical layer is SDH-based, the ATM cells are transmitted by first mapping them into the STM-n frame of the SDH. First, the HEC field is calculated

Table 4.6
Physical Layer ATM Cells with Preassigned Cell Header

ATM Cells	VPI	VCI	PT	CLP
Idle cell	00000000	00000000 00000000	000	1
Reserved cell for physical layer	00000000	00000000 00000000	PPP	1
Physical layer OAM cell (F1)	00000000	00000000 00000000	001	1
Physical layer OAM cell (F3)	00000000	00000000 00000000	100	1

Note: P: bit to be filled in the physical layer.

and inserted into the cells received from the ATM layer, and then idle cells are added to create a signal with the capacity of 149.760 Mbps; afterwards, *self-synchronous scrambling* (SSS) with the characteristic polynomial $x^{43} + 1$ is applied to all the bits in the cell except the cell header, and then the cells are mapped into the VC-4 payload space (that is, the C-4 space) and transmitted in the form of an STM-n frame (a detailed explanation of the scrambling process is given in Section 4.5.7). The receiving side can extract and use clock information from the STM-n frame, and send the corresponding ATM cells up to the ATM layer after performing a reverse of the procedure just mentioned (for more information on the mapping of ATM cells into the STM frame, see Section 3.4.7).

When the ATM cells organized with the 149.760-Mbps cell rate are mapped into the VC-4 payload space, the starting point of the cells for every frame can be recorded in the path overhead's H4. The recorded information consists of the cell offset indication; that is, the interval from the endpoint of the H4 overhead to the starting point of the first ATM cell header that follows can be represented in byte units. The number of bits necessary for this is six, so bits 3−8 of H4 can be assigned for this. At the receiving side in this case, the ATM cell boundaries can be identified by first identifying the STM-n frame and subsequently reading its H4 byte. Of course, the HEC procedure can be applied instead of this.

The section and path overheads of the STM frame provide various maintenance and performance monitoring functions associated with the SDH transmission process. These include STM frame alignment, STM-1 and path identification function, section and path error monitoring, section alarm indication, section and path far-end receive failure, section and path far-end error report, and path signaling indication (for a detailed discussion of the path and section overheads, see Section 3.5). The ATM cell mapping-related overhead also includes the C2 byte, which indicates that the frame in question is carrying ATM cells. The section overhead has a close connection with the OAM information flows F1 and F2, and the path overhead with F3 (the OAM information flow is treated in detail in Section 4.8).

4.5.4 G.702-Based Physical Layer

In the evolution toward BISDN, a capability is required for transmitting ATM cells in the form of existing plesiochronous digital hierarchy signals.[11] Such a capability is referred to as *G.702-based transmission* or *PDH-based transmission*.

---

11. Besides, ATM transmission over the FDDI is of interest. The ATM Forum specified FDDI-based 100-Mbps ATM transmission, according to which ATM cells are transmitted as a byte stream over the FDDI's physical layer without any additional frame structure.

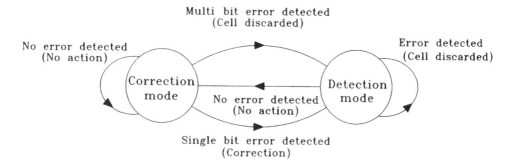

Figure 4.13 Modes of HEC operation.

The physical layer for the G.702-based transmission is similar to that of SDH-based transmission, except that the G.702 tributaries are used instead of the STM-*n* signal. Here, the cell transmission capacity can vary depending on the particular tributary used.

Among the G.702 tributaries employed for ATM cell transmission, those of primary interest are DS-1 (1.544 Mbps), DS-1E (2.048 Mbps), DS-3E (34.368 Mbps), DS-3 (44.736 Mbps), and DS-4E (139.264 Mbps). In the case of the DS-3 signal, the *physical layer convergence protocol* (PLCP) is employed as for the case of the DQDB MAN (refer to Section 5.6.4).[12]

4.5.5 Header Error Control

The role of the HEC function is to correct any single-bit errors found in any part of the cell header and to detect multibit errors. It operates by performing a cyclic redundancy check on the first four header bytes and then recording the result on the fifth byte; the process is repeated at the receiving end to extract relevant error information.

Typical errors that occur on optical fibers are a mixture of single-bit errors and bursty errors. To be able to cope with both forms of errors, the HEC establishes and employs two types of error-checking modes, as shown in Figure 4.13, which are correction mode and detection mode. The correction mode is employed by default and it provides a single-bit error correction function.

Once an error is detected in the correction mode, it is corrected, and the

---

12. In the cases of DS-1, DS-1E, DS-3E, and DS-4E, it is simpler to leave the frame alignment marks for the tributary itself as they are, and instead map ATM cells into the remaining payload space in a continuous manner. But in the case of the DS-3 tributary, the overhead bits are distributed into single bits, which are 85 bits apart; hence, it is simpler to transmit the ATM cells by overlapping a separate frame on top of the DS-3 frame.

receiver moves into the detection mode in the case of single-bit errors; if the error is multibit, the corresponding cell is discarded, and then the receiver moves into the detection mode. When an error is found in the detection mode, the cell is immediately discarded; if no error is detected, the receiver reverts to the correction mode.

Hence, the HEC employed in this manner provides the capability to correct any single-bit errors, along with the ability to detect bursty errors. The receiver moves from the correction mode into detection mode once a multibit error is found, which is due to the fact that the multibit error itself is an advance warning of a bursty error. That is, even though the error that appears after the multibit error might be diagnosed as a single-bit error, there is a greater probability that it is another multibit error.

Figure 4.14 is a flowchart of the possible error control and processing stages. After the HEC procedure, the ATM cell is identified either as a valid cell or an invalid cell. The valid cells can include actual valid cells and cells that carry errors. The valid cells with errors might be created due to an imperfection in the CRC, or because of a flaw in the error correction stage. The cells that are determined and discarded as invalid cells include those with uncorrectable multibit errors or those that are detected as bursty errors. They can also include correctable single-bit errors. The valid cells with errors and invalid (discarded) cells are the primary causes of degradation in the performance of the BISDN. Figure 4.15 is a plot of the probability of their occurrence as a function of the bit error rate (refer to CCITT Recommendation I.432).

The code that is used for HEC is a cyclic code with the generating polynomial $x^8 + x^2 + x + 1$. To elaborate, the first four bytes of a cell header are expressed as a thirty-first degree binary polynomial, which is then multiplied by x^8 and divided by the generating polynomial. The remainder is recorded in the HEC field, which is subsequently confirmed at the receiving side. Here, all of the arithmetic performed is modulo-2.

The confirmation device at the receiving side has to express five header bytes (including the HEC field) as a thirty-ninth degree binary polynomial, and check whether it is divisible by the generating polynomial. In case bit slips are present, the HEC performance can be enhanced if 01010101 is added to the fifth byte before transmission and subtracted at the receiving side before HEC processing.

4.5.6 Cell Delineation

Cell delineation is the function for locating cell boundaries of the incoming data flow. Basically, the method works by observing the degree of correlation between the ATM cell's first four bytes and the fifth byte. That is, after five contiguous bytes are selected and expressed as a thirty-ninth order polynomial, if the polynomial is divisible by the real polynomial $x^8 + x^2 + x + 1$, then for

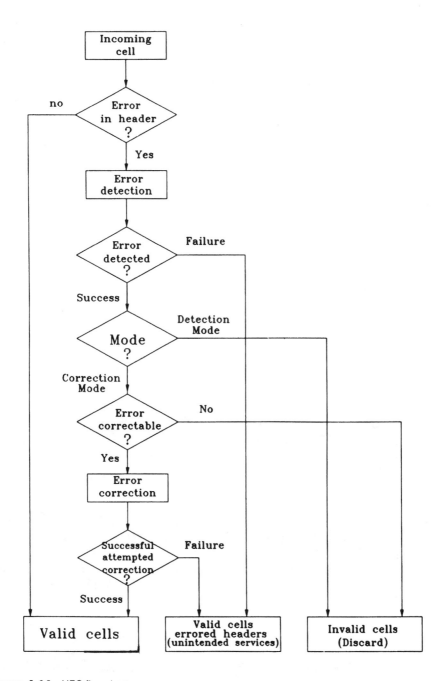

Figure 4.14 HEC flowchart.

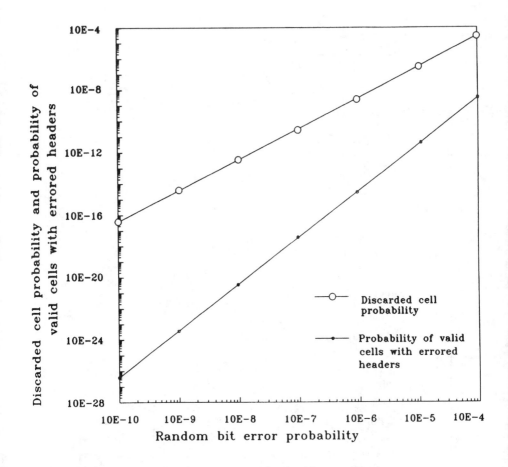

Figure 4.15 Probability of discarded cells and valid cells with errored headers.

the moment the result is regarded as a valid cell header. If the same holds true after the process is applied repeatedly in intervals of 53 bytes, then the result is confirmed as the ATM cell header. The cell delineation technique that uses the HEC has a state diagram shown in Figure 4.16. The three states of the cell delineation procedure are *hunt, presynch,* and *synch,* which are defined as follows:

1. In the hunt state, cyclic redundancy coding is performed on groups of five bytes (five bytes correspond to the length of one header) as they are moved over bit by bit. (It could be byte by byte in the SDH-based case.) If the correct HEC is found, it enters the presynch state.
2. In the presynch state, the cyclic coding is performed as bytes are moved

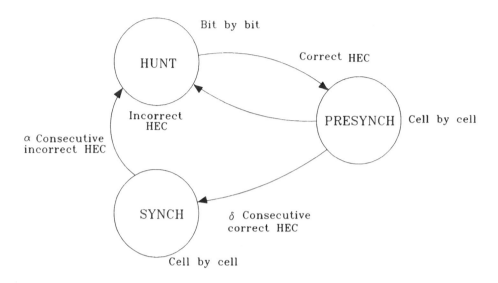

Figure 4.16 Cell delineation state diagram.

over in groups of 53 (length of one cell). If the correct HEC is found δ consecutive times, the procedure moves into the synch state. However, if even a single incorrect HEC appears, the procedure reverts to the hunt state.

3. The synch state also performs cyclic coding on a cell-unit basis. The synch state is maintained even if occasionally incorrect HEC is discovered, but if incorrect HEC is discovered α consecutive times, then the loss of synchronization is assumed and the procedure moves back to the hunt state.

The consecutive header error indicator α, which is the criterion for determining deviation from synchronicity, and the consecutive header confirmation indicator β, which triggers the transition into the synch state, are two of the most important parameters for determining cell delineation performance. The performance evaluation for the 155.520-Mbps signal in terms of these parameters is shown in Figure 4.17 (refer to CCITT Recommendation I.432). It can be seen from the figure that α = 7 and δ = 6 give a satisfactory result.[13]

At an SDH-based physical layer, the STM frame is identified by searching out the bytes A1 and A2 of SOH from the received bit stream, then the AU-4

---

13. In the cell-based physical layer, only six of HEC's eight bits are used for cell delineation purposes; thus, a higher value of δ, which is 8, is used. See Section 4.5.8.

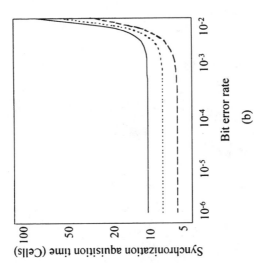

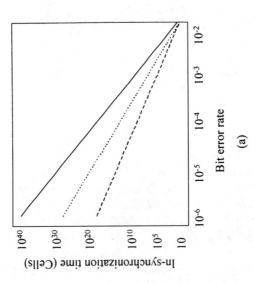

Figure 4.17 Performance of α and δ for 155.520-Mbps signal: (a) α and in-synchronization time and (b) δ and synchronization acquisition time.

pointer bytes within the SOH are read in order to determine the VC-4 starting point, and finally the above HEC is applied to locate the ATM cell boundary. In the final step, the H4-based cell delineation can be applied instead of using the HEC. The performance comparison between H4 byte-based cell delineation and HEC-based cell delineation is displayed in Table 4.7. The former is the case in which the H4 byte is used for delineation and HEC for cell boundary loss confirmation, and the latter is the case in which HEC is used for both the delineation and loss confirmation. It can be inferred from the table that, at low bit error rates, the performances of each method are equivalently good, but as bit error increases, the use of HEC shows a better performance.

4.5.7 Scrambling

In the BISDN physical layer, scrambling is used as a complement and a means to improve the reliability of the cell delineation technique. At the same time, by altering the data of the information field to appear more random, scrambling brings about an enhancement in the transmission performance. The characteristics of the most representative scrambling techniques—*self-synchronous scrambling* (SSS), *frame synchronous scrambling* (FSS), and *distributed sample scrambling* (DSS)—are compared in Table 4.8.

For the SDH-based physical layer, SSS is used with the characteristic polynomial $x^{43} + 1$. The scrambling procedure applies to the entire information field of the 149.760-Mbps cell flow (including idle cells), excluding the header. During the cell delineation procedure at the receiving side, the descrambler does not operate in the hunt state, but does operate in the presynch and synch states.

In SSS, because each state of the *pseudo-random binary sequence* (PRBS) generator depends on the incoming signal, even if the scrambler and the descrambler are slightly out of synch, the synch state can be self-recovered. Therefore, a frame synchronization procedure is not separately required, thus simplifying the implementation of SSS. However, an error in the transmission

Table 4.7
Performance of Cell Delineating Algorithms (unit: byte)

Performance		*H4*	*HEC*
Maximum average synchronization time ($\delta = 6$)	Bit error rate 0.0001	380	340
	Bit error rate 0.001	990	380
Average time for detecting the loss of synchronization ($\alpha = 7$)		400	400

Table 4.8
Comparison of Scrambling Techniques

Scrambling	Characteristics
Self synchronous scrambling (SSS)	No synchronization problem Randomizing effect good Error multiplication problem No synchronization sample transmission problem
Frame synchronous scrambling (FSS)	Synchronization by resetting at each frame Randomizing effect good for large-sized frames No error multiplication problem No synchronization sample transmission necessary
Distributed sample scrambling (DSS)	Synchronization by transmitted samples Randomizing effect is good even for small-sized frames No error multiplication problem Synchronization sample transmission necessary

affects the states of the descrambler's PRBS generator; hence, there is a possibility that one bit of input error magnifies to two bits of output error. A relatively simple characteristic polynomial $x^{43} + 1$ is adopted to limit the effective error multiplication rate to be two; its shortcoming is that its signal randomization performance is not optimal. However, it is complemented by the FSS process during the SDH frame generation, which scrambles the user information field as well as the header.

Such scrambling techniques are not appropriate for the cell-based physical layer, because the cell header is not scrambled again when it is transmitted.[14] Therefore, DSS with the characteristic polynomial $x^{31} + x^{28} + 1$ is used to scramble and transmit both the header and the user information fields.[15]

The DSS method generates a sample of PRBS and performs binary addition (or XOR operation) on the bit streams of header and information fields. Here, the 8 bits of the HEC section are obtained by calculating a CRC-8 code for the 32 bits of the distributed sample scrambled header section.

To synchronize the PRBS generator of the transmitting side with that of the receiver, PRBS samples of the transmitter are added in binary to the bits of the HEC section before they are transmitted. The distributed sample scrambler of the transmitter generates and transmits two such sampled bits per cell. The receiving side searches for the cell boundary using the remaining six HEC bits

---

14. In SDH-based transmission, an FSS is again applied during the process of creating the STM frame; hence, such a problem does not arise.
15. Distributed sample scrambling, unlike the SSS, is a unique scrambling technique that only appears in ATM cell-based transmission. Its basic principles are treated separately in Section 4.5.8.

Table 4.9
Comparison of SSS and DSS

Scrambler	SSS	DSS
Characteristic polynomial	$x^{43} + 1$	$x^{31} + x^{28} + 1$
Error multiplication	2	1
Number of bits used in the header	0	2 (before synchronization) 0 (after synchronization)
Cell transmission possibility	Possible (using SDH based)	Possible
Malicious interference probability	$1/2^{43}$	$1/(2^{31} - 1)$
Synchronization time	43 bits	16 cells
Complexity of scrambler	43 SR* 1 XOR**	31 + 2 SR 2 XOR
Complexity of descrambler	43 SR 1 XOR	31 + 2 SR 19 XOR
Scrambling of header	Excluded from scrambling	Included in scrambling

*SR: Shift Register
**XOR: XOR Gate

until the scrambler and the descrambler are synchronized. Once the cell boundary is found, the other two HEC bits are tracked; after they are found, their values are used to deduce the PRBS sample values. Since the distributed sample scrambler's characteristic polynomial has the order of 31, if cell delineation is performed correctly, then the synchronization between the respective scramblers of the transmitter and the receiver can be achieved after 16 cells.

Table 4.9 gives a comparison of the SSS used in SDH-based transmission and the DSS used in the cell-based transmission. Compared to the SSS, DSS is superior from the error multiplication standpoint, but is inferior in terms of synchronization time and the complexity of the descrambler. However, the chief difference between the two methods is whether or not the header is included as an object to be scrambled.

4.5.8 Distributed Sample Scrambling

DSS is a scrambling technique recently adopted by ITU-T for use in the cell-based physical layer of the BISDN.[16] DSS is basically similar to FSS,

---

16. This subsection gives a detailed description of DSS, which is a unique scrambling method used in the cell-based physical layer of the BISDN, and so it may be skipped without detracting from the overall understanding of the chapter.

which scrambles and descrambles the digital bit streams by adding *shift register generator* (SRG) sequences. But the DSS is different from the FSS in the method of synchronizing the state of the descrambler SRG to that of the scrambler SRG. In the FSS, the scrambler and the descrambler are synchronized by resetting the states of both SRGs to a prespecified state, whereas in the DSS the samples of the scrambler SRG are transmitted to the descrambler SRG for synchronization.

Transmitter Operation

The transmitter SRG sequence is added (modulo-2) to the complete cell bit by bit except for the HEC field. The characteristic polynomial of the SRG in the DSS is $x^{31} + x^{28} + 1$.

The CRC byte for each cell is then modified by modulo-2 addition of the CRC calculated on the first 32 bits of the scrambled header. The first 2 bits of the HEC field are then modified again as follows by two samples from the SRG sequence. To the first HEC bit (HEC$_8$), the value of the SRG sequence that was used 211 bits earlier for scrambling (or s_{t-211}) is added in modulo-2. To the second bit of the HEC field (HEC$_7$), the current value (or s_t) of the SRG sequence is added. These samples are exactly half a cell apart so the first is delayed by 211 bits before being delivered. The SRG sequence and resultant transmitted data structures of the HEC field are shown in Figure 4.18.

Receiver Operation

Receiver operation consists of three basic states: acquisition of scrambler synchronization, verification of scrambler synchronization, and steady-state

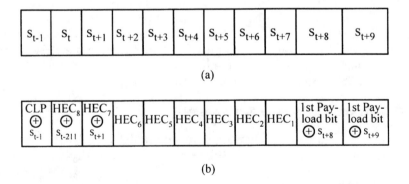

(a)

(b)

Figure 4.18   Data structure at DSS: (a) SRG sequence and (b) resultant transmitted data element.

operation. The transition between these states may be determined by the value of the *confidence counter* (*C*). The state transition diagram for DSS synchronization is shown in Figure 4.19.

A. Acquisition of Scrambler Synchronization

Cell delineation is determined using the last six bits of the HEC field only. The conveyed sample bits are extracted by modulo-2 addition of the predicted values for HEC_8 and HEC_7 from the received value. The descrambler generates its own samples of SRG state in the same manner and compares it to the delivered ones. Scrambler synchronization may, for example, be achieved by applying delivered samples at half-cell intervals to a recursive descrambler in the same interval they were extracted from the source PRBS. The second sample s_{t+1} (derived from HEC_7) is stored for 211 bits before it is used. The resultant descrambler configuration is shown in Figure 4.20. Synchronization principles of the scrambler and descrambler SRGs in relation to this figure are given later in this subsection.

The confidence counter, which initially is set to 0 is incremented by one

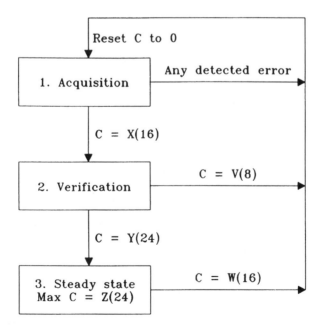

Figure 4.19 State transition of DSS.

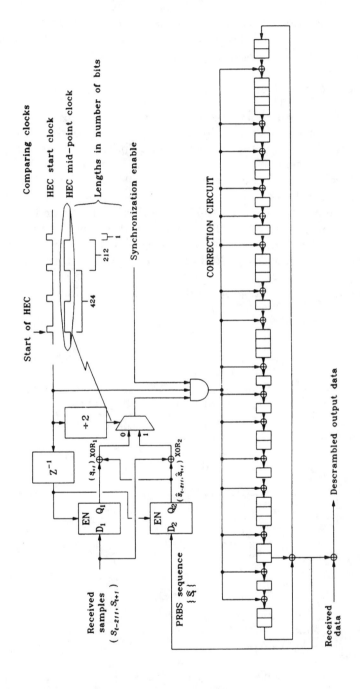

Figure 4.20 Descrambler of DSS.

for every cell received correctly, with no errors detected in the HEC bits 1 to 6. Any error detected in the cell header results in a return to the initial value. Transition to the verification state occurs when the counter reaches X (proposed value = 16).

B. Verification of Scrambler Synchronization

Verification is needed because undetectable errors in the conveyed bits may have occurred during the acquisition phase. The verification state differs from the acquisition in that the descrambler SRG is no longer modified with conveyed scrambler SRG sequence samples.

For every cell received without detected errors, the two conveyed samples are compared to their predicted values. For each cell with two correct predictions received, the confidence counter is incremented. If one or two incorrect predictions are made, then the counter is decremented. If the counter falls below V (proposed value = 8), the system returns to the acquisition initial state and the confidence counter is reset. The transition to the steady state occurs when the counter reaches Y (proposed value = 24).

C. Steady-State Operation (Synchronized Scrambler)

Steady state means that the scrambler and descrambler are completely synchronized. In this state, HEC_8 and HEC_7 bits can both be returned to normal use following their descrambling. Properties of error detection and correction are not affected by this process.

The rules for incrementing and decrementing the confidence counter are the same as for verification state. If the counter drops below W (proposed value = 16), it triggers an automatic return to the acquisition state. The confidence counter has an upper limit of Z (proposed value = 24).

Synchronization Principles of DSS

For a DSS scrambler and descrambler pair, we denote with $\mathbf{d}_k$ and $\hat{\mathbf{d}}_k$ the state vectors of the scrambler and descrambler SRGs, respectively. The DSS scrambler is synchronized to the scrambler if $\mathbf{d}_k = \hat{\mathbf{d}}_k$ for all k.

To synchronize the descrambler SRG whose SR size is L (= 31) to the scrambler SRG state, L or more samples should be transmitted. So, in an efficient synchronization process, we need L times of sampling in the scrambler and L times of correction in the descrambler.

The expression $r + \alpha_i$, $i = 0, 1, \ldots, L - 1$, denotes the sampling time of the i'th sample z_i (or $\hat{z}_i$), and $r + \beta_i$, $i = 0, 1, \ldots, L - 1$, denotes the correction time using the i'th sample, where r indicates a reference time. This is depicted in Figure 4.21 (note that $\alpha_0 = 0$). Note that the ith correction is done after the i'th sampling, but no later than the $i + 1$'th sampling; that is,

$$r < r + \beta_0 \le r + \alpha_1 < r + \beta_1 \le \dots \le r + \alpha_{L-1} < r + \beta_{L-1}.$$

For synchronization of the descrambler, the state of the descrambler SRG is repeatedly corrected until the descrambler SRG sequence $\{\hat{s}_k\}$ becomes identical to the scrambler SRG sequence $\{s_k\}$. For each correction time, the transmitted sample $\hat{z}_i$ is compared to its descrambler counterpart $\hat{z}_i$, and a correction is made to the state of the descrambler SRG in case the two samples are not identical. This process is equivalent to correcting the state vector by adding a correction vector $\mathbf{c}_i$, $i = 0, 1, \dots, L - 1$. Therefore, the correction process can be carried out in the following manner. First, add the two samples z_i and $\hat{z}_i$ (modulo-2 addition), then multiply $\mathbf{c}_i$ to the sum, and finally add the result at the time $r + \beta_i$ to the state vector $\mathbf{d}_{r+i}$. This is depicted in Figure 4.21 also.

We now state the conditions on sampling times, correction times, and correction vectors that provide appropriate guidelines on choosing the terms α_i, β_i, and $\mathbf{c}_i$ to synchronize the descrambler SRG with the scrambler SRG [Kim 94a].

First, the sampling times α_i's must be chosen such that the discrimination matrix

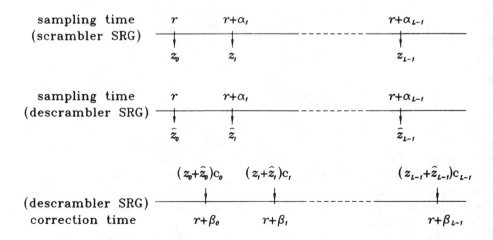

Figure 4.21 Timing diagram for sampling and correction times.

$$D = \begin{bmatrix} \mathbf{h}^t \\ \mathbf{h}^t\mathbf{T}^{a_1} \\ \mathbf{h}^t\mathbf{T}^{a_2} \\ \vdots \\ \mathbf{h}^t\mathbf{T}^{a_{L-1}} \end{bmatrix} \tag{4.1}$$

becomes nonsingular, where $\mathbf{T}$ and $\mathbf{h}$ are, respectively, the state transition matrix and the generating vectors for the SRG.[17] Otherwise, the descrambler SRG states cannot be synchronized with the scrambler SRG states for any choice of correction times and vectors.

Second, if the sampling times a_i's are chosen such that the discrimination matrix $\mathbf{D}$ in (4.1) becomes nonsingular, then the descrambler SRG states can be synchronized with the scrambler SRG states for the correction vector $\mathbf{c}_i$, chosen such that

$$\mathbf{c}_i = \begin{cases} \mathbf{T}^{\beta_i}\Delta^{-1}\left(\mathbf{e}_i + \sum_{j=i+1}^{L-1} u_{i,j}\mathbf{e}_j\right), & i = 0, 1, \cdots, L-2, \\ \mathbf{T}^{\beta_{L-1}}\Delta^{-1}\cdot\mathbf{e}_{L-1}, & i = L-1, \end{cases} \tag{4.2}$$

for the arbitrarily chosen correction time β_i. In the equation, the $L-1$ vector $\mathbf{e}_i$, $i = 0, 1, \ldots L-1$, is the basis vector whose i'th element is 1 and the others are 0; and $u_{i,j}$ is either 0 or 1 for $i = 0, 1, \ldots, L-2$ and $j = i+1, i+2, \ldots, L-1$. Therefore, the sampling times a_i should be chosen such that the resulting discrimination matrix $\mathbf{D}$ in (4.1) becomes nonsingular. Once this is done, we can arbitrarily select correction times β_i and select correction vectors $\mathbf{c}_i$ according to the expression in (4.2).

We apply the two conditions to the DSS for the cell-based ATM transmission in the BISDN. Then the sampling times are $a_i = 212i$, $i = 0, 1, \ldots, 30$, and so the resulting discrimination matrix $\mathbf{D}$ in (4.1) becomes nonsingular. Therefore, if we choose the correction times $\beta_i = 212 + 212i$, we can obtain by (4.2) the common correction vector

---

17. The state transition matrix $\mathbf{T}$ is a matrix representing the relation between the state vectors $\mathbf{d}_k$ and $\mathbf{d}_{k+1}$, and the generating vector $\mathbf{h}$ is a vector representing the relation between the state vector $\mathbf{d}_k$ and the SRG sequence element $\mathbf{s}_k$.

$$c_0 = c_1 = \dots = c_{30} = [0110100110111001100111011000100]^t. \quad (4.3)$$

This provides the descrambler circuit depicted in Figure 4.20.

4.6   ATM LAYER OF BISDN

The ATM layer possesses the processing capability associated with all of the fields of the ATM cell header except the HEC. That is, it performs GFC-related flow control, VPI/VCI-related ATM connection control, and other PT- and CLP-related processing functions. In addition, there are traffic-related ATM layer functions such as network resource management, connection admission control, usage/network parameter control, priority control, and congestion control.

In this section, the BISDN ATM layer-related functions are examined. The related traffic control is handled separately in Section 4.9, and the ATM switching is discussed in detail in Section 4.10.

4.6.1   ATM Layer Connection

A transparent connection provided by the ATM layer to the higher layer is called the *ATM connection*; it is connected end to end through a concatenation of *connection elements* (CE). A CE consists of VP switching and control function, VC and VP switching and control capabilities and connection links. There are two kinds of ATM connections: *VC connection* (VCC) and *VP connection* (VPC). ATM VCC and VPC are illustrated together with physical layer path and sections in Figure 4.22. VC refers to a logical unidirectional connection between two end points for the transfer of ATM cells, and VP implies a logical combination of VCs.

Each VC is assigned a VCI and each VP is assigned a VPI. Inside a VPC, VC links that are different from one another can exist, and each is differentiated through the use of VCI. On the other hand, VCs belonging to different VPs may possess the same VCI. Hence, a VC can be completely identified solely on the basis of its corresponding VCI and VPI.

When switching occurs in a VCC, the value of the VCI is not identically maintained at both ends. Also, when the VP link terminates via a cross-connect equipment, a concentrator, or a switching equipment, the value of VPI can also change accordingly. However, VCI only changes when the VC link terminates, thus, within the same VPC, the same VCI value is maintained. These relationships are illustrated in Figure 4.23. In Figure 4.23(a), S_{VPI} and S_{VCI} represent VP switching and VC switching, respectively, and it can be seen that, in VP switching, the VCI is maintained as it is, and, in VC switching, both sides of VCI and VPI are altered. The corresponding ATM connection formats for the VC and VP are shown in Figure 4.23(b).

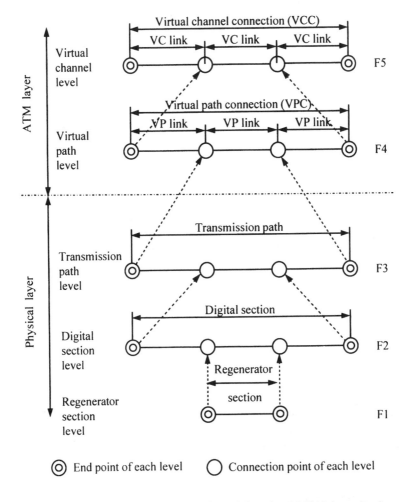

Figure 4.22 Connections in BISDN layer networks and the related OAM information flows.

VC switching and VC/VP switching functions are illustrated in Figures 4.24(a,b). In the figure, VP switching corresponds to the add/drop or cross-connect, and VC/VP switching corresponds to normal switching function.

Twenty-four bits are allocated for VPI/VCI at the UNI, and 28 bits at the NNI. The actual number of bits for the VPI/VCI field used for routing at the UNI is determined through negotiations between the user and the network. Here, the value chosen is the lowest of the values demanded by the user or the network. The allocated VPI values must be consecutive, the values must be selected starting with the least significant bit, and the unused VPI bits should remain at 0. These rules also apply to the VCI. For the indication of meta-signaling VC,

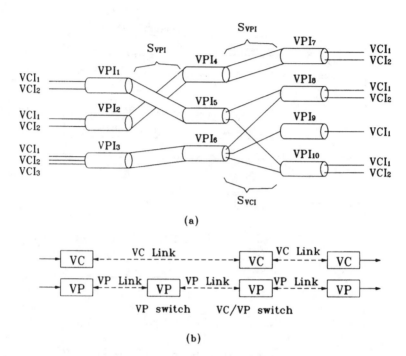

Figure 4.23   ATM layer connection: (a) VPI and VCI assignments and (b) VP and VC connections.

general broadcast signaling VC, point-to-point signaling VC, and OAM VC, the fixed VP and VC identifiers are preassigned as described in Section 4.6.5. The OAM information flows pertaining to the ATM layer are F4 and F5 (see Figure 4.22).

Virtual Channel Connection

VCC refers to a concatenation of VC links for achieving connection between ATM service access points (see Figure 4.22). Here, the term *VC link* implies the unidirectional virtual connection for enabling the transport of ATM cells between points where VCI is assigned and the points at which the VCI gets translated or removed.

A VCC can be provided by the switching equipment, and can be permanent or semipermanent. The integrity of the cell sequence is ensured within the same VCC. A VCC user is provided by the network with a set of QOS parameters such as cell loss and cell delay. At time of VCC setup, user traffic parameters are prescribed through negotiation between the user and the network, and the network monitors the observance of these parameters.

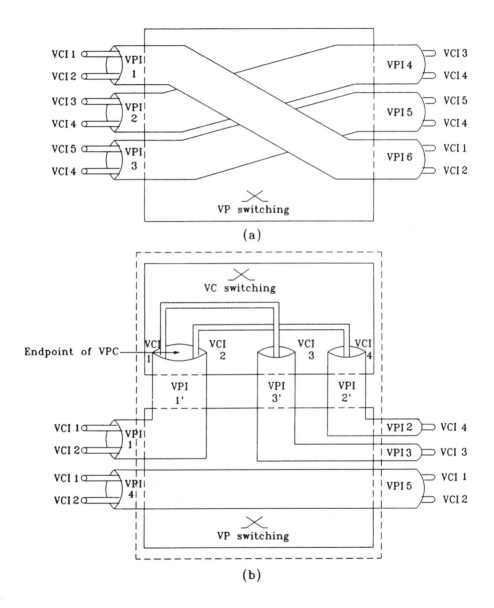

Figure 4.24   VP and VC switching: (a) VP switching and (b) VC/VP switching.

At the UNI, four different methods can be used for establishing or removing VCC. First, the signaling procedure can be bypassed if connection setup or release is achieved through a reservation. This method applies to both permanent and semipermanent connections. The second is the use of a meta-signaling procedure (signaling procedures are described in Section 4.8). That is, a signaling VC is established or removed through the use of a meta-signaling VC. Third is the use of a user-to-network signaling procedure. This involves the use of a signaling VCC to establish or release a VCC for end-to-end communication. The fourth is the use of a user-to-user signaling procedure. This method employs a signaling VCC to establish or release a VCC internal to a VPC preestablished between two UNIs. Accordingly, four methods of assigning VCI values at the UNI are possible: assignment by the network, assignment by the user, assignment through a network-user negotiation, and the use of a standardized method. In general, the assigned VCI value itself is unrelated to the service provided through the corresponding VC. For conveniences such as terminal interchangeability or initialization, it is desirable to assign the same VCI value for some specific set of functions. For example, if a fixed meta-signaling VCI value is used at all UNIs, the initialization of the terminal equipment becomes simple.

Because the cell header is processed at such network elements as the ATM switch, cross-connect equipment, and concentrator, the VCI or VPI translation process is performed at these nodes also. Therefore, when a VCC is established or released in the ATM network, the establishment or release of a VC link can occur at more than one NNI. Here, a VC link is established or released by way of internal signaling or an internetwork signaling procedure at ATM network elements.

Virtual Path Connection

Virtual path connection refers to a concatenation of VP links for connecting the points at which a VPI is assigned with those at which a VPI is translated or removed (see Figure 4.22). Here the term *VP link* implies the VC link groups that join a VPI assignment point with the corresponding setup/removal point.

The VPC can be provided through switching equipment and can be permanent or semipermanent. Cell sequence is ensured for each VCC within the same VPC. QOS parameters such as cell loss rate and cell delay variation are provided for each VPC. Here, VPC QOS must be able to guarantee the best among the VCC service qualities maintained within the VPC. At the time of establishing VPC, the user traffic parameters are determined through a network-user negotiation, and the network monitors the observance of these parameters.

A VPC between VPC end points can be established or released in two possible ways. The first is to establish or release VPC without going through a signaling procedure. In that case, setup or release of the connection is achieved by way of a reservation. Secondly, the VPC is established or removed if the need arises for such purposes as user control and network control.

4.6.2 Generic Flow Control

The GFC field is used to alleviate short-term overload conditions that may occur at the UNI by controlling the flow of traffic submitted to the network by users. The GFC protocol must be able to ensure that the agreed minimum bandwidth capacity is available to each user and that any spare capacity is shared fairly among VBR services. Also, GFC protocol should be able to support all possible configurations such as star, bus, starred-bus, and ring. The GFC protocol should also be insensitive to aggregate traffic, number of terminals, or distance between terminals.

Several examples of protocols can potentially meet these requirements, for instance[18]:

- GFC protocol using counter reset timing;
- GFC protocol using wait-for-transmit queues;
- Modified DQDB protocol using empty slot counter;
- Modified DQDB protocol using priorities;
- ATMR-based GFC protocol.

Among these, the ATMR-based GFC protocol is singled out for more detailed discussion in the following subsection.

ATMR-Based GFC Protocol

Though the original ATMR protocol was developed for constructing a high-speed shared-medium network, it can be easily extended for use as a GFC protocol. The ATMR is based on the slotted-ring architecture, with one slot corresponding to one ATM cell. The ATMR provides such basic functions as *access control* (AC) guarantee of the required bandwidth using *connection admission control* (CAC) based on the AC-Window scheme, distributed monitoring of the TE transmission state using busy address, fair distribution of remaining bandwidth using the AC-Reset scheme, and the prevention of excessive use of cells using UPC-Window.

To guarantee the required bandwidth, the CAC procedure based on the

---

18. These are the GFC protocols proposed to ITU-T for relevant standardization in the 1990s. The proponents are England, the United States, Switzerland, Australia, and Japan, in that order. Although active discussions followed, no specific proposal was selected as a standard. Instead, the *uncontrolled mode*, which does not use the GFC function with the GFC field set to the "all 0" state, was defined.

AC-Window is applied. This procedure relies on the determination of the reference duration T_p for defining the bandwidth of a connection and the calculation of the AC-Window value corresponding to the required bandwidth based on this T_p. Each TE has an AC-Window counter (AC-CTR-W), the initial value of which is set equal to the AC-Window. This indicates the maximum number of cells allowed to be transmitted by the TE during a time interval equal to the T_p. The AC-CTR-W is decremented according to the number of cells transmitted from the TE. When AC-CTR-W becomes zero, the TE stops transmitting cells. The AC-CTR-W is reset to the initial value on receiving the AC-Reset cell. The AC-Reset cell is issued by the TE that has found that all other TEs have completed or stopped cell transmission for that T_p.

The issuing of the AC-Reset cell requires monitoring of the transmission status of other TEs by each and every TE. This is achieved by having every active (or busy) TE (it is active if it has cells to transmit and its AC-CTR-W is not equal to zero) overwrite its own TE *identifier* (TEID) onto the GFC field of each incoming cells. (Such a TE is called *specific address* TE, and specific address refers to the value written in the GFC field. A common address is written in the GFC field in the case of *nonspecific address* TE.) If a TE finds its own TEID in the GFC field, it means that all other TEs are inactive. If a different TEID is found, it means that another TE is still active (i.e., busy). Figure 4.25 illustrates this distributed monitoring scheme.

UPC-Window and UPC-Flag are used to prevent the excessive use of cells by malicious users and to control *cell delay variation* (CDV), from which different services may suffer.. The procedures concerning UPC-Window and UPC-Flag are similar to AC-Window and AC-Reset. That is, both counters are initially set to some initial value and decremented as cells are transmitted, with the condition that the TE can transmit cells only if AC-Window and UPC-Window are not zero. Multiple UPC-Flags may be used to control various long-term and short-term cell usage parameters. Figure 4.26 shows the use of short-term and long-term UPC-Windows and flags. All the flags are issued at the *local exchange* (LEX) periodically. UPC-Window and UPC-Flag, which were not considered for the original ATMR, are employed to alleviate the CDV problem.

4.6.3 Payload-Type Indication Function

The payload type field indicates whether the contents of payload consist of user information or network information, and additionally provides indications of network congestion experience and ATM layer user-to-ATM layer user indication. Among the three bits in the PT field, the first bit is used to indicate whether it is a user information (0) or network information (1). In the case of user information, the second bit is an *explicit forward congestion indication* (EFCI) bit, which indicates whether the relevant cell experienced congestion (1) or not (0) while traversing the network along the specific connection; and

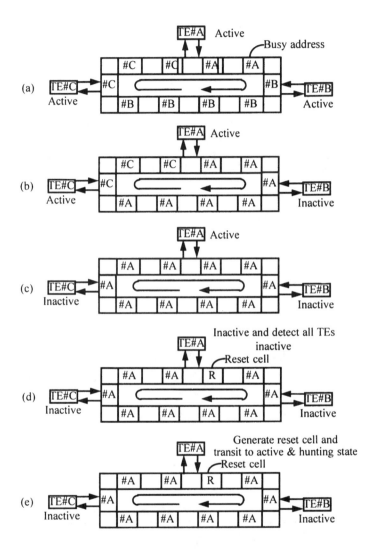

Figure 4.25 Distributed monitor scheme of ATMR: (a) TE#A: Active, TE#B: Active, TE#C: Active; (b) TE#A: Active, TE#B: Inactive, TE#C: Active; (c) TE#A: Active, TE#B: Inactive, TE#C: Inactive; (d) TE#A: Inactive, TE#B: Inactive, TE#C: Inactive; and (e) TE#A: (Generate reset cell) → Active (and hunting state), TE#B: Inactive, TE#C: Inactive.

the third bit is an AAL-indicate bit, which indicates whether the information belongs to the tail part of an AAL-5 PDU (1) or not (0) (see Section 4.7.6). In the case of net work information, the PT field indicates whether the relevant cell is an OAM flow F5 cell or *resource management* (RM) cell. A summary of this is listed in Table 4.10.

4.6.4 Cell Loss Priority Function

Since the VBR services can vary widely in bit rates, at the moment when the various VBR services all manifest their maximum possible bit rates, the network can be severely congested. As a means of resolving such traffic congestion, the CLP function can be used. That is, priority level to be used for cell loss (or cell discard) is recorded in the CLP field of each ATM cell employed for VBR

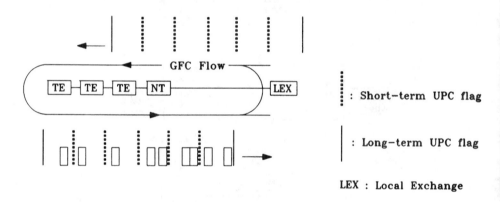

Figure 4.26 Example of UPC-flag usage.

Table 4.10
Usage of PT Field

PT field	Indication
000	User cell, EFCI = 0, AAL-indicate = 0
001	User cell, EFCI = 0, AAL-indicate = 1
010	User cell, EFCI = 1, AAL-indicate = 0
011	User cell, EFCI = 1, AAL-indicate = 1
100	OAM F5 segment associated cell
101	OAM F5 end-to-end associated cell
110	Resource management cell
111	Reserved for future use

services, and when congestion occurs, the cells with lower priority are discarded first. If the CLP bit indicates 1, then it represents a cell with a lower priority that can be abandoned.

The cell loss priority function must be provided in conjunction with the QOSs determined at the time of establishing VPC/VCC. That is, it must be possible to provide the minimum guaranteed bit rate even after the cell loss processing, and the prescribed service quality must be maintained. Consequently, the network must determine the bit rate of the cells with the higher priority at the time of establishing the connection, and the rate must be negotiable even after the connection is completed. The network must constantly monitor via usage parameter control whether the number of cells corresponding to a given connection exceeds the prearranged value. When the cell traffic exceeds the negotiated level, even the cells that have been preassigned with higher priority can be ignored.

4.6.5 Preassigned Cell Headers

Cells reserved for physical layer use have preassigned values in the whole header. Here, the ATM cells used by the physical layer include idle cells and physical layer OAM cells (see Table 4.6).

ATM cells in the ATM layer with preassigned cell headers include unassigned cells, signaling cells, OAM F4 and F5 cells, and resource management cells. Signaling cells consist of meta-signaling cells, general broadcast signaling cells, and point-to-point signaling cells, with VCIs of all 0 but with the last three digits being 001, 010, and 101 preassigned to them, respectively. For the OAM flow F4 cells, the VCIs of all 0 but the last three digits 011 and 100 are preassigned, respectively, to segment OAM F4 cell and end-to-end OAM F4 cell. For the OAM flow F5 cells, the PTIs 100 and 101 are preassigned respectively to segment OAM F5 cell and end-to-end OAM F5 cell. A summary of such cell header preassignment is listed in Table 4.11.

4.7 ATM ADAPTATION LAYER OF BISDN

The AAL of the BISDN is a layer positioned between the ATM layer and the higher user service layer, and its main function is to resolve any disparity between the service provided by the ATM layer and the service demanded by the user. For that purpose, the AAL adapts user service information with the ATM cell format and performs handling of transmission errors, misinserted or lost cells, and errored cells. It also provides flow control function to meet the QOSs demanded by the user, and timing control function to restore the user signal.

The user information field of the ATM cell is fixed at 48 bytes, while the user service information to be adapted to this space is extremely diverse in character. Under such a constraint, the AAL performs various functions, such

Table 4.11
ATM Layer Cells with Preassigned Cell Header

ATM Cells		VPI	VCI	PT	CLP
Unassigned cell		00000000	00000000 00000000	XXX	0
Metasignaling cell		X	00000000 00000001	0A0	C
General broadcast signalling cell		X	00000000 00000010	0AA	C
Point-to-point signaling cell		X	00000000 00000101	0AA	C
OAM F4 flow cell	Segment	Y	00000000 00000011	0A0	A
	End-to-end	Y	00000000 00000100	0A0	A
OAM F4 flow cell	Segment	Y	Z	100	A
	End-to-end	Y	Z	101	A
Fast resource management cell		Y	Z	110	A

Notes: A: Bit to be filled in the ATM layer; C: Bit to be filled by the signaling entity; X: "Don't care"; Y: Any VPI value; Z: Any nonzero VCI.

as ATM cell adaptation, transmission error processing, lost cell and inserted cell processing, flow control, and timing information control. Consequently, it is necessary to consider the following issues.

First, it is necessary to restrict the number of possible protocol types to a minimum. Also, for efficient implementation of these protocols, it is desirable to simplify their structures (as a possible expedient, it is better to construct AAL PDUs or the overheads in byte units as much as possible). Also, it is a fundamental necessity that particular AAL *service data units* (SDU) used for packet-oriented services should have no influence on the design of the AAL.

A flow control function is needed for guaranteeing the user's QOS requirements. The user SDUs should be delivered in a specified time interval according to the user's needs. Also, supporting a multiplexing capability at the AAL should be a simple means of managing a diverse set of services. Furthermore, error detection and correction capability for processing transmission errors must exist, and a means of processing lost cells and inserted cells must be provided as well. The capability to deliver and recover timing information for real-time services must also be provided.

In this section, BISDN service classification is considered from the network's point of view and extended for the AAL classification. Then each type of AAL is examined in the subsequent four subsections, with an example of each. Finally, source clock frequency recovery is discussed in detail.

4.7.1 Classification of BISDN Services

From the communications network's viewpoint, the services can be classified according to the bit rate (constant or variable), timing relations (real time or non real time) or channel connection mode (CO or CL).

Constant-Bit-Rate Services and Variable-Bit-Rate Services

Examined from the bit rate point of view, BISDN services can be broadly categorized into those whose bit rates are constant and those with variable bit rates. The former are CBR services, and the latter are VBR services.

The most representative example of a CBR service is the 64-Kbps PCM voice signal. Video signals or data signals can also be provided in the CBR service format. But since data signals generally manifest VBR characteristics, it is more natural to provide them as a VBR service. On the other hand, both voice and video signals can also be furnished as VBR services.

The bit rate of CBR service is determined through a negotiation between the user and the network, and the same bit rate is maintained as long as the service is continued. In the case of VBR services, the bit rate does vary during the service period. Since excessive variation can obstruct the operation of the network, the network must be notified in advance of the VBR service's characteristic parameters at the time of call setup.

Real-Time Services and Non-Real-Time Services

BISDN services can be categorized into real-time and non-real-time services, depending on the timing relation between the source and the destination. Real-time services can deteriorate in quality or become useless if the associated information transfer becomes delayed; hence, they are sensitive to the time it takes for the unit information entities (ATM cells) to be transferred. On the other hand, the quality of non-real-time services is insensitive to delays in information transfer. Examples of real-time services include video telephony and video conferencing, and non-real-time services are represented by data transmission.

Connection-Oriented Services and Connectionless Services

From the standpoint of the channels within the communications network, services in general can be divided into *connection-oriented* (CO) services and *connectionless* (CL) services. The existing circuit-mode services are all CO services. Among the packet-mode services, those that employ VCs are CO, and those that use the datagram scheme are CL.

Service Classification According to Characteristics

From the viewpoint of processing service information, it is convenient to classify services according to their characteristics. That is, rather than using such criteria as whether the service in question consists mainly of images or sound, it is more useful to consider such service attributes as the character of the bit rate, timing relations, and the channel connection mode, as far as the information processing within the communications network is concerned.

Table 4.12
BISDN Service Classification

Service Class	Timing Relation Between Source and Destination	Bit Rate	Connection Mode	Examples of Services	Relevant AAL
Class A	Required	Constant	Connection oriented	Constant-bit-rate video, DS-1 circuit emulation	AAL-1
Class B	Required	Variable	Connection oriented	Variable-bit-rate video, packet video	AAL-2
Class C	Not required	Variable	Connection oriented	Connection-oriented data transfer, FR, X.25	AAL-3/4, 5
Class D	Not required	Variable	Connectionless	Connectionless data transfer, SMDS, IP	AAL-3/4, 5

Accordingly, in the ATM network, service classification according to characteristics of the services may be more appropriate for the AAL.

According to service attributes, BISDN services are divided into class A, B, C, and D services. Class A services are real time, CBR, and CO services, an instance of which is the constant-rate video signals. The service attributes for other service classes and the respective examples are as listed in Table 4.12.[19]

4.7.2 AAL Classification

Horizontal Classification

To group various user services in an effective manner while keeping the preceding factors in mind, the AAL can be categorized into four types. This involves grouping BISDN services into four classes according to their attributes and dividing the AAL correspondingly. That is, as examined earlier, the services can be categorized into four classes from A to D, depending on the nature of the bit rate, the time-related characteristic, and the connection mode, and in a similar manner the AAL can be divided into four types: AAL-1, AAL-2, AAL-3/4, and AAL-5, as is indicated in Table 4.12.[20]

---

19. This classification is defined in ITU-T Recommendation I.362. Refer to Chapters 5 and 6 for more detailed descriptions of each service examples listed in Table 4.12.
20. In the early stage of ITU-T standardization, four types of AALs—AAL-1, AAL-2, AAL-3, and AAL-4—were defined to match the four service classes of A, B, C, and D, one to one. However, AAL-3 and AAL-4 were later merged into AAL-3/4 due to their similarity, and AAL-5 was newly defined to enable high-speed data transfer.

To elaborate, AAL-1 provides AAL functions to class A services with CBR and real-time characteristics. That is, it provides the capability for delivering CBR SDUs using an identical bit rate, transferring timing information from the information source to its destination, and indicating recoverable and unrecoverable errors. Similarly, AAL-2 provides functions that are suitable for the class B services; AAL-3/4, for class C and D services; and AAL-5, also for class C and D services, but with a simple and efficient protocol. A summary of the most representative functions of AAL-1 through AAL-5 is given in Table 4.13.

Vertical Classification

On the other hand, the AAL can also be divided vertically, into SAR sublayer and CS. That is, the process of converting *user-service data units* (U-SDU) into ATM cells executed by the AAL is divided into two sublayers. The SAR provides the functions associated with the segmentation and reassembly of U-SDUs, and the CS provides the capability for converging specified service-related functions to an upper service layer.

In the direction of transmission, the CS accepts U-SDUs from an upper user layer, to which it adds a header and trailer related to error handling and data priority preservation to create the SAR-PDUs, which are then sent to the ATM layer. In the direction of reception, the SAR sublayer analyzes the SAR-PDUs transported from the ATM layer, and the SAR-PDUs are collected and assembled together with the CS-PDUs and delivered to the CS. Then the CS analyzes the header and trailer of the transported CS-PDUs and extracts the

Table 4.13
Major Functions of AAL-1 Through AAL-5

AAL Type	Major Functions
AAL-1	Transfer of constant-bit-rate SDU with the same bit rate Transfer of timing information between source and destination Error recovery and indication of errored information that is not recovered by AAL-1
AAL-2	Transfer of SDU with variable bit rate Transfer of timing information between source and destination Error recovery and indication of errored information that is not recovered by AAL-2
AAL-3/4	Transfer of class C and D service SDUs from AAL-SAP to AAL-SAP(s) Transfer by CO or CL mode Multiplexing function in the AAL using the MID
AAL-5	Function of AAL-3/4 but with a simple and fast protocol No SAR-PDU header and trailer, no CPCS-PDU header Suitable for high-speed data transfer

original U-SDUs, finally delivering them to the upper user layer. Interentity protocols such as flow control are also handled at the CS. The handling procedures are conceptually illustrated in Figure 4.27. Such vertical classification applies equally to all four AAL types, except that the CS is subdivided into the *service-specific CS* (SSCS) and the *common-part CS* (CPCS) in the cases of AAL-3/4 and AAL 5.

4.7.3 AAL-1 Functions

The delivery of constant-rate U-SDUs, along with the associated timing information using a common bit rate, and the indication of uncorrectable errors are some of the services that AAL-1 provides to the upper layers. AAL-1 provides a function for partitioning and reassembling user information. It also provides a function for handling cell delay variations and lost and inserted cells, and enables the receiver to extract timing information from the information source.

When transferring CBR data, the timing information is delivered by the *synchronous residual time stamps* (SRTS). This arrangement enables the provision of circuit emulation services for the DS-1 or DS-3 signals. For the

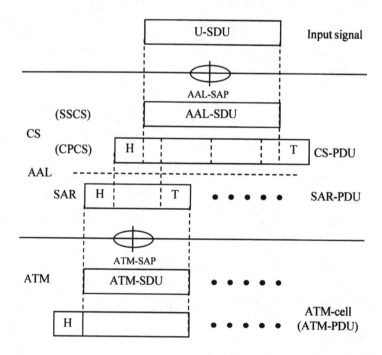

Figure 4.27 Processing of data at AAL sublayers.

octet-structured CBR signals such as $n \times$ DS0 ($n \leq 92$), circuit emulation service becomes possible by employing the *structured data transfer* (SDT) scheme. For error detection and correction, the Reed-Solomon (RS) code is used.

SAR Sublayer

The function of AAL-1's SAR sublayer is to segment the CS-PDUs and then add a header to form the SAR-PDUs and send them to the ATM layer. Also, through a reverse process it reassembles the SAR-PDUs to recover the CS-PDU. An SAR-PDU formed at the SAR sublayer is shown in Figure 4.28. The number of bits assigned to *sequence number* (SN) and *sequence number protection* (SNP) is four each; consequently, the size of SAR-PDU payload space becomes 47 bytes. SN is used for inspecting whether a cell loss or cell insertion has occurred, and the SNP is used for error correction in order to protect SN from errors.

More specifically, the *convergence sublayer indication* (CSI) bit is used for special purposes such as indicating the presence of the CS functions; the *sequence count* (SC) bits indicate a serial SAR-PDU count in modulo-8; the *cyclic redundancy check* (CRC) bits represent the CRC-3 code for the four bits in the SN field; and the *parity* (P) bit denotes the parity check for the preceding seven bits in the SAR-PDU header. The CRC and P bits determine whether or not the SN bits are valid, and this SN checking process is accomplished through the state transition mechanism shown in Figure 4.29, which consists of the two states: correction mode and detection mode.

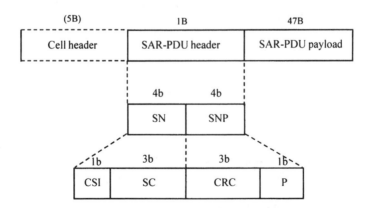

SN: Sequence Number   SC: Sequence Count
SNP: Sequence Number Protection   CRC: Cyclic Redundancy Check
CSI: Convergence Sublayer Indication P: Parity

Figure 4.28 SAR-PDU format for AAL-1.

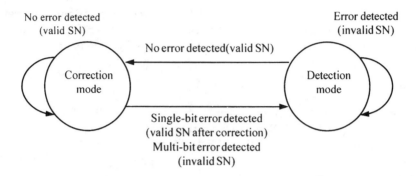

Figure 4.29   SN checking process in the AAL-1 SAR sublayer.

Convergence Sublayer

AAL-1 CS reconstructs the original CBR data stream by eliminating cell jitter through buffering, and by properly handling the lost and misinserted cells through a sequence number checking process. When using SRTS, it inserts and recovers timing information for source clock recovery; and when using SDT, it transfers information on the user data structure. By employing the RS code-based *forward error correction* (FEC), it can also monitor and improve the error status of the end-to-end virtual channel.

Through the SN checking process, AAL-1 CS can determine whether a SAR-PDU is in normal state, or lost, or misinserted. For the lost SAR-PDUs, it can determine the position and the number.

For timing clock recovery, AAL-1 CS can employ the SRTS scheme. If the jitter and wander requirement is not stringent, it can also adopt the adaptive clock recovery method (refer to Section 4.7.7 for a detailed discussion of timing recovery).

When employing the SDT scheme, AAL-1 CS indicates the boundary of user information, for example, the starting point of the $n \times$ DS0 data block, using a one-byte pointer, positioning the pointer at the first byte of the SAR-PDU payload space. In this case, the pointer byte represents, using its latter seven bits, the offset of the data block boundary from the pointer in terms of bytes. A SAR-PDU is called *P-formatted* if it contains such a pointer; and is called *non-P-formatted* otherwise. A SAR-PDU is allowed to be P-formatted only when the corresponding sequence number is 0, 2, 4, or 6, and the P-format is allowed only once in each cycle of eight SAR-PDUs. If there is no boundary information to send during one cycle, the SAR-PDU whose SN is 6 is forced to be a null P-format, having all 1 pointer bytes. The CSI bit is set to 1 for the SAR-PDUs in P-format or null P-format.

When using the RS code, AAL-1 CS provides two different coding schemes depending on whether the data are loss sensitive or delay sensitive. In the loss-sensitive case, it employs the long interleaving format of Figure 4.30(a); and in the delay-sensitive case, it uses the short interleaving format of Figure 4.30(b). In Figure 4.30(a), each 124 octets of input data in the 47×124 octet block is appended by a 4-octet RS code in the row direction, and the resulting 47×128 octet block is read in the column direction for transmission. In contrast, in Figure 4.30(b), each 88 octets of input data in the 8×88 data block is appended by a 6-octet RS code in the row direction, and the resulting 8×94 octet block is read in the diagonal direction for transmission. In each case, the RS-coded data block becomes a CS-PDU, and the CSI bit of the SAR-PDU, which contains the head end of this CS-PDU, is set to 1. The generating polynomials for the two cases are $\prod_{i=0}^{3} (x - \alpha^{i + k})$ and $\prod_{i=0}^{5} (x - \alpha^{i + k})$ respectively, where α is a root of the polynomial $x^8 + x^7 + x^2 + x + 1$. The overhead percentage due to the RS code is 3.1% for Figure 4.30(a) and 6.4% for Figure 4.30(b). The error correction capability is 4 cell loss, 2 cell loss and 1 octet error in each column, no cell loss and 2 octet errors in each column for Figure 4.30(a); and 1 cell loss out of 16 cells and 3 octet errors in each column for Figure 4.30(b). The delay caused by byte interleaving and deinterleaving is about 256 cell times for Figure 4.30(a) and about 16 cell times for Figure 4.30(b).

Illustration of AAL-1 Mapping

Figure 4.31 illustrates the AAL-1 multiplexing process for a CBR service signal. The convergence sublayer does not have any particular format, so user information U-SDU is directly mapped to the SAR data format in units of 47 bytes. The SAR-PDU header is 1 byte, and the RTS generated in the CS for timing information delivery is loaded at the CSI bit of this header alternately. Prefixing a 5-byte cell header to the 48-byte SAR-PDU yields the ATM cell.

4.7.4 AAL-2 Functions

AAL-2 delivers real-time U-SDUs in variable bit rates along with the associated timing information, indicates unrecoverable errors, and provides other such services to the upper layers.

AAL-2, similar to AAL-1, provides a function for segmenting and reassembling user information. It also provides the capability for handling cell delay variation, treating lost or misinserted cells, and, at the receiving end, the capability for recovering the information source clock.[21]

---

21. Standardization of AAL-2 has not progressed very far, with the difficulty in efficient timing information delivery being one of the major obstacles. As a means to handle real-time VBR services, two approaches have been considered. One is based on AAL-1 and the other is based on AAL-5. The AAL-1-based approach has been led by ITU-T, and the goal of it is to provide class B services by taking advantage of the functional commonality between AAL-1 and AAL-2. The AAL-5-based approach has been led by the ATM Forum, and its goal is to provide *video-on-demand* (VOD) and other real-time VBR services by exploiting the fast protocol processing capability of AAL-5.

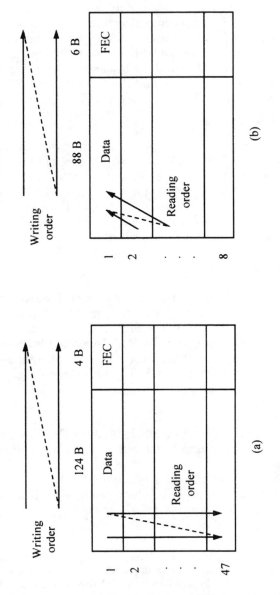

Figure 4.30 RS coding formats for forward error correction: (a) long format and (b) short format.

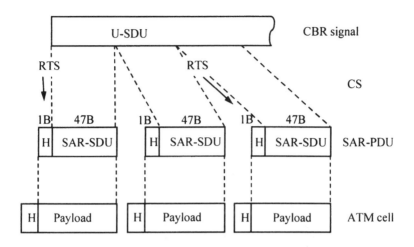

Figure 4.31 Example of AAL-1 multiplexing.

4.7.5 AAL-3/4 Functions

AAL-3/4's function is to establish an adaptation layer connection prior to the transmission, then transport class C and class D service data with VBR characteristics. The services provided at AAL-3/4 can be divided into *message-mode* services and *stream-mode* services. In the message mode, an AAL-SDU passes across the AAL interface in exactly one AAL-IDU (interface data unit), whereas, in the streaming mode, it does so in one or more AAL-IDUs. The two modes are illustrated in Figures 4.32 and 4.33. Here, an internal pipelining function can be applied, and an AAL entity can initiate data transfer to the receiving AAL entity before it has the complete AAL-SDU available.

These two service modes both provide *assured* operation and *nonassured* operation. The assured operation is the operation mode in which all the SDUs are accurately delivered in the order in which they are received from the ATM layer, lost or corrupted cells are retransmitted, and flow control is provided by necessity. The assured operation applies only to point-to-point ATM layer connections. In nonassured operation, lost or corrupted cells are not retransmitted. When the need arises, damaged SDUs are transported to upper layers, and flow control capability is provided for point-to-point ATM connections only.

The services that can be provided based on AAL-3/4 include the connectionless data transfer service (ITU-T), ATM-DXI (ATM Forum), and SMDS (refer to Section 5.6).

SAR Sublayer

The AAL-3/4 SAR sublayer receives variable-length CS-PDUs from the CS and then segments and appends a header and trailer to form the SAR-PDUs, which

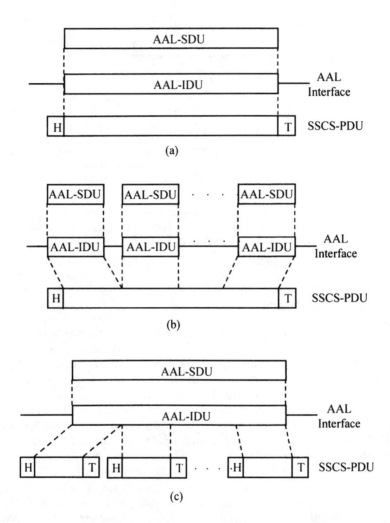

Figure 4.32 Illustration of message mode: (a) when one AAL-SDU forms one SSCS-PDU; (b) when multiple AAL-SDUs form one SSCS-PDU; and (c) when one AAL-SDU gets divided to form multiple SSCS-PDUs.

are then sent to the ATM layer. It can also reassemble SAR-PDUs through a reverse of the process and recover the CS-SDUs.

The structure of the SAR-PDU of AAL-3 is shown in Figure 4.34(a). In the figure, the *segment type* (ST) indicates whether the corresponding payload is BOM (beginning of message), COM (continuation of message), EOM (end of message), or SSM (single segment message), and SN indicates the serial number of each message. The multiplexing identification (MID) field is used when

(a)

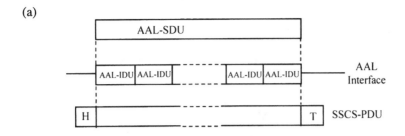

(b)

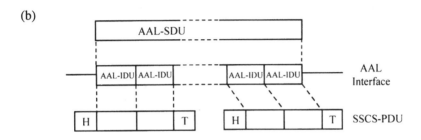

Figure 4.33 Illustration of streaming mode: (a) when one AAL-SDU forms one SSCS-PDU and (b) when one AAL-SDU gets divided to form multiple SSCS-PDUs.

multiple CPCS (see the CS section) connections are multiplexed through one ATM layer connection; length indication (LI) indicates the length of the SAR-PDU payload in octets; and CRC is the CRC code for the entire SAR-PDU including the header.

Convergence Sublayer

The AAL-3/4 CS provides various functions for AAL-3/4 service users, including transparent delivery of AAL-SDUs, mapping between AAL-SAP and ATM layer connections, error detection and treatment (CS-PDU damage detection and appropriate treatment procedure), message segmentation and reassembly, information identification, and buffer allocation. The AAL-3 CS also provides special functions specific to class C and class D AAL-3/4 services.

The CS functions of AAL-3/4 can be rearranged into the *common part CS* (CPCS), which is common to all class C and class D services, and the

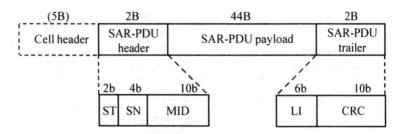

CRC : Cyclic Redundancy Check code
ST : Segment Type   MID : Multiplexing Identification
SN : Sequence Number   LI : Length Indicator

(a)

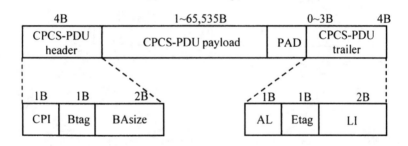

CPI : Common Part Indicator
Btag : Begin tag   AL : Alignment
BAsize : Buffer Allocation size   Etag : End tag
PAD : PAdding   LI : Length Indication

(b)

Figure 4.34   (a) SAR-PDU and (b) CPCS-PDU structures for AAL-3/4.

service-specific CS (SSCS) services, which differ depending on each specific service (see Figure 4.27).[22]

The structure of the CPCS-PDU is as shown in Figure 4.34(b). In the figure, CPI indicates whether the corresponding PDU belongs to a common part, B/ Etag is a tag attached to the header and trailer of a CPCS-PDU so that they are

---

22. The formats in Figures 4.32 and 4.33 are two special examples of SSCS. In general, however, null SSCSs are used in most applications, which merely provides primitive mappings between a higher layer and the CPCS.

identical: BAsize indicates the size of buffer to be allocated in the receiver and PAD is a padding to create a CPCS-PDU payload sized in multiples of four bytes. Finally, LI indicates the length of CPCS-PDU payload, and AL is a filler to make the CPCS-PDU trailer size 32 bits.

Illustration of AAL-3/4 Multiplexing

Figure 4.35 illustrates the AAL-3/4 multiplexing process for two simultaneously arrived VBR data packets of lengths 78 bytes and 121 bytes each. At first, the packet lengths are up-sized to be multiples of 4 bytes by appending 2-byte and 3-byte paddings, respectively. Then a 4-byte header and a 4-byte trailer are attached to each to form 88-byte and 132-byte CS-PDUs, respectively. Next, the CS-PDUs are segmented in units of 44 bytes, and a 2-byte header and a 2-byte trailer are attached to each 44-byte SAR-SDU (or SAR-PDU payload) to make a 48-byte SAR-PDU (or ATM cell payload). Finally a 5-byte cell header is prefixed to each ATM payload to produce an ATM cell. In this process, each CS-PDU is independently treated, and the MID and the segment type (BOM, COM, EOM) fields are marked based on each CS-PDU. The SAR-PDU multiplexing capability associated with the MID field is a unique feature of AAL-3/4 that is not available for AAL-5.

4.7.6 AAL-5 Functions

AAL-3/4, standardized for connection-oriented and connectionless services, is not adequate for high-speed data communication because of its heavy protocol overhead. AAL-5 is a derivative of AAL-3/4, which is endowed with a simplified protocol and is thus capable of high-speed processing. This is the key feature of AAL-5.

AAL-5 protocol has a simple header structure, and this enables simple protocol processing and, consequently, fast and efficient data communication. AAL-5, like AAL-3/4, supports connection-oriented and connectionless VBR data services, and consists of SAR, CPCS, and SSCS sublayers. In reality, however, the SAR sublayer of AAL-5 is a blank layer that has no particular header functions. As a consequence, the SAR-PDU multiplexing capability associated with the MID field is no longer available for AAL-5, and the SNP and CRC functions are not supported. Instead, a CRC-32 function is provided at the CPCS sublayer.

The services that can be provided based on AAL-5 include *frame relay* (FR) of ITU-T, ATM-*data exchange interface* (DXI) of ATM Forum, and *Internet Protocol* (IP) of the ATM Forum and IETF (Internet Engineering Task Force) (refer to Section 5.9). It appears also possible to transfer MPEG-2 coded video data stream over AAL-5 (see Section 6.5).

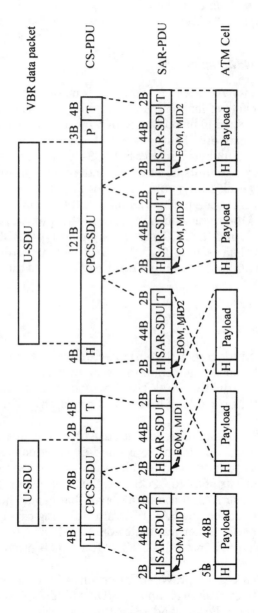

Figure 4.35   Illustration of AAL-3/4 multiplexing.

SAR Sublayer

The AAL-5 sublayer segments the CPCS-PDU in units of 48 bytes to produce SAR-PDUs, and, conversely, reassembles the original CPCS-PDU out of SAR-PDUs. The structure of SAR-PDU is shown in Figure 4.36(a). As can be seen from the figure, the SAR sublayer has no protocol overhead, so the protocol processing becomes very simple. Instead, an indication is put on the ATM header to indicate whether or not the particular ATM cell carries the rear-end portion of a CPCS-PDU. More specifically, the *AAL-indicate* bit in the PT field of the ATM cell header is set to 1 if the corresponding ATM cell contains the rear-end portion, and to 0 otherwise (refer to Section 4.6.3).

Convergence Sublayer

The AAL-5 CS consists of AAL-5 CPCS and AAL-5 SSCS. AAL-5 CPCS provides the mapping function between CPCS-SDU and CPCS-PDU, the error detection

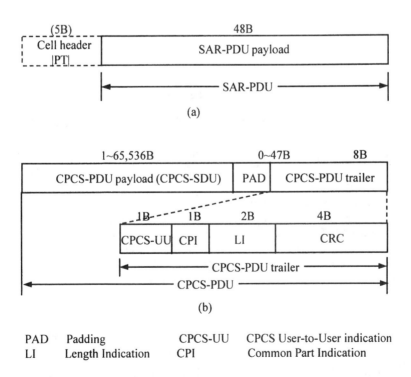

| PAD | Padding | CPCS-UU | CPCS User-to-User indication |
| LI | Length Indication | CPI | Common Part Indication |

Figure 4.36 (a) SAR-PDU and (b) CPCS-PDU structures for AAL-5.

and correction function, and so on. AAL-5 SSCS is used in the form of null SSCS in most cases.[23]

The structure of AAL-5 CPCS-PDU is shown in Figure 4.36(b). It has no header but has a trailer of 8 bytes. PAD is the padding field to fill 0 to 47 null bytes to make the length of CPCS-PDU a multiple of 48 bytes.

The CPCS-UU (user-to-user) in the CPCS trailer conveys user-to-user information transparently and CPI indicates whether or not the corresponding PDU is a common part and can also indicate the unit of the LI count. LI indicates the length of the CPCS-PDU payload, whereas CRC carries the CRC-32 code for the CPCS-PDU (payload, PAD, and the first 4 bytes of the trailer).

Illustration of AAL-5 Multiplexing

Figure 4.37 illustrates the AAL-5 multiplexing process for the two VBR data packets that were considered for the illustration of AAL-3/4 multiplexing in Figure 4.35. The 78-byte data packet is appended by a 10-byte padding and an 8-byte trailer; and the 121-byte data packet is appended by a 15-byte padding and an 8-byte trailer. The resulting lengths of the CS-PDUs are 96 bytes and 144 bytes, respectively. These are segmented in units of 48 bytes to produce SAR-PDUs, which become the ATM cell payloads themselves without further processing. The AAL-indicate bit in the PT field of ATM header is set to 1 only for the ATM cell, which contains the rear-end portion of a CS-PDU, as is illustrated in the figure.

4.7.7 Source Clock Frequency Recovery

One of the important functions of the AAL for real-time services is the source clock frequency recovery at the receiver.[24] Due to the statistical nature of the ATM-based network, the ATM cells arriving at the destination will not be periodic, even in the case of CBR services. If the ATM-based network is asynchronous (i.e., different timing references for the nodes in the network), then the only information on the source clock frequency available at the receiver is the long-term average cell throughput. In this case, PLL can be used to regenerate the bit clock. However, if the ATM-based network is synchronous (i.e., the reference timings of the nodes are traceable to a single timing source), then it is possible to synchronize the transmission and service rate with the network timing. In this case, a synchronization pattern or time stamp can be used to solve the source clock frequency problem.

---

23. In the ITU-T standard, SSCS is defined only for the frame relay service. FR-SSCS performs the multiplex and demultiplex function, FR-SSCS PDU length indication, congestion control, and so on.
24. This section describes only the source clock frequency problems, so the reader may skip it without losing the overall continuity.

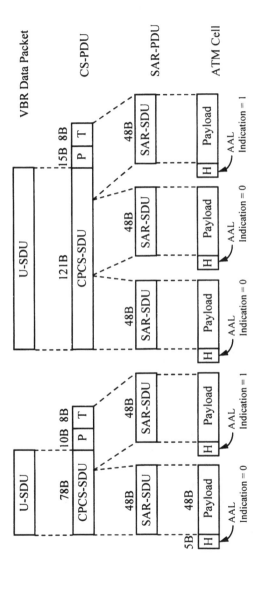

Figure 4.37 Illustration of AAL-5 multiplexing.

In this section, we examine the source clock frequency recovery methods in synchronous ATM-based networks. For this, CBR timing recovery methods are briefly discussed first, followed by a detailed description of the *synchronous residual time stamp* (SRTS) method. Lastly, source frequency clock recovery for the VBR services is briefly considered.

CBR Timing Recovery Methods

For CBR timing recovery, two methods, the *synchronous frequency encoding technique* (SFET) and the *time stamp* (TS), have been introduced, which were later merged into the SRTS method.

The fundamental concept of SFET is that in a synchronous optical network, common clocks that are available at both the transmitter and receiver can be used as the timing reference. The source clock, which is asynchronous with respect to the network timing, is compared to the network clock, and the frequency difference information, together with the common network clock, is transmitted and then used to reconstruct the source clock at the receiver.

The TS method uses the common network clock and a 16-bit TS to convey source clock information. The TS is a 16-bit binary number representing the number of network clock cycles corresponding to a fixed number of service clock cycles. Since a common network clock is available at the receiver, the TS conveys necessary information for the reconstruction of the source clock. The TS was proposed to be carried in the CS overhead, which occurs once every 16 cells.

Each of the two methods has a mixture of advantages and disadvantages in terms of overhead efficiency and service adaptability. The advantage of SFET is that very little overhead is needed to convey the frequency difference information. The major concern in SFET, however, is that, for every new service, a new network-derived clock needs to be defined. The TS method can relax this constraint, but at the price of a larger overhead. These considerations have led to a solution based on a modified TS, referred to as the SRTS.

The SRTS method uses the *residual time stamp* (RTS) to measure and convey information on the frequency difference between a common reference clock derived from the network and a service clock. The same derived network clock is assumed to be available at both the transmitter and the receiver. The SRTS method is also capable of meeting the jitter requirements specified in CCITT Recommendations G.823 and G.824. This method is described in detail in the following subsection.

Aside from these three CBR timing recovery methods, there is an adaptive clock method. The *adaptive clock method* is a conventional method, widely used in existing terminals and the network. The receiver writes the received information into a buffer and then reads it with a local clock. The fill level of the buffer is used to control the frequency of the local network. The control is performed by continuously measuring the fill level around its medium position, and by using this information to drive the PLL, which provides the local clock. The fill level of the buffer may be maintained between two limits in order to prevent buffer overflow and underflow. Compared to the two previous methods, the adaptive clock method requires larger buffer size, but its response time is comparatively long.

Synchronous Residual Time Stamp Method

Two clock frequencies are involved in the SRTS method: the reference clock frequency f_r, which is derived from the network frequency f_n, and the service frequency f_s, which depends on the service in question. The reference clock is required to be larger than or equal to the service clock, but smaller than or equal to twice the service clock or, equivalently, $f_s \leq f_r \leq 2f_s$. If the reference clock is derived from the network clock f_n such that

$$f_r = f_n / 2^k \qquad (4.4)$$

for an integer k, then it is always possible to meet this requirement.

The fundamental idea of the SRTS method is to send to the receiver the information on the difference appearing between f_r and f_s. Since f_r is also available to the receiver, it is possible to determine f_s in the receiver based on the difference. As a time reference to measure the difference, we specify the period of RTS, which is T seconds long and which corresponds to N cycle times for the service clock f_s.

If M denotes the number of cycles the reference clock has during the RTS period, then

$$M = \frac{f_r}{f_s} N \qquad (4.5)$$

So M is not an integer in most cases (see Figure 4.38). Let the integer part of M be M_q. Then M_q is actually made up of a nominal part and a residual part. The nominal part is obtained from the nominal value of f_r and is well known to the receiver also. The residual part conveys the frequency difference information and the quantization effect, which are unknown to the receiver. Therefore, the residual part is the information to be sent to the receiver for the delivery of the difference information.

The residual part of M_q is conveyed over the RTS, which is P bits long. The size P of RTS can be determined by considering the tolerance of the service

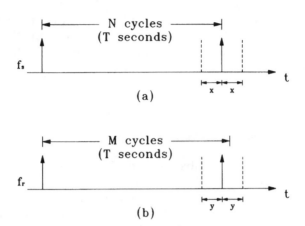

Figure 4.38 The principle of the SRTS method: (a) service clock and (b) reference clock.

clock f_s. If the tolerance of f_s is x, then the corresponding clock deviation y of the reference clock f_r has the expression

$$y = x \frac{f_r}{f_s} N \tag{4.6}$$

in view of (4.5) and Figure 4.38. Since the RTS should be capable of absorbing twice this deviation, it is necessary to choose the counter size P such that it meets the relationship

$$2^P > 2\lceil y \rceil \tag{4.7}$$

where $\lceil y \rceil$ denotes the smallest integer larger than or equal to $\lceil y \rceil$.

The RTS can be generated using the process shown in Figure 4.39. In the figure, counter C is a P-bit counter, which is continuously clocked by the reference clock f_r. The output of counter C is sampled every RTS period, and this period can be generated by counter A through the divide-by-N operation.

With a knowledge of the RTS and the nominal part of M_q, the value of M_q is completely determined at the receiver. This M_q can be used to produce the reference timing signal for a PLL to finally reconstruct the service clock.

For a practical application within the synchronous ATM network, we may choose f_n in (4.4) to be 155.52 MHz; N of 3,008, which corresponds to the number of bits in eight SAR-SDUs; tolerance x of 200×10^{-6}; and the size of RTS of four bits. The four RTS bits can be transmitted in the serial bit stream provided by four CSI bits in four successive odd sequence-numbered SAR-PDU headers (refer to Section 4.7.3 for a more detailed description of the CSI bit).

The clock recovery process in the receiver is depicted in Figure 4.40. The received RTS first passes through the FIFO, which can absorb the accompanied network jitter. The FIFO-output RTS is then compared with the P-bit counter driven by the clock f_r. The pulse generated in this process has period P, as illustrated in Figure 4.41. The gating circuit inside the dotted block then selects appropriate RTS pulses to pass to the PLL out of this long pulse stream. The gating circuit employs an M_l counter to measure the nominal value M_{nom}, where M_l is related to M_{nom} through the relation

$$M_l = \lceil M_{nom} \rceil - 2^{P-1}. \tag{4.8}$$

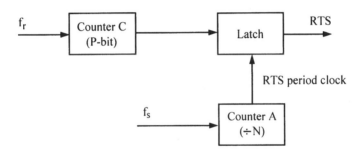

Figure 4.39 RTS generation process.

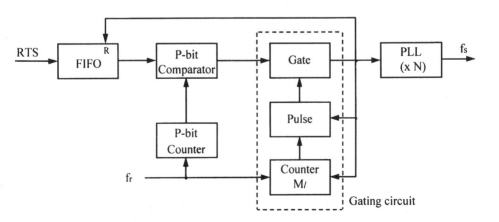

Figure 4.40 Block diagram for source clock recovery in the receiver.

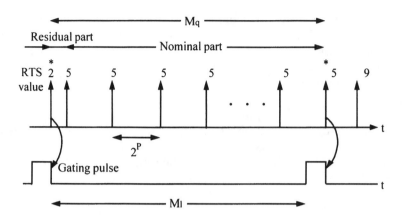

Figure 4.41 Timing diagram describing the gating function (* indicates the RTS pulses to be sent to the PLL).

The gating pulse generated by the M_l counter, which is driven by f_r, selects the first RTS pulse appearing after M cycles, neglecting all pulses that appeared before the Mth cycle. At the moment this desired RTS is selected, the M_l counter is reset, and the next RTS value comes in from the FIFO. Then the duration between a gating pulse and the first RTS pulse indicates the residual part of M_q, while the duration between this RTS pulse and the subsequent gating pulse indicates its nominal part (see Figure 4.41). This process enables the reconstruction of M_q in the receiver, and the gated RTS pulses are used as the reference signal in the PLL for recovering the source clock.

Source Clock Frequency Recovery for VBR Services

The aforementioned clock recovery methods, such as SFET, TS, SRTS, and adaptive clock methods are all for AAL-1 CBR services. For AAL-2 time-related VBR services, however, no fully reliable clock recovery methods are available yet. The CBR clock recovery methods can be modified for use in VBR environments, but the traffic characteristics of the VBR services are a critical factor for the modification. Since there is no fixed period or time reference in the cell stream of VBR services, the user must implant synchronization patterns within a layer above the AAL to aid source clock frequency recovery. But the randomness of the VBR traffic characteristics again constrains the performance and applicability of this method. Figure 4.42 illustrates one possible arrangement to apply the SRTS method for VBR services, in which RTS carrying cells are accompanied by their indication cells. The four indicating cells, starting with an even sequence number, indicate the location of the RTS carrying cells. It is requisite in this application to secure a reliable traffic shaping mechanism that

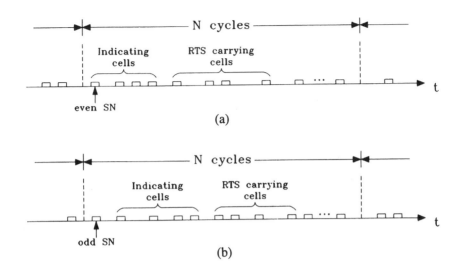

Figure 4.42 An illustration of SRTS for VBR services: (a) when the first cell has even SN and (b) when the first cell has odd SN.

can spread the traffic pattern such that the minimum number of cells is guaranteed at all times [Kim95].

4.8 CONTROL AND MANAGEMENT IN BISDN

All the discussions that have been made so far are related to user information. To transport user information in the BISDN, however, network control and management functions must be supported. In terms of the BISDN protocol reference model shown in Figure 4.12, these two functions belong to the control and management planes.

The control plane performs the functions associated with the virtual connection, including addressing and routing. The related topics such as a higher layer signaling protocol and the signaling AAL have been standardized in the first level. The management plane is associated with the operation and maintenance of the network, and is composed of layer management and plane management.

In this section, topics related to BISDN control and management, such as the signaling and OAM functions, are discussed on the basis of ITU-T Recommendation I.610.

4.8.1 Signaling and Control Plane

Signaling, in general, refers to the procedures and messages for establishing, maintaining, and releasing ATM layer connections. Because various different types of services exist in BISDN, it is desirable to use an out-of-band signaling method, which enables us to manage ATM call connections efficiently. So, in BISDN, the signaling information is carried via dedicated signaling VCs, separate from the user information.

BISDN Signaling Requirements and Functions

Because BISDN is supposed to support NISDN application services, BISDN signaling specifications are designed to include the NISDN signaling specifications. In addition, they require the following ATM specific features.

First, the BISDN signaling scheme must be capable of controlling the ATM VCCs and VPCs for information transfer. This entails the capability to establish, maintain, and release VCCs and VPCs, and provide a semipermanent or permanent setup when required. Point-to-point, point-to-multipoint, and broadcast communication configurations must be also supported. It must allow for the negotiation of traffic parameters at the time a connection is established, and also renegotiate the traffic parameters for the readily established connections.

Second, the BISDN signaling scheme must be capable of supporting simple multiparty and multiconnection calls. For this purpose, it must support symmetric or nonsymmetric simple calls and allow for the possibility of establishing or removing multiconnections associated with a single call simultaneously. It should also allow for the addition or removal of groups. It should possess the ability to correlate connections making up a multiconnection call and to reconfigure a multiparty call or separate an unspecified number of calls.

In addition, the capability to reconfigure readily established connections is required. Also, the accommodation of systems with different coding schemes must be supported, as well as the interworking with non-BISDN services.

These signaling capabilities, however, are too complex to implement all together before the initial deployment of the BISDN. So, ITU-T has taken a phased approach that divides the required signaling capabilities into three groups according to the degree of difficulty in realization, and plans to proceed with the standardization in that order.

The first phase of standardization, which is called *capability set-1* (CS-1), includes the features of point-to-point connection, interworking with NISDN, negotiation of VPI/VCI, peak cell rate based bandwidth allocation, and so on.

The next phase, *capability set-2* (CS-2), supports the point-to-multipoint connection and the multiconnection call, which is made possible by separating the call and the connection control. It can also optimally allocate the bandwidth using the user declared information, i.e., the QOS level and the traffic descriptor. Thus, it will be possible to support VBR services in the second stage. The functions of CS-2 are further divided into two groups, CS-2.1 and CS-2.2.

Finally, *capability set-3* (CS-3) supports distributed multimedia services and broadcasting connections. At this stage, QOS is decided not by the user but by negotiation between the user and the network.

Signaling Virtual Channel

In BISDN, signaling information is transferred, separate from the user information, through a dedicated channel, a *signaling virtual channel* (SVC). There are four types of SVCs: meta SVC, general broadcast SVC, selective broadcast SVC, and point-to-point SVC.

Meta SVC is used for establishing, confirming, and releasing the point-to-point SVCs or selective broadcast SVCs. The meta-signaling procedure, which is a layer management function, is in charge of the transfer of control information only. Meta SVC has predetermined VPI/VCI values (see Table 4.11). Meta-signaling performs the functions of allocating the capacity of SVC and of associating call setup requests to appropriate service profiles. It also provides a means to resolve simultaneous call requests, which is similar to the procedure that allocates the *terminal equipment identifier* (TEI) in NISDN.

Broadcast SVC is employed to transfer signaling messages to all or some signaling end points. *General broadcast SVC* exists in every signaling end point regardless of the service profile, and the VPI/VCI values are preassigned (see Table 4.11). *Selective broadcast SVC* is allocated to each service profile of each signaling end point.

Point-to-point SVC is used for establishing, maintaining, and releasing a VC to convey user information, and its VPI/VCI values are also preassigned (see Table 4.11).

Signaling Protocol Architecture

The CS-1 signaling protocol architecture is shown in Figure 4.43. Higher layer signaling protocol uses the Q.2931 at the UNI and the *broadband ISDN user part* (B-ISUP) at the NNI. The signaling AAL, which lies below the signaling protocol, accomplishes the adaptation function between the signaling protocol and the ATM layer.

To remain compatible with NISDN and in order to save time for standardization, ITU-T has adopted the existing signaling protocols for BISDN with only minor modification. Consequently, the Q.2931 protocol at the UNI is determined based on the NISDN signaling protocol Q.931, and the B-ISUP protocol at the NNI is specified on the basis of the NISDN signaling protocol ISUP.

Signaling Protocol at UNI

The signaling protocol at the UNI, Q.2931, describes in detail the specifications of messages and information elements that are used for signaling procedures,

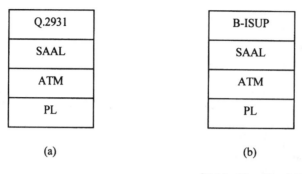

SAAL : Signaling AAL
PL : Physical Layer
B-ISUP : Broadband ISUP

Figure 4.43 Signaling protocol architecture: (a) UNI and (b) NNI.

and the signaling procedures between the signaling end points at the UNI. Although based on Q.931, the third layer signaling system of NISDN, Q.2931, also includes new features appropriate to BISDN whose key technology is ATM. The information elements newly added in Q.2931 include broadband bearer capability, ATM traffic descriptor, AAL parameters, and connection identifier, which are mainly associated with ATM technology. In addition, the allocation and negotiation of the VPI/VCI, compatibility test at the receiving end and the interworking procedures between Q.931 and Q.2931 are also included. Q.2931 has a structure that will result in an easy evolution toward CS-2 and CS-3. Major signaling procedures described in Q.2931 can be summarized as follows:

1. Point-to-point call/connection setup procedure;
   - VPI/VCI allocation and selection procedure;
   - QOS and traffic parameter selection procedure;
2. Point-to-point call/connection release procedure;
3. Call/connection restart procedure;
4. Error state recovery procedure;
5. State monitoring procedure;
6. Interworking procedure with 64-kbit circuit-mode ISDN.

The information needed in the Q.2931 signaling procedure is conveyed through the general message whose format is shown in Figure 4.44. A general message consists of four essential elements whose order must be kept: protocol discriminator, call reference value, message type, and other information elements. The major functions of each element are discussed below.

The *protocol discriminator* is located in the frontmost position. It

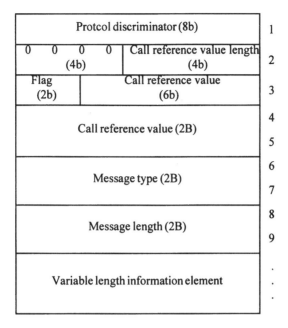

Figure 4.44 General message format.

differentiates the UNI call control messages from other proposed messages. For example, the protocol discriminator of the Q.2931 message has the value of 10001001.

The *call reference value* discriminates the call to which the message belongs. The value is meaningful only for the local area and meaningless for end-to-end link. The call reference value assigned by the calling party remains unchanged during the duration of the call and is returned at call termination for other callers' use.

The *message type* indicates the function of the message, which includes call setup message, call release message, and messages for other uses.

The *message length* indicates the length of the message contents in octets. This field itself is made up of 2 octets.

The *variable-length information element* consists of the *mandatory information element* and the *optional information element*. The length of this field varies according to the number and length of the information element.

A user can invoke a call setup procedure by sending a SETUP message after determining the destination address, QOS level, traffic parameters, and so on. As for call release, either the user or the network can invoke the procedure using the RELEASE message. An example of call setup and release is shown in Figure 4.45.

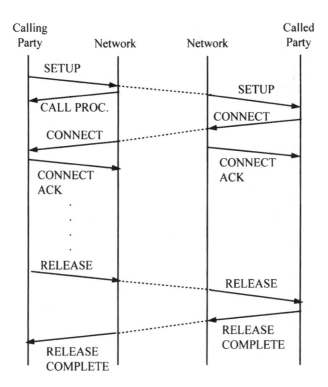

Figure 4.45 Point-to-point call setup and release procedure.

Signaling Protocol at NNI

The signaling protocol at the NNI is B-ISUP, which is specified in Q.2761 through Q.2764. B-ISUP is restructured out of ISUP so that it can fit to the ATM environment and can easily accommodate newly arising services. Since it provides the control procedures required for supporting CS-1 applications, B-ISUP differs from ISUP in both structure and functions.

B-ISUP constructs an *application service element* (ASE) for each function and thus ensures independence among different elements. If a new service is introduced, it can flexibly accommodate it by adding a new ASE. It provides the basic framework of architecture for separate control of call and connection, which is the ultimate goal of CS-2 and beyond.

B-ISUP interworks with Q.2931 to perform the call setup and release procedures through transparent information transfer between end users, and through access to information elements in the network. Call connection control procedures are carried out by exchanging a series of messages between intermediate switches and the switches at the source and destination sides.

Signaling AAL

Signaling AAL (SAAL), outlined in Q.2100, refers to the AAL layer of the control plane, which plays an important role of adapting the signaling protocols such as Q.2931 and B-ISUP to the ATM layer protocol. The standardization for SAAL has progressed keeping the principle in mind that it should use a common protocol for both UNI and NNI in order to manage the network efficiently and to save time for standardization.

SAAL uses the same SAR and CPCS protocols as AAL-5 does, but it newly defines SSCS protocols suitable for higher layer protocols. SSCS is composed of the *service specific connection-oriented protocol* (SSCOP) and the *service specific coordination function* (SSCF), as depicted in the layered architecture of Figure 4.46.

SSCOP, specified in Q.2110, performs the following functions: sequence integrity, error detection, error recovery through retransmission, receiver-based flow control, assured and nonassured transmissions of the user information, and establishment, release, and synchronization of the SSCOP connection.

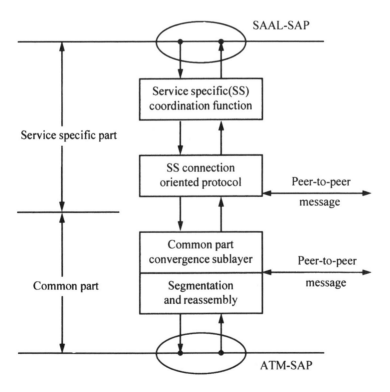

Figure 4.46 SSAL layer protocol structure.

SSCF is for the mapping of service primitives between the higher layer signaling protocol and the SSCOP. It is impossible to use a common SSCF at the UNI and NNI since the higher layer signaling protocols are different each other. The SSCF at the UNI, which is specified in Q.2130, converts the SSCOP services to the services requisite for Q.2931, and the SSCF at the NNI, specified in Q.2140, converts the SSCOP services to the services requisite for B-ISUP.

4.8.2 BISDN Operation, Administration, and Maintenance

Basically, the BISDN's OAM functions are required to be capable of performance monitoring, defect and failure detection, system protection, failure of performance information transfer, fault locating, and so on. As a systematic way to meet these requirements, OAM information flow is assigned and monitored at every network level. The OAM levels and OAM information flows are introduced in this section, and the OAM functions are discussed.

OAM Levels

The BISDN transport network can be layered into an ATM layer network and a physical layer network. The ATM layer network can be further divided into virtual channel level and virtual path level, and the physical layer network can also be divided into a transmission path level, digital section level, and regenerator section level. The concept of a layered network is applicable for both cell-based networks and SDH-based networks.

Table 4.14 represents the layered structure of the ATM network, and the relationship among the layers is shown in Figure 4.22.

At the VC level, a VCC is provided for user-to-user, user-to-network, and network-to-network information transfer. As can be seen from Figure 4.22, a VCC is composed of a concatenation of VC links. User-to-user VCC is estab-

Table 4.14
The Layer and Level Discrimination of
the ATM Network

Layer	Level
ATM layer	Virtual channel
	Virtual path
Physical layer	Transmission path
	Digital section
	Regenerator section

lished between T_B or S_B reference points, and the cells associated with the same VCC are transported through the same path.

At the VP level, a VPC is provided for user-to-user, user-to-network, and network-to-network information transfer. As shown in Figure 4.22, a VPC consists of a concatenation of VP links. User-to-user VPC is established between T_B (or S_B) reference points, and the cells associated with the same VPC are transported via the same path.

The physical layer network corresponds to a union of the transmission medium layer network and the path layer network of the synchronous transmission network (refer to Section 3.9.1). In other words, in SDH-based systems, the transmission path level is equivalent to the VC-4 transmission path, and the digital section level and regenerator section level, respectively, represent the multiplexer section and regenerator section.

OAM Information Flows

To perform the OAM function in a systematic manner, the OAM function itself can be divided into five OAM hierarchical levels. This is identical to the division of the ATM network layer. Consequently, the OAM function can be represented by five information flows: F1, F2, F3, F4, and F5, as shown in Figure 4.22. Because such division is not always necessary, in case one of the levels is omitted, its respective OAM function can be performed by an upper level.

The OAM function associated with each level is independent of that of other levels. In order for a level to obtain information on performance quality and condition, it has to perform the necessary procedure itself. The result is delivered to the management plane, and also to the next higher level as the occasion arises. However, the higher order layer function is not separately required to support the lower order layer's OAM.

A. Physical Layer OAM Flows

The physical layer encompasses the regenerator section level, the digital section level, and the transmission path level, and the corresponding information flows are defined as F1, F2, and F3, respectively. The method of providing the OAM function required to generate an OAM flow for each case depends on the particular physical layer transmission technique chosen.

In the SDH-based transmission, F1 is conveyed via RSOH, F2 via MSOH, and F3 via POH. Parts of F3 are sometimes transported over physical layer OAM cells.

In the cell-based case, the multiplexer section is not applicable and hence OAM information flow F2 does not exist. Both F1 and F3 are conveyed via the physical layer's OAM cells, and the headers are assigned with the bit patterns indicated in Table 4.6. These physical layer OAM cells are not sent up to the ATM layer.

The physical layer's OAM cells are inserted repeatedly into the ATM cell flow. Here, the insertion of physical layer cells must not hinder the transfer capability of the ATM layer. Consequently, the maximum frequency of physical layer OAM cells allowed is limited to 1 per 27 ATM cells. The minimum frequency possible is 1 physical layer OAM cell per 512 ATM cells.

In G.702's PDH-based transmission, the OAM flow is conveyed through the maintenance function possessed by the system (refer to Section 3.7.4). In this case, the capability to deliver OAM information other than bit messages is extremely limited.

B. ATM Layer OAM Flows

The ATM layer encompasses the information flows F4 and F5. Here, F4 and F5 are associated, respectively, with the VP level and the VC level. These information flows are delivered to the VPC and VCC using the cells that are responsible only for the ATM layer OAM function. These cells can be used to achieve communication between peer layers residing in the same management plane of different transmission equipment.

The OAM information flow F4 provides, in support of the VPC OAM function, such capabilities as VPC alarm monitoring, VPC continuity check, and VPC performance monitoring. In case a VPC failure is detected at the VPC point, it sends VP-AIS in the direction of the downward termination point, and if a VP-AIS or a VPC failure is detected at the VP termination point, it sends VP-RDI in the upward direction. Also, if no user information cells have been sent for a fixed duration of time, it creates and sends continuity check cells in order to verify the continuity of VPC. In addition, information related to error blocking and cell loss/insertion is loaded onto the cells and delivered to the other party for the purpose of end-to-end monitoring.

In support of the VCC OAM function, the OAM information flow F5 provides such capabilities as VCC alarm monitoring, VCC continuity verification, and VCC performance monitoring. The details of their operation are analogous to those of F4.

If the OAM flow of the BISDN user access is realized in terms of several physical configurations, the result is as shown in Figure 4.47. As can be seen from the figure, F1 terminates at B-NT1 and the regenerator; F2 terminates at B-NT1, B-NT2, and LT; and F3 terminates at B-NT2, ET, and VP-XC. It can also be inferred that F4 terminates at B-NT2 and ET; and F5 at B-NT2 or B-TE.

OAM Functions

The BISDN OAM function is divided into five types as follows. The first is to monitor, either continuously or periodically, all the entities managed by the

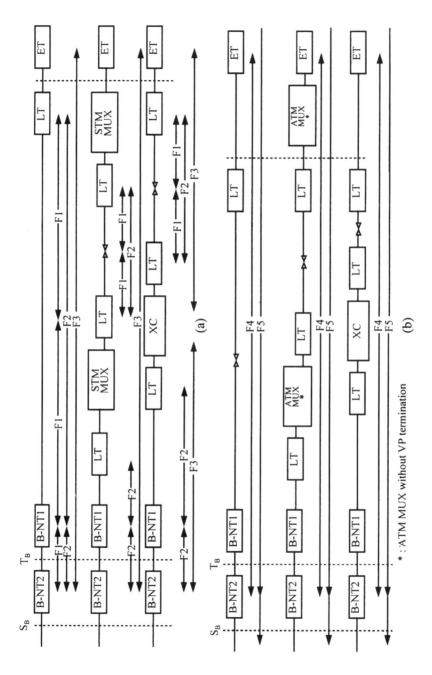

Figure 4.47 Physical structure and OAM flow: (a) OAM flow F1, F2, F3 and (b) OAM flow F4, F5.

network in order to verify their normal operation. As a result of such performance monitoring, maintenance event information can be generated. The second is to detect failure conditions through a continuous or a periodic inspection. As a result of defect detection, maintenance event information of various alarms can be generated. The third is to minimize the effect of the failure of the managed entity by blocking or replacing it. As a result of such system protection measures, the failed entity is excluded from operation. The fourth is to deliver performance information or impairment information to other management entities. As a result, alarm indications can be delivered to other management planes, and a report on the ongoing status can be given. The fifth is to use an internal or an external test system to determine the impaired entity if the given impairment information proves insufficient. As a result of such impairment locating, the impaired entity can be isolated or replaced.

To support the above OAM functions, loss of frame, loss of cell synchronization, and header error must be detected, and the error performance must be monitored, and finally either an *alarm indication signal* (AIS) or *remote defect indication* (RDI) signal must be generated (refer to Tables 4.15 and 4.16).

If physical layer OAM functions are divided according to each OAM flow type, then the result is as listed in Table 4.15 (refer to ITU-T Recommendation I.610). In the table, the entries indicated with "S" and "C" represent SDH-based and cell-based transmissions, respectively, and "-" denotes the absence of any applicable data. Also, PLOAM denotes the *physical layer OAM* and CN the *customer network*. The ATM layer's OAM functions are summarized in Table 4.16.

OAM Cell Format

Figure 4.48 depicts the format of the OAM cell, which is common to both F4 and F5 OAM flows. The VPI field in the F4 OAM cell header is coded according to the VPI value of the VP to be managed. The VCI field is used to differentiate whether the OAM information carried by the cell is for the link (VCI = 3) or for the end-to-end operation (VCI = 4). Other VCI values are reserved for future use. As for F5 OAM flow, both VPI and VCI are used to identify the VC to be managed, and the PTI field is used for the discrimination between the link OAM (PTI = 100) and the end-to-end OAM (PTI = 101).

The first octet in the payload field of the ATM OAM cell is composed of *OAM type* and *function type*. There are three OAM types: fault management, performance management, and activation/deactivation, and for each OAM type two or three function types are assigned. The OAM types and function types are listed in Table 4.17 together with their codes.

The *function-specific field* is used to transfer the function-specific OAM information in a predetermined format (refer to ITU-T Recommendation I.610). The octets that are not in use are all stuffed with "01101010."

Table 4.15
Physical Layer OAM Functions

Level	Function	Flow	Defect/Failure Detection	System Protection and Failure Information		
				B-NT2-B-NT1 Section	B-NT1-LT Section	B-NT2 Transmission Path Termination
Regenerator section	Frame alignment (S*)		Loss of frame	Section AIS/RDI		—
	Section error monitoring (S,C**)		Degraded error performance	Section AIS/RDI		
	Section error reporting (C)	F1	Degraded error performance	Section AIS/RDI		
	PLOAM cell recognition (C)		Loss of PLOAM cell recognition	Section AIS/RDI		—
	Cell delineation		Loss of cell sync	Section AIS/RDI		
Digital section	Frame alignment (S)		Loss of frame	Section AIS/RDI		—
	Section error monitoring (S)		Degraded error performance	Section AIS/RDI		—
	Section error reporting (S)	F2	Degraded error performance	Section AIS/RDI		—
Transmission path	VC-4 offset (S)		Loss of AU PTR			
	CN status monitoring (S,C)		CN-AIS			Path AIS/RDI
	Cell delineation (S,C)		Loss of cell sync			Path AIS
	Header error detection/correction (S,C)	F3	Uncorrectable header	—	—	Path RDI
	Header error performance (S,C)		Degraded header error			(Path management message)
	Cell rate decoupling (S,C)		Failure of insertion and suppression of idle cells			(Path management message)
	Path error monitoring (S,C)		Degraded error performance		—	
	Path error reporting (S,C)		Degraded error performance		—	Path AIS/RDI
	PLOAM cell recognition (C)		Loss of PLOAM cell recognition		—	Path AIS/RDI Path RDI

*S: SDH-based
**C: cell-based

Table 4.16
OAM Function of the ATM Layer

Level	Function	Flow	Detect/Failure Detection	System Protection and Failure Information
Virtual path	Monitoring of path availability	F4	Path not available	—
	Performance monitoring		Degraded performance	
Virtual channel	Performance monitoring	F5	Degraded performance	—

5B	4b	4b	45B	6b	10b
Header	OAM type	Function type	Function specific fields	Reserved	CRC-10

Figure 4.48 ATM OAM cell format.

4.9 TRAFFIC AND CONGESTION CONTROL IN BISDN

The advantages of ATM lie in the efficient use of network resources and the flexibility to support various services. But if these points are to be fully exploited, the problems of traffic control and congestion control must be resolved.

Many traffic control mechanisms have already been developed for existing packet communication networks. But these mechanisms are perceived to be minimally effective in controlling congestion in BISDN, and the reasons are as follows:

- ATM networks should support various services with probably significantly different bit rates.
- A single connection should accommodate heterogeneous traffic streams with different statistical characteristics and performance objectives.
- High transmission speed and large propagation delay limit the use of the window-based flow control.
- Existing and foreseeable services require qualitatively and quantitatively different QOSs.
- Delay-related performances such as the maximum delay and cell delay variation become important as real-time services emerge.

Table 4.17
OAM Types and Function Types

OAM Types		Function Types	
Fault management	0001	AIS	0000
		RDI	0001
		Continuity check	0100
Performance management	0000	Forward monitoring	0000
		Backward monitoring	0001
		Monitoring report	0100
Activation/deactivation	0000	Performance monitoring	0000
		Continuity check	0001

In this section, the basic concepts of traffic control and congestion control in BISDN are reviewed first. Then the traffic parameters that specify a service's statistical properties are discussed. Last, several traffic and congestion control functions are considered.

4.9.1 Basic Concepts

In BISDN, *congestion* refers to the state when the negotiated QOS can no longer be guaranteed by the network elements—a situation caused by unpredictable statistical fluctuation of traffic flows and fault conditions within the network. Congestion should be distinguished from the case when buffer overflow is causing cell losses, yet meeting the negotiated QOS.

Traffic control, which is also called preventive control, is defined as the set of actions taken by the network to avoid congestion. *Congestion control*, which is also called reactive control, is defined as the set of actions taken by the network to minimize the intensity, spread, and duration of congestion.

The two main objectives of traffic control and congestion control are to protect the network and the user in order to achieve network performance objectives and to optimize the use of network resources.

To understand the structure of traffic control and congestion control in the B-ISDN, each step involved in setting up, maintaining, and releasing a connection needs to be understood. In the following, a simple overview of these procedures is given, with an emphasis on traffic control.

When a user wants to set up a connection, first the traffic parameters representing the statistical characteristics of the source and the required level of the QOS are passed to the network. Next, the *connection admission control* (CAC) decides whether the call may be accepted without affecting the QOSs of other connections in progress. Once a connection is admitted, the source may send cells into the network at the rate specified in the traffic contract. While the connection is in use, the *usage parameter control* (UPC) polices the traffic emitted from the source at the UNI to make sure that the source is complying with its traffic contract. Throughout all these processes, the network may manage network resources using a virtual path to separate traffic flows according to service characteristics, or may use feedback controls to control the traffic flow injected into the network by the user. Also, the user may generate different priority traffic flows by using the CLP bit so that a congested network element can drop the cell as necessary.

Figure 4.49 shows the reference model for traffic control and resource management as defined by ITU-T. CAC is performed over the entire network, while the UPC and the *network parameter control* (NPC) are located at the UNI and NNI, respectively.

4.9.2 Traffic Descriptors and Parameters

Definitions

A *traffic parameter* as defined by ITU-T is a specification of a particular traffic aspect of a source. An *ATM traffic descriptor* is the generic list of traffic parameters that can be used to capture the traffic characteristics of an ATM connections, while a *source traffic descriptor* is a subset of traffic parameters belonging to the ATM traffic descriptor. It is used in the connection setup phase to capture the intrinsic traffic characteristics of the connection requested by a particular source. The *connection traffic descriptor,* which specifies the traffic characteristics of the ATM connection at the UNI, includes the *cell delay variation* (CDV) tolerance[25] and the conformance definition[26] in addition to the source traffic descriptor.

CAC and UPC/NPC procedures require a knowledge of the characteristics of ATM layer connection. This information is called the traffic contract, which consists of a connection traffic descriptor and a requested QOS class for each direction of the ATM connection.

GCRA Algorithm

The ITU-T recommendations provide two examples of algorithms—the virtual scheduling algorithm and the continuous-state leaky bucket algorithm—that are useful in specifying and monitoring the *peak cell rate* (PCR) in an operational

---

25. The issue of CDV tolerance is discussed later in this section.
26. The conformity of cells of an ATM connection at the UNI is defined according to the GCRA algorithm in relation to the corresponding traffic parameters. For details, refer to the ATM Forum's UNI specification.

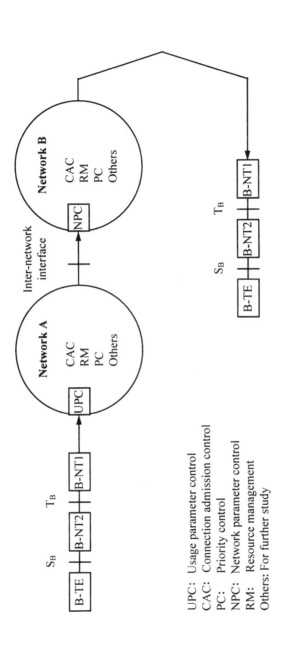

UPC: Usage parameter control
CAC: Connection admission control
PC: Priority control
NPC: Network parameter control
RM: Resource management
Others: For further study

Figure 4.49 Reference configuration for traffic control and congestion control. (Note: NPC may apply as well at some intranetwork NNIs. The arrows indicate the direction of the cell flow.)

manner, while taking into account a certain CDV tolerance. These operationally equivalent algorithms are both named the *generic cell rate algorithm* (GCRA) by the ATM Forum.

For each cell arrival, the GCRA determines whether the cell is conforming to the traffic contract of the connection, and thus the GCRA is used to provide the formal definition of traffic conformance to the traffic contract.

The GCRA formally defined in Figure 4.50 depends on two parameters: the increment I and the limit L. The virtual scheduling algorithm updates the *theoretical arrival time* (TAT) assuming equally spaced cell arrival with an interarrival time of I and compares it with the actual arrival time t_a. If the actual

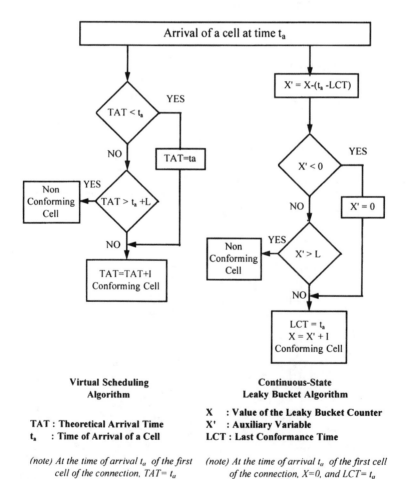

Figure 4.50 Two equivalent GCRA algorithms.

arrival is too early relative to the TAT, that is, $t_a < TAT\text{-}L$, the relevant cell is nonconforming; otherwise the cell is conforming. In the case of the continuous-state leaky bucket, the value of the leaky bucket counter X increases by increment I, and decrease at a continuous rate. If the cell arrives while the value of X is greater than the limit value of L, the cell is nonconforming; otherwise the cell is conforming. The two algorithms in Figure 4.50 are equivalent in the sense that for any cell stream the two algorithms reach the same conclusion as to its traffic conformance to the relevant traffic contract.

Traffic Parameters

The only traffic parameter currently standardized by ITU-T is the PCR. The PCR in the source traffic descriptor specifies an upper bound on the traffic that can be submitted on an ATM connection.

The equivalent terminal configuration for the definition of the PCR is given in Figure 4.51. The PCR is defined at the physical (PHY) layer *service access point* (SAP) of an equivalent terminal based on the basic event, that is, the request to send an ATM-PDU. The PCR (R_p) of the ATM connection is the inverse of the minimum interarrival time T between the preceding two basic events, where T is called the peak emission interval of the ATM connection.

The output of the traffic shaper at the PHY SAP is supposed to conform to GCRA $(T,0)$. But due to various operations in the equivalent terminal and other CPEs, a certain amount of CDV characterized by τ is generated. The value τ is chosen such that the cell flow at the T_B point conforms to GCRA (T,τ). The value τ is called the CDV tolerance and represents a bound on the cell clumping phenomenon at the UNI (T_B point). Users are urged to select explicitly or implicitly a value for the CDV tolerance at the UNI from a set of values supported by the network.

Due to the need for flexible mapping of the traffic parameters of some existing services such as frame relay, the ATM Forum has defined two new parameters, the *sustainable cell rate* (SCR) and the *burst tolerance* (BT). The SCR specifies an upper bound on the possible conforming average rate of an ATM connection, whereas the BT together with the SCR and PCR determine the *maximum burst size* (MBS) that may be transmitted at the peak rate and still be conforming. These parameters enable the end-user to describe the traffic characteristics in greater detail than just the PCR. Consequently, the network provider will be able to utilize the network resources more efficiently, resulting in possible benefits such as a reduced charge for the end user.

4.9.3 Traffic Control Functions

If congestion occurs anywhere in the network, the network performance is severely degraded, resulting in poor service quality for the user. The main goal of

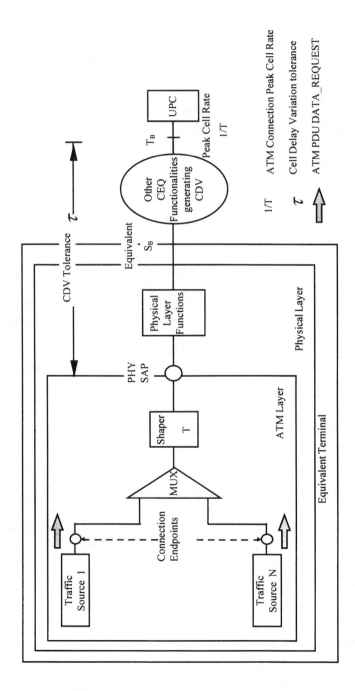

Figure 4.51   Equivalent terminal for the definition of the PCR.

traffic control is to prevent such congestion. In the BISDN, large propagation delay and high transmission speed may render reactive control functions ineffective. As a consequence, preventive control mechanisms become more dependable in ATM networks. For this reason, ATM networks provide several traffic control functions, such as CAC, UPC/NPC, and *priority control* (PC).

Connection Admission Control

One of the main methods for traffic control in ATM networks is CAC, which limits the number of active calls so that QOS requirements for accepted calls can be met. The CAC, as defined by ITU-T, refers to the set of actions taken by the network during the call setup phase (or during the call renegotiation phase) in order to establish whether or not a virtual channel/virtual path connection request can be accepted. Based on the traffic characteristics, QOS requirements, current network load, and the amount of network resources, the CAC function carries out the following: It decides whether to grant or to refuse the connection; determines the traffic parameters for UPC; and allocates network resources.

It has been shown that a so-called effective bandwidth exists that is bounded by the peak and the average rate of the source.[27] The most important property of the effective bandwidth is that it is simply additive for several sources. The acceptance of a new call can be decided by testing whether or not the effective bandwidth of the new call is greater than the available link capacity.

The effective bandwidth method is often difficult to use in a real environment because it usually relies on a particular traffic model, which may be hard to relate to the traffic descriptor, and also it usually requires considerable computation to get reliable results. A possible solution can be found in the *class-related rule* (CRR), which divides virtual channels into a small number of distinct classes, with all channels of the same class assumed to possess similar traffic characteristics and QOS requirements.[28] This method requires a table that lists the bandwidth to accommodate a number of VCs of the same class. For example, the (i,j)'th element of the table lists the bandwidth required for accommodating j of class i connections. Once the table is constructed off-line, CAC may be carried out by looking up the appropriate entry in the CRR table. To build the table, however, numerous complicated simulations or complex analyses need to be carried out for all possible combination of connections.

27. Elwalid and Mitra [Elwalid93] obtained an effective bandwidth for general Markovian traffic sources via the spectral decomposition method. In this case, the effective bandwidth was the maximal real eigenvalue of a matrix that is directly obtained from the source characteristics and the admission criterion. Kesidis and Walrand [Kesidis93] provided another method for obtaining an effective bandwidth by taking a large deviation approach. They have revealed the existence of an effective bandwidth for more general source models.
28. For a more detailed description of CRR, refer to [Gallassi90].

Another instance of the simple CAC method is the fixed boundary method, which preallocates a fixed amount of each network resource, such as bandwidth and buffer space, to each class. This CAC algorithm is advantageous in terms of implementation, but is likely to suffer from considerably low bandwidth utilization because the unused resources of one class cannot be used by other classes.

The CAC algorithms we have discussed are based on the traffic descriptor specified by the user. However, it is difficult, in general, for the user to specify its traffic characteristics accurately. This observation has led to the notion that a CAC scheme relying on real-time traffic measurements rather than any user-declared traffic descriptors may be a better solution. The *dynamic CAC algorithm* is an embodiment of this notion, in which the call acceptance decision is made based on the on-line evaluation of the upper bound of the cell loss probability, which is estimated out of cells arriving in a fixed interval [Saito91]. This algorithm can be used without modeling the input traffic, and thus enables the network to manage its resources tightly using only the peak cell rate. If the relevant estimation can be made accurate and cost effective, this algorithm can become the most promising one in the early stage of the BISDN in which only the peak cell rate is defined by ITU-T.

Policing

Users may violate the traffic contract negotiated during the call setup phase, either deliberately or due to malfunctioning of the user equipment. To minimize any negative effects on other conforming cells, a policing action should be taken. This function is referred to as the UPC or NPC depending on the location of application. The UPC/NPC algorithm should be able to detect the precise violation as quickly as possible and should be transparent[29] to conforming cells.

A common example of the UPC/NPC algorithm is the *leaky bucket algorithm*. The basic form of the leaky bucket is depicted in Figure 4.52(a), and there are a number of variants as well. The most generalized version among them is the buffered leaky bucket shown in Figure 4.52(b), which has both the token pool of size M and the user buffer of size K. It uses a token generator that generates tokens periodically with the period of slots. The generated tokens go into the token pool, and a token is removed from the pool whenever a cell is transmitted. Cells arriving while the token pool is empty are stored in the user buffer until the matching tokens are generated. The leaky bucket algorithm is a simple traffic policing algorithm but it can control various traffic parameters by controlling the size of the token pool, buffer size, and token generation rate.

---

29. In general, the cell blocking probability at the UPC scheme should be less than 10^{-10}, regardless of the cell loss probability requested by a user.

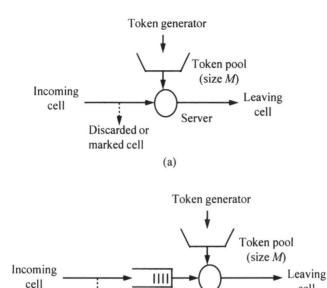

Figure 4.52 Leaky bucket algorithm: (a) basic LB and (b) buffered LB.

Dimensioning a leaky bucket algorithm is a rather complex task. First, the values a and $M + K$ need to be determined, since the blocking probability at the buffered leaky bucket is known to depend only on the sum of the token pool size and the buffer size (refer to [Rathgeb 91] for a more detailed description). Once the value of $M + K$ is evaluated, the individual values are determined according to the delay tolerance. That is, a small delay at the leaky bucket can be achieved by decreasing the buffer size. But this increases the token pool size and thus allows for the leaky bucket to emit a large burst of data into the network. Therefore, it is desirable to take the largest possible buffer size as long as the delay constraint can be satisfied.[30]

Another example of UPC/NPC algorithm is the *jumping window mechanism,* which checks the number of cells transmitted during a constant interval called window. At the call setup phase, a set of parameters (T,X) is defined, where X is the number of cells allowed to be transmitted during the interval T.

---

30. In fact, the CDV originated at the CPN could further complicate the leaky bucket dimensioning procedure. In this situation, it is necessary to analyze the CDV mathematically, applying the result to leaky bucket dimensioning. Refer to [Guillemin91, Lee94a].

This method is also simple to implement, and various traffic parameters can be controlled by using different pairs of (T,X).

The preceding traffic policing techniques enable the detection of nonconforming cells. The detected nonconforming cells are then disposed according to the operation policy. Possible actions of the UPC/NPC function, as defined by ITU-T, are as follows: Pass or reschedule[31] cells identified by UPC/NPC as conforming; and tag or discard the nonconforming cells. Another action, which may be optionally taken at the connection level, is to release the noncompliant[32] connection.

Priority Control

In BISDN, where different classes of traffic requiring different QOSs share the same network resources, it is relatively easy for network elements to discriminate cells in different connections, or traffic classes having different QOS requirements. So it is useful to take advantage of this aspect in setting priority among different QOS traffic as a means for traffic control. Such *priority control* can be classified into two categories: time priority control and space priority control. The former scheme rearranges the service order and thus changes the delay performance, whereas the latter scheme selectively discards cells and thus changes the loss performance. The two priority control schemes can also be combined into a single queue service discipline. In a congested traffic environment, the space priority control scheme is advantageous because it can alleviate the congestion by selectively discarding cells.

The queueing strategies developed for a data-oriented network are tailored for reliable data transmission, not for real-time data transmission. So, new queueing strategies are necessary for ATM networks where real-time services may be dominant. Among the schemes possibly applicable in high-speed real-time communication, there are work-conserving ones and non-work-conserving ones.[33] A *work-conserving* server never goes idle even when there is a backlog, whereas a *non-work-conserving* server may become idle even when there are cells to send. A work-conserving discipline is advantageous in terms of network efficiency but has the drawback that it can hardly minimize the delay jitter and requires a larger buffer space than a non-work-conserving strategy. In the

---

31. Rescheduling refers to the cell spacing action taken when traffic shaping is combined with the UPC function.

32. The term *compliance* is applied to the connection, whereas *conformity* is applied to the individual cell.

33. For example, *generalized processor sharing* (GPS) [Parekh93] and Virtual clock [Zhang90] are work-conserving ones, and Stop-and-Go [Golestani90], *hierarchical round-robin* (HRR) [Kalmanek90], *rate-controlled static priority* (RCSP) [Zhang93], and *residual service interval-based priority* (RSIP) [Lee95b] are non-work-conserving ones.

conventional data networks, non-work-conserving queueing strategies have been seldom studied, because the real-time related features were of no concern. In ATM networks, however, the advantages of the non-work-conserving strategies need to be fully exploited.

The queueing strategies developed for supporting real-time communications can essentially guarantee a bounded delay, for deterministically characterized traffic sources, for which a typical example is the traffic constrained by leaky buckets.[34] The schedulability condition is, in general, an essential part of the queueing strategy, which, when applied to admission control, can significantly facilitate the CAC function. In addition, the queueing strategy can be used to police misbehaving users. These observations indicate that the CAC and the UPC functions are closely related to intelligent queueing strategies, and thus an integrated framework can be devised for traffic control based on such queueing strategies.

4.9.4 Congestion Control Functions

Even with all the available traffic control functions, congestion could still occur due to malfunctioning of those functions, which can be caused by unpredictable statistical fluctuations of traffic flows or by network failures. Therefore, congestion control functions are necessary in addition to traffic control functions in order to protect effectively the network and guarantee the required QOS. In this section, some congestion control functions are considered.

Selective Cell Discarding

As previously mentioned, a congested network element can provide different loss performances among different traffic classes by selectively discarding low-priority cells. This enables the protection of high-priority cell flows and minimizes the spread and duration of congestion. The CLP field in ATM cell header can be used for indicating the cell loss priority such that CLP = 0 for high priority and CLP = 1 for low priority.

A well-known selective cell discarding (or space priority) scheme is the *pushout policy,* which discards low-priority cells when the queue is full. If a high-priority cell arrives while no buffer space is available, one of the low-priority cells in the queue is pushed out. This policy, even though complicated to implement, has been considered to be optimal with respect to the throughput of low-priority flows while providing a desired level of performance to high-priority flows. But recent studies have revealed that this policy has quite limited capability in protecting high-priority cells [Cidon94].

---

34. This is also called *linear bounded arrival process* (LBAP) [Cruz91].

Another popular policy is the *threshold policy,* which accommodates the low-priority cells if and only if the buffer occupancy is below the threshold value. From the implementation's point of view, this scheme is much simpler than the pushout scheme because only the buffer occupancy needs to be monitored in this scheme. There could be many other variations to these policies.

Explicit Congestion Notification

When congestion occurs at an intermediate network node, the node may notify the relevant end nodes of the congestion status so that the end nodes can take some appropriate actions. This congestion control scheme is called the explicit congestion notification scheme. Depending on the direction of the notification, this scheme is divided into two types: *explicit backward congestion notification* (EBCN or BECN) and *explicit forward congestion notification* (EFCN or FECN).

In the case of the EBCN, when an intermediate node gets congested, a special cell is generated to notify the status of the congestion at the congested node and is then sent back to all the connected source nodes. In contrast, in the case of the EFCN, the congestion notification is delivered forward to all the destination nodes. Usually, it takes a shorter time to notify backward to the source than to notify forward to the destination. So the EBCN scheme has been widely used in the legacy data networks as a back-pressure mechanism. However, it is not appropriate for ATM networks since it imposes a considerable burden on the intermediate nodes.

In the case of the EFCN, which has been standardized by the ITU-T for use in the ATM network, the destination node is notified of the congestion status over the PTI field of the ATM cell. If congestion occurs, the congested node sets the second bit of the PTI field in the user information cell to 1 (refer to Section 4.6.3). On recognizing this PTI markup, the destination node takes appropriate action by signaling the source node to slow down the upcoming transmission. This EFCN operation is illustrated in Figure 4.53. For this scheme to be effective, end-users should cooperate with the network, and the round-trip delay should be substantially smaller than the expected congestion duration.

Feedback Flow Control for ABR Services

The ABR service defined by the ATM Forum gives the network the opportunity to allocate resources preferentially to high-priority traffic such as the real-time services and share the remaining bandwidth fairly among all ABR connections. To support the ABR service effectively, two feedback flow control algorithms have been considered: rate-based flow control and credit-based flow control.

Rate-based flow control is an end-to-end flow control, which controls the source rate directly according to the congestion status of the network. The

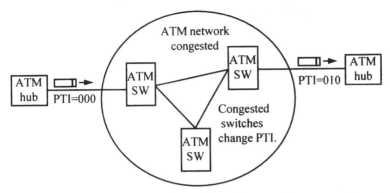

- PTI=000 : user data cell, congestion not experienced
- PTI=010 : user data cell, congestion experienced
- Receiving device may take appropriate action.

Figure 4.53   Operation of the EFCN scheme.

congestion in the network may be notified either is the backward direction (i.e., EBCN) or forward direction (i.e., EFCN). The congestion indication, which is usually represented by a single bit, may increase (positive polarity) or decrease (negative polarity) the source cell rate. The problem with negative polarity is that the rate would keep increasing if the resource management cells indicating the congestion status get lost, and the network congestion would get worse. Therefore, positive polarity is adopted in the *proportional rate control algorithm* (PRCA). The flow control would be faster and its implementation would be more flexible, if the explicit available network capacity, rather than just the existence of the congestion, is informed. This notion has motivated the *enhanced PRCA* (EPRCA).[35]

Credit-based flow control is a link-by-link, per-VC flow control in which each node manages a fixed-size buffer for each VC. The receiver periodically sends credits to the sender indicating the available buffer space for receiving data cells of the VC. Since the sender transmits cells only when allowed with the credits, buffer overflow is not likely to occur. When congestion occurs, back-pressure will propagate backward along the congested VCs quickly. The source of the congested VC, which encounters back-pressure, would be throttled, blocking the excessive traffic at the boundary of the network. The per-VC

35. Rate-based flow control has been opted for over credit-based flow control as the ATM Forum's standard for its simplicity, but it is often argued to be inappropriate for supporting the bursty and interactive data traffic. For a more detailed discussion on rate-based flow control, refer to [Bonomi95].

queueing strategy of credit-based flow control requires a considerable amount of buffer in each node. This problem is partly resolved by adaptively allocating the buffer for each VC. Credit-based control has an advantage in that it can guarantee fairness and loss-free performance, and it has a much smaller ramp-up time than the rate-based one. However, it has the critical disadvantage that it requires excessively large buffer space.[36]

4.10 ATM SWITCHING

Rapid evolution of technological changes has brought about new switching concepts such as multirate circuit switching, fast circuit switching, and fast packet switching. All these have been considered as possible switching technologies in the BISDN environment, and the fast packet switching technology has been finding its embodiment in ATM switching. ATM switching can contribute much toward accommodating versatile services, effectively utilizing resources, and efficiently accommodating bursty services.

From the functional point of view, an ATM switch is practically the same as the packet switch used in today's computer networks. However, there are two main differences in the incoming packet size and the required switching speed. In ATM switches, the incoming packets (cells) have the fixed size of 53 bytes and are transmitted at or above the 155-Mbps rate. So, in contrast to packet switching, new switching techniques such as parallel processing and self-routing[37] are required in ATM switching. Also, in contrast to circuit switching, in ATM switching more than one cell arriving at the same time slot may be destined for the same output port because cells are issued without call setup processing. This causes packet conflicts, and thus packet buffering must be provided within the switch. As a consequence, an ATM switch can be considered to be a box that provides switching and buffering functions together.

In this section, ATM switching mechanisms and architectures are discussed in detail.[38] First, basic switching techniques and the architectures of moderate-sized ATM switches are described, followed by discussions on several large-scale ATM switch architectures. Based on this, several methods to accommodate point-to-multipoint services in the ATM switching systems are

36. For more details about credit-based flow, refer to [Kung95].
37. Self-routing is a new routing scheme in which each incoming packet is transferred to its destined output port based only on its own routing information without assistance from the centralized processes.
38. For simplicity of presentation, each incoming and outgoing line of ATM switches is assumed to have the same transmission capacity, and the packet arrival times of all input lines are assumed to be synchronized in this section. Under this assumption, time is divided into the units of *time slots*, which is the transmission line of an ATM cell. Further, all ATM switches in this section are assumed to have *N* inputs and *N* outputs.

discussed. Finally, other related issues for ATM switching systems are briefly considered.

4.10.1 ATM Switch Architecture

In general, an ATM switch consists of a set of *line interfaces* (LI), a *call processor* (CP), a *signal processor* (SP), and a switching network as depicted in Figure 4.54. The LI performs optical-to-electrical signal conversion, cell synchronization, VPI/VCI header translation, traffic congestion control, and insertion and extraction of routing information. That is, it performs all the processing required for incoming packets before being transferred to the switching network. The switching network routes incoming packets using the routing information of the packets. The CP and the SP perform the functions concerned with the ATM connection setup and release operations and can be connected to the LIs through two different configurations: front signal processing and rear signal processing, as shown in Figures 4.54(a,b), respectively. In the front signal processing, signaling packets are transferred through a separated bus to the CP and the SP. In the rear signal processing, however, there is no separate bus for signaling packet transfer, but signaling and control information is handled in a unified way together with the user traffic. Therefore, the hardware configuration of the rear signal processing case could be simpler than the front signal processing case.

ATM switches can be classified into shared-buffer switch, shared-medium switch, and space-division switch depending on the switching mechanism employed [Tobagi90]. In the following subsections, ATM switch architectures are examined according to the preceding categories.

Shared-Buffer Switch

In shared-buffer switches [Kuwahara89] a single buffer memory is shared by all input and output lines as shown in Figure 4.55, where it is used as the central component for the switching operation. Packets arriving at all input lines are multiplexed into a single stream and then fed to the shared memory for storage according to their destined output addresses. At the same time, an output stream of packets is created by retrieving packets in the output queues sequentially. This output stream is then demultiplexed to distribute packets to each individual output line. So, in shared-buffer switches, high-speed controllers and memories are necessary that can process N input packets and N output packets during one slot time. As a consequence, the switch size is determined by the available memory speed and achievable processing speed. Shared-buffer switches have some advantages such as efficient buffer utilization, easy accommodation of point-to-multipoint services, and priority-control-based buffer management.

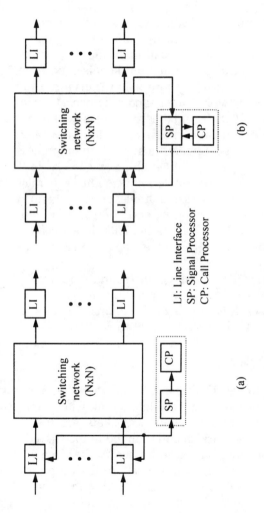

Figure 4.54 General structure of ATM switch: (a) front signal processing and (b) rear signal processing.

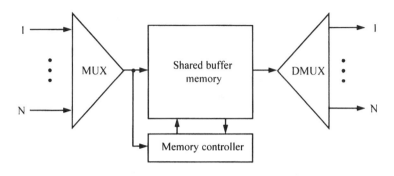

Figure 4.55 Basic structure of shared-buffer switch.

To perform the required switching function, the shared memory in a shared-buffer switch must operate like *N* buffers logically. The controller that performs the memory management controls addresses each output port individually or stores the addresses in the shared buffer in conjunction with the packets in linked list form. It is also possible to simplify buffer management by employing the *content addressable memory* (CAM) as the shared buffer.

Shared-Medium Switch

In shared-medium switches [Suzuki89], all packets arriving at the input lines are synchronously multiplexed onto a common high-speed medium of bandwidth equal to *N* times the rate of a single input line. Each output line is connected to the bus via an interface consisting of an *address filter* (AF) and an output FIFO queue as shown in Figure 4.56. The AFs determine whether or not the packet observed in the bus is to be written into the FIFO queue. Functionally, this structure is similar to that of the shared-buffer switches, so the point-to-multipoint services can be easily accommodated. However, because independent queues are used at the output lines, efficient buffer utilization is not achieved.

Space-Division Switch

Space-division switches can provide a path between an input line and an output line in the space-division fashion. So, in this type of switch, several independent paths can be set up simultaneously. This is quite different from the case of the shared-buffer switch or the shared-medium switch in which paths are set up sequentially in a time-division fashion. This difference means that the

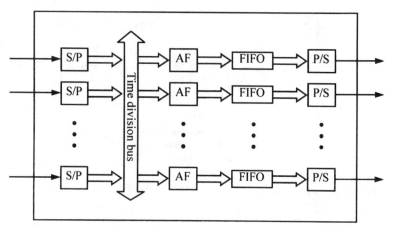

AF: Address Filter

Figure 4.56 Basic structure of shared-medium switch.

space-division switch may operate at a lower speed than the others types.[39] However, the situation that all incoming packets cannot be transmitted to their desired destination may occur, and therefore buffers are necessary to resolve the packet conflicts caused by such resource limitations.

The space-division switches can be classified into four categories depending on the buffer arrangements: input buffer switch, output buffer switch, input/output buffer switch, and internal buffer switch. In the following subsections, the operation and structure of these four categories are described in detail.

The switching network of space-division switches can be divided into two types, i.e., a *blocking* and *nonblocking* switching networks. If a switching network can switch every different-destined input packet to the desired output port then it is a nonblocking switching network; otherwise it is a blocking switching network. In general, blocking switching networks are simpler in structure than, but inferior in performance to, nonblocking switching networks.

The *multistage interconnection network* (MIN) structure is one example of the blocking switching networks. It consists of multiple stages of networks with each stage consisting of a column of 2 × 2 switching elements that performs a fixed pattern of permutation functions. A MIN structure has different names depending on the interconnection pattern between the switching stages, namely, banyan network, baseline network, shuffle-exchange (or OMEGA) network, and flip network. Figure 4.57 depicts the interconnection patterns of these

---

39. Note that this advantage comes from the constituent hardware, which is much heavier in the space-division switch than in the other switches.

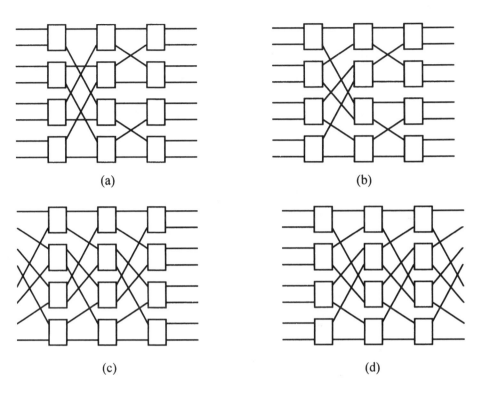

(a)

(b)

(c)

(d)

Figure 4.57 Interconnection patterns of multistage interconnection networks: (a) banyan network; (b) baseline network; (c) shuffle-exchange network; and (d) flip network.

four MIN structures. Note that the differences in interconnection patterns do not necessarily lead to any performance difference in the packet-switched environment.

Basically, every MIN has the unique path property that only one path exists from each input to each output. Further, the route that an incoming packet passes within the network can be completely determined by the destination address affixed to the packet. This implies that the connection state of each switching element in the network is completely determined by the destination of the incoming packet. This observation leads to the self-routing scheme, which can be realized by adding a routing header that contains the destination address and other information necessary for the routing to the front end of the packet, and by letting each switching element operate as directed in this header. For example, in the case of the 8×8 baseline network shown in Figure 4.58, the three-bit header affixed to each packet determines the route in such a way that the switching elements in the first stage are controlled by the *most*

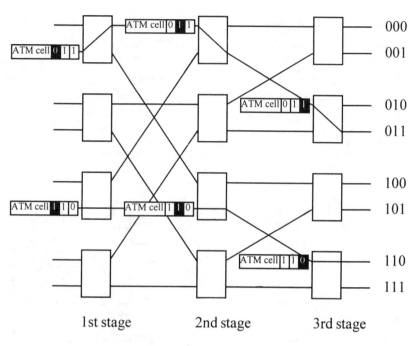

Figure 4.58   A routing example of an 8×8 baseline network.

significant bit (MSB) of the routing header, those in the second stage by the next bit, in that order.

Examples of the nonblocking switching network include crossbar switch, Benes network, and Batcher-banyan network. The crossbar switch consists of N^2 cross-point switches, one for each input-output pair, so the realizable switch size tends to be limited. In addition, it has the problem that the transit time is not constant over all input/output pairs unless artificial delays are introduced at the inputs and outputs of the switch. The Benes network has a structure in which two baseline networks are cascaded back to back. So, the number of required switching elements is $\frac{N}{2}(2\log_2 N - 1)$, which is much less than that of the crossbar switch. However, the self-routing capability is not applicable to this network. The Batcher-banyan network is a nonblocking network that can keep its self-routing capability while overcoming the internal blocking that is the drawback of the banyan network. The Batcher-banyan network can avoid internal blocking by sorting the incoming packets based on their destination addresses first and then routing through the banyan network [Narasimha88].

Since the Batcher-banyan network requires $\frac{N}{2}\log_2 N(\frac{1}{2}(1 + \log_2 N) + 1)$

switching elements, it lies in between the crossbar and Benes network as far as hardware complexity is concerned. Figure 4.59 illustrates the basic structure and routing operation of an 8×8 Batcher-banyan network.

4.10.2 Space-Division Switch Architectures

As discussed in the previous subsection, space-division switches can be classified into input buffer switch, output buffer switch, input/output buffer switch, and internal buffer switch depending on the buffer arrangements. We discuss each of these four categories now.

Input Buffer Switch

In input buffer switches, a nonblocking switching network is used as the switching network and a dedicated buffer is allocated for each input port. Because the switching network can transfer only one packet to each output in each time slot, an arbiter is needed to avoid packet conflicts that may occur in the switching network. Figure 4.60(a) shows the structure of input buffer switches.

The performance of input buffer switches depends on the operation of the input buffers. If the buffers behave like simple FIFO queues, then at the beginning of each time slot only the packets at the *head of line* (HOL) of the buffers contend for access to the switch outputs. If every packet is addressed to a different output, the nonblocking switching network allows each packet to reach the desired output. If k packets at the heads of the N input buffers are addressed to the same output, only one of them is allowed to pass through the switching network, and the other k-1 packets must wait until the next available time slot. While a packet in the HOL is waiting for its turn to access the next available slot, other packets heading for idle output ports must be queued behind it in the buffer. This is known as HOL blocking. Due to this blocking the maximum throughput is limited to 0.586 for a uniform traffic[40] and a large N. The contention among the HOL packets can be resolved by employing the ring-reservation scheme as indicated in Figure 4.60(b). The ring reservation scheme is actually a token ring scheme in which each HOL packet can make a reservation when it catches the token.

To improve the performance of input queueing, it is necessary to eliminate the requirement that the buffers behave like FIFO buffers, thus removing the HOL blocking problem. Examples of non-FIFO buffering can be found in the scheme that includes non-HOL packets for output contention or in the scheme that preschedules the output timing of each cell in the input buffer.

---

40. The uniform traffic refers to the traffic whereby the process describing the arrival of packets at an input line is a Bernoulli process, independent of all other input lines, and whereby the requested output port for a packet is uniformly chosen among all output ports, independently of all arriving packets.

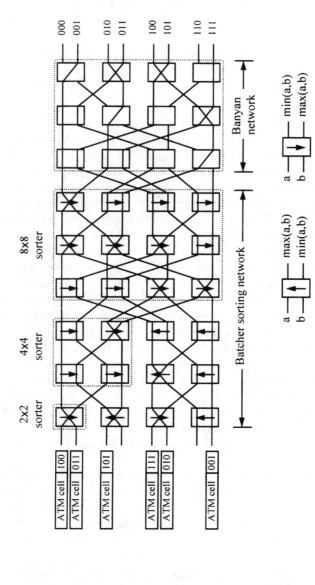

Figure 4.59 Basic structure of an 8×8 Batcher-banyan network.

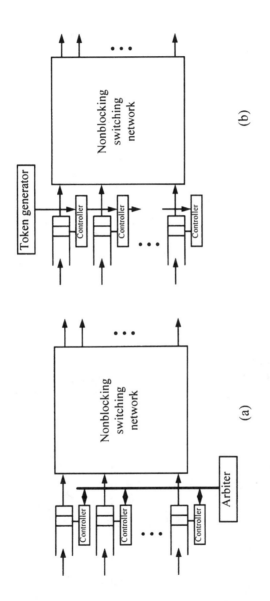

Figure 4.60 Structure of input buffer switch: (a) input buffer switch with arbiter and (b) input buffer switch with ring-reservation scheme.

Output Buffer Switch

In space-division switches, if the incoming packets are the uniform traffic with input load rate p and if each output port can accommodate L packets simultaneously, then the packet loss probability of the internal nonblocking switching network is as follows [Hluchyj88]:

$$P_{loss}(L) = \frac{1}{p} \sum_{k=L+1}^{N} (k - L)\binom{N}{k}\left(\frac{p}{N}\right)^k \left(1 - \frac{p}{N}\right)^{N-k} \tag{4.9}$$

We can show, using this equation, that for the acceptable cell loss probability 10^{-6}, the required L value is 8 regardless of the switch size N when p is 0.9. This means that an ATM switch that requires the above level of packet loss probability can be constructed by using a nonblocking switching network that can transfer up to 8 packets to each output port simultaneously in a time slot. This observation leads to the design of an output buffer switch structure that has dedicated buffers at each output port.

Figure 4.61 shows the structure of output buffer switches. In the figure, the interconnection fabric broadcasts each incoming packet to all interface modules simultaneously, and each interface module selects the packets that are destined to itself. That is, a broadcast-and-select mechanism is adopted in the switch. The broadcasting function can be realized by a parallel bus, binary trees, or a high-speed transmission bus. Examples of the output buffer switches include the knockout switch, the sunshine switch, and the tandem banyan switch.

In the *knockout switch* [Yeh87], N inputs form N buses that are directly connected to each of the N interface modules as shown in Figure 4.62. This

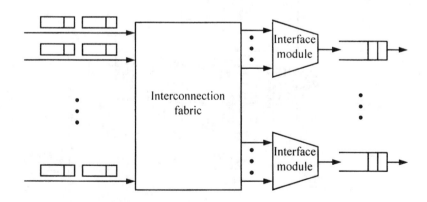

Figure 4.61 Structure of output buffer switch.

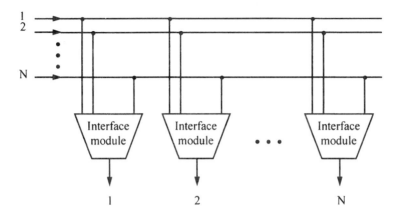

Figure 4.62   Basic structure of the knockout switch.

simple structure causes no packet conflicts among the packets bound for different outputs, and brings forth good modularity and broadcast and multicast capabilities. The interface module consists of three major components: packet filter, concentrator, and shared buffer. The packet filter examines the address of every packet on each of the N buses and filters out those addressed to itself. The concentrator then concentrates the input lines to L output lines such that up to L packets can emerge at the output of the concentrator. The shared buffer secures FIFO buffers equivalent to a single queue of L inputs and one output for each interface module to store the concentrator output packets. The knockout switch has low latency and a simple switching mechanism. However, because each bus has a large number of fanouts, it is difficult to implement large-size knockout switches.

In the *sunshine switch* [Giacopelli91], a Batcher network and L banyan networks are used as the interconnection fabric as shown in Figure 4.63. In the structure, all incoming packets are first sorted by the front-end Batcher network according to their destination addresses and then are transferred to their destined output ports via L parallel banyan networks simultaneously. This interconnection fabric can transfer up to L packets to each output port simultaneously, which becomes possible because the Batcher network and the banyan networks are interconnected in such a manner that L consecutive outputs of the Batcher network can be transferred to L different banyan networks. So, a maximum of L packets destined to the same output can be serviced by the switching fabric without loss. However, if the number exceeds L, then the number of packets that exceed L are transferred to the delay circuit to rejoin the sorting operation in the next time slot.

In the structure, the trap network resolves output port contention by selecting L packets for each destination address, the concentrator separates

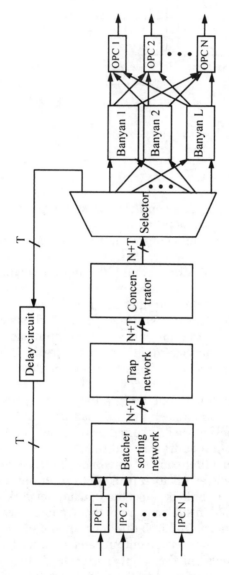

Figure 4.63 Basic structure of the sunshine switch.

packets for recirculation from those for banyan routing, and the selector directs packets to their destinations as determined by the concentrator. The rear *output port controllers* (OPC) terminate the banyan networks and queue the output packets.

The *tandem banyan switch* structure [Tobagi91] consists of banyan networks and *L* interface modules as shown in Figure 4.64. All incoming packets are fed to the first banyan network. The correctly routed packets at the end of the first banyan network exit to their respective interface modules. The misrouted packets carry on into the next banyan network where the same process is repeated. This repeats up to the *L*th banyan network, and the packets that are still misrouted even at the output of the last network are regarded as lost. As the load on successive banyan networks decreases, so does the likelihood of conflicts. This implies that it is possible to reduce the packet loss rate to the desired low level by attaching a sufficiently large number of banyan networks.

Various methods exist to resolve conflict if it occurs between two packets in a 2×2 switching element. The random selection method is one simple example. It picks up an arbitrary one of the two conflicted packets and routes it to the correct output port, while misrouting the other packet to the wrong output port with a mark put on its collision bit. Whenever a conflict occurs between a properly routed packet and a misrouted one, the misrouted packet is doomed to lose the conflict. Therefore, it is possible that a packet that gets misrouted at some stage within a banyan network does not influence the routing of the properly routed packets in that banyan network. When a misrouted packet begins routing afresh at the next banyan network, the collision bit is reset to that of a properly routed packet. The packets that arrived at its destined interface module are accepted via the address filters and then stored at its buffer after multiplexing.

Input/Output Buffer Switch

The input buffer switch has a simple switching structure, but requires a complex arbiter in order to obtain the desired switching performance. On the contrary, the output buffer switch does not require an arbiter, but its hardware complexity is high. As a compromise, a switch structure that has a simple arbiter and moderate hardware complexity can be devised by introducing buffers both at the input ports and at the output ports as shown in Figure 4.65. In such an input/output buffer switch, the switching network that can transfer three or four packets to a particular output simultaneously is normally used.

In this switch, a back-pressure mechanism between the input buffers and the output buffers can be used not only to overcome packet loss at the output buffers but also to reach higher throughput with a quite limited number of output buffers, at the expense of a large number of buffers installed at the input ports. Buffering at the input ports costs less, as the input buffer operates at the line speed with a single input line feeding each buffer. On the other hand, an

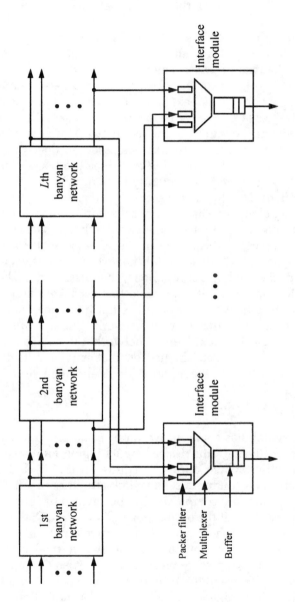

Figure 4.64   Basic structure of the tandem banyan switch.

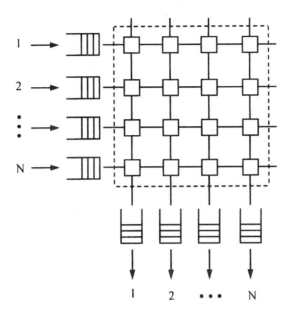

Figure 4.65   Structure of input/output buffer switch.

output buffer has to have multiple inputs or run at a speed faster than the output line in order to be able to accept packets from multiple inputs during the same time slot. So, it is advantageous to have a small amount of buffering at the output ports and to have a sufficient amount of input buffering to achieve the desired performance.

Internal Buffer Switch

In the previous three buffering schemes, packet contention within the switching network is avoided by employing an arbitration function outside the switching network, so buffers are located external to the switching network, that is, at the input and/or output ports. In the internal buffer switch, the packet contention resolution functions are distributed internally throughout the switching network and, consequently, the buffering functions are located inside the switching network. The buffered banyan network is a typical example of the internal buffer switch. In this switch, if two packets at both input ports of a switching element are addressed to the same output at the same time slot, only one is allowed to pass, whereas the other is stored in the internal buffer for one time slot.

4.10.3 Large-Scale Switching Systems

In practice, it is estimated that a central office of a broadband network would require switching fabrics with tens of thousand of high-speed ports and connections for hundreds of thousand of terminals. Most switch architectures discussed so far cannot be easily scaled up to that size. For example, a Batcher-banyan network of ten thousand ports requires synchronization of up to ten thousand packets over a network of about one hundred stages. The knockout switch also encounters significant difficulties for ten thousand ports, because each bus in the architecture must have fanouts of more than ten thousand. So, a popular approach to designing a large-scale switching system has been to interconnect many independent small switch modules in such a way that the overall system can meet the switching requirements.[41] In the following, some of architectures proposed for large-scale ATM switches are introduced.

Knockout Switch-Based Large-Scale Structure

As discussed before, the knockout switch is difficult to implement as the network size becomes larger due to the increasing fanout problem. This problem can be worked out by partitioning the output ports into a number of groups and providing a common interface module that has a switching function for each output group [Chao91]. Figure 4.66 shows the switch structure based on this output grouping concept. The operation of the switch is similar to that of the knockout switch. Each incoming packet is first transferred to an appropriate interface module via the interconnection fabric and then transferred to its destined output port at the module.

Multiplexer-Based Large-Scale Structure

This structure consists of a set of switch modules and multiplexers, as shown in Figure 4.67 [Lee90]. In this structure, the incoming packets are first transferred to their destined output ports by the switching modules and then multiplexed into a single stream by the back-end multiplexers. The switching modules, whose size is $M \times N$ (with $M = N/K$), consist of a Batcher sorting network, a stack of binary trees, and a bundle of banyan networks.

In this structure, a set of switch modules is interconnected at the outputs by multiplexers. Thus, no interference occurs between switch modules. This

---

41. Another approach can be found in implementing large-scale ATM switches using high-speed switching modules. Although this approach raises many problems for LSI devices (especially memory) and packaging technologies, it has the advantages that fewer interface modules may be required due to multiplexing, and may be more efficiently utilized due to statistical multiplexing.

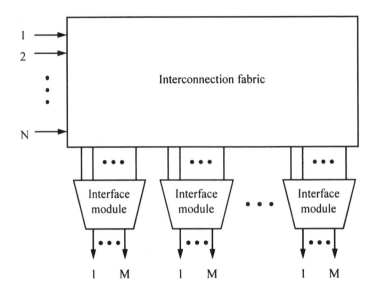

Figure 4.66 Large-scale structure based on the knockout switch.

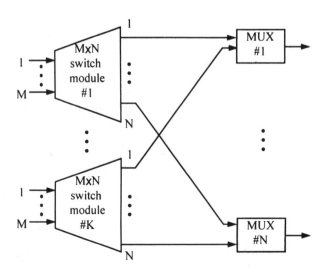

Figure 4.67 Large-scale structure based on the multiplexer.

architecture allows for independent clocking of modules, which substantially simplifies timing. It also allows for simple fault tolerance by providing a spare module, not a duplication of the entire switch. However, the number of interconnections between the switching modules and the multiplexers is very large.

Clos Network Based Large-Scale Structure

This switch structure has the three-stage structure [Eng92, Chiussi92] depicted in Figure 4.68, where a partition is inserted between the front-end interconnection fabric, which consists of two stages of switch modules and the column of output packet switch modules. The outputs are divided into groups of n lines each. All incoming packets are routed through the front-end interconnection fabric for instantaneous delivery based on their destined output group addresses. For each output group, the corresponding output packet switch module has m ($m > n$) inputs, meaning that up to m packets can be accepted for that output group in each time slot.

In this structure, an input packet can always be transferred to its destination via m different second-stage switch modules. This comes from the fact that the interconnection patterns between the switching stages are the same as those of the Clos network. So, the overall performance depends on the packet distribution algorithm that distributes incoming packets to the second-stage switch modules. In addition, the structure of each switch module is important to get the desired performance.

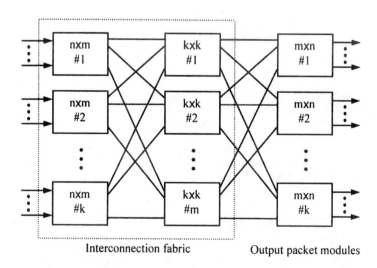

Interconnection fabric   Output packet modules

Figure 4.68   Large-scale structure based on the Clos network.

Banyan Network-Based Large-Scale Structures

Figure 4.69 shows one example of a large-scale ATM switch structure that is based on the banyan network [Liew90, Kim95c]. Similar to the previous scheme, this structure is composed of a set of switching stages. However, the interconnection patterns between the switching stages in this structure follow that of the banyan network. So, the switch modules to which each incoming packet should be routed is uniquely determined by the routing information of the packet.

In this structure, two adjacent switch modules are interconnected by more than one physical link, called a *channel group*. So, multiple paths exist between any input and any output in the overall switch architecture. This means that packets of an ATM connection may travel over different paths within the switch and arrive at their outputs out-of-sequence, due to delay differences of the paths. This implies that in order to guarantee in-order packet sequencing throughout the switch, it is necessary to maintain the time ordering among the packets not only in the same link but also in all links that belong to the same channel group.

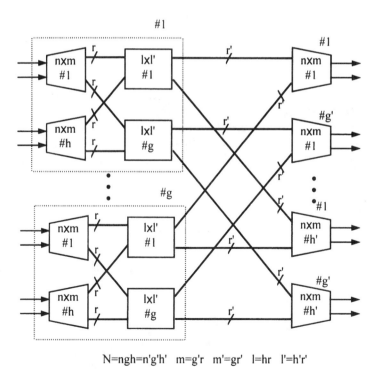

$$N=ngh=n'g'h' \quad m=g'r \quad m'=gr' \quad l=hr \quad l'=h'r'$$

Figure 4.69 Large-scale structure based on the banyan network.

An efficient way to achieve this is to equate the time ordering among the packets of an ATM connection with their ordering among the packets and maintain this relative ordering throughout packet propagation within the switch. Then, within a channel group, a lower indexed link has the higher time ordering than does a higher indexed link at each time slot.

4.10.4 Accommodation of Point-to-Multipoint Services

To accommodate point-to-multipoint services such as teleconferencing, distributed data processing, LAN bridging, and video distribution, a function that can transfer an incoming packet to a set of output port groups simultaneously, that is, a packet replication function, should be added to ATM switching systems.

Addition of the packet replication function to the ATM switch depends largely on the ATM switching system architecture. For example, it can be easily accommodated in the shared-medium switch, because all incoming packets are broadcast to a shared-medium to which all output ports are connected in the switch. However, in the switches based on the banyan network, the replication function cannot be accommodated without altering the switch structure. In this case the packet replication function is provided by cascading an additional copy network in the front stage as shown in Figure 4.70, in general. According to the structure, a point-to-multipoint packet is delivered to its destined outputs through the following three steps: First, the packet is replicated in the front-end copy network. Second, the destined outputs of the replicated packets are inquired after and an appropriate routing field is added to each of them in the *trunk number translators* (TNT). Third, the replicated packets are routed to their destination ports through the point-to-point switching network.

Because the copy network is in charge of the packet replication function only, it can be easily implemented by using a *broadcast banyan network* (BBN),

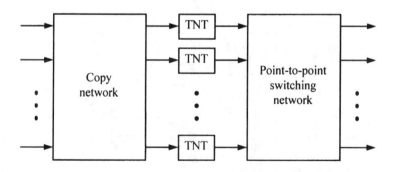

TNT: Trunk Number Translator

Figure 4.70 Accommodation of point-to-multipoint services based on a copy network.

which is a banyan network equipped with the packet replication function. One example [Lee88] of a copy network based on the BBN is shown in Figure 4.71. The packet replication process of this network can be divided into the address encoding process and address decoding process. The address encoding process transforms the set of copy numbers specified in the headers of incoming packets into a set of monotone address intervals, which form new packet headers. This process is performed by the concentrator, the *running adder* (RA), and the *dummy adder encoders* (DAE). The running adder adds the copy numbers of incoming packets sequentially from the top and transfers the results to the DAEs. Then, the DAEs assign a set of monotone output addresses to each packet based on the results. These addresses indicate the outputs to which each packet should be transferred in the following BBN. In the address decoding process, the packet replications are performed at the BBN. Finally, the destinations of the replicated packets are determined by the TNTs.

4.10.5 Other Related Issues

A number of issues need to be considered, other than the switching fabric itself, in order to implement an efficient ATM switching system. In the following, some of important issues are considered, such as routing information placement, switch test and supervision, and switch control schemes.

Routing Information Placement

The self-routing scheme is a switching technique that is normally employed in ATM switching systems. In an ATM switch that employs the self-routing scheme, incoming packets are routed to their destined output ports based on only their routing information. With respect to the routing information

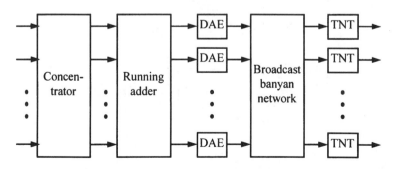

DAE: Dummy Address Encoder
TNT: Trunk Number Translator

Figure 4.71 Example of copy network with the broadcast banyan network.

placement, the information can either be carried by each packet in the so-called routing header, or it can be stored in routing tables in the basic switching elements of the switching network. In the former case, a routing header is added to each incoming packet at the line interfaces in general, and the packet is self-routed to its destination based on this routing header. This added header requires an increased speed, compared to the external speed. In the latter case, the routing table in each basic switching element provides the necessary routing information, so neither routing header nor operating speed increase is required. Instead, internal routing tables are necessary for switching elements. Figures 4.72(a,b), respectively, illustrate these two examples of routing information placements.

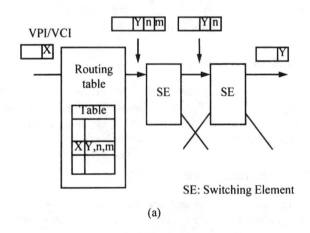

(a)

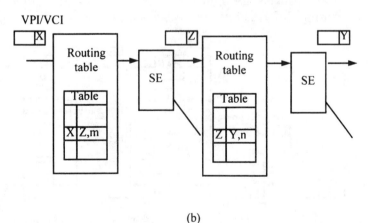

(b)

Figure 4.72 Routing information placement: (a) in routing header and (b) in routing table.

Switch Test and Supervision

Since the ATM switching system manages the logical channels of the VP and VC, the system should have the OAM functions to supervise VP/VC channels. Also, it must be capable of testing itself to determine if failure has occurred in switching paths or if misrouting has occurred in ATM cells. In general, such functions can be performed by using test cells. That is, it is possible to perform the functions by comparing the original test cells with the cells that have traveled through the path under test. Test cells can be generated by attaching a test generator as shown in Figure 4.73, with a special VCI assigned in the VCI field for easy distinction from other cells, and with a cell sequence number and pseudorandom binary bits assigned in the payload field.

Switch Control Schemes

For control of the switching system and call processing, switching systems, in general, utilize a mixture of central control and distributed control schemes. In the BISDN environment, the distributed control scheme appears more popular due to its flexible and modular realizability. Figure 4.74 illustrates ATM switch structures that employ the two typical distributed control schemes: a *multi-module structure* and a *multiprocessor structure*.

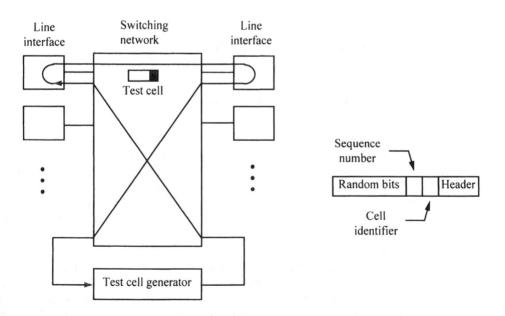

Figure 4.73 Example of switch test and supervision function implementation.

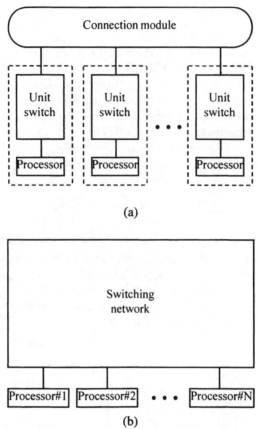

Figure 4.74 ATM switching system structure based on the distributed control scheme: (a) multi-module-based structure and (b) multiprocessor-based structure.

The ATM switch employing the multimodule-based control scheme consists of a connection module and multiple unit switch modules. Each switch module has a switching network and a processor for call processing. Multiple-unit switches are connected through the connecting module to attain the desired switching capacity. The multimodule-based architecture renders control simple because call processing and system control are independently done in each switch module.

The ATM switch employing the multiprocessor-based control scheme consists of a large switching network and a set of processors for call processing. This architecture can be flexibly designed to satisfy the need for BISDN as well as low-bit-rate services. If lower bit rate services dominate, then a small switching network can be used together with a large number of processors; and if

BISDN services dominate, then a large switching network can be used with a reduced number of processors.

Selected Bibliography

ATM Forum 93-215(R8), "Broadband interconnection interface specification document," 1993.

ATM Forum 93-590, "Data exchange interface specification document," 1993.

ATM Forum, "ATM user-network interface specification version 2.0," 1992.

ATM Forum, *ATM user-network interface specification: Version 3.0,* Prentice Hall, 1993.

ATM Forum, "Network compatible ATM for local network applications," Phase 1, Version 1.0, 1992.

ITU-T Rec. F.811, "Broadband connection-oriented bearer services," 1992.

ITU-T Rec. F.812, "Broadband connectionless data bearer service," 1992.

ITU-T Rec. I.113, "Vocabulary terms for broadband aspects of ISDN," 1993 (Rev).

ITU-T Rec. I.120, "Integrated services digital networks (ISDN)," 1993 (Rev).

ITU-T Rec. I.121, "Broadband aspects of ISDN," 1991 (Rev).

ITU-T Rec. I.140, "Attribute technique for the characterization of the telecommunication services supported by an ISDN and network capability of an ISDN," 1992 (Rev).

ITU-T Rec. I.150, "BISDN asynchronous transfer mode functional characteristics," 1993 (Rev).

ITU-T Rec. I.211, "BISDN service aspects," 1993 (Rev).

ITU-T Rec. I.311, "BISDN general network aspects," 1993 (Rev).

ITU-T Rec. I.321, "BISDN protocol reference model and its application," 1991.

ITU-T Rec. I.327, "BISDN functional architecture," 1993.

ITU-T Rec. I.356, "BISDN ATM layer cell transfer performance," 1993.

ITU-T Rec. I.361, "BISDN ATM layer specification," 1993 (Rev).

ITU-T Rec. I.362, "BISDN ATM adaption layer (AAL) functional description," 1993 (Rev).

ITU-T Rec. I.363, "BISDN ATM adaptation layer (AAL) specification," 1993 (Rev).

ITU-T Rec. I.364, "Support of broadband connectionless data service on BISDN," 1993.

ITU-T Rec. I.371, "Traffic control and congestion control in BISDN," 1993.

ITU-T Rec. I.374, Framework recommendation on "network capabilities to support multimedia services," 1993.

ITU-T Rec. I.413, "BISDN user-network interface," 1993 (Rev).

ITU-T Rec. I.432, "BISDN user-network interface—physical layer specification," 1993 (Rev).

ITU-T Rec. I.610, "BISDN operation and maintenance principles and functions," 1993 (Rev).

Abdelaziz, M., and I. Stavrakakis, "Some optimal traffic regulation schemes for ATM networks: A markov decision approach," *IEEE/ACM Trans. on Networking,* Vol. 2, No. 5, Oct. 1994, pp. 508–519.

Anderson, J., and M. D. Nguyen, "ATM-layer OAM implementation issues," *IEEE Commun. Mag.,* Vol. 29, No. 9, Sep. 1991, pp. 79–81.

Aoyama, T., I. Tokizawa, and K. Sato, "ATM VP-based broadband networks for multimedia services," *IEEE Commun. Mag.,* Vol. 31, No. 4, Apr. 1993, pp. 30–39.

Armbruster, H., and G. Arndt, "Broadband communication and its realization with broadband ISDN," *IEEE Commun. Mag.,* Vol. 25, No. 11, Nov. 1987, pp. 8–19.

Armbruster, H., "The flexibility of ATM: Supporting future multimedia and mobile communications," *IEEE Commun. Mag.,* Feb. 1995, pp. 76–84.

Armitage, G. J., and K. M. Adams, "Packet reassembly during cell loss," *IEEE Network,* Sep. 1993, pp. 26–34.

Babson, M., D. Buster, G. De Val, and J. S. Xavier, "ATM switching and CPE adaptation in the north carolina information highway," *IEEE Network,* Vol. 8, No. 6, Nov./Dec. 1994, pp. 40–47.

Basch, B. E., et al., "VISTAnet: A BISDN field trial," *IEEE LTS Mag.,* Vol. 2, No. 3, Aug. 1991, pp. 22–30.

Bauch, H., "Transmission systems for the BISDN," *IEEE LTS Mag.*, Vol. 2, No. 3, Aug. 1991, pp. 31–36.

Berenbaum, A., J. Dixon, A. Iyengar, and S. Keshav, "A flexible ATM-host interface for XUNET II," *IEEE Network*, Vol. 7, No. 4, Jul. 1993, pp. 18–23.

Bianchi, G., and J. S. Turner, "Improved queueing analysis of shared buffer switching networks," *IEEE/ACM Trans. on Networking*, Vol. 1, No. 4, Aug. 1993, pp. 482–490.

Bonomi, F., and K. W. Fendick, "The rate-based flow control framework for the available bit rate ATM service," *IEEE Network*, Mar./Apr. 1995, pp. 25–39.

Boudec, J. Y. L., "The Asynchronous transfer mode: A tutorial," *Computer Networks and ISDN Systems*, Vol. 24, 1992, pp. 279–309.

Breuer, H. J., "ATM layer OAM: Principles and open issues," *IEEE Commun. Mag.*, Vol. 29, No. 9, Sep. 1991, 75–78.

Burgin, J., and D. Dorman, "BISDN resource management: The role of virtual paths," *IEEE Commun. Mag.*, Vol. 29, No. 9, Sep. 1991, pp. 44–49.

Byrne, W. R., B. L. Kilim, and M. D. Soneru, "Broadband ISDN technology and architecture," *IEEE Network*, Vol. 3, No. 1, Jan. 1989, pp. 7–13

Byrne, W. R., et al., "Evolution of metropolitan public network and switch architecture," *IEEE Commun. Mag.*, Vol. 29, No. 1, Jan. 1991, pp. 69–82.

Byun, J. W., and T. T. Lee, "The design and analysis of an ATM multicast switch with apaptive traffic controller," *IEEE/ACM Trans. on Networking*, Vol. 2, No. 3, Jun. 1994, pp. 288–298.

Cao, X. -R., and D. Towsley, "A performance model for ATM switches with general packet length distributions," *IEEE/ACM Trans. on Networking*, Vol. 3, No. 3, Jun. 1995, pp. 299–309.

Chao, H. J., "A recursive modular terabit/sec ATM switch," *IEEE JSAC*, Vol. 9, No. 8, Oct. 1991, pp. 1161–1172.

Chao, H. J., and B. -S. Choe, "Design and analysis of a large-scale multicast output buffered ATM switch," *IEEE/ACM Trans. on Networking*, Vol. 3, No. 2, Apr. 1995, pp. 126–138.

Chen, D. X., and J. W. Mark, "SCOQ: A fast packet switch with shared concentration and output queueing," *IEEE/ACM Trans. on Networking*, Vol. 1, No. 1, Feb. 1993, pp. 142–151.

Chen, T. M., and S. S. Liu, "Management and control functions in ATM switching systems," *IEEE Network*, Vol. 8, No. 4, Jul./Aug. 1994, pp. 27–40.

Chiussi, F. M., and F. A. Tobagi, "A hybrid shared-memory/space-division architecture for large fast packet switches," in *Proc. ICC'92*, pp. 905–911.

Cidon, I., I. Gopal, and R. Guerin, "Bandwidth management and congestion control in planet," *IEEE Commun. Mag.*, Vol. 30, No. 10, Oct. 1991, pp. 54–65.

Cidon, I., I. Gopal, and A. Segall, "Connection establishment in high-speed networks," *IEEE/ACM Trans. on Networking*, Vol. 1, No. 4, Aug. 1993, pp. 469–481.

Cidon, I., R. Guerin, and A. Khamisy, "On protective buffer policies," *IEEE/ACM Trans. Networking*, Vol. 2, No. 3, Jun. 1994, pp. 240–246.

Clarke, D. E. A., and T. Kanada, "Broadband: The last mile," *IEEE Commun. Mag.*, Vol. 31, No. 3, Mar. 1993, pp. 94–100.

Coppo, P., M. D'Ambrosio, and R. Melen, "Optimal cost/performance design of ATM switches," *IEEE/ACM Trans. on Networking*, Vol. 1, No. 5, Oct. 1993, pp. 566–575.

Coudreuse, J. P., "Network evolution towards BISDN," *IEEE LTS Mag.*, Vol. 2, No. 3, Aug. 1991, pp. 66–70.

Cruz, R. L., "A calculus for network delay, part I: Network elements in isolation," *IEEE Trans. Information Theory*, Vol. 37, No. 1, Jan. 1991, pp. 114–121.

Daddis, G. E., Jr., and H. C. Torng, "A Taxonomy of broadband integrated switching architectures," *IEEE Commun. Mag.*, Vol. 27, No. 1, May 1989, pp. 32–42.

Day, A., "International standardization of BISDN," *IEEE LTS Mag.*, Vol. 2, No. 3, Aug. 1991, pp. 13–21.

Decina, M., C. Mossotto, and A. Roveri, "The ATM test-bed: An experimental platform for broadband communications," *IEEE Commun. Mag.*, Vol. 32, No. 10, Oct. 1994, pp. 78–83.

De Prycker, M., R. Peschi, and T. V. Landegem, "BISDN and the OSI protocol reference model," *IEEE Network,* Vol. 7, No. 2, Mar. 1993, pp. 10–19.

De Zhong, W., Y. Onozato, and J. Kaniyil, "A copy network with shared buffers for large-scale multicast ATM switching," *IEEE/ACM Trans. on Networking,* Vol. 1, No. 2, Apr. 1993, pp. 157–165.

Domann, G., "Two years of experience with broadband ISDN field trial," *IEEE Commun. Mag.,* Vol. 29, No. 1, Jan. 1991, pp. 90–96.

Eckberg, A. E., B. T. Doshy, and R. Zoccolillo, "Controlling congestion in BISDN/ATM: Issues and strategies," *IEEE Commun. Mag.,* Vol. 29, No. 9, Sep. 1991, pp. 64–74.

Eigen, D. J., "Narrowband broadband ISDN CPE directions," *IEEE Commun. Mag.,* Vol. 28, No. 4, Apr. 1990, pp. 39–46.

Elwalid, A. I.,and D. Mitra, "Effective bandwidth of general Markovian traffic sources and admission control of high speed networks," *IEEE/ACM Trans. Networking,* Vol. 1, No. 3, June. 1993, pp. 329–343.

Eng, K. Y., M. J. Karol, and Y. S. Yeh, "A growable packet (ATM) switch architecture: Design principles and applications," *IEEE Trans. Commun.,* Vol. 40, No. 2, Jan. 1992, pp. 423–439.

Filipiak, J., "M-architecture: A structural model of traffic management and control in broadband ISDNs," *IEEE Commun. Mag.,* Vol. 27, No. 5, May 1989, pp. 25–31.

Fischer, W., et al., "Data communications using ATM: Architectures, protocols, and resource management," *IEEE Commun. Mag.,* Vol. 32, No. 8, Aug. 1994, pp. 24–33.

Fowler, H. J., "TMN-based broadband ATM network management," *IEEE Commun. Mag.,* Vol. 33, No. 3, Mar. 1995, pp. 74–79.

Frame, M., "Broadband service needs," *IEEE Commun. Mag.,* Vol. 28, No. 4, Apr. 1990, pp. 55–58.

Gallassi, G., G. Rigolio, and L. Verri, "Resource management and dimensioning in ATM networks," *IEEE Network,* May 1990.

Garcia-Haro, J., and A. Jajszczyk, "ATM shared-memory switching architectures," *IEEE Network,* July/Aug. 1994, pp. 18–26.

Gechter, J., and P. O'reilly, "Conceptual issues for ATM," *IEEE Network,* Vol. 3, No. 1, Jan. 1989, pp. 14–16.

Giacopelli, J. N., et al., "Sunshine: A high-performance self-routing broadband packet switch architecture," *IEEE JSAC,* Vol. 9, No. 8, Oct. 1991, pp. 1289–1298.

Gilbert, H., O. Aboul-Magd, and V. Phung, "Developing a cohesive traffic management strategy for ATM networks," *IEEE Commun. Mag.,* Vol. 30, No. 10, Oct. 1991, pp. 36–45.

Goeldner, E. H.,and M. N. Huber, "Multiple access for BISDN," *IEEE LTS Mag.,* Vol. 2, No. 3, Aug. 1991, pp. 37–43.

Golestani, S. J., "Congestion-free transmission of real-time traffic in packet networks," in *Proc. INFOCOM,* 1990, pp. 527–536.

Guillemin, F.,and J. W. Roberts, "Jitter and bandwidth enforcement," in *Proc. GLOBECOM,* Dec. 1991, pp. 261–265.

Haas, Z., "A protocol structure for high-speed communication over broadband ISDN," *IEEE Network,* Vol. 5, No. 1, Jan. 1991, pp. 64–70.

Habib, I. W., and T. N. Saadawi, "Controlling flow and avoiding congestion in broadband networks," *IEEE Commun. Mag.,* Vol. 30, No. 10, Oct. 1991, pp. 46–53.

Habib, I. W., A. A. Tarraf, and T. N. Saadawi, "Intelligent traffic control for ATM broadband networks," *IEEE Commun. Mag.,* Vol. 33, No. 10, Oct. 1995, pp. 76–85.

Handel, R., "Evolution of ISDN towards broadband ISDN," *IEEE Network,* Vol. 3, No. 1, Jan. 1989, pp. 7–13.

Handel R., and M. N. Huber, *Integrated Broadband Networks: An Introduction ATM-Based Networks,* Addison Wesley, 1991.

Hluchyj, M. G.,and M. J. Karol, "Queueing in high-performance packet switching," *IEEE JSAC,* Vol. 6, No. 9, Dec. 1988, pp. 1587–1597.

Hong, D., and T. Suda, "Congestion control and prevention in ATM networks," *IEEE Network*, Vol. 5, No. 4, Jul. 1991, pp. 10–17.

Imai, K., T. Honda, H. Kasahara, and T. Ito, *ATMR: Ring Architecture for Broadband Networks*, 1990.

Kalmanek, C. R., and H. Kanakia, "Rate controlled servers for very high-speed networks," in *Proc. GLOBECOM*, 1990, pp. 12–20.

Kesidis, G., J., Walrand, and C. -S. Chang, "Effective bandwidths for multiclass Markov fluids and other ATM sources," *IEEE/ACM Trans. on Networking*, Vol. 1, No. 4, Aug. 1993, pp. 424–428.

Kim, S. C., and B. G. Lee, "Synchronization of shift register generators in distributed sample scramblers," *IEEE Trans. on Commun.*, Vol. 42, No. 3, Mar. 1994, pp. 1400–1408.

Kim, H. S., "Design and performance of multinet switch: A multistage ATM switch architecture with partially shared buffers," *IEEE/ACM Trans. on Networking*, Vol. 2, No. 6, Dec. 1994, pp. 571–580.

Kim, B. G., and P. Wang, "ATM network: Goals and challenges," *Communications of The ACM*, Vol. 38, No. 2, Feb. 1995, pp. 40–44.

Kim, K. S., and B. G. Lee, "Three-level traffic shaper and its application to source clock frequency recovery for VBR video services in ATM network," *IEEE/ACM Trans. on Networking*, Vol. 3, No. 4, Aug. 1995, pp. 450–458.

Kim, Y. M., and K. Y. Lee, "KSMINs: Knockout switch based multistage interconnection networks for high-speed packet switching," *IEEE Trans. on Commun.*, Vol. 42, No. 8, Aug. 1995, pp. 2391–2398.

Kishimoto, R., and I. Yamashita, "HDTV communication systems in broadband communication networks," *IEEE Commun. Mag.*, Vol. 29, No. 8, Aug. 1991, pp. 28–35.

Kitawaki, N., H. Nagabuchi, M. Raka, and K. Takahashi, "Speech coding technology for ATM networks," *IEEE Commun. Mag.*, Vol. 28, No. 1, Jan. 1990, pp. 21–27.

Kodama, T., and T. Fukuda, "Customer premises networks of the future," *IEEE Commun. Mag.*, Feb. 1994, pp. 96–98.

Kung, H. T., and R. Morris, "Credit-based flow control for ATM networks," *IEEE Network*, Vol. 9, No. 2, Mar./Apr. 1995, pp. 40–48.

Kuwahara, H., et al., "Shared buffer memory switch for an ATM exchange," in *Proc. ICC'89*, pp. 4.4.1–4.4.5.

Kwok, T., "A vision for residential broadband services: ATM-to-the-home," *IEEE Network*, Vol. 9, No. 5, Sep./Oct. 1995, pp. 14–29.

La Porta, T. F., et al., "BISDN: A technological discontinuity," *IEEE Commun. Mag.*, Vol. 32, No. 10, Oct. 1994, pp. 84–97.

Lazar, A. A., and G. Pacifici, "Control of resources in broadband networks with quality of service guarantee," *IEEE Commun. Mag.*, Vol. 30, No. 10, Oct. 1991, pp. 63–73.

Lee, T. T., "Nonblocking copy networks for multistage packet switching," *IEEE JSAC*, Vol. 6, No. 9, Dec. 1988, pp. 1455–1467.

Lee, T. T., "A modular architecture for very large packet switches," *IEEE Trans. on Commun.*, Vol. 38, No. 7, Jul. 1990, pp. 1097–1106.

Lee, T. H., "Design and analysis of a new self-routing network," *IEEE Trans. on Commun.*, Vol. 40, No. 1, Jan. 1992, pp. 171–177.

Lee, J. W., and B. G. Lee, "MMPP-model based analysis of cell delay variation in ATM networks," in *Proc. ICCCN*, 1994, pp. 12–16.

Lee, B. G., and S. C. Kim, *Scrambling Techniques for Digital Transmission*, Springer Verlag, 1994.

Lee, B. G., and S. C. Kim, "Low-rate parallel scrambling techniques for today's lightwave transmission," *IEEE Commun. Mag.*, Vol. 33, No. 4, Apr. 1995, pp. 84–95.

Lee, J. W., and B. G. Lee, "New queueing strategies employing service interval-based priority for real-time communications in ATM networks," *Annales Des Telecommunications*, Vol. 50, No. 7–8, Jul./Aug. 1995, pp. 617–623.

Li, S. Q., "Performance of a nonblocking space-division switch with correlated input traffic," *IEEE Trans. Commun.*, Vol. 40, No. 1, Jan. 1992, pp. 97–108.

Liew, S. C., and K. W. Lu, "A 3-stage interconnection structure for very large packet switches," in *Proc. ICC'90*, pp. 316.7.1–316.7.7.

Lyles, J. B.,and D. C. Swinehart, "The emerging gigabit environment and the role of local ATM," *IEEE Commun. Mag.*, Vol. 30, No. 4, Apr. 1992, pp. 52–59.

Mckinney, R. S.,and T. H. Gordon, "ATM for narrowband services," *IEEE Commun. Mag.*, Apr. 1994, pp. 64–72.

Mesiya, M. F., "Implementation of a broadband integrated services hybrid network," *IEEE Commun. Mag.*, Vol. 26, No. 1, Jan. 1988, pp. 34–43.

Minzer, S. E., "Broadband ISDN and asynchronous transfer mode (ATM)," *IEEE Commun. Mag.*, Vol. 27, No. 9, Sep. 1989, pp. 6–14.

Minzer, S. E.,and D. R. Spears, "New directions in signalling for broadband ISDN," *IEEE Commun. Mag.*, Vol. 27, No. 2, Feb. 1989, pp. 6–14.

Mitra, N. and,S. D. Usikin, "Relationship of the signalling system No. 7 protocol architecture to the OSI reference model," *IEEE Commun. Mag.*, Vol. 5, No. 1, Jan. 1991, pp. 26–37.

Miura, H., K. Maki, and K. Nishihata, "SDH network evolution in Japan," *IEEE Commun. Mag.*, Feb. 1995, pp. 86–92.

Murano, K., et al., "Technologies towards broadband ISDN," *IEEE Commun. Mag.*, Vol. 28, No. 4, Apr. 1990, pp. 66–70.

Narasimha, M. J., "The Batcher-banyan self-routing network: Universality and simplification," *IEEE Trans. on Commun.*, Vol. 36, No. 10, Oct. 1988, pp. 1175–1178.

Natarajan, N.,and G. M. Slawsky, "A framework architecture for multimedia information networks," *IEEE Commun. Mag.*, Vol. 30, No. 2, Feb. 1992, pp. 97–104.

Nederiof, L., et al., "End-to-end survivable broadband networks," *IEEE Commun. Mag.*, Vol. 33, No. 9, Sep. 1995, pp. 63–71.

Nishio, M., S. Suzuki and K. Kaede, "A new architecture of photonic ATM switches," *IEEE Commun. Mag.*, Vol. 31, No. 4, Apr. 1993, pp. 62–69.

Okada, T., H. Ohnish, and N. Morita, "Traffic control in asynchronous transfer mode," *IEEE Commun. Mag.*, Vol. 29, No. 9, Sep. 1991, pp. 58–63.

Oki, E., N. Yamanaka, and F. Pitcho, "Multiple-availability-level ATM network architecture," *IEEE Commun. Mag.*, Vol. 33, No 9, Sep. 1995, pp. 80–89.

Olshanksy, R., "Subscriber multiplexed broadband service network: a migration path to BISDN," *IEEE LTS Mag.*, No. 3, Aug. 1990, pp. 30–34.

Padmanabhan, K., "An efficient architecture for fault-tolerant ATM switches," *IEEE/ACM Trans. on Networking*, Vol. 3, No. 5, Oct. 1995, pp. 527–537.

Parekh, A. K.,and R. G. Gallager, "A generalized processor sharing approach to flow control in integrated services networks: the single-node case," *IEEE/ACM Trans. on Networking*, Vol. 1, No. 3, Jun. 1993, pp. 344–357.

Pattavina, A. and G. Bruzzi, "Analysis of input and output queueing for nonblocking ATM switches," *IEEE/ACM Trans. on Networking*, Vol. 1, No. 3, Jun. 1993, pp. 314–328.

De Prycker, M., "ATM switching on demand," *IEEE Network*, Vol. 6, No. 2, Mar. 1992, pp. 25–29.

Ramakrishanan, K. K.,and P. Newman, "Integration of rate and credit schemes for ATM flow control," *IEEE Network*, Vol. 9, No. 2, Mar./Apr. 1995, pp. 49–56.

Ramanathan, S. and,P. V. Rangan, "Adaptive feedback technique for synchronized multimedia retrieval over integrated networks," *IEEE/ACM Trans. on Networking*, Vol. 1, No. 2, Apr. 1993, pp. 246–260.

Rathgeb, E. P., "Modeling and performance comparison of policing mechanisms for ATM networks," *IEEE JSAC*, Vol. 9, No. 3, Apr. 1991, pp. 325–334.

Rider, M. J., "Protocols for ATM access networks," *IEEE Network*, Vol. 3, No. 1, Jan. 1989, pp. 17–22.

Rigolio, G., and L. Verri, "Resource management and dimensioning in ATM networks," *IEEE Network*, Vol. 4, No. 3, May 1990, pp. 8–17.

Roberts, J. W., "Variable bit rate traffic control in BISDN," *IEEE Commun. Mag.*, Vol. 29, No. 9, Sep. 1991, pp. 50–57.

Saito, H.,and K. Shiomoto, "Dynamic call admission control in ATM networks," *IEEE JSAC*, Vol. 9, No. 7, 1991.

Sato, K., H. Ueda, and N. Yoshikai, "The role of virtual path crossconnection," *IEEE LTS Mag.*, Vol. 2, No. 3, Aug. 1991, pp. 44–54.

Smouts, M., *Packet Switching Evolution from Narrowband to Broadband ISDN*, Artech House, 1992.

Stallings, W., *Advances in ISDN and Broadband ISDN*, IEEE Computer Society Press, 1992.

Suzuki, H., et al., "Output-buffer switch architecture for asynchronous transfer mode," in *Proc. ICC'89*, pp. 99–103.

Suzuki, T., "ATM adaptation layer protocol," *IEEE Commun. Mag.*, Apr. 1994, pp. 80–83.

Tassiulas, L., Y. C. Hung and S. S. Panwar, "Optimal buffer control during congestion in an ATM network node," *IEEE/ACM Trans. on Networking*, Vol. 2, No. 4, Aug. 1994, pp. 374–386.

Thekkath, C. A., et al., "Implementing network protocols at user level," *IEEE/ACM Trans. on Networking*, Vol. 1, No. 5, Oct. 1993, pp. 554–565.

Tobagi, F. A., "Fast packet switch architecture for broadband integrated services digital networks," in *Proc. IEEE*, Vol. 78, No. 1, Jan. 1990, pp. 133–167.

Tobagi, F. A., T. Kwok, and F. M. Chiussi, "Architecture, performance, and implementation of the tandem banyan fast packet switch," *IEEE JSAC*, Vol. 9, No. 8, Oct. 1991, pp. 1173–1193.

Toda, I., "Migration to broadband ISDN," *IEEE Commun. Mag.*, Vol. 28, No. 4, Apr. 1990, pp. 55–58.

Vakil, F.,and H. Saito, "On congestion control in ATM networks," *IEEE LTS Mag.*, Vol. 2, No. 3, Aug. 1991, pp. 55–65.

Vecchi, M. P., "Broadband networks and services: architecture and control," *IEEE Commun. Mag.*, Vol. 33, No. 8, Aug. 1995, pp. 24–33.

Vetter, R. J., "ATM concepts, architectures, and protocols, *Communications of the ACM*, Vol. 38, No. 2, Feb. 1995, pp. 31–38.

Walters, S. M., "A New Direction for Broadband ISDN," *IEEE Commun. Mag.*, Vol. 29, No. 9, Sep. 1991, pp. 39–43.

White, P. E., "The Role of the Broadband Integrated Services Digital Network," *IEEE Commun. Mag.*, Vol. 29, No. 3, Mar. 1991, pp. 116–119.

Wolf, S., et al., "How will we rate telecommunications system performance," *IEEE Commun. Mag.*, Vol. 29, No. 10, Oct. 1991, pp. 23–30.

Yeh, Y. -S., M. G. Hluchyj, and A. S. Acampora, "The knockout switch: A simple, modular architecture for high-performance packet switching," *IEEE JSAC*, Vol. SAC-5, No. 8, Oct. 1987, pp. 1274–1283.

Yoneda, S., "Broadband ISDN ATM layer management: operations, administration, and maintenance considerations," *IEEE Network*, Vol. 4, No. 3, May 1990, pp. 31–35.

Zhang, L., "Virtual clock: A new traffic control algorithm for packet switching networks," in *Proc. SIGCOMM*, 1990, pp. 19–29.

Zhang, H. and D. Ferrari, "Rate-controlled static-priority queueing," in *Proc. INFOCOM*, 1993, pp. 227–236.

High-Speed Data Networks and Services

5

Spurred by rapid advances in semiconductor, transmission, switching, protocol, and computer technologies, information networks are becoming faster, more diverse, and more broadband oriented, taking us ever closer to satisfying all the communication needs of human beings. The scope of communication needs ranges from the transfer of simple voice sounds to the transfer of data, graphics, and images, and from low transmission speeds of a few kilobits per second to high transmission speeds of hundreds of megabits per second. Its area of coverage is globalizing, expanding from local area to worldwide usage.

The LAN systems, which were installed to integrate a multiple number of computers and to share data and resources on the premises of universities, research institutions, and business sectors, used to be mostly medium- to low-speed at around 10 Mbps, as exemplified by Ethernet. But recently, the appearance of powerful, high-performance workstations, possessing network operating systems for LAN applications, has led to the emergence of the FDDI. The FDDI is a high-speed backbone LAN designed to accommodate distributed processing environments that integrate mainframes, and to link low-speed LANs scattered inside a building.

With increasing speed and distributed processing, the communication capacity of such a LAN environment is going beyond local areas to wide areas, and the need for high-speed, wide-area networking, with the internetworking of LANs as its main objective, is being recognized. Networks already exist that can interconnect LANs, such as the *packet-switched data network* (PSDN), *circuit-switched data network* (CSDN), and high-speed digital leased line. However, the frame relay technology that is based on frame multiplexing at the data link layer enables high-speed transmission by minimizing network functions for error control and flow control to meet high-speed access requirements for LAN interconnection.

Whereas a LAN covers a local area of small radius, a MAN has a structure and function that can transmit high-speed multimedia data over a metropolitan area of large radius. DQDB was adopted as the standard technology for the MAN by the IEEE 802.6. MANs are expected to take the roll of LANs to the BISDN in the early stages of BISDN.

Figure 5.1 displays the above-mentioned data communication networks and technologies in terms of bandwidths and application areas.

To realize high-speed networks that can meet the requirements of real-time multimedia communication services, research and standardization activities have been carried conducted during the past decades. The fields where such efforts have been exerted include IEEE high-speed LAN, DQDB, SMDS, and ATM data services. In this chapter, these topics are investigated one by one. For this, backgrounds and fundamentals are introduced on data communications and networks first, then the open system interconnection, TCP/IP protocol suite, and LAN, which are the bases for all computer communication networks, are reviewed. Then various high-speed data networks and services, such as FDDI, IEEE's high-speed LANs, DQDB, SMDS, and ATM data services are

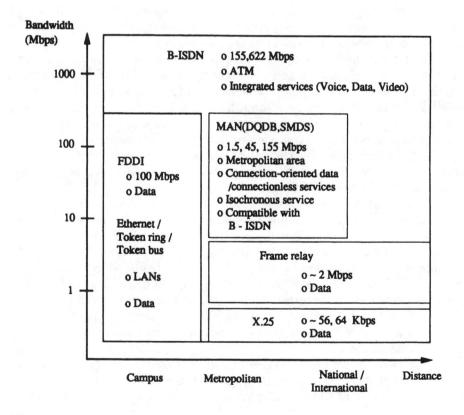

Figure 5.1 Data communication networks and technologies.

described, along with various high-speed network protocols. Finally, Internet and information infrastructures for information superhighways are discussed.

5.1 INTRODUCTION

As a preliminary to high-speed data networks and services, the fundamental concepts of data communications, such as communication protocols and architectures, *open system interconnection* (OSI), the TCP/IP suite, and fundamental data networks, are briefly reviewed in this introductory section. However, more detailed descriptions on OSI and TCP/IP follow in the subsequent two sections. With regard to the *local-area network* (LAN), which is the most fundamental form of data network, a separate section is dedicated to it, following the sections on OSI and TCP/IP.

5.1.1 Communication Protocols and Architectures

Due to its indispensable role in everyday human life, today's computer has been introduced in almost every part of the home and the office, and is being used in various applications such as word processing, business, customer management, automation management, and point-of-sale management. Until recently, most computers were used as standalone devices. However, the unification of computer and data communication fields in the 1970s and the early 1980s brought about a large change in the technologies and products of computer/communication and related industries, leading to the conceptual separation of data and information. *Data* refers to a syntax that has been formatted in a way suitable for communication, processing, and analysis by computers or human beings, while *information* refers to the semantics assigned to data by human beings according to defined conventions.

If the unification of computers and communications is illustrated from the data and information point of view, then, as shown in Figure 5.2, computer communications can be categorized into the network service that provides *bearer* services (i.e., data transport service) and the *telematic* service that provides end-to-end information services.

For two entities to communicate with each other they must speak the same language and must define what, how, and when to communicate through a common convention acknowledged by both. The basic elements of protocol, which is a rule for the exchange of data between two entities, are syntax, semantics, and timing. *Syntax* includes such items as data format, coding, and signal levels; *semantics* includes control information for coordination and error handling; and *timing* includes speed matching and sequencing. In most cases, the task of communicating between two entities on different systems is too complicated to be handled by a single process or module. Figure 5.3 conceptually illustrates a structural set of protocols and represents the case in which

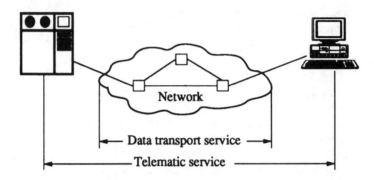

Figure 5.2   Concepts of computer communications.

two stations are connected through a multiswitched network. Between stations 1 and 2 of the figure, an application protocol is required that links the respective operations of two application processes and defines the appropriate syntax and semantics. Such an application protocol does not require information about the communication facilities existing between, and it possesses an interentity process-to-process protocol corresponding to the network service entity. This protocol executes such tasks as flow control or error control. Also, a protocol is required between station 1 and network A, and between station 2 and network B, and the access protocol allows a device to access the network. Other protocols are required within the network itself, such as node-to-node protocol and entry-to-exit protocol. Finally, a protocol is required between networks, which is the internetwork protocol.

Figure 5.3 is a general classification of protocols, and protocol architecture can be subdivided according to specific requirements. The structure of hardware and software required for defining functions of communications in this manner is called the *communication architecture*. In the following section, these concepts are used as a focal point to describe the open systems architecture standards and the OSI reference model.

5.1.2 Open Systems Interconnection (OSI)

The standards established by the international bodies for the computer industry of the past were mainly concerned with the internal operations of computers or connection with various types of peripheral devices. As a result, the hardware and software communications systems that were in circulation in the early days were plug-compatible systems that could only exchange information at the computers themselves. Such systems are called *closed systems*. Consequently, to exchange information with a computer from a different manufacturer, the respective standards of the other manufacturer must be followed.

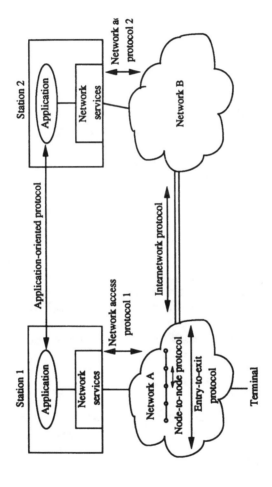

Figure 5.3 Relations of communication protocols.

In contrast, in the past few years several international standards bodies associated with common carriers have adopted standards that allow interlinking of network devices. Recommendations adopted in this manner provide compatibility among equipment from different vendors, allowing a purchaser to select equipment most suitable to his or her needs among the diverse equipment categories.

Initially, the services provided by most common carriers were mainly associated with data transmission; hence, standards were only concerned with interfacing a device to the network. Recently, however, common carriers have begun providing extensive distributed information services, such as teletex and videotex. Accordingly, standards bodies associated with the telecommunications industry have enacted high-level standards related not only to network interfacing, but also to information exchange control and the data format among systems, enabling mutual information exchange among equipment from different manufacturers. Such a system is called an *open system*, and such an environment is called the *open systems interconnection environment* (OSIE).

In the mid 1970s, different forms of distributed systems rooted in public and private data networks became widespread, magnifying the necessity for an open system. From then on, various categories of standards have been developed, and the very first to be announced at ISO was the *OSI reference model*, which was concerned with the overall structure related to the complete communication system inside each computer.

The objective of the OSI reference model is to provide a framework for standards development, as well as support existing and evolving standards activities within the framework, and eventually enable communication between application processes of different computers supporting the same standards. In other words, OSI is associated with information exchange between processes and allows an application process to carry out the distributed information processing task independently of the computer operation. (See Section 5.2 for a detailed discussion on OSI).

5.1.3 TCP/IP Suite

Although the OSI reference model standardizes the communication system structure as a template with the standards activities associated with each layer as the basis, it was not intended to standardize protocols associated with each layer. Rather, each layer is related to a set of standards, each of which provides different levels of functionality. Consequently, in a specific open systems environment, a set of standards is defined for all systems under that environment.

Before the OSI standards activities, the U.S. *Department of Defense* (DOD), through its own *Defense Advanced Research Projects Agency* (DARPA), has carried out basic research on computer communications and networking since 1969. As a part of the research, computer networks associated with several universities and research institutes were linked with DARPA's computer network, and this internetwork was called ARPANet.

The original aim of the ARPANet was to study the possibility of communications between computers. But, by the mid 1970s, this experiment has evolved into a system for interconnecting various networks of computers and into an architecture with different manufacturer's proprietary networks, data transmission protocols, hardware, and operating systems software. In response to this demand, the *transmission control protocol/internet protocol* (TCP/IP) suite was developed. In 1978, it was officially adopted by the U.S. government as the "preferred" networking protocol.

The TCP/IP protocol suite includes both the network-oriented protocols and application support protocols (see Section 5.3). TCP/IP is broadly used in the existing Internet, and a considerable portion of the TCP/IP protocols was used as the basis for OSI standards. Furthermore, since TCP/IP does not require a license fee, all of the associated protocol specifications are in the public domain. To create open systems networking environments, they are widely used by the commercial and public authorities as well. In practice, therefore, there are two major vendor-independent open systems standards: the TCP/IP protocol suite and those based on the OSI standards (see Sec. 5.3 for a detailed discussion on TCP/IP suite).[1]

5.1.4 Fundamental Data Networks

A *local-area network* (LAN) is the most fundamental data communication network formed among computers and peripheral equipment in a local area. In the LAN, computers and equipment are connected using coaxial cables or twisted-pair copper wires as the transmission medium. So the transmission speed of the LAN is usually about 10 -Mbps or below, and the transmission coverage is usually limited to a radius of about 2 km.

In general, the communications characteristics and performance of a LAN depend on the network topology and the transmission medium. The LAN's topology is usually categorized into ring, star, tree, and bus, with the most frequently used among them being the bus and ring topologies. Depending on the used transmission medium and transmission speed, LANs are divided into baseband systems and RF band systems.

Standards for various LAN protocols are well organized by the IEEE 802

---

1. A range of standards is associated with each layer in the case of ISO/ITU-T standards, and such standards allow an administrative authority to choose the set of standards most suitable for its application. The resulting protocol suite is known as the *open systems interconnection profile* (OSIP). A number of such profiles have now been defined, including *technical and office product system* (TOP), *manufacturing automation protocol* (MAP), and U.S. and U.K. GOSIP, for use in U.S. and U.K. government projects, respectively.

committee. In particular, the basic standards IEEE 802.3 through 802.5 prescribe the protocols, respectively, for CSMA/CD bus, token bus and token ring. These three protocols form the foundation of data communications for the LAN, which are further developed to encompass a wide area of coverage and high-speed data transmission (see Section 5.4 for detailed discussions of LANs).

As the transmission medium has evolved to fiber-optic cables, high-speed LANs have emerged. A typical example is the *fiber distributed digital interface* (FDDI) LAN. An FDDI LAN can transmit data at the 100-Mbps rate and can cover a much wider area. It is also possible to provide 100-Mbps rate data services on the copper-based medium by exploiting advanced transmission techniques, and typical examples can be found in the IEEE 802.3 and 802.12 high-speed LANs (see Section 5.5 for details about FDDI and other high-speed LANs).

As computers performance evolved and as users demanded even higher speed data communications and wider area connections, existing LANs evolved to satiate these needs. *Metropolitan-area networks* (MANs) and *wide-area networks* (WANs) emerged in answer to those demands.

MANs have the structure and function appropriate for a metropolitan area. A MAN can transmit high-speed data in the range of 2 to 155 Mbps, and can cover a metropolitan area with a radius of about 50 km. For the MAN, the preferred medium of transmission is optical cable, and the preferred topology is the bus, although the ring topology is also used. For example, a *distributed-queue dual-bus* (DQDB) MAN uses the bus topology, while the FDDI LAN uses the ring topology (FDDI LAN is sometimes categorized in the MAN's domain). Although a MAN is a high-speed data network from a functional point of view, it is a private network from the user's point of view (refer to Section 5.6 for a detailed discussion on DQDB).

WANs are networks that interconnect computers and equipment spread over a very wide area, such as nationwide public networks. The X.25-based *packet-switched public data network* (PSPDN) is a representative WAN with a limited transmission capacity of the DS-0 rate. Frame relay technology is the next evolutionary step from X.25 technology that has increased the transmission capacity to the DS-1 rate. *Switched multimegabit data service* (SMDS), which is the next generation of technology based on the DQDB MAN, can provide wide-area data services at the DS-3 rate (refer to Section 5.4.6 for the frame relay services and Section 5.6.5 for the SMDS information).

5.2 OPEN SYSTEM ARCHITECTURE

For today's data communication, the concept of open systems interconnection assumes a very fundamental role. It enables compatibility among data equipment from different venders, thus allowing for interconnection among independent data systems. In this section, the open systems interconnection and the related protocol reference model, services, and protocols are considered.

5.2.1 OSI Reference Model

In the early stages, the software of communication systems was implemented in the form of single, complex, and unstructured programs. As a result, the software developed in this fashion was extremely difficult to test and modify. Also, communication protocols were too complicated to implement as single protocols and could not support different physical networks.

To overcome these problems, ISO adopted a layered approach for the reference model. According to the layered principle, a complete communication system was subdivided into several layers, each performing a well-defined task and implementing any new level of abstraction as a well-defined function.

Concepts of layered architecture cover layer entities, protocols, and services as a whole, as indicated in Figure 5.4. Every entity at a given layer communicates through the service access point within a system. Layer $(N + 1)$ communicates with another peer layer $(N + 1)$ by way of a service supported at layer (N), and layer (N) is called a service provider and layer $(N + 1)$ is called a service user.

Conceptually, the layers of a communication system perform one of two generic functions: network-dependent functions and application-oriented functions. As shown in Figure 5.5, the operational environment of a communication system can be divided into three distinct operational environments: the *network environment*, which is concerned with the protocols and standards relating to the different types of underlying data communication networks; the *OSI*

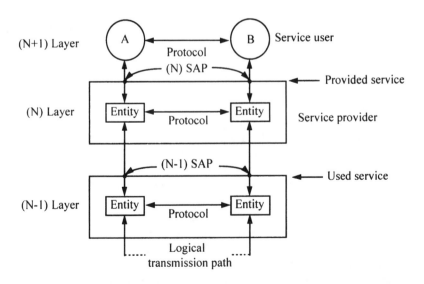

Figure 5.4 Concept of layered architecture.

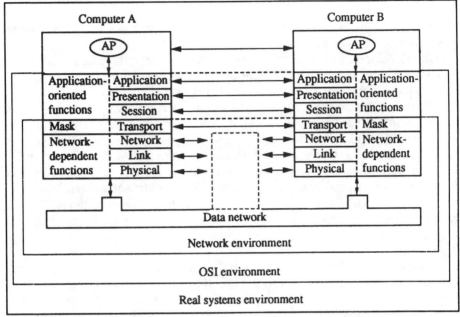

AP : Application Processor

Figure 5.5 Operational environments and OSI reference model.

environment, which embraces the network environment and adds additional application-oriented protocols and standards to allow end systems (computers) to communicate with one another in an open way; and the *systems environment*, which is built on the OSI environment and is concerned with a manufacturer's own proprietary software and services to perform a particular distributed information processing task.

The selection of functions executed at each layer and at boundaries between the layers is based on the experience gained from the early standardization activities. Each layer performs a single well-defined function in an overall communication system interconnection, and achieves communication with the peer layer of a remote system according to the protocol defined in exchanging message units composed of user data and additional control information. In this case, each layer possesses a single well-defined interface with the adjacent upper layer as well as with the adjacent lower layer. In conclusion, the implementation of the optional protocol layer is *independent* of the implementation of other layers.

As indicated in Figure 5.5, the logical structure of the OSI reference model is composed of seven protocol layers. The lower three layers (1–3) are *network dependent* and represent the protocols associated with the data communication network that links a pair of communicating computers. On the other hand, the upper three layers (5–7) are *application-oriented* protocols that allow a pair of end-user application processes to interoperate through services provided at the local operating system. The transport layer in the middle (4) assumes the role of a mask that conceals the detailed operations of the lower network-dependent layers from the upper application-oriented layers. Basically, this layer is implemented above the network-dependent layers in order to provide network-independent message interchange services to the application-oriented layers. The transport layer is sometimes considered as an extension of the network-dependent layers, which then encompasses layers 1 through 4.

5.2.2 Functions of Seven Layers

As stated earlier, the function of each layer in the OSI reference model is formalized into conventions for communication with a peer layer of a remote system and formalized into protocols that define a set of rules. Consequently, each layer provides a defined set of services to the adjacent upper layer, as indicated in Figure 5.4, and uses the services provided by the adjacent lower layer in order to transmit protocol-related message units to a remote peer layer.

The functions of each of the seven layers are summarized in Figure 5.6, which provides a categorized display of the seven layers. The lower four layers including the transport layer belong to the category of network-dependent layers, and their functions differ depending on the type of physical network. Basically, the network-dependent layers provide connection setup and message transfer functions between communication entities, thus providing reliable message exchange services to the application-oriented layers above. In contrast, the application-oriented layers, which consist of the upper three layers, provide network-independent functions necessary to support various distributed information services for the end-user application processes.

The *physical layer* provides physical access between host computer and network termination equipment. In particular, the physical layer is closely related to the mechanical and electrical attributes of the physical access. Therefore the physical layer standard defines the mechanical and electrical characteristics of the relevant devices in the physical network termination. Examples of the standard include RS-232-C/V.24, X.21, and various LAN-related interface standards, and can include the ATM physical layer standard also.

The *data link layer* provides reliable data transmission functions to the network layer by utilizing the physical access functions of the physical layer. For this, it creates data packets, synchronizes the data packets, detects and corrects errors, and controls the flow of the packet stream. The data link layer provides data link control functions that consist of the *medium access control* (MAC) function and the *logical link control* (LLC) function.

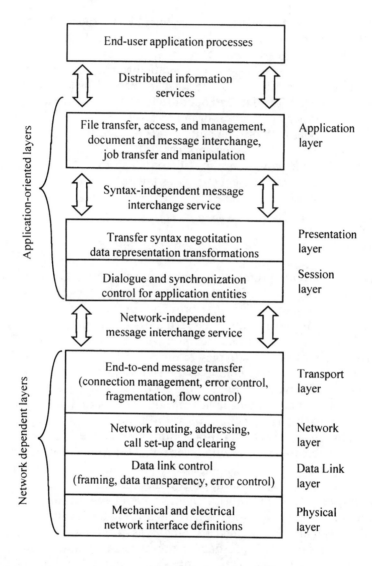

Figure 5.6   Functions of each layer protocol of OSI reference model.

The *network layer* provides routing (addressing) and call setup and release functions between two transport layer protocol entities and sometimes provides a flow control function to the computers connected to the network. If the OSI protocol-based networks are connected through intermediate networks, the network layer protocols are arranged such that network layer services can be provided regardless of the number of intermediate networks. As such, the network layer can provide an end-to-end service to the transport layer in the internetworked environment.

The *transport layer* provides network-independent message transport services to the session layer, which is the adjacent upper layer. For this, it performs the end-to-end message transfer functions such as connection management, error control, flow control, and message segmentation. It supports both connection-oriented and connectionless operation modes.

The *session layer* manages dialogue, assembles data in sequence, and provides the synchronization required to transfer data between two application layer protocol entities in two different computers. For this it provides the means to set up and release communication channels between two presentation layer protocol entities, and also the means to insert checkup points into the data stream for use in retransmitting data after interruption.

The *presentation layer* provides the means to negotiate the data format to be exchanged between two end-user application processes. The mutually agreed-on format is called the *transfer syntax* or concrete syntax. In contrast, the format within the application process that systematically presents the data structure is called the *abstract syntax*, which could be different from the agreed-on transfer syntax. In this situation, the presentation layer protocol entity of the corresponding application process provides the format translation function.

The *application layer* provides user interfaces to various protocol entities so that the application process can perform versatile distributed information processing functions in an open fashion. The application layer services include, in addition to the information transfer, such services as destination name (or address) configuration, encryption technique negotiation, destination connectability check, destination validity check, dialogue method selection, negotiation of error recovery responsibility, and restriction on data format.

5.2.3 OSI Services and Protocols

To describe the functions and operations of an OSI protocol layer, it is necessary to define the service that a layer provides to its adjacent upper layer, the internal processing (or the protocol) of the layer, and the service provided to the layer by its adjacent lower layer.

Service and Service Access Point

A service is accessed at the *service access point* (SAP) as shown in Figure 5.4, and the SAP of the (N) layer becomes the access point of the service provided by the (N) layer to the (N+1) layer. Such a SAP possesses a unique address that identifies itself. For example, in a telephone system, the SAP corresponds to the socket to which a telephone module jack is connected, and the SAP address corresponds to the telephone number of the socket. Therefore, in order to make a telephone call, the caller must know the SAP address of the called party.

Likewise, in the communication systems it is important to differentiate the identity of the user *application process* (AP) and the location of the process within the network. In general, the identity of the user AP is represented by the *name*, while the location in the network is represented by the *address*.

Since the name is used to identify the user, the name allocated to the AP of each user should be unique in a particular OSI environment. In general, an AP in a network need not know the physical location of another AP because one AP can communicate with another simply by designating its name. Therefore, so as to guarantee a unique name to each individual AP in a particular OSI environment, it is necessary to be equipped with a means to manage the name assignment among all the user and service provider APs. Such a means is called the *name server*, and it is usually owned and operated by the organization that manages the corresponding OSI environment.

In the case of a relatively small-scale network environment, as small as a single LAN-based computer network, one name server can serve for the overall system. However, in the case of a large-scale network environment that consists of several LANs and WANs with thousands of computer systems, it is difficult to use one name server to cover them all. In this situation, it is desirable to use separate name servers among different subnetworks and additionally affix the subnetwork name in front of the user AP name, so that a unique name can be allocated to each user AP within the overall OSI environment.

As the name is used in the user level, so is the address used in the OSI environment. It first determines the location of the computer in which the requested AP resides, and then confirms the identity of the AP protocol used by the corresponding AP. In this case, the OSI environment takes the role of interrelating the name of the counterpart requested by the user AP to a specific network address, and the relations between the names and addresses are listed in the system index.

The address used in the OSI environment is composed of a SAP or a cascade of multiple subaddresses. For example, let PSAP represent the SAP between the application layer and the presentation layer; SSAP the SAP between the presentation and the session layers; TSAP the SAP between the session and the transport layers; and NSAP the SAP between the transport and the network layers, all within the system in which the corresponding AP resides. Then the AP address becomes the cascade of PSAP, SSAP, TSAP, and NSAP, or equivalently, AP address = PSAP + SSAP + TSAP + NSAP.

Service Classification

Each layer can provide two modes of services to the adjacent upper layer: connection-oriented service and connectionless service. Connection-oriented service is modeled after the telephone system, and is thus provided in such an order that a connection is set up first, then data transmission follows, and finally the connection is released. In contrast, a connectionless service is modeled after the postal system, and is provided by creating data packets and then transmitting the data without establishing connection. Consequently, it can happen that the packet issued later reaches the destination earlier than the packet issued earlier, which never occurs in connection-oriented services.

Table 5.1 lists a classification of services, together with their application examples. As can be seen, services can be characterized according to the service quality. In particular, some services require reliable lossless data transmission, which can be secured by an arrangement such that the receiving and the transmitting devices exchange confirmation messages on receipt of each data packet, but at the cost of delayed transmission and reduced transmission throughput.

Service Primitives

Service is prescribed as a set of primitives used by the user or the corresponding upper layer entities, and the primitive describes the jobs to be carried out for the corresponding service. In the OSI reference model, primitives are divided into four categories, as listed in Table 5.2 and illustrated in Figure 5.7. For example, if a connection request is made by the service user in the $(N+1)$ layer to the service provider (i.e., protocol entity) in the (N) layer, the peer service provider indicates it to the peer service user. Then the peer service user responds regarding whether or not to accept the connection, and this message is confirmed to the original service user.

Table 5.1
Service Classification

Modes	Services	Examples
Connection oriented	Reliable message stream	Page sequence integrity
	Reliable byte stream	Remote log-in
	Reliable connection	Digital voice
Connectionless	Unreliable datagram	Electronic mail
	Confirmed datagram	Confirmed electronic mail
	Request-response	Database retrieval

Table 5.2
Service Primitives

Primitives	Interpretation
Request	Requests the service layer to perform a job
Indication	Indicates the job requested by the peer entity
Response	Reponds to the job requested by the peer entity
Confirm	Confirms the response to the requested job

If viewed from the service primitive's point of view, there are two types of services, as illustrated in Figure 5.7(b): confirmed service and nonconfirmed service. The confirmed service consists of all four primitives, but the nonconfirmed service consists of request and indication only. For example, connection setup should be processed as a confirmed service, whereas data transmission may be handled as a nonconfirmed service.

Most primitives are accompanied by some parameters. For example, the connection request can have parameters specifying destination address, source address, service type, maximum message size, and so on, and the connection indication can use the parameters for source entity identification, requested service type, requested maximum message size, and others. If the maximum message size for a peer service user differs from that requested by the original service user, negotiation can be made through repeated request-response primitive exchanges.

Interlayer Relations

For two adjacent layers to exchange information, they should mutually agree on the interface. In general, the $(N+1)$ layer entity transmits an *interface data unit* (IDU) to the (N) layer through the (N) SAP (see Figure 5.8). IDU is composed of the *service data unit* (SDU) and the *interface control information* (ICI): SDU is the information finally delivered to the peer $(N+1)$ layer, and ICI is the information added to control the interface. In general, the (N) SDU is equivalent to the $(N+1)$ PDU and, likewise, the (N) PDU is equivalent to the $(N-1)$ SDU.

To transmit the (N) SDU, the (N) layer entity segments it and then attaches the (N) *protocol control information* (PCI) to each segment as the overhead. The PCI is the information to be used by the peer entity to process the received protocol and includes such information as the type of contained data, sequence number, counter and so on.

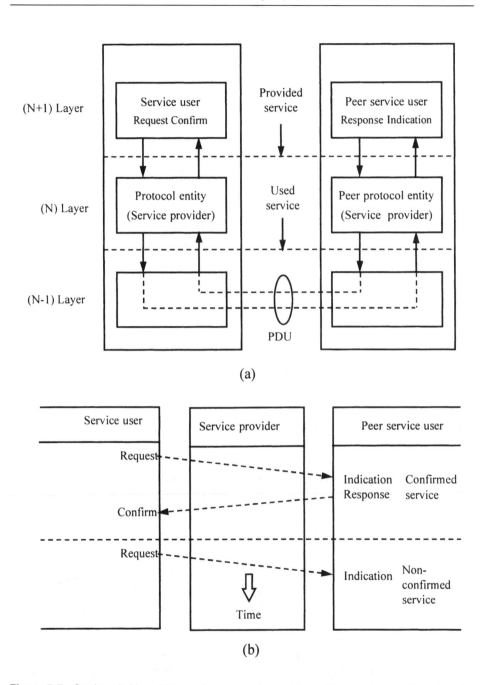

Figure 5.7 Service primitives: (a) layered representation and (b) confirmed and nonconfirmed service primitives.

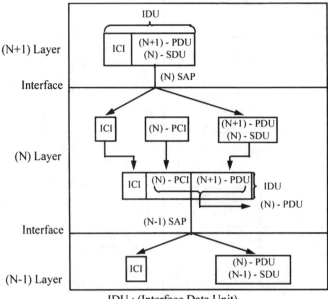

IDU : (Interface Data Unit)
ICI　: (Interface Control Information)
SDU: (Service Data Unit)
PCI : (Protocol Control Information)

Figure 5.8　Interrelation of the layers.

In summary, the PDU generated in a layer is transferred to the adjacent lower layer through the SAP and the interfacing of the two layers is controlled by the added information, ICI. PCI is attached to help the counterpart peer layer correctly process the received PDU. This inter-layer relation is depicted in Figure 5.9 for the OSI seven layers.

The terms *service* and *protocol* are frequently confused for each other although they are conceptually different. The service is represented by a set of primitives that a layer offers to its adjacent upper layer. It defines the job to be performed on behalf of a user, but does not specify how to realize the job. As a consequence, the service pertains to the interface of the two adjacent layers within the same system. On the other hand, the protocol is a set of regulations that treats the formats and meaning of the frames, packets and message to be exchanged between a layer entity and its counterpart peer layer entity, and each entity employs protocols to realize the assigned service. Therefore, the protocol entity of each layer can be freely modified as long as the service to be offered to its upper layer remains unchanged.

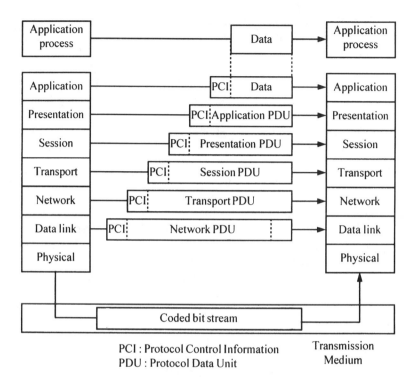

Figure 5.9 OSI reference model and interlayer relationships.

5.3 THE TCP/IP PROTOCOL SUITE

In this section the basic TCP/IP communication protocol suite, which is the basis of the Internet[2] architecture, is examined in detail. First, the features and design philosophy of the TCP/IP protocols are explained. Then the actual protocol architecture is described, following which the network and transport layers are described in detail. Finally, the IP routing mechanism and name services are more closely examined.

5.3.1 Features and Design Philosophy of the TCP/IP Protocols

The Internet has grown exponentially due to several attractive features of the TCP/IP protocol suite [Hunt92]. The distinctive features can be roughly summarized by the following three points.

---

2. Note that an *internet* differs from the *Internet*. An internet means any group of networks interconnected by a common protocol, whereas the Internet (with a capital I) refers to the worldwide interconnection of networks based on the TCP/IP protocol suite.

First, TCP/IP protocols are open protocol standards. They are open in the sense that they are not under the ownership of any one company or group, they are freely available to anyone interested, and anyone can contribute to the standardization process.[3]

Second, TCP/IP protocols are independent of the actual physical network, and computer hardware and software. They can run over Ethernet, token ring, X.25, modems, UNIX workstations, IBM PCs, and so on.

Third, TCP/IP uses a common global address scheme that enables worldwide connectivity. Since the address is globally unique, anyone can connect to anyone else on the Internet worldwide.

The first and second features have resulted in full support from both academic and industrial sectors. This made TCP/IP an ideal basis for interconnecting heterogeneous computer systems. For example, SUN workstations, IBM PCs, and VAX minicomputers could be all easily interconnected with TCP/IP. It also gave birth to many popular applications based on the TCP/IP protocols, such as *ftp, telnet, e-mail, usenet,* and the WWW. The third feature is important in that it allows for the growth of the Internet's connectivity in line with the growth in size. This has increased the usefulness of the Internet for all those readily connected, resulting in a positive feedback effect. This can be viewed as the underlying reason for the Internet explosion of today.

The TCP/IP suite is designed based on the following two principles: The first is the "IP over everything" principle, which enables a single network layer protocol IP to connect any communicating entities. The second is the "end-to-end" principle, which is arranged such that most communication-related processing can be carried out at the communicating end points.

The "IP over everything" principle is simply the decision to base the TCP/IP protocol suite on an internetworking protocol layer that would overlay all networks to be interconnected.[4] Because of this principle it has become extremely easy to adapt new network technologies. Furthermore, a unique global addressing method could be defined so that all hosts can be easily reached.

---

3. Even though the TCP/IP protocols were already available as open standards and widely deployed when the ISO started to develop its own suite, the ISO did not adopt the TCP/IP protocols as they were mainly for two reasons: First, enhancements were to be added to the basic TCP/IP protocols. Second, non-U.S. manufacturers were concerned about the dominance of U.S. manufacturers, who would have had an advantage. But basically the TP4 was based on TCP, and CLNP was also based on IP. The TCP/IP protocol suite is currently much more widely deployed than the OSI protocol suite.

4. There are basically two methods for interconnecting networks: *translation* and *overlay.* Interconnecting by translation means that gateways are used to map the data and control information of similar services from one network to another. For example, mail gateways between TCP/IP networks and OSI networks translate TCP/IP mail into OSI mail, and vice versa. Obviously, translation has many of the problems observable in human language translations. Consequently, the Internet architects have chosen the overlay method, which uses a single unifying network, on which all higher layer protocols are based, to interconnect all networks.

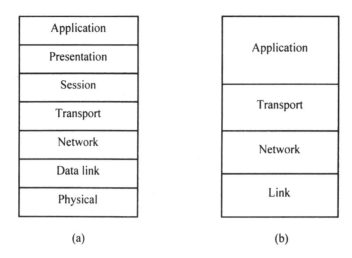

Figure 5.10 Comparison of (a) OSI and (b) TCP/IP protocol layers.

The "end-to-end" principle states that the network should offer only the basic transport functions, and any final decisions regarding any extra communication functions should always be made by the end users themselves. That is, the networking functions should thus be delegated as much as possible outside the network, not inside the network. For example, for error checking and control functions, which can be executed either hop-by-hop or end-to-end, it is natural to use only end-to-end error control in TCP.[5]

5.3.2 TCP/IP Protocol Structure

The TCP/IP protocol suite does not have an exact layered model as does the OSI seven layer reference model, but it can be divided into four layers, namely, the *link layer*, *network layer*, *transport layer*, and *applications layer*, as shown in Figure 5.10. While a direct comparison of the TCP/IP suite with the OSI seven layer model is not easy, the internal layers of the two are matched in the following way: The TCP/IP link and network layers correspond to the physical, data link, and network layers of the OSI reference model; the TCP/IP transport

---

5. Note that this principle does not apply to all network problems. The answer may also depend on the network technologies used. For example, according to the end-to-end principle, error correction should only be performed internally if it would result in a critical improvement in performance compared to not performing it internally. So hop-by-hop error correction makes sense over an extremely lossy wireless link, but should be avoided over a low bit error rate fiber link.

layer to the OSI transport and session layers; and the TCP/IP applications layer to the OSI presentation and application layers. Figure 5.11 shows the main protocols related to these four layers.

In this subsection the main characteristics of each protocol layer are examined based on Figure 5.11. Among the four layers, the network and transport layers are discussed in more detail in the subsequent two sections.

Link Layer

The link layer is the lowest layer of the TCP/IP protocol suite. It provides the means for the transmission and receiving of data over a physically connected transmission media. As such, encapsulating IP datagrams into the actual transmission frames of the link layer network and mapping of the global IP address to the link layer network address are two main functions of the link layer.

Due to its strong dependence on the actual physical network, the link layer protocols must be developed in accordance with the corresponding physical networks. For example, the standards for encapsulating IP datagrams in

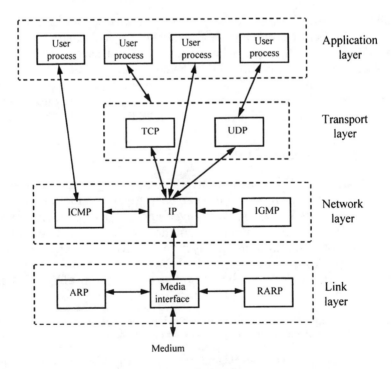

Figure 5.11   Layers and related protocols of the TCP/IP protocol suite.

Ethernet frames are defined in RFC894 and RFC1042, while the *address reso-lution protocol* (ARP) and *reverse ARP* (RARP) protocols for converting between IP addresses and Ethernet addresses are defined in RFC826 and RFC903, respectively.

Network Layer

Whereas the link layer is used to transfer data between IP nodes, the network layer transfers data between IP end nodes by using those link layer functions. There are three protocols defined in the TCP/IP network layer: the *Internet protocol* (IP), the *Internet control message protocol* (ICMP) and the *Internet group management protocol* (IGMP).

Among the three protocols, IP is the most important protocol in the TCP/IP network layer. IP routes (i.e., transports) the IP datagrams over various networks from the sender system's transport layer to the receiver system's transport layer.[6] ICMP and IGMP protocols are used to provide other functions in support of this basic transport function (for details refer to Section 5.3.3).

Transport Layer

The transport layer services are the actual end-to-end data transfer services seen by the end user. Depending on the type of user services two types of protocols—the *transmission control protocol* (TCP) and the *user datagram protocol* (UDP)—are used. UDP is used for a simple datagram transfer service, while TCP is used for a reliable connection-oriented service. Because the service offered by UDP is not complicated, it is realized by adding a minimal amount of control information to the basic IP protocol. In contrast, the service offered by TCP is more complicated than that that of UDP, so more functions such as error detection and control must be provided (for details refer to Section 5.3.4.).

Applications Layer

The application layer, which is located above the transport layer, is composed of application programs. Applications can be divided into two types—user applications and infrastructure applications, which are used by other applications to carry out their functions. The DNS name service and the SNMP network management service are examples of infrastructure applications, whereas *ftp, telnet, rlogin, e-mail,* and the WWW are the examples of user applications.

---

6. Note that the end user of the datagram is an application process above the transport layer, not the transport layer itself.

5.3.3 The Network Layer Protocols

Among the three protocols defined in the TCP/IP network layer (i.e., IP, ICMP, and IGMP), the IP protocol is detailed in this subsection.

IP Network Layer Characteristics

The IP network layer has the characteristics that it is connectionless and unreliable. The fact that the IP protocol is connectionless means that there is no need to exchange any signaling or control information before the source sends a datagram to the destination. The IP end user just sends the packets with the destination's address attached. If a connection-oriented mechanism is needed, a higher layer protocol such as TCP must be added. The fact that the IP protocol is unreliable means that there is no error correction, retransmission of lost datagrams, or guarantee of in-order delivery. That is, the IP layer tries its best to deliver datagrams correctly, but it does not guarantee anything for it (i.e., best effort service).

These characteristics can be seen to reflect the "end-to-end" design principle. That is, in both cases the end user may or may not need the functions related to these characteristics, so only the minimal requirements are specified in the IP layer, with more specific functions left for higher layer protocols.

IP Datagram Format

The IP datagram consists of a variable-length header and a data field as shown in Figure 5.12. The fields of the header are aligned to 32-bit words for ease of handling in a computer. The *version* field contains the version number of the protocol.[7] The *Internet header length* (IHL) field contains the variable length of the header. The *type of service* (TOS) field was originally defined for allocating network resources according to the needs of the user, but in fact is almost never used. The total length field indicates the whole length of the IP datagram in bytes. The *identification, flags,* and *fragmentation offset* fields are used in the fragmentation of IP datagrams (see below). The *time to live* (TTL) field is used to ensure that no packet loops forever in the network. Basically it is a hop count and the packet is dropped when it becomes zero. The *protocol* field is used to differentiate end-user transport layers. The *header checksum* is a 16-bit one's complement sum of the IP headers.[8] The *source IP address* and the *destination IP address* are 32-bit IP addresses. The *options* field is of variable length and can contain various options.

---

7. The current version is IPv4. The next-generation IP version has been termed IPv6.
8. A "16-bit one's complement sum" means that the data are divided into 16-bit unit slices, then the one's complements is calculated for each 16-bit slice, and finally all the one's complement numbers are summed together in modulo-2 operation.

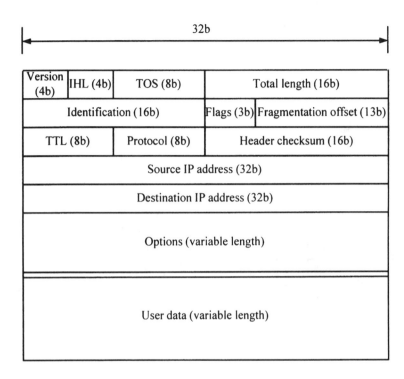

IHL : Internet Header Length
TOS : Type of Service
TTL : Time to Live

Figure 5.12   IP datagram format.

Routing of IP Datagrams

A TCP/IP network is composed of two types of nodes: *hosts* and *routers*. A router[9] is a node with two or more network interfaces that are capable of forwarding a datagram from one network to another, and a host is a node that is not a router.

A simple example of a TCP/IP internetwork is shown in Figure 5.13. In this example there are three networks, namely, the InmacNet, the SnuNet and the EngNet. The InmacNet is a 10-Mbps Ethernet network, the SnuNet is a

---

9. Originally, TCP/IP networks were defined to be composed of hosts and gateways, with gateways being the routers defined above. Today the term *gateway* is usually reserved for the devices that translate data between similar services in different protocol families, such as a mail gateway between TCP/IP and OSI networks.

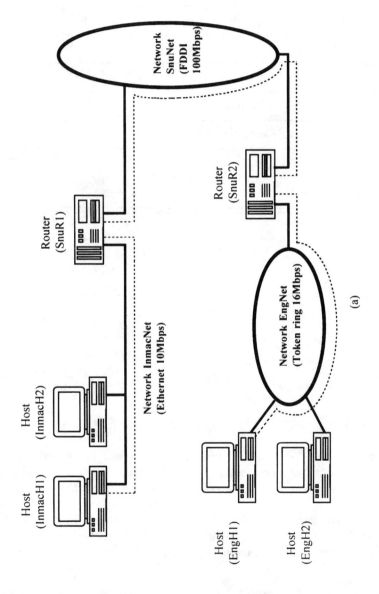

Figure 5.13 Example of a TCP/IP network: (a) physical connections and (b) abstract protocol connections.

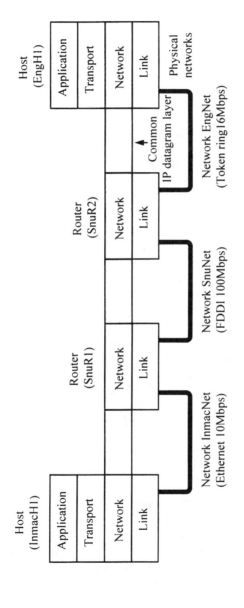

(b)

Figure 5.13 (continued)

10-Mbps FDDI backbone network, and the EngNet is a 16-Mbps token ring net-work. The hosts on each network are named InmacH1, SnuH1, and EngH1. The routers connecting InmacNet and EngNet to the backbone SnuNet are called SnuR1 and SnuR2, respectively. In this example, if the host InmacH1 wants to send a datagram to the host EngH1 on EngNet, it must send the IP datagram to the router SnuR1 first. SnuR1 then sends the datagram to the router SnuR2, which then forwards the datagram to EngH1. For EngH1 to send data to any host on SnuNet, it would send the datagram to SnuR2, which would then send it to the appropriate host.

In all cases the routing of the datagram is based on the destination IP address in the IP header and the routing information in the routing tables that are maintained by each host and router. The hosts do not know anything about the networks beyond their own. If the destination is on their network, it is forwarded there. Otherwise, based on the information in its routing table, the host or router just sends the datagram to the next router on the route to the destination. (For explanations on the use of routing tables and the routing op-eration, refer to Section 5.3.5.)

Once the data arrive at the destination host, they must be reassembled into the application data units and passed on to the appropriate users. This means that the datagrams must be demultiplexed and sent to the appropriate software or protocol modules. Multiplexing and demultiplexing occur in three places in the TCP/IP protocol suite: At the three boundaries between the link layer and the IP layer, the IP layer and the transport layer, and the transport layer and the application layer. This is indicated in Figure 5.14.

At each boundary an identification field in the lower layer's protocol header is used to differentiate between the higher level users. If Ethernet is used as the link layer, the *frame type* field in the Ethernet header is used to differ-entiate IP datagrams, ARP packets, and RARP packets. The *protocol* field value in the IP header is used to identify the type of protocol used by its payload among TCP, UDP, ICMP and IGMP. The *destination port* number in the TCP or UDP header is used to differentiate various user processes (see Figure 5.15).

Fragmentation of IP Datagrams

In the course of IP datagram routing, IP datagram fragmentation can also occur. This is because the various networks over which a TCP/IP internet operates may use different size transmission frames.

Each link layer network has a maximum transmission frame size. For ex-ample Ethernet has a maximum frame size of 1,518B including the header and trailer and the maximum-size data unit that can be transmitted over it is 1,500B [RFC894]. This is the *maximum transmission unit* (MTU) of Ethernet. The MTU differs among link layer networks. For example, the FDDI MTU is 4,352B, the X.25 MTU is 576B, and the SLIP line MTU is 256B.

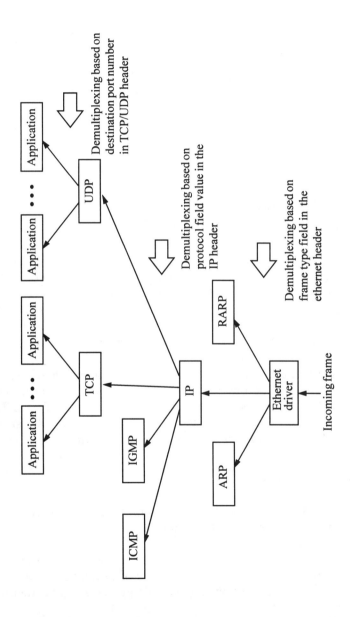

Figure 5.14 Demultiplexing procedure in the TCP/IP protocol stack (with an Ethernet link layer).

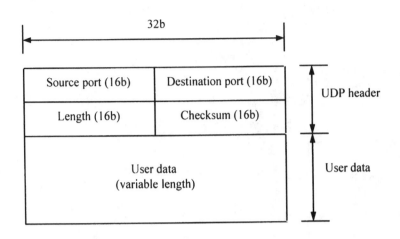

Figure 5.15 UDP datagram format.

Fragmentation occurs when an IP datagram passes over a network whose MTU size is smaller than the size of the datagram. The IP datagram must then be broken up into smaller pieces (fragmented), to be transported over this network, and then reassembled at the receiver system. This will result in the original IP datagram being delivered to the destination host as fragments. The identification, flag, and offset fields in the IP header are used for this IP fragmentation mechanism.

Because fragmentation usually results in the degradation of network performance, its use is not recommended in the future. Recently a *path MTU* (PMTU) discovery method has been developed for finding the MTU of the whole path to the destination. The source can then send datagrams as large as this PMTU without fragmentation on the way to the destination.

Other Network Layer Protocols

Besides the IP protocol, two more protocols are defined in the network layer, the ICMP and IGMP protocols. Both protocols use the basic IP datagram with special protocol field numbers in the header.

The ICMP datagrams are used primarily for diagnosis or control. For example, the echo message is used to check whether or not the destination node is operating, while the destination unreachable message signals that the destination node is unreachable.

The IGMP datagram also uses the IP datagram. The IGMP protocol is used to form and maintain multicast groups.

5.3.4 The Transport Layer Protocols

Two transport layer protocols, UDP and TCP, are discussed next.

UDP

Basically UDP is the IP datagram service having the minimal control information needed for an end-to-end transport protocol. Because no special functions are added other than those offered by the IP layer, UDP is an unreliable connectionless datagram service.

The structure of the UDP datagram format is as shown in Figure 5.15. The 16-bit *source port* and *destination port* numbers are used to identify the user application at the source and the destination, respectively. The *length* field contains the length of the whole UDP packet. The *checksum* is a 16-bit one's complement sum of the UDP header, the pseudo UDP header,[10] and the UDP data.

Though UDP supplies only an unreliable datagram service, there are still three types of applications: The first type of applications uses UDP because the application is based on a very simple model of exchange such as query and reply, where a connection setup and release procedure would be burdensome. An example would be the ping service. A second type uses UDP because very low protocol overhead is needed, that is, the amount of resources consumed by the protocol must be minimized. An example is diskless workstation booting. The third type is for its own type of transmission control such as *simple network management protocol* (SNMP).

TCP

Applications that require reliable delivery of data use the TCP instead of the UDP. It is a reliable, connection-oriented, byte-stream transport protocol. The structure of the TCP segment is shown in Figure 5.16. The *checksum, source port,* and *destination port* fields are the same as those in the UDP datagram. The *sequence* and *acknowledgment* fields are used for reliable data transmission and acknowledgment, and the *window* field is used to indicate the receiver window size in the sliding window algorithm. The *urgent pointer* field is used to indicate the position of data, which must be processed immediately at the receiving station.

TCP is connection oriented because a logical end-to-end connection must be set up between the two end hosts before data transmission. This means that a three-way handshake must take place between the two end points to

---

10. The pseudo UDP header consists of the source and destination IP addresses of the IP datagram header and the actual UDP header. It is used only for calculating the UDP checksum.

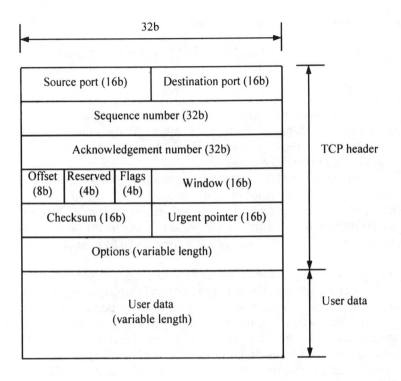

Figure 5.16   TCP segment format.

synchronize them before data transfer can occur. This is illustrated in Figure 5.17. In this figure SYN is the sequence number to be synchronized, and ACK is the acknowledgment number. End points are uniquely identified by their IP address and port number pairs.[11]

TCP is reliable because it relies on positive acknowledgments with retransmission. This means that every data segment sent by the sender must be checked and acknowledged by the receiver. If the receiver does not acknowledge the segment within a timeout, the transmitter assumes that the segment has been lost and retransmits the segment.

TCP is a byte-stream protocol, because, unlike UDP, it regards the data it sends as a continuous stream of bytes. A TCP host assigns a number to each byte it sends, and the positive acknowledgment and retransmission mechanisms are executed based on these numbers.

---

11. To be more accurate, the protocol field value in the IP header is also needed, but this is only needed for differentiating between the users of TCP and UDP protocols. So a TCP service user would actually need only the IP address and port number to uniquely identify its corresponding host, since they both must be using TCP.

TCP also contains a flow control mechanism modeled on the sliding window method. It is realized by using the window and acknowledgment fields in the TCP header. The amount of data that a host can send without receiving an acknowledgment is set by the last window advertised by the receiver. An example is shown in Figure 5.18. In the figure, the transmitter sends data bytes 1–6, but receives ACKs for only bytes 1–3. Because the advertised window size

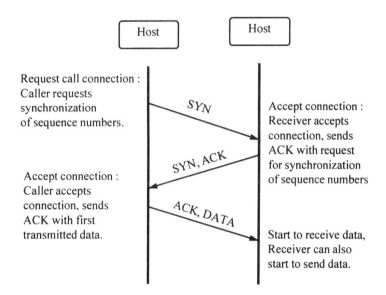

Figure 5.17 TCP call connection setup.

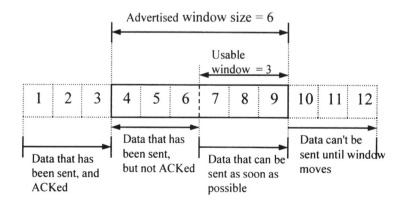

Figure 5.18 Example of the flow control mechanism used in TCP.

is six bytes, the sender may send only three more bytes, 7–9, waiting for a new acknowledgment.

TCP also contains sophisticated congestion control algorithms based on implicit probing of the network congestion status. Recently methods of extending TCP to Gbps rates have also been developed.

5.3.5 Routing

One of the most important parts of the IP architecture is the mechanics of how IP datagrams are routed. In this subsection, the basic routing mechanism is explained. First the steps involved in routing are explained through a basic example. Then IP addresses, routing tables, address resolution protocols, and routing protocols are examined.

Example of Datagram Routing

Figure 5.19 depicts an example of how routing is done in IP interconnected networks. This example is an extension of the example shown in Figure 5.13. Assume that a user wants to send data from the host InmacH1 on the network

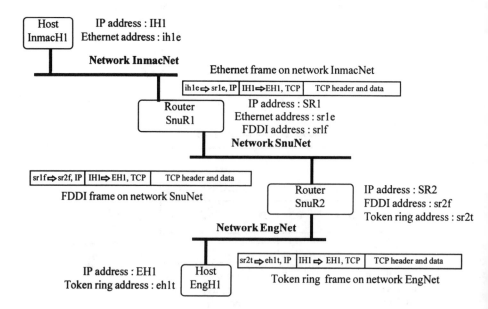

Figure 5.19   Example of routing in TCP/IP network.

InmacNet to the host EngH1 on the network EngNet. Then the following processes will occur for routing the data:

1. The host InmacH1 finds out the IP address EH1[12] of EngH1.
2. InmacH1 makes an IP datagram with the source IP address IH1, and the destination IP address EH1.
3. InmacH1 searches for the host or router on InmacNet to which it must send the datagram in order to route it to EngH1. It then finds that the datagram must be sent to the router SnuR1, which has an IP address SR1.
4. To actually send the data to SnuR1, InmacH1 then finds out the Ethernet address, sr1e, of SnuR1.
5. InmacH1 encapsulates the IP datagram in an Ethernet packet addressed to sr1e, and transmits it on the InmacNet network.
6. The router SnuR1 receives the Ethernet packet, and checks the destination IP address. SnuR1 must also find out (just as InmacH1 did) to which host or router on SnuNet it has to send the datagram to route it to EH1. It finds out that it has to send it to router SnuR2.
7. SnuR1 finds out the FDDI address of SnuR2, which is sr2f, and sends the datagram to the router SnuR2 encapsulated in an FDDI frame.
8. SnuR2 receives the FDDI packet, and checks the destination IP address. SnuR2 determines which host or router on SnuNet to which it has to send the datagram to route it to EH1. It then finds that SnuR2 must send the datagram directly to the host EngH1, because both are directly connected to the same network, EngNet.
9. SnuR2 finds the token ring address of EngH1, which is eh1, and encapsulates the datagram in a token ring packet transmitted to EngH1.
10. The host EngH1 receives the packet and passes the data to the appropriate user application.

In the preceding steps, various problems arise. For example, in item 1, how does the host find the IP address EH1 of EngH1 from its name EngH1? In item 3, how does the host find out that it must send to the router SnuR1 in order to send the datagram to EngH1? In item 4, how does the host find the Ethernet address sr1e of SnuR1? Variations of these same problems occur in the other steps as well.

Before discussing how these problems are solved in TCP/IP, the IP address structure must be explained.

---

12. The IP address is a 32-bit number, but for ease of explanation the names EH1, IH1, SR1, and SR2 are used for the addresses. The same also applied to the Ethernet, token ring, and FDDI addresses in the example.

IP Address

The 32-bit IP address uniquely identifies a host or router on the Internet. The IP address is made up of *netid* (network number) and *hostid* (host number). This hierarchical structure is used because it simplifies routing.

Variously sized networks may exist depending on the usage and objective of the network. So, as shown in Figure 5.20, five types of IP addresses, classes A, B, C, D, and E, are defined. The first few bits of the classes A, B, and C addresses are fixed as 0, 10, and 110, respectively, with the rest of the address bits, D and E, being divided into *netid* and *hostid*. The first 4 bits of the classes D and E addresses are fixed as 1110 and 1111, respectively. The class D addresses use the rest of the address bits as a multicast *groupid* (group number), while the usage of the remaining bits is not defined for class E addresses. In the figure, the group of numbers to the right represents the range of possible 32-bit IP addresses in that class. Note that the normal notation for IP addresses is four decimal numbers separated by dots, with each decimal number representing 8 bits of the IP address.[13]

Due to the spread of local-area networks and private company networks during the 1980s, it was found that a more structured address space was desirable. So, as shown at the bottom of Figure 5.20, *subnet* fields were added to the address structure to ease these management and routing problems. This is accomplished by using 32-bit subnet masks to divide the IP address *hostid* field into a number of subnets. The bits that are 1's in the 32-bit subnet mask mark the bits that can be used as the *netid* or *subnetid*. Examples of subnets and subnet masks are shown in Table 5.3. In the first example, the subnet mask is 255.255.0.0, while the address is of class A since the first bit is 0. Therefore the *netid* is 18, the *subnetid* is 18.20, and the last two bytes 16.91 represent the *hostid* on that subnet. In the third example, the subnet mask is 255.255.255.192, so 26 bits out of the 32 bits are used as the *netid* or *subnetid*, while only the last six bits are used as the *hostid*. Also as the first three bits are 110, the address is a class C type, and the first three bytes 192.90.88 are the *netid*, 192.90.88.128 is the *subnetid*, and the *hostid* is 4.

Routing Tables

The routing table contains information on where the IP datagram must next be sent in order to get to the desired destination. When an IP datagram is received, the destination IP address is used to look up the relevant routing table entry

---

13. Organizations are given with class A, B, or C addresses according to their needs. A class C network can have 254 hosts, a class B network can have roughly 64,000 hosts, while a class A network can have about 16,000,000 hosts. Very few organizations are given class A addresses, and most are given class B or class C network addresses.

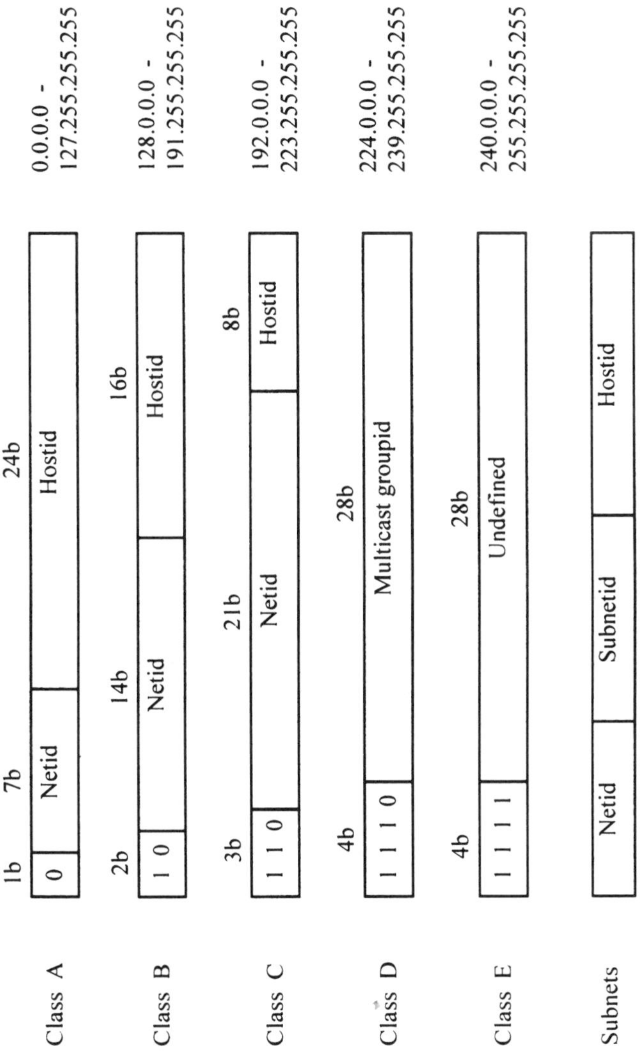

Figure 5.20 Internet address classes.

and the datagram is sent to the next hop destination (host or router) in that entry.

The destination host may be on a network directly connected to the host or the router that has just received the datagram. (As explained later, this is easily found out by examining the routing table.) In this case, the datagram can be routed to the destination host by just using the physical network address of the final destination. This is called *direct routing*. But usually the destination host will be on a different network. In this case, the next hop router to which the IP datagram must be sent in order to reach the destination must be found. The routing table contains the IP address of this next hop router. The datagram is sent to this next hop router, which will then forward the datagram according to its own routing table to the appropriate network or the host. This is called *indirect routing*.

In the following, we describe the routing function in detail through an example. Table 5.4 is a sample routing table. The first column of the table lists the address of the host or network to which the datagram must finally be sent. The second column lists the address of the next hop router or host to which we must actually send datagrams destined for the networks or hosts in the first column. The third column contains information on columns 1 and 2. The fourth column indicates the actual network interface to which the datagram has to be sent. This is necessary because routers usually have more than one network

Table 5.3
Example of Subnets

IP Address	Subnet Mask	Interpretation
18.20.16.91	255.255.0.0	Host 16.91 on subnet 18.20.0.0
147.46.66.19	255.255.255.0	Host 19 on subnet 147.46.66.0
192.90.88.132	255.255.255.192	Host 4 on subnet 192.90.88.128

Table 5.4
Example of a Routing Table

Destination	Gateway	Flags	Interface
127.0.0.1	127.0.0.1	UH	lo0
Default	147.46.66.1	UG	emd0
147.46.66.0	147.46.66.19	U	emd0
147.46.148.0	147.46.66.1	UG	emd0

interface. In the table H means that the destination address type in column 1 is a single host, otherwise it is a network; U means that the router or host in the second column is in operation; and G means that the next hop destination in column 2 is a router, otherwise it is a host.

Next, the first entry of Table 5.4 is for local loopback. It is used by the host to send datagrams back to itself for such purposes as diagnostics and debugging. The second entry is the default entry. If no match for the destination address is found in the routing table, the datagram is forwarded to this router. The third entry is for hosts on the subnet 147.46.66.0.[14] Because the host is directly connected to this network, the destination gateway entry is itself. This means that any datagrams destined for hosts on this network should be sent by just using the link layer network. The fourth entry is for hosts on subnet 147.46.148.0. Since these hosts are on a different network, the destination is a router (with an IP address 147.46.66.1) connected to the host's own network 147.46.66.0.

Figure 5.21 exemplifies how to perform a table-based routing in relation to Figure 5.13. In the figure routing tables for each host and router on the path from InmacH1 to EngH1 are shown. Figure 5.13 used IH1 and EH1 to signify their host addresses, but Figure 5.21 uses their actual numerical IP address values, which are 147.46.66.19 and 147.46.148.12 respectively.

As in the original example, a datagram must be sent from InmacH1 to EngH1. The IP address of EngH1 is 147.46.148.12 (how InmacH1 knows this is discussed later in this section). InmacH1 examines its routing table, and finds that to send the datagram to hosts on network 147.46.148.0, it must forward the datagram to the router with an address of 147.46.66.1, which is the router SnuR1. After SnuR1 receives the datagram, it examines its routing table, and finds that to forward the datagrams destined for network 147.46.148.0, it must send the datagrams to the router 147.46.80.99 (i.e., the router SnuR2). Router SnuR2 finds that it is directly connected to the network 147.46.148.0, so it just needs to send the datagram to its interface 147.46.148.1. The interface then transmits the datagram on the EngNet to reach its final destination EngH1.

Address Resolution Protocol

In the preceding example, InmacH1 knows that it must forward the IP datagram to the router SnuR1 on the InmacNet network. InmacH1 also knows the destination IP address of SnuR1 to be 147.46.66.1, as it was in the routing table. But for InmacH1 to send the datagram to SnuR1, it must first know the physical

---

14. The fact that this is a subnet is easily recognized as follows. Since the first byte is 147, this is a class B address, which uses the first two bytes for *netid*, and the last two bytes for *hostid*. But because there is no H flag in the third column of the routing table, the corresponding entry in the first column must be a network address. This means that three bytes are used to identify this network, so the third byte must be a *subnetid* that is used to define 256 subnets.

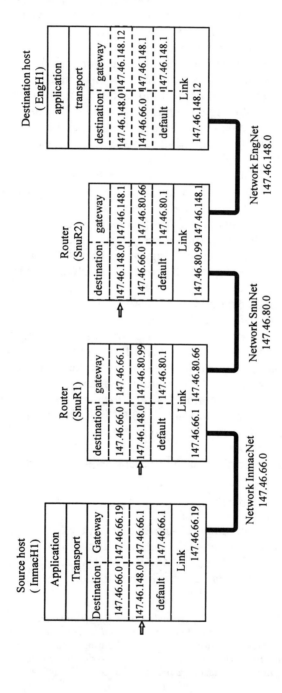

Figure 5.21 Example of table-based routing.

network (i.e., Ethernet) address of SnuR1. This is done by using the ARP protocol.

The ARP protocol is a method for finding the link layer address of a host or a router when only its IP address is known. In the preceding example, the protocol operates as follows: First, InmacH1 broadcasts a packet containing the IP address 147.46.66.1 on the InmacNet network to determine the Ethernet address of the host or router. Every host on the Ethernet will hear the broadcast and the correct one, that is, SnuR1, will answer with an ARP reply message containing the corresponding Ethernet address. Usually the results of an ARP request and reply sequence are kept in an ARP cache, so in most cases the ARP protocol is used only a few times and most link layer addresses are obtained from the ARP cache maintained by the host.

Routing Protocols

In some networks, routes are calculated centrally and distributed to all nodes, either automatically or manually. But in the Internet all routes are calculated in a distributed fashion. Two algorithms, *routing information protocol* (RIP) and *open shortest path first* (OSPF), are used in TCP/IP networks. But all routers in a single routing domain use the same routing protocol, whether it is RIP or OSPF. Each router exchanges routing data and calculates the routes to all destinations based on the data it receives.

As a routing domain becomes larger, routing protocols such as RIP and OSPF become unusable due to problems with scaling. So when the Internet started to become too large for any single routing protocol, the concept of autonomous systems and *exterior gateway protocols* (EGPs) was developed to handle the problems of routing in the Internet. A single autonomous system is operated under a single organization, and usually uses a single routing algorithm. The routing protocols within an autonomous system are called *interior gateway protocol* (IGP), and RIP and OSPF are popular IGP's. Autonomous systems are interconnected through exterior gateways. Exterior gateways must exchange routing data among themselves and calculate routes so that datagrams can be routed to destination hosts on different autonomous systems. Exterior gateways use EGPs to exchange this data.

5.3.6 Name Service

All Internet interfaces have a numerical 32-bit IP address. This is the actual numerical value used inside the Internet to route packets. But IP addresses are hard to remember and to use. So usually a nonnumeric host name is given to each host interface. For example, a host may have an IP address, 147.46.66.19, and a host name, *tsp7.snu.ac.kr*. In this case when a user uses *telnet*, instead of typing "*telnet 147.46.66.19*," the user types "*telnet tsp7.snu.ac.kr*." The

TCP/IP software will then automatically translate *tsp7.snu.ac.kr* into 147.46.66.19. When the user uses *e-mail*, in the same manner the user types "*mail blee@tsp 7.snu.ac.kr*" and the software translates this into "mail the message to user *blee* at the host 147.46.66.19."

But two problems arise with the use of names. First, how will the system translate the nonnumeric name into the numerical IP address? Second, how can host names be globally known? Basically there are two ways to solve these problems: One is by using host tables, and the other is to use the *domain name service* (DNS). The host tables method is the intuitive but naive solution. Each host maintains a host table that contains a simple mapping between host names and host IP addresses. Whenever a name is used for communication, the system looks up the name in the host table, and uses the IP address in there. There are problems with this method, with scalability being the most obvious one. Also this does not define a global host name to IP address translation method, so we can only connect to hosts defined in the host table. But if DNS is used, both problems can be solved, and consequently DNS is the Internet standard name service.

Basically DNS is a name service that uses a very large distributed database containing all Internet host names and their corresponding IP addresses. All internet hosts have a domain name such as *tsp7.snu.ac.kr*. DNS is based on a hierarchy. There are "top-level domains" based on countries such as *.kr, .uk, .jp*, and also some based on classes such as *.com, .gov, .org, .edu*. Each top-level domain can then assign subdomains such as *ibm.com, dod.gov, ac.kr, ntt.jp*. This is repeated at each layer in the hierarchy. In the previous example, the host's top-level domain is *kr*, its subdomain is *ac.kr*, and its subsubdomain is *snu.ac.kr*, with the name of the actual host being *tsp7.snu.ac.kr*. Due to this hierarchical structure, a unique domain name may be obtained for everyone, and their IP addresses can also be easily found by searching the corresponding DNS name servers. Note that there is no deterministic relationship between an IP address and a domain name. Even if the domain names look similar, their IP addresses may be completely different.

5.4 LOCAL AREA NETWORKS

A *local-area network* (LAN) is a data communication network that is formed among computers and peripheral equipment in a local area. A traditional LAN uses coaxial cable or twisted-pair as the transmission medium. It has a transmission speed of 10 Mbps or below and a transmission distance that is limited to a radius of about 2 km. Several transmission media have been developed as a result of recent technological advances, and high-speed LANs that employ optical fiber as the transmission medium have emerged. As LANs spread widely on campuses and in business areas, *metropolitan-area networks* (MANs) have come into being, as have *wide-area networks* (WANs), which are remote extensions via the public network.

In this section, the topologies, protocols, and services of traditional LANs are discussed. High-speed LANs such as FDDI are handled separately in Section 5.5. Fundamentals of LAN are introduced first, followed by the description of each LAN topology. Next, the data link protocols, namely, the MAC and LLC protocols, are examined. Then we discuss LAN interconnection issues and introduce wireless LANs. Finally, the evolution of LANs to MANs and WANs is considered, in conjunction with a brief introduction of the frame relay service. (DQDB MANs are discussed in detail in Section 5.6.)

5.4.1 Fundamentals of LANs

In general, LAN characteristics vary depending on the network topology and transmission medium, and the transmission rate and transmission efficiency also vary accordingly. LAN network topology can be categorized into ring, star, tree, and bus, and the most widely used today are the bus and the ring topologies. The standards for these LAN protocols, such as *logical link control* (LLC), *carrier-sense multiple access with collision detection* (CSMA/CD), token bus, and token ring, were announced in 1985 by the IEEE 802 Committee, and were also adopted by the ISO. LAN standards correspond to layers 1 and 2 of the OSI reference model, described in Sec. 5.2, and are configured as shown in Figure 5.22.

The 802.1 *high-level interface* (HLI) deals with issues related to network architecture, internetworking, and network management. The 802.2 LLC is employed at the upper layers of MAC standards, and its objective is to provide a means of data exchange between different MAC users. The MAC protocol is a

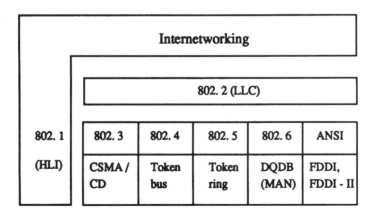

Figure 5.22   Organization of IEEE 802 and ANSI's FDDI standards.

key function of the LAN standards and regulates procedures required for data transmission. It is specified in detail in IEEE 802.3, 802.4, and 802.5 standards.

Network topologies of main interest in LANs are bus and ring, and depending on the applied access control schemes, they can form CSMA/CD bus, token bus, and token ring structures.

Network Topology

The four categories of LAN's topology, namely, star, ring, bus, and tree, are depicted in Figure 5.23.

In the case of the *star* topology, there is a switching node in the center that relays the connection among all stations [see Figure 5.23(a)]. A station that wants to transmit data requests a connection to the switching node, and then the switching node activates the switching function to connect it to the desired destination station. Once the connection is established, the two stations can communicate as if they were on a point-to-point dedicated line. The configuration is robust to the failure of stations, but the failure of the switching node

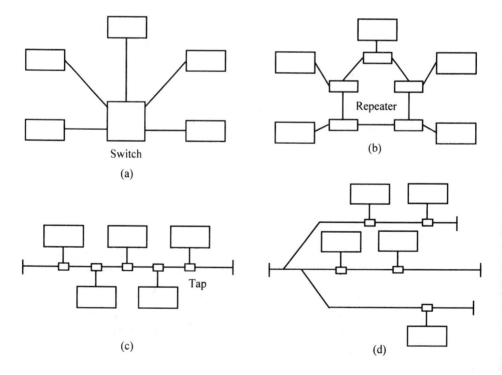

Figure 5.23   Network topologies of LAN: (a) star, (b) ring, (c) bus, and (d) tree.

can cause a breakdown in the entire network operation. Further the star configuration entails high costs in deploying cables in the initial staging and experiences difficulty in developing high-performance switching node. Therefore the star topology is not often used in LANs.

The *ring* topology forms a closed loop by connecting repeaters to tied stations [see Figure 5.23(b)]. A station that wants to transmit data waits for a token, which authorizes its turn to transmit, and if it catches one it transmits the data to the ring in the form of a packet that contains the source and destination addresses. The packet then circulates on the ring until the destination station copies it down and the source station removes it.

In contrast to the ring topology, in which stations are separated from the transmission medium because of the interposed repeaters, the *bus* topology provides a direct connection from each station to the medium through a tap [see Figure 5.23(c)]. As a consequence, in the bus topology the data transmitted by one station is delivered to all the other stations at the same time, and only one station can transmit data at a moment. As in the case of the ring topology, the transmitted data are in the form of a packet that contains the source and destination addresses, and the destination station copies the packet.

The *tree* topology is a combination of multiple bus topologies [see Figure 5.23(d)].

Transmission Medium

A LAN uses copper wires, coaxial cables, and optical cables as its transmission media, and the transmission rate varies depending on the network topology.

In the case of copper wires, a transmission rate of a few megabits per second is usually obtained. Copper medium is weak to interferences and noise, such as the crosstalk from neighboring copper wires. Nonetheless, it is adequate for use within buildings or in low-traffic applications, because it is easy and cheap to deploy, and because plenty of copper wire is readily deployed.

If a higher rate of transmission is desired, coaxial cables are used. Coaxial cables, in general, can provide connections to many stations and equipment, and can support long-distance transmission. Coaxial cables can be used for baseband as well as for RF band transmission systems (see Section 5.4.2).

Optical cables, though small and light, have a much larger transmission capacity than coaxial cables. Optical fibers are immune to noise and do not interfere with neighboring fibers. However, optical cables have the drawback that the optical devices are expensive and cable-to-cable connection is comparatively cumbersome. Nevertheless, optical cables have been deployed widely in recent years, and the high-speed LANs such as FDDI are based on these optical cables.

On the other hand, wireless LANs have been actively deployed recently.

Wireless LANs are useful when moving an office frequently, or installing LANs for temporary use (see Section 5.4.5).

5.4.2 Characteristics of LAN Topologies

Among the four network topologies mentioned in the last section, the star topology is not included in the IEEE 802 LAN standards due to the drawbacks discussed in Section 5.4.1, and the tree topology can be viewed as an extension of the bus topology. Therefore, in this subsection only the characteristics of bus and ring topologies are considered.

Bus Topology

The bus topology is the most general type of network topology, with the most typical example being the Ethernet. The bus topology has a multipoint configuration with multiple stations connected to the common medium; consequently, only one station can transmit data each time.

When a station on the bus transmits data to the medium, the signal level should be kept within some particular range. It should be large enough for the receiver on the far end to detect even with the attenuation of the medium. However, if the signal level is too high, it imposes a power burden on the transmitter and generates spurious signals also. It may be comparatively easy for a point-to-point system to find some balanced signal level, but it is complicated for multipoint systems to balance signal levels among all stations. It is even more difficult in RF band systems due to the interference among RF signals. To resolve this signal level balancing problem, the bus topology LAN divides the medium into small segments, balances the signal level within each segment, and connects the individual segments using repeaters or amplifiers.

There are two ways to transmit data to the medium: Use either the baseband system or the RF band system. The baseband system transmits data as a digital pulse stream, whereas the RF band system modulates the baseband data to an RF band analog signal. For baseband transmission, both copper wire and coaxial cable can be used as the transmission medium, but for RF band transmission, only coaxial cable can be used as the medium because the frequency band expands quite a bit in this case.

A. Baseband System

Figure 5.24(a) depicts the network configuration of a bus topology that employs baseband transmission. Since the baseband system transmits a digital pulse stream, the transmission distance is limited to about 1 km due to severe attenuation and harmonic distortion. The baseband signal propagates to both directions, but cannot penetrate the coupler or amplifier in the tree topology network.

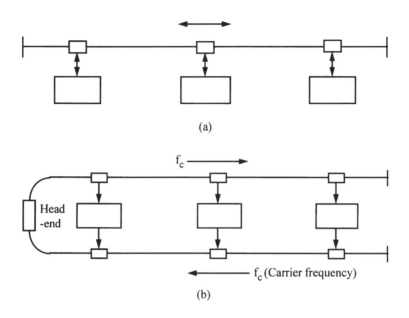

(a)

(b)

Figure 5.24 Transmission techniques in bus topology LANs: (a) baseband transmission and (b) RF band transmission.

In baseband bus networks, copper wire or coaxial cable is used, and a LAN can be constructed simply by using only one cable segment of 500m at maximum. A termination is attached at each end of the cable to prevent signal reflection, and each station is connected to the cable by affixing a tap. Each tap is positioned at the distance of multiples of 2.5m from its adjacent tap to reduce reflection, and a maximum of 100 taps can be connected in a cable segment.

The transmission distance can be extended by placing a repeater between two LAN segments. A repeater consists of a paired transmitter and receiver, and it amplifies and regenerates the signal in one segment, then passes it to another segment.

For example, the standard Ethernet LAN 10BASE-5 has the following specifications: It supports the transmission speed of 10 Mbps; it employs baseband transmission; and its segment length is 500m at maximum. These specifications are indicated, respectively, by 10, BASE, and 5 in the notation for the name. The maximum distance between two neighboring stations within a segment is 50m, the maximum length of an Ethernet is 2.5 km, and a maximum of 1,024 stations can be connected to an Ethernet. There are other types of Ethernet LANs, namely, 10BASE-2, which is a 10-Mbps thin coaxial cable LAN; 1BASE-5, which is a 1-Mbps copper wire LAN; and 10BASE-T, which is a 10-Mbps unshielded twisted-pair LAN.

B. RF Band System

Figure 5.24(b) depicts the bus network configuration that employs RF band transmission. The RF band system transmits RF band analog signals, which are obtained by modulating the original baseband signals to the analog carrier of frequency f_c. It can additionally employ *frequency-division multiplexing* (FDM) to divide the RF band into multiple subbands for separate use among various different signals such as audio, video, and data. The RF band system uses analog amplifiers to compensate for the attenuation that occurs while propagating through the cable, and thus the amplifier can be used to extend the transmission distance. Because it is difficult to implement bidirectional amplifiers, the RF band system employs unidirectional transmission. Therefore, each station requires two unidirectional transmission paths, which meet at the headend as shown in the figure. The headend does the unidirectional relaying function: A transmitting station transmits modulated data to the headend, and the receiving station receives data from the headend.

IEEE 802 prescribes the 10BROAD-36 system, which is a thick coaxial-cable-based 10-Mbps RF band LAN.

In the RF band system, adding or dropping signals is simple because it is an analog system. It is possible to transmit signals farther than the baseband system. However, it is difficult to design and maintain the related analog circuits due to the required fine tuning. Further, as the transmission rate of the baseband system grows higher, the RF band system becomes less attractive.

Ring Topology

The ring network consists of multiple repeaters, which jointly form a closed unidirectional transmission path. Data are transmitted in a bit stream and each repeater regenerates the received bit stream. The repeater carries out the three basic functions required for data transmission in the ring network: data transmission, data receiving, and data removal. Each repeater provides a connection point to its relevant station.

In contrast to the bus network, the ring network requires the data removal process. As opposed to the bus network, where data packets are absorbed by the terminations in both ends, data packets in the ring network are regenerated by the repeaters and therefore can circulate the ring endlessly. There are two possible ways to remove the data packets: Either to let the repeater of the destination station remove them, or to let the transmitting station remove them when they return after the circulation trip. The latter scheme is usually preferred because it can check if the data packets have made a normal circulation trip on the ring, and also can allow for a station to transmit data to multiple stations at the same time.

Since the repeaters placed on the ring separate the transmission medium into multiple segments, the RF band analog signals cannot successfully propagate on the ring. Therefore it is not adequate to employ the RF band system in the ring network.

5.4.3 Data Link Layer Protocols

The data link layer consists of two sublayers, namely, the *medium access control* (MAC) and *logical link control* (LLC) layers. The MAC layer is concerned with the medium access functions such as frame structure and error checking, which vary depending on the network topology. In contrast, the LLC layer controls the transmission and receiving of frames, which is independent of the network topology.

MAC Protocol

Stations connected to a LAN share the transmission medium and its transmission capacity. To share transmission capacity among stations, it is necessary to control the access of the stations to the common medium. The protocol designed for this purpose is the MAC protocol.

In general, two ways are available for controlling medium access: centralized control and distributed control. A LAN employs the distributed control, which is subdivided into contention scheme and token control scheme. In *contention scheme* each station can attempt to access the medium at any time on a contention basis, but in *token control scheme* a station can access the medium at a preregulated time and order. The most typical contention scheme is the *carrier sense multiple access with collision detection* (CSMA/CD) scheme, and the token control scheme includes the token bus and token ring schemes. The CSMA/CD and the token bus schemes are used in bus topology networks, and the token ring scheme is used in ring topology networks. In both the token bus and token ring schemes, medium access is controlled by a special frame called a *token*: The station that catches the token owns the exclusive privilege to transmit frames for a preregulated duration of time. If the station finishes transmitting data, or if the allowed time is consumed, the station passes the token to the next station.

CSMA/CD Scheme

CSMA/CD is currently the most widely used scheme, and its most representative product is the Ethernet. In this CSMA/CD scheme, a station that wants to transmit can transmit a data frame if the medium is not being used by another station. The transmitted frame has the format as shown in Figure 5.25.

The media access algorithm of the CSMA/CD scheme is as follows:

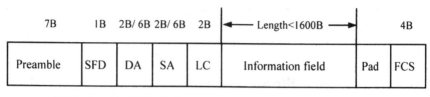

SFD : Start Frame Delimiter   LC   : Length Count
DA : Destination Address   FCS : Frame Check Sequence
SA   : Source Address

Figure 5.25 IEEE 802.3 CSMA/CD MAC frame format.

- Step 1: If the medium is "idle," then commence transmission; otherwise, go to step 2.
- Step 2: If the medium is "busy," then monitor the medium until "idle" is detected; if "idle" is detected, then commence transmission immediately.
- Step 3: If collision is detected during transmission, then send a jamming signal that notifies all the stations of frame collision and discontinue transmission.
- Step 4: Reattempt transmission sometime after the jamming signal has been transmitted. (Go to step 1.)

After a given frame has collided with another frame, the source station that has transmitted the colliding frame must continue the transmission until the frame returns in order to discern whether the frame has been in a collision.

The collision detection time becomes maximum when a collision occurs between the two frames emitted by the two stations located on the opposite ends of the network. In this case if the frame emitted by one station collides with the frame emitted by the other station in front of the other station, then the collision detection time reaches the maximum. Therefore, the maximum collision detection time is twice the transmission delay time between the two farthest stations in the network (see Figure 5.26). This implies that a transmitting station has to continue transmitting data at least for the maximum collision detection time before it can detect whether or not the transmitted data collided. Consequently, the minimum frame size of the MAC protocol should be determined such that its equivalent time length is twice the maximum propagation delay within the network.

Token Bus Scheme

The token bus scheme has a physical bus structure, but adopts a logical ring structure in transmitting frames between stations. To form a logical ring

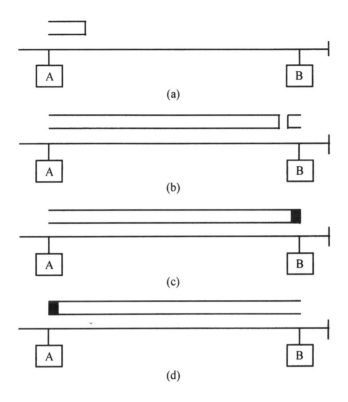

Figure 5.26 Detection of the frame collision: (a) A starts transmission; (b) B starts transmission; (c) B detects collision; and (d) A detects collision before one frame ends.

structure, each station possesses a predecessor parameter, which denotes the station that has sent the token to that particular station, and a successor parameter, which denotes the station that is to receive the token. Figure 5.27 is a representation of the token bus structure, and a logical ring is formed through the path A-D-B-C. When a station is newly registered or removed, the ring structure is reconstructed by altering the *next station* (NS) parameter contained inside the station contiguous to the station in question, as well as the *previous station* (PS) parameter of the next station.

For a station to transmit data, it has to acquire a token, and once the token has been obtained, it can transmit one or more data frames over a specified period of time. A station that has acquired the token hands over the right to transmit to the next station if there are no more data frames to be transmitted or if the *token holding time* (THT) that it has been assigned has expired. A station on the ring performs such functions as ring initialization, registration

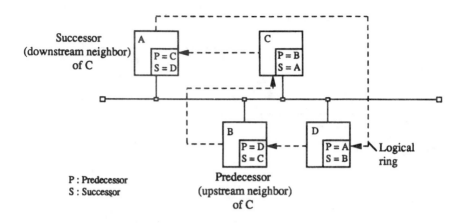

Figure 5.27 Token bus operation.

and removal of stations, recovery when the token gets lost or duplicated, and priority control.

In the token bus scheme, four priority classes are defined: 6, 4, 2, and 0. Each station can possess more than one priority class. At the time of ring initialization, each station is assigned a THT, which allows the station to send Class 6 frames, and *token rotation time* (TRT) values are specified, which are used in transmitting class 4, 2, and 0 frames. Once a station acquires a token, it can transmit the highest priority frame (class 6) during the preassigned THT interval, and in the case of low-priority frames (class 4, 2, 0), the rotation time of the circulating token and the already designated $TRTn$ ($n = 0, 2, 4$) value are compared. If the $TRTn$ value is greater (i.e., if the token arrives sooner than the $TRTn$ of the frame to be transmitted), the frame can be transmitted only during the time difference.

Token Ring Scheme

The token ring is the oldest ring control technique. IBM was the first to develop a product based on the token ring standard, and since then numerous companies have brought out compatible products. Also, the token ring scheme subsequently became the base frame of the FDDI standards. Figure 5.28 represents the format of the frame used.

In a token ring structure, a station can transmit a data frame when it receives a token; the I-bit of the *end delimiter* (ED) field is set to 1 during transmission, and to 0 when the final frame is being transmitted. If all of the frames transmitted by a given station have completed their rotation around the ring and returned to the source station, the station releases the token so that another

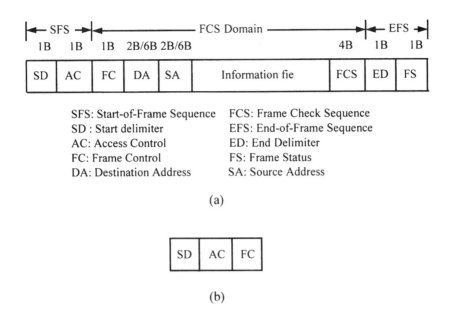

Figure 5.28 IEEE 802.5 token ring frame format: (a) frame format and (b) token format.

station can transmit data. Each station checks the *destination address* (DA) when it receives the frame transmitted by the previous station, and if DA is the same as its own, it copies the frame and transmits the frame to the next station. The station that transmitted the frame verifies whether the returned frame has been correctly transmitted and whether any error has occurred, and then removes the frame from the ring. Such a data transmission procedure of the token ring scheme is depicted in Figure 5.29.

The priority control scheme of the token ring is controlled using the AC field inside the token. The AC field's priority bits (PPP) and reserved bits (RRR) can be used to designate eight priority classes. A token can have several priority classes, and a station is allowed to transmit if it has a frame with a higher priority than the token. Also, when a data frame passes by, a station possessing a high-priority frame can alter the RRR value of the AC field to its own priority value, thus reserving the right to transmit. After a station has completed the transmission of the data frame, it stores the RRR value (contained in the data frame) in the buffer, and hands over the token after replacing the token's PPP value with the data frame's RRR value. When the token rotates around the ring, if there are no stations with a priority higher than the PPP value of the token, the token arrives at the station that has made a reservation. After the transmission of high-priority data, the station that has accepted the token returns the token that has the same PPP value as when the token was received. The station

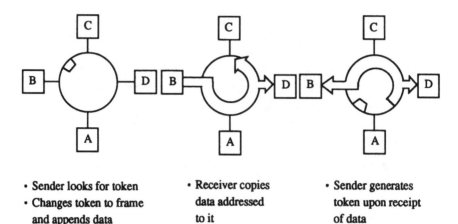

• Sender looks for token	• Receiver copies	• Sender generates
• Changes token to frame	data addressed	token upon receipt
and appends data	to it	of data

Figure 5.29   Token ring operation.

that has transmitted by increasing the token's PPP value compares the priority value of the token (when it arrives) with the value stored in the buffer, and if the two values are the same, it assumes that there is no higher priority station and releases the token. In other words, the station that releases the token after increasing its priority has the responsibility of restoring the original priority of the token when it returns.

LLC Protocol

The LLC protocol is the protocol that controls the transmission and receiving of frames independently of the network topology. The relation of the LLC frame to the MAC frame is illustrated in Figure 5.30. An LLC service is defined by the service primitives and the service parameters exchanged by the user and the service entity. IEEE 802.2 LLC provides the following three types of services to the upper layer: type 1, which provides unacknowledged connectionless service; type 2, which provides connection-oriented service; and type 3, which provides acknowledged connectionless service.

There are several types of LLC frames, namely, the unnumbered frame to indicate the start and end of communication, the supervisory frame to confirm the information frame and to control errors, and the information frame to transmit user data and higher layer protocols. These frames are assigned by different values at the control field. Each station on the LAN has multiple *service access points* (SAP) to provide LLC services to upper layers, and the application process in a station communicates with the application process in another station through the SAPs.

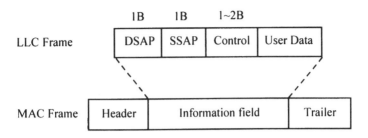

DSAP: Destination Service Access Point
SSAP: Source Service Access Point

Figure 5.30 LLC frame format.

5.4.4 LAN Interconnection

As stated earlier, a LAN is suitable for use as the customer-premise network on campuses or in business areas. As the number of LANs in these applications increases, the demand for their interconnection also increases. If LANs located at a distance are to be interconnected, repeaters, bridges, routers, and gateways can be used, but two separate LAN segments located in the same building can be interconnected using repeaters or bridges (see Figure 5.31).

The *repeater* is a layer 1 interconnection device that simply regenerates electrical signals for LAN interconnection. The *bridge* is a layer 2 interconnection device that looks at the source and destination addresses of each packet to decide whether or not to pass it to the other LAN segment. Because the data of each station is broadcast within a LAN, efficiency could drop significantly in a large-scale LAN due to the resulting enormous amount of traffic. In this situation, the LAN bridge installed between two LAN segments can increase the efficiency because it passes only the traffic that is destined for other LAN segments. There are two types of bridges: the local bridge for a customer premise, and the remote bridge for interconnecting LANs at long distance via public networks.

The *router* is a layer 3 interconnection device that sets up the necessary routes to deliver packets to their final destination. The router can be programmed to be able to transfer packets according to the desired objectives, and is usually used in complex internetworking.

The *gateway* operates in the OSI transport layer or higher layers, and provides the protocol conversion function that converts the protocol of the received packets to that of the destination network. The gateway is used when interconnecting LANs to heterogeneous networks having different protocols such as SNA, DECNET, and X.25.

LAN layers correspond to layers 1 and 2 of the OSI reference model,

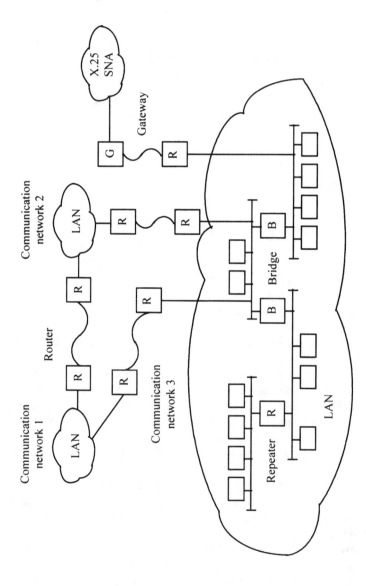

Figure 5.31   Interconnection of LANs using repeaters, bridges, and routers.

which are intended to transfer data exactly to the destination station over the LAN. However, when multiple LANs are interconnected via router or gateway, the layer 1 and 2 protocols are not sufficient. Higher layer protocols are also required: Layer 3 and 4 protocols are necessary for two end stations to communicate; and layer 5 to 7 services may also be necessary depending on applications.

5.4.5 Wireless LANs

In the early stages of LANs, wireless LANs were used for the purpose of saving the cost of installing and relocating wired LANs, and their major applications were as extensions of the wired LANs. But as the number of personal portable computers increased, the demand for wireless services such as ad hoc networking, nomadic access, and mobile computing has also increased. This facilitated the development of wireless LANs and the fusion of computer and wireless communications.

Physical Layer

Wireless LANs perform packet-mode communications using the air as the shared transmission medium. This transmission requires a medium frequency allocation for wireless communications, and this differentiates wireless communications from the wired communications.

The physical layer of a wireless LAN is composed of the *physical medium dependent* (PMD) sublayer and the *transmission convergence* (TC) sublayer. The PMD sublayer depends on the transmission medium such as infrared or RF, and the TC sublayer converges various different physical media to a common MAC layer. For medium access control, wireless LANs use the methods used in wired LANs for interworking with traditional wired LANs. Wireless LANs also require additional medium access controls because they also support the specialized medium access control functions for real-time services, in addition to the basic functions for data communications.

Wireless LAN systems can be classified according to wireless medium, technology, and frequency bands. In terms of the medium, the systems are divided into the *infrared* (IR) and RF systems. The infrared systems are subdivided into *diffused infrared* (DF-IR) and *direct-beam infrared* (DB-IR) systems; and the RF systems are subdivided into the licensed *non-spread-spectrum* (NSS) systems and the unlicensed *spread-spectrum* (SS) systems that operate in the ISM (industrial, scientific, and medical) bands. A DF-IR LAN provides a 1-Mbps data transmission rate and moderate size of coverage, and it is easier to install than a DB-IR LAN. A DB-IR LAN provides high-speed transmission data in a small area by aligning the transmitter and the receiver. An NSS LAN supports high-bit-rate data transmission because this operates in a high-frequency band. But the radio transmitters and receivers are expensive. An SS LAN is popular today. It has wider coverage than the IR and the NSS systems, is robust to interference and fading, and does not need to have its bands licensed. But its data transmission rate is usually low. An SS LAN can use either a *direct sequence* (DS) or *frequency hopping* (FH) scheme for spectrum spreading. The characteristics of wireless LAN systems are summarized in Table 5.5.

Medium Access Control

For the medium access control of wireless LANs, the following should be considered: Since the wireless bandwidth is very limited, the medium access control should be able to provide high throughput and acceptable delay performances. It should be able to support different physical layers transparently, and to support data, audio, and video services altogether. Further, fairness of access, low battery power consumption, reconfigurability of the network robustness, and handover are other important aspects to be supported.

The medium access control mechanisms of wireless LANs include CSMA, polling, and TDMA, and reservation schemes are also used to support various real-time services. While wired LANs commonly use the CSMA/CD, wireless LANs instead use a *collision avoidance* (CA) variation, or CSMA/CA, because reliable carrier sensing is extremely difficult in wireless LANs. The polling method used in wireless LANs transfers the token to the mobile nodes sequentially and permits data transmission only to the node possessing the token. The DB-IR LANs adopt the polling method. The TDMA method is modified to the *reservation TDMA* (R-TDMA) in wireless LANs to support the time-bounded services. This method employs a fixed-length frame structure and a channel use time reservation scheme.

Network Configurations

The configurations of wireless LANs are largely divided into two categories: *infrastructured* networking and *ad hoc* networking, as respectively shown in Figure 5.32(a,b). An infrastructured wireless LAN has central nodes (i.e., base stations) that provide various control and connections between the backbone network and the mobile station at the same time. Major applications of this configuration are LAN extension, building-to-building communication, nomadic access, and so on. An ad hoc wireless LAN is an independent private communication network formed among computer users. Its typical applications can be found in wireless classes or wireless battlefields in which ad hoc networking is formed among the participating mobile computers.

In Figure 5.32(a), the contact point between the backbone network and the wireless link is called the access point, which corresponds to a base station or a repeater. The traffic in such an infrastructured wireless LAN is either the

Table 5.5
Comparison of Wireless LANs

| | Infrared | | Radio-Frequency | |
	DF-IR	DB-IR	NSS	SS(DS, FH)
Data transmission speed	1~4 Mbps	10 Mbps	5~20 Mbps	1~20 Mbps
Mobility	Good	None	Better	Best
Range	20~60m	25m	10~40m	30~240m
Wavelength(λ)/frequency(f)	$\lambda = 0.8$–$0.9\mu m$	$\lambda = 0.8$–$0.9\mu m$	$f = 5.2, 17.1, 18$ GHz	$f = 0.9, 2.4, 5.7$ GHz
Transmission power	—	—	25 mW	<1W
Available systems	PhotoLink	InfraLAN	Altair	WaveLAN, RangeLAN, Freeport

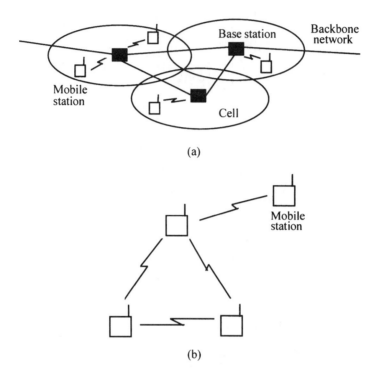

Figure 5.32   Wireless LAN architecture: (a) infrastructured networking and (b) ad hoc networking.

downlink traffic traveling toward the mobile station from the access point or the uplink traffic toward the backbone network from the mobile station. The downlink traffic is broadcast on a common channel, and the uplink traffic is transferred through the wireless medium access control. Since the access point to the backbone network changes as the mobile station moves, the access control function should provide a handover function for sustained communication. In Figure 5.32(b), the networking supports the connections among the mobile stations directly within a reachable area. Each mobile station in such an ad hoc LAN should use the medium access control to communicate to other stations. It is possible to expand the coverage of the network by implementing the forwarding function in one of the mobile stations transferring packets in multiple hops.

Wireless LAN Evolution

Standardization activities on wireless LANs have been concentrated mostly on the unlicensed frequency bands, and there are two major approaches to regulate the unlicensed bands: One is to set some interoperable rules for all equipment to achieve mutually dependable communications, and has been promoted by the IEEE 802.11 and HIPERLAN. The other is to provide the so-called spectrum etiquette: the minimum sets of rules that enable wireless LAN equipment manufactured by different vendors to fairly share the wireless resources. The latter approach does not preclude the former, and has been promoted by WINFORUM. IEEE 802.11, HIPERLAN, and WINFORUM are three major standards bodies for wireless LANs. IEEE 802.11 develops standards for DF-IR, DSSS, and FHSS operating in the ISM band. HIPERLAN develops standards for operation in the 5.2- and 17.1-GHz frequency bands. It studies new techniques such as multicarrier modulation and adaptive equalization, instead of spread spectrum, as a means to enhance wireless LAN performance. WINFORUM works toward standards and spectrum etiquette on data and voice services in the PCS band.

The currently available wireless LANs provide comparatively low data transmission rates of 1 to 20 Mbps. However new wireless LANs that can support a 20- to 150-Mbps data transmission rate will soon emerge, and the related standardization is in progress. If such broadband wireless LANs come into service, it will be possible to obtain such mobile broadband services as remote information search, video transfer, and mobile computing.

5.4.6 LANs, MANs, and WANs

As the performance of computers evolved and became higher and higher, and as users found needs for wider area connections, the need to expand LANs and to connect distant LANs increased. Demands for long-distance connections can be partially satiated by leasing dedicated lines, by connecting to X.25 packet networks, or by connecting to the public telephone networks through dial-up modems. However, network efficiency is low in the case of leased lines, and transmission capacity is small in the case of public-network-based connections. An ultimate solution to these problems could be found by resorting to broadband public networks such as BISDN, but it is a long-term solution. As such, interim practical solutions have been found in metropolitan-area networks (MANs).

A MAN has a structure and function that are appropriate to the metropolitan area and it can transmit high-speed data, high-definition images, and various multimedia information at the speed of 2 to 155 Mbps. A MAN, whose coverage is between LANs and WANs, encompasses a 50-km metropolitan area, whereas a LAN covers only a radium of about 2 km. A MAN can support switched or dedicated line services and can reduce transmission costs and improve transmission efficiency by allocating the bandwidth at users' requests. A MAN is a high-speed data network from the functional point of view, and a private network from the user's point of view. MANs are expected to serve as *customer-premises networks* (CPNs) in the early stage of BISDN. In 1990, DQDB was adopted as the standard technology for MANs by the IEEE 802.6 subcommittee (refer to Section 5.6).

WAN refers to a network used for interconnecting terminals that are spread across a nation. The X.25-based PSPDN is a representative WAN, which has an inherent limitation on its transmission capacity. With the increasing demand for high-speed data services, frame relay, which can provide higher performance than the traditional X.25 protocol, has appeared. SMDS is a rival technology of frame relay, which can support internetworking services at the DS-3 rate (refer to Section 5.6.5). Figure 5.33 indicates the position of X.25, frame relay, and SMDS among several categories of internetworking technologies.

The X.25 packet standard, recommended by ITU-T in 1976, was developed under the assumption that the transmission medium is error prone, and to ensure end-to-end quality, error management is performed at every node through a resource-intensive high-level data link control (HDLC) protocol. Accordingly, error detection and recovery functions are executed link by link, which becomes the major cause of network delay, thus hindering the network from operating at higher speeds.

Frame relay technology is the next evolutionary step in the X.25 technology and is designed to improve the efficiency of packet networks, as well as accommodate wide-area interconnection of LANs at the DS-1 rate. In 1988 the frame relay technology was first introduced by ITU-T and was defined by ANSI

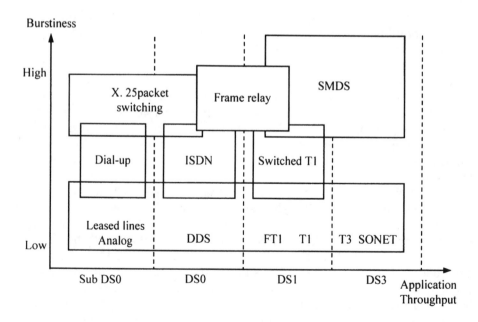

Figure 5.33 Transmission speed and burstiness of WAN services.

as *frame relaying bearer services* (FRBS) in 1990. And in 1992 ITU-T recommended frame relay services as ISDN's *frame mode bearer services* (FMBS) and defined the frame relay transmission technique as LAP-F. In frame relay, error correction and flow control functions are handled at the end users' CPE to enhance performance. As a consequence, bandwidth requirements can be lowered, which in turn curtails communications costs and reduces the amount of packet handling equipment in the network.

Frame relay, like X.25, regulates the interface between computer customers and networks and is implemented in such products as the LAN bridge, router, and T1 multiplexer. To achieve the inherent benefits of frame relay, fast packet switches must be used. Fast packet switches employ statistical multiplexing techniques to adapt a channel's entire bandwidth for transmission, and can guarantee high transmission efficiency for the users with bursty traffic.

Frame relay technology, when coupled with fast packet switches, has advantages over the traditional X.25 in terms of transmission speed and delay. And it guarantees higher transmission efficiency than leased line since it allows dynamic bandwidth allocation. In Table 5.6, the general features of frame relay are compared with X.25 and the leased line.

At this stage, frame relay services can only be provided by PVC, and frame switching capability through SVC is not yet implemented. For an efficient provision of frame relay services, a frame switching function should be added and various interfaces for CPEs such as workstations should be developed.

5.5 FDDI AND IEEE HIGH-SPEED LANS

Most LANs standardized in the 1980s provided a link capacity of 10 Mbps and all the stations in a LAN shared this capacity. As the number of stations

Table 5.6
Comparison of Frame Relay Service with Other Services

Comparison Features	Frame Relay	X.25 Packet Network	Leased Line
Transmission rate	56 Kpbs–1.544 Mbps	Up to 56 Kbps	Up to 1.544 Mbps
Bursty traffic	Accommodatable	Rather accommodatable	Inappropriate
Transmission delay	Short	Long	No delay
Routing	Yes	Yes	Predetermined
Network management	Yes	Yes	Restricted
Call setup	PVC	Yes	No
Layer	Data link	Network	Physical
Charging system	Access rate	Usage	Access rate/distance

Note: PVC: permanent virtual connection.

increased, the shared bandwidth decreased, and thus the transmission efficiency degraded. Further, as the stations evolved to possess high-speed processing and communication capabilities, the 10-Mbps link rate itself turned out to be a serious limitation. The emergence of highly advanced multimedia services including data, audio, still pictures, and moving images also required much faster data transmission speeds. As a consequence, LANs in the range of 100-Mbps link rate have been sought and are collectively called *high-speed LANs* in contrast to the conventional 10-Mbps range low-speed LANs. There are two representative approaches to high-speed LANs: the FDDI, which was promoted by the ANSI, and the IEEE 802 high-speed LANs, which have been standardized by the IEEE.

ANSI recognized the need for high-speed data communications in the early 1980s and began to develop high-speed LAN technology since then. In 1982, the X3T9 committee, which was responsible for the standardization on interconnections among computers, organized the X3T9.5 subcommittee to develop and standardize protocols for high-speed data communications. The X3T9.5 subcommittee generated in the late 1980s a series of standards in regard to the LAN technology that provided a 100—Mbps transmission capacity using optical fiber. This gave birth to the FDDI LAN. The FDDI LAN, which could also be classified as a MAN depending on its uses, is a 100-Mbps high-speed LAN that has been deployed mainly in private networks.

The IEEE P802 committee started the standardization of high-speed LANs rather late. In the early 1990s it began to plan for 100-Mbps high-speed LANs, but utilizing the existing low-speed LAN equipment and facilities. The IEEE's approaches can be classified into two categories: The first category is the IEEE 802.3 subcommittee's approach, which increases the transmission capacity to 100 Mbps while maintaining the conventional MAC protocol CSMA/CD. The second category is the IEEE 802.12 subcommittee's approach that develops a new MAC protocol to replace the CSMA/CD protocol, which is viewed as inappropriate for high-speed LANs.

In this section the FDDI and the IEEE's high-speed LANs are considered. The structure and functionality of the FDDI are discussed first, followed by a brief discussion of FDDI-II, which is an expanded version of the FDDI supporting isochronous services. Then the structures and operations of two IEEE high-speed LANs—(IEEE 802.3 subcommittee's high-speed LAN and IEEE 802.12 subcommittee's high-speed LAN) are considered.

5.5.1 FDDI Configurations and Protocols

The FDDI uses optical fiber as the transmission medium and is a LAN protocol based on the token ring access method of IEEE 802.5, providing a transmission speed of 100 Mbps. The FDDI was proposed by the X3T9.5 Working Group of the *American National Standards Institute* (ANSI) and is being applied in back-end local networks (which are used for connecting mainframe computers and large-capacity storage devices), high-speed office networks, and LAN backbone networks of the IEEE 802 party.

The FDDI has a dual-ring structure and can link up to 500 stations within a radius of 100 km, with the distance between stations being limited to 2 km. The ring topology used in the FDDI is similar to that of the IEEE 802.5 token ring, but is different in data coding and medium access protocols.

FDDI Configurations

The FDDI has a layered architecture, as shown in Figure 5.34, and is composed of the following layers corresponding to OSI layers 1 and 2: *physical medium dependent* (PMD), *physical* (PHY), LLC, MAC, and *station management* (SMT). The PMD layer regulates the characteristics of a fiber-optic medium, the connector for connecting each station to the medium, the wavelength to be used for transmission, the power requirement of the transmitter, and the method of optically bypassing a nonoperational node. The PHY layer regulates the 125-MHz clock speed, clocking scheme, and control symbol, whereas the MAC layer regulates token passing, frame formation, addressing, error detection and recovery, and the bandwidth allocation method between nodes. Finally, the SMT layer provides such network management services as station insertion and removal, ring configuration, and error logging. In the FDDI standards, the protocol of the LLC layer is not separately prescribed; hence, the IEEE 802.2 protocol of Figure 5.22 can be used.

The FDDI ring is composed of a primary ring and a secondary ring that have opposite data transmission directions, as shown in Figure 5.35. In normal operation, data flows in the primary ring only, while the secondary ring remains in the idle state. A station in the ring is categorized into class A or class B, depending on how it is connected to the ring. These two types of stations are

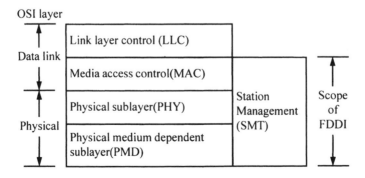

Figure 5.34   Four key standards of FDDI specification.

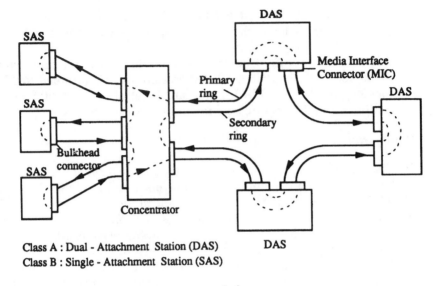

Class A : Dual - Attachment Station (DAS)
Class B : Single - Attachment Station (SAS)

(a)

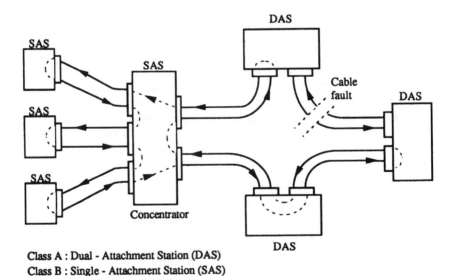

Class A : Dual - Attachment Station (DAS)
Class B : Single - Attachment Station (SAS)

(b)

Figure 5.35   FDDI ring configuration: (a) normal operation and (b) during a fault.

named, respectively, *dual-attachment station* (DAS) and *single-attachment station* (SAS) in Figure 5.35. A class A station is a station that is connected to both the primary ring and secondary ring and can change the configuration of the ring when a fault occurs in the link or station [see Figure 5.35(b)]. On the other hand, a class B station is connected only to the primary ring through a concentrator and cannot reconfigure the ring, even when a link fault occurs. The concentrator is a special form of a class A station and is a device that concentrates multiple class B stations by connecting them to its lower part.

Ring Failure Recovery

FDDI ring topology is advantageous in that its initial configuration as well as its reconfiguration is simple. It enables the insertion or elimination of stations without affecting the ring traffic, and supports sustained operation even when failure occurs. Therefore, it is easy to isolate and recover failures in the FDDI ring.

There are three ways to handle failures in the FDDI ring. First, stations can handle failures, as illustrated in Figure 5.36. In normal operation dual rings are in operation with the secondary ring in standby [Figure 5.36(a)]. If failure occurs on the link, the adjacent stations reconfigure the connection to form a single ring, which is free from the failure [Figure 5.36(b)]. If failure occurs in a station, the neighboring stations take similar actions to eliminate the failed station from the reconfigured single ring [Figure 5.36(c)].

Second, the concentrator can also handle failures. The concentrator can reconfigure the connection to eliminate the failed station and link. For example, if an SAS fails in Figure 5.35, the concentrator can eliminate it from the connection while maintaining the connection.

Third, the bypass switch can be used to handle the failure in the physical medium level. Optical switches can be installed in class A stations. If failure occurs in a station, the optical bypass switch in the station can directly bypass the incoming signal to the next station in the physical medium level without tossing it up to the physical layer.

FDDI-MAC Layer

The MAC protocol of FDDI is analogous to the token ring of IEEE 802.5, but differs in its token handling, priority, and management mechanisms.

The frame structures of the information frame and the token frame of the FDDI are shown in Figure 5.37. (Compare those with the token ring in Figure 5.28.) The term *symbol* in the figure refers to a four-bit unit.

Preamble (PA) consists of a maximum of 16 symbols, and functions to acquire and maintain the clock synchronization. *Start delimiter* (SD) indicates the start of the frame, and uses the symbols J and K (see Table 5.7). *End delimiter*

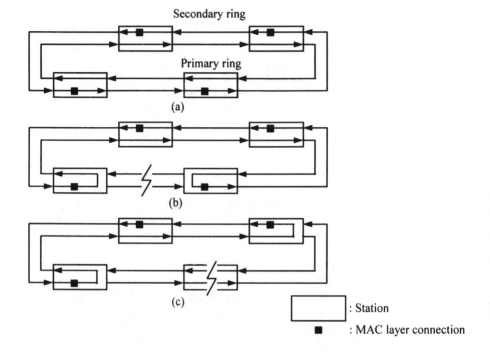

Figure 5.36 Reconfiguration of FDDI ring in failed state: (a) normal operation; (b) failure on a link; and (c) failure in a station.

(ED) indicates the end of the frame and uses the symbol T. *Frame control* (FC) denotes the frame type (i.e., synchronous frame or asynchronous frame), the frame field length, and frame category (i.e., LLC frame or MAC frame). *Destination address* (DA) and *source address* (SA) are address fields, and *frame clock sequence* (FCS) contains the 32-bit CRC calculated using the IEEE 802 prescribed polynomial. *Frame state* (FS) indicates the state of the frame with regard to error detection, destination address clock, and frame copy.

Token is a fixed sized frame consisting of only PA, SD, FC, and ED as shown in Figure 5.37(b).

A station with a token possesses the right to transmit data and can transmit a multiple number of frames during a designated time interval. Each station that is connected to the ring releases the token to the next station if there is no more data to transmit or after the designated data transmission time has elapsed. Also, each station repeats the frame transmitted from the previous station to the next station, and when the frame arrives at the destination station designated by the frame, the destination station obtains a copy of the frame. Here, the destination station establishes frame error detection status, address acknowledgment status,

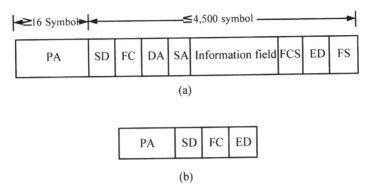

PA : Preamble   FCS : Frame Check Sequence
SD : Start Delimiter   ED : End Delimiter
FC : Frame Control   FS : Frame Status
DA : Destination Address
SA : Source Address

Figure 5.37   FDDI frame format: (a) FDDI Information frame format and (b) FDDI token format.

Table 5.7
4b/5b Code

Symbol	Encoded Bits	Symbol	Encoded Bits
0	11110	C	11010
1	01001	D	11011
2	10100	E	11100
3	10101	F	11101
4	01010	Q (Quiet)	00000
5	01011	I (Idle)	11111
6	01110	H (Halt)	00100
7	01111	J (Start 1)	11000
8	10010	K (Start 2)	10001
9	10011	T (Terminate)	01101
A	10110	R (Reset)	10111
B	10111	S (Set)	11001

and frame duplication status in the status bit, and retransmits the frame to the next station. The station that has transmitted a frame must remove the frame after it has rotated around the ring and returned. This is called *stripping*. Whether the frame had been properly transmitted can be determined here by checking the frame's status information. But error recovery is handled by an upper layer.

In contrast to the token ring of IEEE 802.5, the token in the FDDI is released after a station has transmitted a frame, and hence frame transmission can continue even if the frames transmitted prior to the release of the token still exist in the ring. Thus, multiple frames transmitted from multiple stations can exist above the ring at specific periods of time. The diagram in Figure 5.38 illustrates the frame flow mechanism in the FDDI.

Physical Layer

The physical layer is composed of PHY and PMD, with PHY regulating upper layer protocols, and PMD regulating hardware parts associated with optical fibers composing the FDDI link.

The FDDI node uses the optical wavelength of 1.3 µm and uses multimode fiber with a core diameter of 50 µm, cladding diameter of 125 µm, and a bit error rate of 10^{-9}.

The PHY transmitter converts the four-bit symbols sent from the MAC into a five-bit code group and transmits it to the medium, and the PHY receiver examines the SD from the encoded data stream, separates the symbol boundaries, and delivers the decoded symbols to the MAC. Additional symbols such as QUIET, IDLE, and HALT are acknowledged at the PHY and are used in supplementing *single-mode fiber* (SMF) interface functions.

In token ring or Ethernet, the Manchester coding method is used in data line coding. In Manchester coding, two transition bauds are generated in marking one bit. In the FDDI, the 4b/5b coding method is used, which uses five bauds to indicate four bits (see Table 5.7). The 4b/5b coding method has the virtue of limiting run lengths and allowing small transitions, thus enabling the transmission of more bits using the same bandwidth. Consequently, the data transmission speed of FDDI is around 100 Mbps, and the clock speed of each node can be as high as 125 MHz.

In the token ring of IEEE 802.5 an active monitor exists for the entire ring. In contrast, a distributed clock scheme is used in the FDDI. Also, a FIFO elastic buffer is employed when a station retransmits a bit in order to compensate for the difference between the input and output clock speeds that results from the jitter of the multimode fiber. The receiver employs such techniques as PLL in order to recover the clock of the previous station from the received data, while the transmitter uses a fixed local clock.

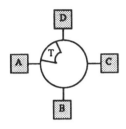

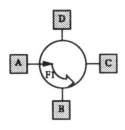

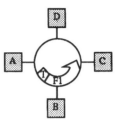

1) A waits for token to arrive.

2) A seizes token, begins transmitting frame 1.

3) A adds token to the end of frame 1.

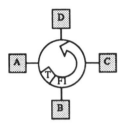

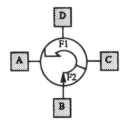

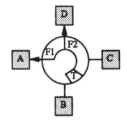

4) C copies frame 1, which is addressed to it.

5) While C continues to copy, B seizes token from the ring and begins transmitting frame 2.

6) B emits token. D copies frame 2, which is addressed to it, and A removes frame 1 from the ring.

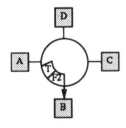

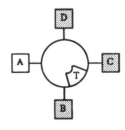

7) A removes frame 1 completely, but lets frame 2 and token pass. B removes frame 2.

8) B removes frame 2 completely and lets token pass.

Figure 5.38 FDDI token operation.

SMT layer

SMT of the FDDI is a part of the FDDI management functions possessed by each station, and it enables the station to control a process that goes on at several different FDDI layers. That is, SMT regulates system management functions, including the control function required for the proper operation of an FDDI node in the ring. Other functions provided by SMT include link management, insertion and removal of a station, station initialization, configuration management, fault isolation and recovery, scheduling policy, and statistical resources accumulation. Services defined in FDDI SMT can be separated into three types: service access management employing SMT frame, the service for connecting PHY and MAC, and the service for reporting the MAC condition via *ring management* (RMT) and for resolving error conditions.

5.5.2 FDDI-II

FDDI-II evolved from FDDI, which provides real-time circuit-switched services in addition to the original packet services. FDDI-II establishes isochronous channels to provide real-time services, and employs the cycle structure to transmit isochronous data. The cycle is generated by the station called the *cycle master*. The cycle master generates and distributes a cycle every 125 μs (i.e., 8 kHz), so that isochronous data can be transmitted at a constant rate.

An FDDI-II station operates either in basic mode or in hybrid mode, depending on the presence of isochronous data. Basic mode operates as in the case of the FDDI. When a user asks SMF for the transmission of isochronous data through appropriate signaling, SMT transmits the ring to the hybrid mode.

In FDDI-II, the bandwidth of 100 Mbps is dynamically separated into *wideband channel* (WBC) units, and up to 16 WBCs can be allocated. Each WBC can transmit 6.144 -Mbps, and once a station uses a WBC, WBC can be divided into subchannels in multiples of 8 Kbps, for example, 16, 32, 64, 384, 1,920, and 2,048 Kbps. In the FDDI-II ring, various service channels can coexist, and various service rates are possible depending on applications.

In FDDI-II, a *hybrid ring control* (HRC) layer is added to the existing FDDI structure, and the extended FDDI-II functions are mostly controlled by the HRC (see Figure 5.39). The MAC layer of FDDI-II is divided into *packet MAC* (P-MAC) and *isochronous MAC* (I-MAC), and the HRC consists of the IMAC and a *hybrid multiplexer* (H-MUX). The H-MUX controls the 125-μs cycle synchronization using the *latency adjustment buffer* (LAB) and manages the transition between the basic mode and the hybrid mode. It also controls the data flow between the MAC and PHY layers. The I-MAC provides the P-MAC-like functions to isochronous services.

The station management functions of the FDDI-II include all the functions for hybrid mode operation as well as for the basic mode token ring operation.

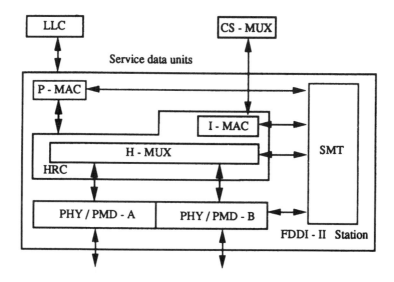

L L C   : Logical Link Control
CS-MUX : Circuit-Switched Multiplexer
P - MAC   : Packet Medium Access Control
I - MAC   : Isochronous Medium Access Control
H - MUX : Hybrid Multiplexer
H R C   : Hybrid Ring Control
P H Y   : Physical Layer
P M D   : Physical Medium Dependent
S M T   : Station Management

Figure 5.39   FDDI-II architecture.

This extended station management function consists of initialization and recovery procedures of the hybrid mode, hybrid mode ring control, bandwidth management, and so on.

5.5.3 IEEE 802.3 High-Speed LAN

When the IEEE P802 committee initiated the standardization process for high-speed LANs in the early 1990s, it considered the following requirements:

1. Support of the 100-Mbps transmission capacity;
2. Maximum distance of over 100m between hub and station;
3. Support of the previous 802.3 CSMA/CD frame structure;
4. Support of the Cat 3 transmission medium of EIA 568;

5. Error rate of 802 committee standard;
6. Observance of the FCC and European microwave emission standards;
7. Support of both the 10- and 100-Mbps rates;
8. Support of the interface independent of the transmission medium;
9. Support of multiple hubs;
10. Support of the RJ-45 connector of UTP (unshielded twisted-pair).

The IEEE's objective for high-speed LANs basically differs from that of ANSI's in that IEEE intended to spread high-speed LANs at low cost by reusing the existing low-speed LAN equipment and facilities, while satisfying the preceding requirements. With regard to the medium access control, two different approaches were considered: One uses the existing CSMA/CD method and the other employs a new MAC protocol suitable for real-time communications. The former approach, proposed by LAN Media and Grand Junction, uses the CSMA/CD medium access control protocol as well as the CSMA/CD frame structure, and thus it can easily interwork with the widespread low-speed LANs. The latter approach, proposed by Hewlett-Packard (H-P) and AT&T, takes the polling and *demand priority access* (DPA) based MAC protocol instead of the contention based CSMA/CD, so that it can guarantee the maximum delay bound and support real-time services. These two approaches are compared in Table 5.8.

Among the two approaches, the former one that is based on the CSMA/CD is called the 100BASE-T LAN, and the IEEE 802.3u subcommittee handles the standardization. The latter one that employs the polling and DPA for the MAC protocol is called the DPA LAN in this section, and the IEEE 802.12 subcommittee is responsible for its standardization. In this section, the 100BASE-T LAN is considered with the DPA LAN left for discussion in the next section.

Table 5.8
Comparison of IEEE High-Speed LANs

	802.3 (100BASE-T)	*802.12 (DPA)*
Frame structure	802.3 CSMA/CD	802.3 CSMA/CD
MAC protocol	CSMA/CD	Polling and DPA
Signaling scheme	8B6T or 4b5b	5b6b quartet
Transmission medium	2, 3, or 4 pairs Cat 3,4,5 and STP	Cat 3 of voice quality 4 pairs (UTP, STP, fiber)
Maximum length	Up to 250 m	Depends on the transmission medium

Note: DPA: demand priority access.

Classification of 100BASE-T LAN

The 100BASE-T LAN is a 100-Mbps baseband LAN, which employs the CSMA/CD scheme and the IEEE 802.3 MAC protocol. The 100BASE-T is composed of several sets of standards, namely, 100BASE-T4, 100BASE-TX, 100BASE-FX, and so on, which differ in the PMD layer. The 100BASE-T can support 1, 10, and 100-Mbps rates, but only the 100BASE-T with its 100-Mbps rate is considered in this section.

Since the 100BASE-T LAN employs the 802.3 CSMA/CD MAC protocol, it renders smooth networking with conventional IEEE 802.3 LANs, but at the same time it suffers from a problem inherent to the CSMA/CD protocol: As the transmission capacity increases, the propagation delay becomes more dominant than the transmission delay, thus decreasing the transmission efficiency. Since the 100BASE-T LAN uses the frame structure of a minimum of 512 bits, its network size should be limited to a tenth of that for the 10-Mbps LAN in order to detect the collision properly. Therefore, the end-to-end length of the 100BASE-T LAN is limited to about 250m maximum.

The 100BASE-T4 provides the 100-Mbps transmission capacity through four pairs of Cat3, Cat4, or Cat5 cables. It supports these voice communication cables so as to maximize the reuse of the readily deployed cables. Voice communication cables are simple and easy to deploy but their transmission quality is comparatively poor. As a consequence, the cable lengths are limited in practical applications. For example, the distance is less than 100m, and the overall network length is limited to about 200m even with two repeaters inserted for extension. The 100BASE-T4 need not transmit signals during the idle time between frame transmissions. This helps to reduce power consumption, so the 100BASE-T4 is adequate for battery-operated systems.

The 100BASE-TX and the 100BASE-FX both use the PMD sublayer and the physical layer of the FDDI, so they have much in common. (These two are called 100BASE-X together.) However, they differ in the transmission medium level: The 100BASE-TX uses the ANSI X3T9.5 TP-PMD/312 two pairs of Cat 5 cables or a 150-Ω shielded cable in full-duplex mode but the 100BASE-FX uses two multimode optical fibers. In the 100BASE-TX LAN, the distance between a hub and a local port is less than 100m, and the overall network length is limited to about 200m even with two repeaters inserted, but in the 100BASE-FX LAN, the distance between a hub and a local port can be extended to 300m and the overall network length can be extended to as long as 400m.

Structure of 100BASE-T Protocol

Even if the physical layer is different among 100BASE-T4, 100BASE-TX, and 100BASE-FX, the data link layer is common among all components of the 100BASE-T, because they use the common CSMA/CD MAC layer. Figure 5.40

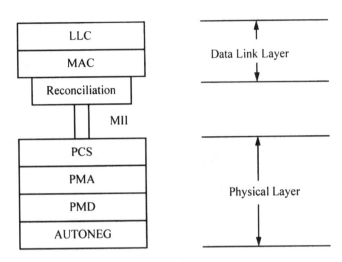

MII : Medium Independent Interface
PCS: Physical Coding Sublayer
PMA: Physical Medium Attachment
PMD: Physical Medium Dependent
AUTONEG : Auto-Negotiation

Figure 5.40   Protocol structure of 100BASE-T high-speed LAN.

shows the structure of the 100BASE-T protocol. The physical layer is composed of the *auto-negotiation* (AUTONEG) function, the PMD, the *physical medium attachment* (PMA), and the *physical coding sublayer* (PCS); the data link layer is composed of the MAC and LLC; and the two layers are connected through the *media independent interface* (MII) and the reconciliation sublayer. Among them, we examine some selected functions.

   A. Auto-Negotiation Function: Auto-negotiation refers to the negotiation function performed by two stations to notify the other of the functional status of each station for the purpose of establishing the best possible communication condition. This function is available among the network equipment that shares a common communication link. The AUTONEG function operates in the following procedure: In a link, a station that wants to communicate with another station in the same link notifies its own functional status to the other through broadcasting. Then the counterpart equipment receives the information and determines whether the sender's functional status conforms with its own. If so, it sends a coincidence signal to the sender; otherwise, it transmits a reject signal. In case the two stations have more than one function in common, they can select one according to some prespecified priority.

   B. MII and Reconciliation Sublayer: The MII and reconciliation sublayer provides a simple and easily implementable means by which to connect the MAC sublayer to the physical layer. The MII provides the following functions:

1. Supports both 10- and 100-Mbps transmission rates;
2. Synchronizes the reference clock of data to that of the delimiter;
3. Provides the four-bit-wide independent transmitting and receiving path;
4. Uses the TTL signal level that is suitable for general digital CMOS ASIC devices;
5. Provides a simple management interface;
6. Drives shielded cables of limited length.

Even if the MII supports two different transmission rates of 10 and 100 Mbps, the provided functions and the timing relation are the same in both cases. Data transmission in each direction is done over seven-bit signals, which consist of data (four bits) and delimiter, error, and clock bits. In addition, the MII provides two medium state signals to indicate the existence of the carrier and the occurrence of a collision, and it provides a signal to the management system.

C. Physical Coding Sublayer: As shown in Figure 5.40, the PCS lies between the MII and the PMA and provides the data encoding/decoding function to them. It receives data nibbles from MII, encodes them using a proper encoding scheme (i.e., 4b5b or 8B6T), and then sends the coded data to PMA. It performs the reverse function in the opposite direction. In addition, the PCS performs PCS carrier sensing, PCS error detection, PCS collision detection, and management system interface functions. In the PCS, the coding scheme differs between the 100BASE-T4 and the 100BASE-X systems. The 100BASE-T4 uses the 8B6T coding scheme, but the 100BASE-X uses the 4b5b coding scheme, which is also used in FDDI. The 8B6T coding scheme converts eight binary digits into six ternary digits, but the 4b5b coding scheme converts four-bit data into five-bit symbols (see Table 5.7). In this case, the internal data rate of 100 Mbps is converted to the 125-Mbps rate for transmission, accordingly.

5.5.4 IEEE 802.12 High-Speed LAN

In the case of the DPA high-speed LAN, which is being standardized by the IEEE 802.12 subcommittee, the repeater to which all stations are connected in the star topology plays the central role. If local stations make transmission requests to the repeater, the repeater selects a station among them using the round-robin polling and demand priority. The physical medium connecting the repeater and each station is UTP, STP, or optical fiber and through this medium a 100-Mbps bandwidth is provided between repeater and each station. By using these media, the DPA-based LAN can utilize the existing transmission resources. In addition, by supporting the 802.3 and 802.5 frame structures, it can smoothly interwork with the legacy low-speed LANs.

Structure of IEEE 802.12 High-Speed LAN

Figure 5.41 illustrates the structure of the 802.12 high-speed LAN using only one repeater. As shown in the figure, the 802.12 network is composed of the three elements: repeater, link, and station.

A repeater has several *local ports* (LPs) and a *cascade port* (CP): Each LP is used when connecting to a station, and the CP is used when connecting to another repeater. A link provides connection between a repeater and a local station or another repeater. The station in the figure may be a computers or network equipment such as a LAN analyzer, bridge, or router.

In all types of IEEE LANs, data transmission is done in such a way that data packets are broadcast among all stations and each station inspects them all but takes only the packet destined to itself. But the IEEE 802.12 high-speed LAN allows for the *privacy mode* in addition to this *promiscuous mode*. The privacy mode refers to the transmission scheme that each station notifies its address to the repeater and then the repeater transmits the packet only to the destined station. This mode enables a secure communication for each station and also reduces each station's frame processing overhead. Each station can select either operation mode as it desires. However, the repeater, bridge, and router should operate in promiscuous mode, and the broadcast and multicast packets should be transmitted to all stations.

Among the physical media used in the links of the LAN, the UTP supports a 30-Mbps transmission rate and four pairs of 100Ω UTP cables are used jointly to yield the transmission rate of 120 Mbps. When based on the UTP, the link

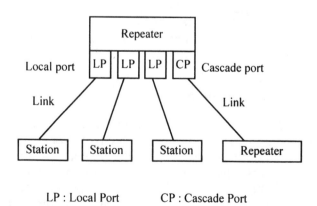

LP : Local Port   CP : Cascade Port

Figure 5.41 Configuration of 802.12 high-speed LAN having a single repeater.

can extend up to 100m. The STP link is composed of two pairs of 150Ω STP cables, and can also support a 120-Mbps transmission rate and 100m link length. The optical fiber link is composed of two multimode fibers of length 2,000m at maximum and supports the transmission rate of 120 Mbps also.

A large size IEEE 802.12 high-speed LAN can be constructed by attaching multiple repeaters in a tree structure as depicted in Figure 5.42. In this structure, the repeater at the highest level is called the level 1 repeater or root repeater, and the repeaters at lower levels are called level 2 repeater, level 3 repeater, and so on, in that order. In connecting repeaters, the port in the upper level repeater may be a local port but the port in the lower level repeater must be a cascade port.

Figure 5.43 shows the protocol structure of the IEEE 802.12 high-speed LAN. Since the repeater reconciles the transmission requests coming from each station, its MAC protocol, namely, the *repeater MAC* (RMAC) protocol, is different from that of the station. The physical layer consists of the *physical medium independent* (PMI), PMD, MII, and *medium dependent interface* (MDI) sublayers. The PMI uses an identical protocol regardless of the transmission medium, whereas the PMD requires different protocols for different transmission media. The MDI defines the physical specification of the interface contacting the transmission medium and the MII enables us to access heterogeneous transmission media. The PMI transmits the byte stream from MAC to the PMD after encoding it using the 5b6b coding scheme, which converts each five-bit piece of data into a six-bit symbol code with a balanced number of 0's and

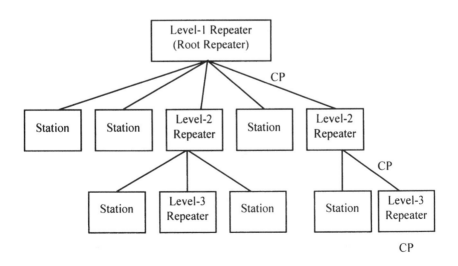

Figure 5.42 Configuration of multiple repeater 802.12 high-speed LAN.

1's. In addition, the PMI performs other functions such as scrambling, preamble generation, and data delimiting.

MAC Protocol of IEEE 802.12 High-Speed LAN

The DPA-based MAC protocol can be explained well by considering the IEEE 802.12 high-speed LAN shown in Figure 5.44, in which a repeater and three

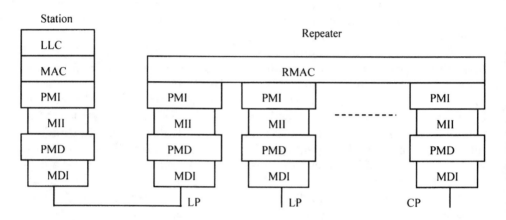

LLC   : Logical Link Control
MAC   : Medium Access Control
RMAC : Repeater Medium Access Control
PMI   : Physical Medium Independent
MII   : Medium Independent Interface

PMD   : Physical Medium Dependent
MDI   : Medium Dependent Interface
LP   : Local Port
CP   : Cascade Port

Figure 5.43   Protocol structure of IEEE 802.12 high-speed LAN.

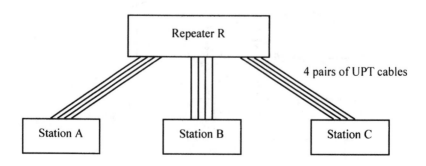

Figure 5.44   Example of IEEE 802.12 high-speed LAN configuration.

local stations are interconnected through four pairs of UTP cables. When transmitting the control signals, two out of the four pairs of UTP cables are used for the transmission from the repeater to the station and the other two pairs are used for the transmission in the other direction. When transmitting the user data, however, all four pairs are used for data transmission.

The operation of the high-speed LAN in Figure 5.44 follows the procedure illustrated in Figure 5.45. In the initial state when each station has no data to transmit, each station sends the Idle-Up signal to the repeater and the repeater sends an Idle-Down signal to each station [Figure 5.45(a)]. When a station has some newly generated data to transmit, it sends a Request-Normal or Request-High signal to the repeater [Figure 5.45(b)]. The Request-Normal signal indicates that the data block has normal priority, and the Request-High signal indicates that the data block has high priority. If two or more local stations send the Request-Normal signal simultaneously as shown in Figure 5.45(b), the repeater selects one according to the priority and round-robin scheme. For example, if the round-robin order is A, B, C, A, . . . , station A is selected. Once station A is selected, the repeater sends a Grant signal to station A and an Incoming signal to the other stations to notify them that data will arrive soon [Figure 5.45(c)]. The local station, which has received the Incoming signal, suspends

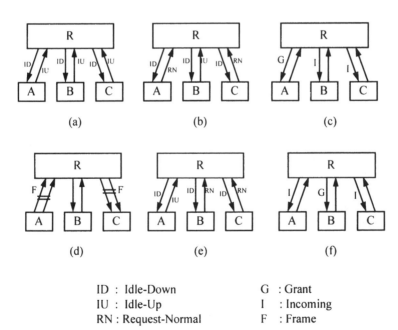

ID : Idle-Down   G : Grant
IU : Idle-Up   I : Incoming
RN : Request-Normal   F : Frame

Figure 5.45 Operation of the DPA protocol: (a) idle state; (b) A and C request; (c) A is granted; (d) cut-through; (e) B and C request; and (f) B is granted.

transmissions and gets ready for data reception. Station A transmits data to the repeater and the repeater translates the destination address of the frame and then transmits the frame to the destination (or to all stations if in promiscuous mode). The frame is transmitted directly to the destination station in a cut-through manner, not in a store-and-forward manner [Figure 5.45(d)]. Such a cut-through scheme maximizes the network efficiency.

Once the transmission of station A is completed, then the next request is processed. If stations B and C send the Request-Normal [Figure 5.45(e)], station B is selected for transmission according to the round-robin order even if station C requested transmission a little earlier than station B [Figure 5.45(f)].

If two requests of different priorities are made simultaneously, then the one with higher priority is selected. If two or more requests are of the same priority, one is selected according to the round-robin scheme. So the repeater should be aware of the point where the service is being made for both the high- and the normal-priority requests. Since the service on the normal-priority requests can be delayed excessively if many high-priority requests are made successively, the normal priority is arranged such that it is automatically upgraded to high-priority status after a predetermined period of time (e.g., 250 ms).

In the multilevel network including multiple repeaters, the service order is determined on the assumption that all the stations are connected to a single repeater, as illustrated in Figure 5.46.

5.6 DQDB AND SMDS

As LANs gradually became more widely distributed, the necessity for interconnecting the LANs that are scattered about locally increased. Accordingly, the IEEE 802.6 Committee has been working on MAN technology standardization since 1981 for provision of integrated services such as data, voice, and video in large metropolitan areas, and in December 1990 it approved the DQDB protocol as the IEEE 802.6 standard.

DQDB has a dual-bus structure and features a distributed queue that depends on the reservation scheme. Accordingly, DQDB accommodates data, voice, and video signals at high speeds, and can provide integrated services efficiently, without any waste in bandwidth. Broadly speaking, DQDB can provide three types of services: connectionless data services, connection-oriented data services, and isochronous services.

On the other hand, the expansion of LAN and high-performance computers necessitated the connection of distant LANs and MANs over the public network. *Switched multimegabit data service* (SMDS) is a connectionless packet switching (datagram) standard that is designed to meet exactly these objectives. It was standardized by Bellcore in 1989 as a MAN construction plan in order

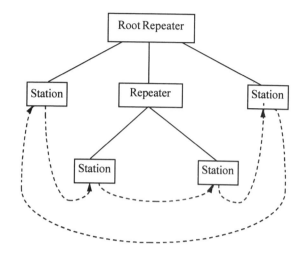

Figure 5.46 Service order in multilevel 802.12 high-speed LAN.

to accommodate high-speed switched data services inside the *local access and transport area* (LATA) of regional BOCs (RBOC). The SMDS protocol is rooted in DQDB.

In this section, the DQDB architecture and each layer function of the DQDB protocol are considered, followed by a discussion of the SMDS.

5.6.1 DQDB Architecture

A DQDB network possesses a dual-bus structure that consists of two unidirectional buses with opposite transmission directions. Such a bus is shared by multiple access nodes, as depicted in Figure 5.47, each node is linked to the transmitter module and receiver module, respectively, of the two unidirectional buses. Therefore, full-duplex mode communication is possible between the nodes. Also, the operations of the two buses are independent of one another with regard to data transmission; hence, a DQDB network has twice the capacity of a comparable system that employs only a single bus.

The first node of each bus performs the function of *head of bus* (HOB), which entails periodically producing empty slots or management information and releasing it into the bus. Other nodes transmit information by loading it onto the slots sent by the HOB. The bus terminates with the last node, and thus it is not necessary to erase data sent from the bus intentionally. DQDB is basically a bus structure, and hence it can be configured so that a malfunctioning node can be easily disconnected from the bus so as not to affect the operation of the overall network.

While DQDB has a dual-bus structure, it can be adapted into the looped bus topology shown in Figure 5.48. Here, DQDB has a ring shape physically but has a bus structure logically.

A DQDB network with the looped bus topology provides reinforced fault tolerance. That is, if the bus becomes disconnected, as shown in Figure 5.49, the beginning and end points of the bus are joined at the HOB node, and the two nodes adjacent to the broken point are made into the HOB so that the network can regain normal operation.

The DQDB functional architecture consists of the LLC layer, the DQDB layer, and the physical layer. The LLC and DQDB layers correspond to the OSI data link layer. The internal structure of the three layer functions are shown in

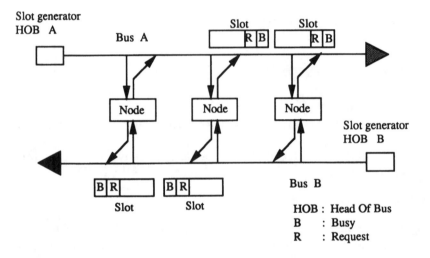

Figure 5.47   DQDB network architecture.

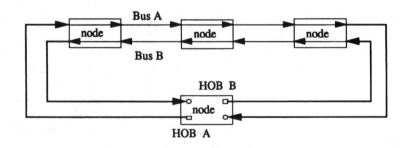

Figure 5.48   DQDB network of looped bus topology.

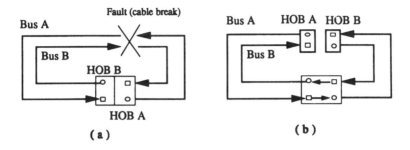

Figure 5.49 Fault tolerance of looped bus topology: (a) fault loop and (b) regained operation.

Figure 5.50. In the following three subsections, DQDB access control is considered first, followed by discussions of the DQDB and the physical layer functions.

5.6.2 DQDB Access Control

In a DQDB network with a dual-bus structure, the HOB node at the beginning of each bus sends 53-octet empty slots into the bus. A node that has data to transmit does so by loading such empty slots with its own information through the employment of the distributed queuing method. Slots are divided into *queue-arbitrated* (QA) slots for nonisochronous traffic, and *pre-arbitrated* (PA) slots for isochronous traffic. QA slots are used for either connectionless MAC services or connection-oriented data services. PA slots are used for isochronous services such as voice and video (see Figure 5.50).

QA Access Control: Distributed Queue-Arbitrated Access Protocol

DQDB network access through the QA slots is controlled according to the distributed queuing principle. The distributed queuing concept involves controlling the order of the nodes attempting to access the network in terms of the overall network, and a virtual *first-come first-served* (FCFS) queue exists for the entire network. That is, the request by a node to access the network gets in the distributed queue, and the network handles the access requests one by one from the head of the distributed queue. Controlling the distributed queue at each node to operate as if it were a single queue is the essence of the distributed queuing concept.

In the QA slot, a BUSY bit and a *request* (REQ) bit are placed in the *access control field* (ACF) in order to control the usage of each channel. The REQ bit of each QA slot actually consists of three bits in order to support three priority queues. In the following we discuss the basic distributed queuing method, which controls the access at the same priority level, the distributed queuing

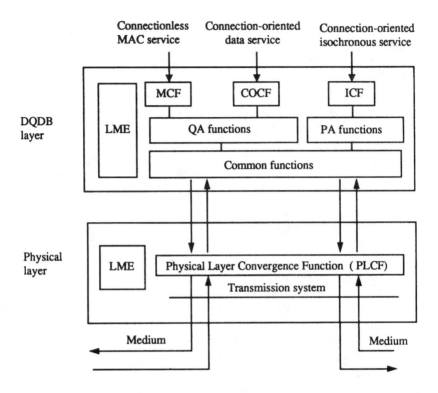

Connectionless   Connection-oriented   Connection-oriented
MAC service   data service   isochronous service

MCF : MAC Convergence Function
COCF : Connection-Oriented Covergence Function
ICF   : Isochronous Convergence Function
LME : Layer Management Entity

Figure 5.50 Functional structure of DQDB node.

method to set the priority depending on the connection-oriented or connectionless services, and bandwidth balancing, in that order.

Basic Distributed Queuing

The distributed queue control method of the QA slot at the single priority level
is depicted in Figure 5.51. An internal node of the DQDB network possesses a
request (RQ) counter corresponding to each bus, as well as a *countdown* (CD)
counter. When a node has data to send, it waits for a slot whose REQ bit is 0
among the slots that pass through the reverse direction bus. When such a slot
is found, its REQ bit is altered to 1 to notify an upper node of the request for

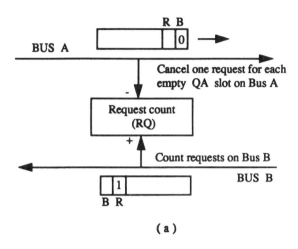

(a)

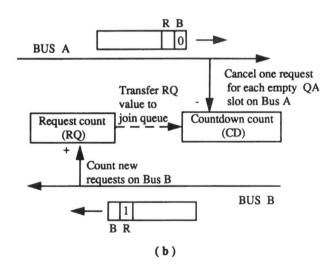

(b)

Figure 5.51 Distributed queuing protocol: (a) operation of request counter before joining distributed queue and (b) operation of request and countdown counter after joining distributed queue.

transmission. The node without data to transmit increments the RQ counter if the REQ bit of the slot that is passing through the reverse direction bus is 1, and decrements the RQ counter if the BUSY bit of the slot transported to the forward direction bus is 0 [see Figure 5.51(a)]. In this case, the value of the RQ counter belonging to a node indicates the number of slots waiting to be transmitted that have been placed in the queue by the nodes located downstream of the forward direction bus.

After setting the REQ bit to make notification of the request for transmission, the node with data to transmit copies the RQ counter value onto the CD counter and goes into the countdown state after setting the RQ counter value to 0 [see Figure 5.51(b)]. The node in the countdown state increments the RQ counter if a slot whose REQ bit is 1 passes through the reverse direction bus, and decrements the CD counter if an empty slot with a 0 BUSY bit passes through the forward direction bus. Here, if the CD counter value becomes 0, data are transmitted by loading them onto the next arriving empty slot. In the countdown state, the CD counter value represents the total number of requests that had been queued by the downstream nodes on the forward bus before the moment in which the node queued the new request. The RQ counter keeps track of the accumulated number of requests that have been queued by the downstream nodes on the forward bus since the moment when the node queued the request. Therefore, transmitting data in an empty slot when the CD counter value becomes 0 is equivalent to operating the distributed queue as if it were a single queue. To transmit a slot, an empty slot with 0 BUSY bit is awaited after the CD counter value becomes 0, and the slot is transmitted after BUSY bit is converted to 1.

Priority Distributed Queuing

The distributed queuing protocol supports the assignment of priority to QA segment access. However, priority requests are not supported in the base standard; all connectionless data segments must be sent at the lowest priority. For upward compatibility, distributed queues for each level of priority are required in this implementation.

The operation of separate distributed queues for each level of priority is achieved by using a separate REQ bit at the access control field on the reverse bus for each level of priority, and separate RQ and CD counters for each priority level. The counters operate similarly to the single priority case, except that account must be taken of REQs at higher levels. That is, for an access unit that does not have a QA segment queued at a particular priority level, the counter operating at that level counts REQs at the same and higher priority levels. Thus, the RQ counter records all queued segments at equal and higher priorities.

If the access unit does have a QA segment queued at a particular priority level, then the RQ counter operating at that level only counts REQs at the same priority level. However, the operation of the CD counter at that priority level is slightly altered. The CD counter, in addition to counting down the received empty slots, increments for REQs received at higher priority levels. This allows the higher priority segments to claim access ahead of already queued segments.

Note that decrementing the RQ counters and CD counters at all priority levels occurs when an empty QA slot is received at the function within the node operating the distributed queue, not when it is sent by this function. This ensures that the correct counter values are maintained if the highest priority segment queued by the access unit gains access to the empty slots and the access unit marks the slot as busy.

The intent of the priority scheme is for the access performance of the highest priority traffic to be unaffected by lower priority traffic. Such a feature is very important in network signaling. However, note that priority should be decided depending on subnetwork distance, speed, and loading conditions.

The IEEE 802.6 standard specifies three levels of access priority. The definition of how these levels are used is a matter for the operator of the particular DQDB subnetwork, although it is likely that the highest level would be used for network management and signaling. The DQDB layer assumes that the user of the QA access requests the appropriate priority level for a particular segment transfer.

Bandwidth Balancing

If the speed of an electromagnetic wave is assumed to be infinite, then the DQDB queuing method processes data segments arriving at each node in the ideal FCFS manner. In reality, however, a propagation delay exists that might cause the REQ bit transmitted by a lower node on a bus to be delivered to an upper node after a finite time delay, which allows the upper node, if it has data to send, to use the available empty slot first. This creates a problem of unfairness, which might become more severe with the increase in the network size and transmission speed. To resolve this problem, DQDB employs a bandwidth balancing counter for each of the buses. Each node increments the counter every time a slot is transmitted, and when the counter reaches the fixed bandwidth balancing modulus (BWB_MOD) value, its value returns to zero and the CD counter and RQ counter values increment by one. In this arrangement, a node that has transmitted BWB_MOD number of slots concedes one empty slot to a low-priority node. This has the effect of reducing unfairness to some extent. There are other ways to handle the unfairness problem also.

Pre-Arbitrated Access Control

PA slot access is typically used to provide for transfer of isochronous service octets. PA slots are distinguished from QA slots as shown in Table 5.9 by a

Table 5.9
DQDB Slot States

BUSY Bit	SL_Type Bit	Slot State
0	0	Empty QA slot
0	1	Reserved
1	0	Active QA slot
1	1	PA slot

combination of the BUSY bit and the slot type (SL_type) bit of the access control field, which is the first octet of the slot's frame.

In contrast to QA slots, a PA slot can be shared by several nodes inside the DQDB network. In other words, the payload space inside a PA slot is composed of several octets, and the node sending isochronous data is allotted a place to load its data (among the PA slot's payload space) according to a DQDB layer management procedure. Each node uses VCI to identify its respective slots among the PA slots passing through the channel. To ensure provision of isochronous service, the HOB node of each bus must produce PA slots at regular intervals and also designate the VCI corresponding to the service provided in the PA slot's header.

5.6.3 DQDB Layer Services

As depicted in Figure 5.50, DQDB maintains functions for provision of three types of services: MAC services for the LLC layer, connection-oriented services, and isochronous services.

MAC Services for LLC Layer

In the DQDB layer, the header and trailer for transmission are appended to the *MAC service data units* (MSDU) delivered from the LLC layer to produce the *initial MAC protocol data units* (IMPDU). The IMPDU header section contains B/Etag for verifying the beginning and end points of IMPDU, IMPDU length information, address of the transmitter/receiver, and quality of service. An IMPDU is divided at the MCF block into segmentation units with the fixed size of 44 octets, and each segmentation unit is appended with two bytes, respectively, of header and trailer to form a 48-octet *derived MAC protocol unit* (DMPDU). Each DMPDU is transmitted through the payload space of QA slots and reassembled at the receiving node. Figure 5.52 depicts the process of creating an IMPDU as well as a DMPDU.

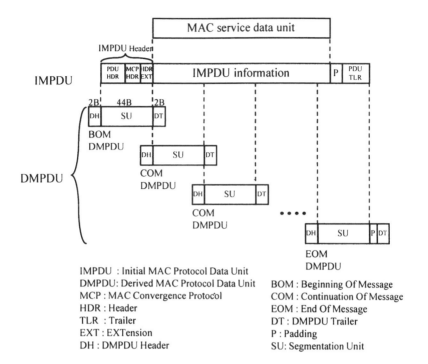

IMPDU : Initial MAC Protocol Data Unit
DMPDU: Derived MAC Protocol Data Unit
MCP : MAC Convergence Protocol
HDR : Header
TLR : Trailer
EXT : EXTension
DH : DMPDU Header

BOM : Beginning Of Message
COM : Continuation Of Message
EOM : End Of Message
DT : DMPDU Trailer
P : Padding
SU: Segmentation Unit

Figure 5.52 Generation of IMPDU and DMPDU of DQDB protocol.

The QA segment is divided into the QA segment header of four octets and the QA segment payload space, which contains the 48-octet DMPDU, as shown in Figure 5.53. The QA segment header includes the VCI to which the QA segment belongs; payload type, which denotes the characteristics of the data being transmitted; priority; and the *header check sequence* (HCS) for detecting bit errors within the QA segment.

The header section of the DMPDU segment payload contains the segment type, which denotes the location of the information of 44 octets transmitted by DMPDU within the IMPDU; sequence number, which indicates the order of a DMPDU within the IMPDU; and *message identifier* (MID), which is used in identifying the DMPDUs derived from a single IMPDU. The DMPDU trailer section contains the length of IMPDU information transmitted by the DMPDU and the payload CRC for detecting DMPDU transmission errors.

The production of a QA slot is complete with the addition of a 1-octet ACF to the 52-octet QA segment. The ACF consists of the BUSY, SL_type, and REQ bits mentioned earlier, as well as the *previous slot release* (PSR) bit, which is used for raising the utilization rate of DQDB. Figure 5.53 shows the overall format of the DQDB slot.

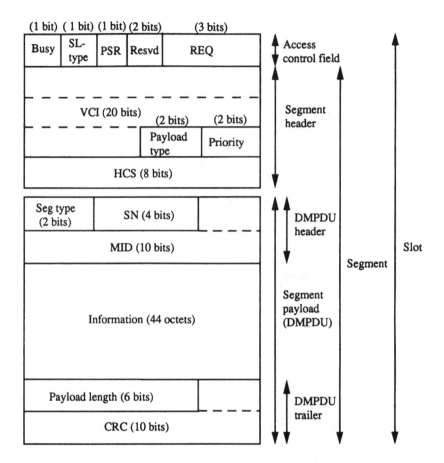

S L　　: Slot　　　　　　　　　　　　PSR : Previous Slot Release
REQ　: Request　　　　　　　　　　VCI : Virtual Channel Identifier
HCS　: Header Check Sequence　　　S N　: Sequence Number
MID　: Message Identifier　　　　　CRC : Cyclic Redundancy Check
DMPDU : Derived MAC Protocol Data Unit

Figure 5.53　DQDB slot structure.

The operation at the receiving unit is as follows. Each node of DQDB node checks the VCI of every slot that passes through the bus, and if a slot is found that corresponds to itself, it examines the DMPDU existing in the payload space of the slot. If the DMPDU indicates BOM, the destination address is examined; and if the DMPDU is intended for itself, it makes a copy of the data, memorizes the sequence number and message identifier, and goes on to accept the slots with that particular MID until the message terminates. After it has finished receiving the message, it removes the information of the MID corresponding to that message. This is due to the possibility that the transmitting unit might use the same MID for transmitting a different message. As described, a single message is transmitted through several slots; hence, the receiving unit can receive several messages concurrently.

The QA function block of Figure 5.50 transmits 48-octet nonisochronous connectionless MAC service and connection-oriented data segments.

Connection-Oriented Data Services

The *connection-oriented convergence function* (COCF) block is responsible for transmitting connection-oriented data, but the procedure associated with the maintenance or disassembly of data links is something that should be performed at an upper layer rather than at the DQDB layer. Like the MCF, the COCF exchanges data through the QA function block of Figure 5.50 via a data disassembly/assembly procedure.

Isochronous Services

In order for a DQDB network to provide isochronous services, the *access unit* (AU) of each node must be inherently capable of controlling the access of PA slots, and the HOB of each bus must generate PA slots periodically.

In case the HOB is incapable of producing PA slots periodically, the PA function block of Figure 5.50 does not operate in actual isochronous mode. Therefore, the *isochronous convergence function* (ICF) block performs buffering in order to provide sufficient isochronous services.

Common Function and Layer Management

The common function block performs transmission of PA and QA slots through PA and QA function blocks, as well as the provision of DQDB layer management information octets. The DQDB *layer management entity* (LME) block performs header maintenance functions and functions for assigning a MID page to each node.

5.6.4 Physical Layer

As shown in Figure 5.50, the physical layer of a DQDB layer consists of a transmission system, a *physical layer convergence function* (PLCF) block, and a physical LME block. In the DQDB standard, the type of transmission system is not specified, and convergence procedures for the physical layers of DS-3 (44.736 Mbps), DS-3E (34.386 Mbps), DS-4E (139.264 Mbps), and the SDH (155.520 Mbps) are of main interest.

The PLCF block is provided so that the DQDB layer can perform its functions independently of the particular transmission system used, and it executes the function of converting the format of timing information, slot octets, and management information so that they can be delivered through the given transmission system. Consequently, each transmission system requires its own appropriate PLCF protocol. The LME block of the physical layer executes management-related functions, and hence performs such functions as fault detection of nodes and links and isolation of impaired nodes.

As a specific example of DQDB transmission systems, we consider the physical layer convergence procedure of the DS-3 transmission system. The DS-3 signal operates at the 44.736-Mbps rate and consists of 699 octets per 125-µs time period. One bit in 85 is used for DS-3 overhead functions providing a nominal information payload rate of 44.210 Mbps ($84/85 \times 44.736$ Mbps) and leaving approximately 690.78 octets available for use by the DS-3 *physical layer convergence procedure* (PLCP). As shown in Figure 5.54, the PLCF frame format consists of 12 rows by 57 octets, with the last row containing a trailer of either 13 or 14 nibbles. The PLCP frame has a nominal duration of 125 µs, and each frame transmits 12 DQDB slots to the DQDB layer. The PLCP frame format is asynchronously mapped into the DS-3 information payload space using a nibble-stuffing technique (i.e., stuffing done in a four-bit unit). A nibble-stuffing opportunity occurs once every 375 µs to maintain a nominal frame repetition

	1B	1B	1B	1B	53B		
1	A1	A2	P11	Z6	DQDB slot		
2	A1	A2	P10	Z5	DQDB slot		
3	A1	A2	P9	Z4	DQDB slot		
4	A1	A2	P8	Z3	DQDB slot		
5	A1	A2	P7	Z2	DQDB slot		
6	A1	A2	P6	Z1	DQDB slot		
7	A1	A2	P5	F1	DQDB slot		
8	A1	A2	P4	B1	DQDB slot		
9	A1	A2	P3	G1	DQDB slot		
10	A1	A2	P2	M2	DQDB slot		
11	A1	A2	P1	M1	DQDB slot	13-14 nibbles	
12	A1	A2	P0	C1	DQDB slot		

A1, A2 : Frame   125 µs
P11-P0 : Path overhead identifier
Z6-Z1, F1, B1, G1, C1 : PLCP path overhead octets

Figure 5.54   DQDB physical layer convergence protocol frame.

rate of 125 µs. Within the stuffing opportunity cycle, there are three PLCP frames. The first frame contains a trailer of 13 nibbles, the second frame contains a trailer of 14 nibbles, and the third frame contains either 13 or 14 nibbles, depending on whether a nibble stuffing has occurred. Figure 5.54 shows the PLCP frame format for the DS-3 signals.

5.6.5 Switched Multimegabit Data Services

SMDS is a service standard designed to provide connectionless packet-switched data services among distant computer terminals and LANs via public networks, and its basic structure is illustrated in Figure 5.55. In the figure, *customer-premise equipment* (CPE), or a LAN consisting of CPEs, is connected to the *MAN switching system* (MSS) through a *subscriber network interface* (SNI). Inside an SMDS network, several MAN switching systems are interconnected through the *interswitching system interface* (ISSI).

The *SMDS interface protocol* (SIP) employs the IEEE 802.6 DQDB protocol, and the terminal based on the IEEE 802.6 protocol is connected through the SMDS network via an access DQDB, which is a DQDB-based subscriber network

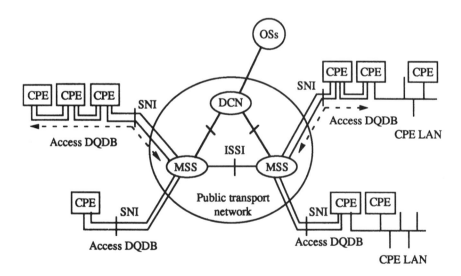

SNI : Subscriber Network Interface	DCN : Data Communications Network
MSS : MAN Switching System	CPE : Customer Premises Equipment
ISSI : Inter-Switching System Interface	CPE LAN : CPE Local Area Network
OSs : Operations Systems	DQDB: Distributed Queue Dual Bus

Figure 5.55 Network in support of SMDS.

unit connection protocol. Here, the DS-3 level access network allows either a single or multiple terminals, but the DS-1 level access network allows the connection of single terminals only. The DQDB subnetwork itself can operate in bus configuration or loop configuration. However, access DQDB must operate in the bus configuration.

SMDS Interface Protocol

The SIP is a connectionless data transmission protocol based on the MAC service protocol of the DQDB discussed earlier. As shown in Figure 5.56, the SIP is composed of three levels. SIP level 3 produces L3-PDUs by adding a header and trailer to the SMDS service data units from an upper layer user. The header section contains the source/destination address, as well as information for verifying the loss of segments to be transmitted, and the trailer section contains such information as the length of the transmitted service data. SIP level 3 corresponds to the functions associated with IMPDU creation and reception (handling) among the MCFs of the DQDB protocol. SIP level 2 performs functions associated with cell-unit segmentation and reassembly of L3-PDUs, which correspond to the IMPDU segmentation/reassembly and DMPDU creation functions of the DQDB protocol. Also, slot transmitting/receiving functions are also performed, such as the access control function for cell-unit transmission, and these functions correspond to the QA function block of the DQDB protocol. SIP level 1 is a physical layer and corresponds to the PLCF of the DQDB protocol. Figure 5.56 illustrates the relationship between the SIP and the IEEE 802.6 DQDB protocol.

The procedure to append a header and a trailer to the SMDS SDU to generate the L3-PDU, and the procedure to segment the L3-PDU to generate L2-PDUs are, respectively, the same as the procedures to generate the IMPDU and the DMPDU in the case of the DQDB protocol shown in Figure 5.52. In other words, if IMPDU, DMPDU, and PLCP in Figure 5.52 are replaced,

L3_PDU	IMPDU
L2_PDU	Segmentation & Reassembly
	DMPDU
	Slot
L1_PDU	PLCP
SIP	DQDB

Figure 5.56 Correlation of protocol layers between SIP and DQDB.

respectively, with L3-PDU, L2-PDU, and L1-PDU, then the result is the procedure to generate the corresponding PDUs of the SMDS (but the detailed names differ).

The L3-PDU is formed with the addition of a header and a trailer to the user data of 9,188 octets or less. The L3-PDU's header consists of BEtag, BAsize, destination address, source address, *high-level protocol identifier* (HLPI), and *header extension* (HE) length indication, and the HE field is appended to it. HLPI is for use in aligning the SIP format and the DQDB protocol format. The L3-PDU's trailer includes BEtag and length indication.

The L2-PDU is formed by dividing an L3-PDU into 44-octet segments and adding a 7-octet header and a 2-octet trailer to each of the segments. The structure of the L2-PDU is the same, in size, as that of the DQDB slot (i.e., the DMPDU) in Figure 5.53, and is similar in usage. The first octet of the L2-PDU provides the access control function as in the case of the DQDB slot but the use of the internal fields differs. The next 4 octets are used for network control information, and the remaining 53 octets are the same as for the DQDB slot both in size and use.

SIP level 1 is a physical layer protocol that operates between the subscriber terminal and the switched network at either 44.736 Mbps (DS-3) or 1.544 Mbps (DS-1) and is divided into the PLCP and the transmission system sublayer. The transmission system sublayer defines the method and characteristics of the connection with the transmission link. The PLCP assumes the role of mapping SIP level 1 control information and L2_PDUs in a format suitable for the transmission system sublayer. The DS3 PLCP frame structure for the SMDS physical layer is the same as that shown in Figure 5.54, but with the DQDB slots replaced with L2-PDUs.

SMDS Services

The SMDS is a high-speed packet switching service that possesses public or logical private network functions. Each data unit contains source/destination addresses and is transported to its destination together with user information. The addressing system is analogous to the telephone numbering system, and it employs a group address scheme to enable the transport of information to a multiple number of destinations.

SMDS possesses address screening functions. Destination address screening is the scheme in which data is transmitted only when its destination address is an approved network address. In source address screening, the data to be sent to the SNI is transmitted there only if it has an approved source address. By applying such address screening functions, an SMDS network can be used as a logical private network.

CPE can access an SMDS network via a DS-1 or DS-3 class transmission path, and a single CPE or multiple CPEs can exist at a single SNI. The important point here is that a single SNI should always be shared by way of a terminal belonging to one user only. This is essential for ensuring the privacy of subscriber information.

Since SMDS subscribers have several traffic requirements, SMDS provides several access classes. Each access class regulates a number of different transfer classes according to the allowed sustained information transfer and burstiness, and the traffic of each class is controlled by way of a credit manager algorithm. That is, the credit value is determined according to the specified class, and it gets lowered every time information is transmitted. The credit value increases periodically by a specified amount, but cannot exceed the maximum credit value. Therefore, the attributes of each class are represented by three parameters: credit increment value, credit increase period, and maximum credit value.

Five access classes are defined in the DS-3 path, while only one access class is defined in the DS-1 path. Such access classes are defined for the user-to-network direction of SNI, and the packets from the CPE to the network that exceed the access class are not transmitted. But in case a packet from the network to the CPE exceeds the access class, the packet is preserved in the network buffering and retransmitted when the credit is allowed.

SMDS specifies transmission delay objectives for high-speed data transmission in many different application areas. In the case of individually addressed L3_PDUs, 95% of the transmission delay of the delivered L3_PDUs should be less than 20 ms in the case of the DS-3-based access path, and it should be less than 140 ms in the case of the DS-1-based access path.

In general, SMDS subscribers have already invested substantially in their computers, hardware, and software for their LANs; hence, in order for SMDS to be competitive, it must offer maximum additional advantages at the lowest possible cost.

The anticipated SMDS subscribers must be assumed to possess several corporate networks, including LANs, leased-line circuits, and X.25 packet networks, and in order to connect these various corporate networks as if they are operating as a single network, an internetworking protocol, situated between the host network and gateway (which connects the support networks), must operate above several different support network protocols.

Currently, the most widely used internetworking protocols are connectionless, and they rely on the overlay transport protocol in order to secure end-to-end reliability and control. Typical examples of such an internetworking/transmission protocol pair are DARPA TCP/IP, ISO 8073/ISO 8473, and Xerox, DEC. Figure 5.57 depicts the role of SMDS as a high-speed subnetwork between subscriber networks that employ the TCP/IP protocol. Figure 5.57(a) depicts the method of utilizing a router with the IP function to transmit, if needed, IP data from a LAN to a network that provides SMDS. In contrast, in Figure 5.57(b), a bridge rather than a router is used between the LAN and the SMDS subnetwork. The bridge does not participate directly in handling internetworking protocol, and it operates only at the subnetwork protocol layer.

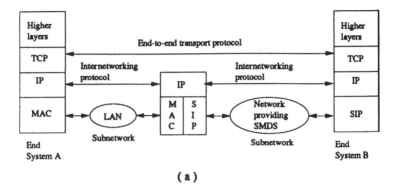

(a)

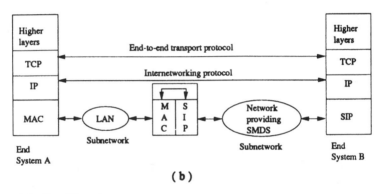

(b)

TCP : Transmission Control Protocol
MAC : Medium Access Control
IP   : Inter net Protocol
SIP   : SMDS Interface Protocol

Figure 5.57   Example scenario of the use of SMDS: (a) router functionality and (b) bridge functionality.

For SMDS to execute its role efficiently, compatibility with the protocols of the subscriber networks that it is linked with must be considered, and the access device between SMDS and the subscriber network must be simple so that subscribers can be connected to the network easily.

5.7 DATA SERVICES IN ATM NETWORKS

Most existing local-area networks are based on shared media interconnection, which is likely to become a bottleneck in accommodating the rapid growth of high-speed data communications and multimedia services. ATM is considered to be the technology capable of resolving this problem, so it has been attracting network providers and computer manufacturers. However, in the initial stage of its introduction to LANs or MANs, ATM will be used in conjunction with the existing network layer and MAC layer protocols. Therefore, it has become an important issue to support existing connectionless data services without change in the ATM networks.

Existing networks for the most part have been providing connectionless data services. For example, broadcast or shared medium networks (e.g., Ethernet), token networks (e.g., token ring, token bus, FDDI), or slot networks (e.g., DQDB) are connectionless networks, which broadcast or transmit data using common MAC, LLC, or bridging protocols without setting up connections. Connectionless wide-area networks (e.g., Internet) are also connectionless service networks that use the datagram routers. Connectionless networks are advantageous in that end systems have no burden on connection management or routing decisions.

In contrast, ATM networks are connection-oriented networks. In ATM networks virtual circuits are established between two end systems and cells are switched according to their connection identifiers. Since resources are statistically allocated per connection basis, ATM networks can provide guaranteed quality-of-service for connections. Also since routing is done only at the connection setup phase, ATM networks can support cell sequence integrity.

Connectionless networks often provide connection-oriented network services as well, in order to take advantage of connection-oriented networks. For instance, the IEEE 802.2 LLC layer emulates connection-oriented services on a hop-by-hop basis, in which LLC functions to establish and release logical link connections for guaranteed data delivery. Some higher layers can also provide connection-oriented services: For example, transport layer protocols such as TCP provide connection-oriented services on an end-to-end basis. In these cases, connection-oriented services are implemented in protocols because they are not the originally intended services in connectionless networks.

ATM networks stand in the opposite direction. They are connection-oriented networks by nature but are supposed to provide connectionless services for the purpose of interoperating with other existing connectionless networks and services. Recently international standards organizations and forums have been studying methods to support connectionless data services in ATM networks, which can be categorized into the following three: The first is to provide *connectionless broadband data services* (CBDS) at the UNI by deploying *connectionless service functions* (CLSF). The second is to put the MAC layer protocol on the ATM layer, which is called *LAN emulation*. The third is to put the IP protocol on the ATM layer, which is called *IP over ATM*. The first approach is suitable for interconnecting connectionless LANs and MANs via ATM networks, and the latter two approaches focus on small- and large-area networks, respectively. The CBDS approach was recommended by ITU-T; LAN emulation by the ATM Forum; and IP over ATM by IETF.

In the following section, we consider the CLSF approach first, followed by discussions of the LAN emulation and IP over ATM approaches.

5.7.1 CLSF-Based Data Services

The ITU-T recommends two different configurations to support connectionless data services in BISDN: the indirect and direct methods. In the indirect method, CLSFs and the associated adaptation layer entities are located outside the BISDN; in the direct method, CLSFs are located inside the BISDN. Figure 5.58 shows connectionless service reference models for these two methods.

Indirect Method

In the indirect method, connectionless services are provided through the virtual connections that connect ATM *interworking unit* (IWU) pairs. Each IWU provides an interface between a connectionless LAN and the ATM network. Virtual connections connecting IWUs form a dense mesh of connections in the network. Such connections may be established all together by using *permanent virtual circuits* (PVCs) or may be established only when necessary by using *switched virtual circuits* (SVCs). The choice of a PVC versus an SVC depends on the network size and the service requirements.

SVCs can support large-size networks because it is not necessary to maintain connections when there are no data to transmit. So, PVCs are used only when the size of the network is small enough to be able to interconnect fully all IWUs with one mesh network. On the other side, SVCs require some connection setup overhead: Each IWU must buffer packets until the connection is established, which causes long transmission delays. But the PVCs help to eliminate such delays.

The indirect method has the drawback, whether it uses PVCs or SVCs, that it cannot efficiently utilize the network resources, especially the bandwidth. When establishing a connection for a connectionless datagram, if the allocated bandwidth is too large, it means that bandwidth is being wasted. In contrast, if the allocated bandwidth is too small, it causes excessive delays in data transmission. Further, the indirect method makes it difficult to scale up the network since an increase of the number of end systems accompanies a rapid increase of the number of connections to support them all.

Direct Method

The direct approach can resolve the problems of scalability and bandwidth utilization. The direct method implements the CLSF using *connectionless* (CL)

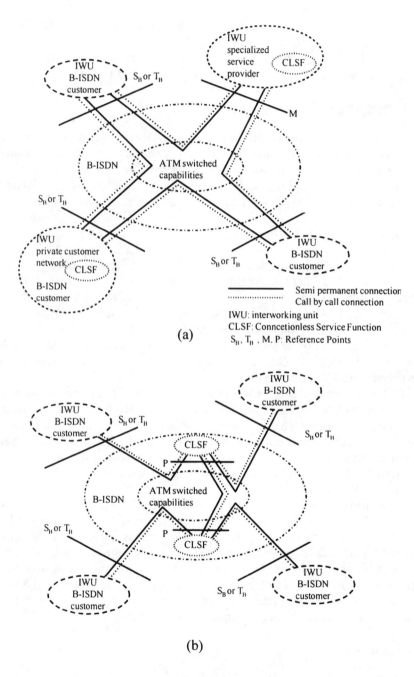

(a)

(b)

Figure 5.58   Connectionless service reference models: (a) indirect method and (b) direct method (ITU-T Rec.I.327) [ITU-T93].

servers and IWUs. An IWU interconnects connectionless networks and ATM networks, and segments and reassembles the connectionless data. Each CL server may be integrated as a part of an ATM switch or may be attached to an ATM switch within the BISDN. In general, CL servers are interconnected through PVCs so that connection setup delays can be reduced. Each CL server makes routing decisions to have each connectionless datagram packet delivered to the next-hop CL server or the destination IWU.

Figure 5.59 depicts the protocol architecture of the connectionless service using CL servers. The *connectionless network access protocol* (CLNAP) in the source IWU encapsulates connectionless datagrams before delivering them to the ATM adaptation layer. The CLNAP frame is encapsulated in an AAL-3/4 convergence sublayer PDU, and then is segmented into many ATM cells. The ATM cells are delivered to the CL servers through ATM switches, and are forwarded, with or without the reassembly/processing/segmentation treatment, to the next CL server or to the destination IWU. At the NNI each CLNAP frame is encapsulated with an additional four-octet header by the *connectionless network interface protocol* (CLNIP). The *mapping entity* (ME) is responsible for the necessary encapsulation and decapsulation processes. The destination IWU reassembles the received cells into CLNAP frames and then decapsulates them into the connectionless datagrams. The connectionless datagrams are finally delivered to the appropriate end systems.

The direct method has advantages over the indirect method in delivering connectionless data over public networks. First, each IWU in the direct method requires only one connection to deliver connectionless data to ATM networks, so the IWUs are not required to make the routing decisions. Second, all connectionless data traffic is aggregated in some connections between CL servers, which can increase the statistical multiplexing gain and can make the network management simple. Third, the number of required connections is much smaller in the direct method than in the indirect method because only the CL servers are interconnected, so the direct method is scaleable.

On the other side, the direct method has some potential weak points: CL servers and connections between CL servers may become bottlenecks. The direct method has not resolved the complicated routing job but, instead, has shifted the job to the ATM network. There may be an interoperability problem when interoperating with LAN emulation or IP over ATM, which has decided to use AAL-5, because the direct approach is likely to utilize AAL-3/4.

The direct method is likely to use AAL-3/4 because cell-based forwarding is much simpler than frame-based forwarding, among the two forwarding schemes that the CL servers support. In the case of the cell-based scheme, a CL server forwards cells to the next CL server or IWU as soon as it receives them, but in the case of frame-based scheme, cells are reassembled into frames at each CL server. The reason that the cell-based forwarding scheme is required to use AAL-3/4 is as follows: When a CL server receives the first cell of a frame, it

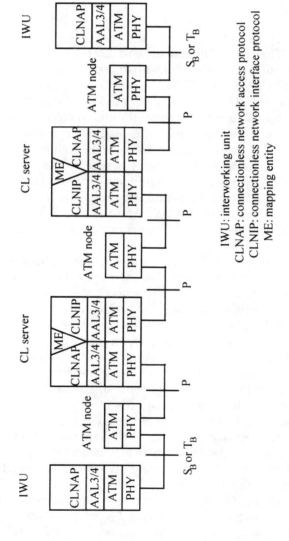

Figure 5.59   Protocol architecture of the connectionless service using CL servers [Vickers94].

finds out the output VPI/VCI and MID in the routing table using the input VCI/VPI and MID stored in the cell. Then the mapping information is preserved and utilized so that the other cells of the same frame, which have the same VPI/VCI and MID, can be forwarded as soon as they arrive. To keep pace with the flow of the traffic, a cell-based CL-server must be able to perform the three-phase process—receive a cell, look up the routing table, and forward the cell—within one cell transmission time. This implies that for an STM-1 155-Mbps interface all processes must be finished within 2.7 ms, and for STM-4 622-Mbps, within 680 ns.

On the other hand, in the case of the frame-based forwarding scheme, a CL server buffers all cells to be accommodated in a frame, and then reassembles them into a frame. Then it makes the routing decision or carries out other processes, and finally resegments the frame before forwarding. Since the frame is processed in one time, the MID field is not necessary, so AAL-5 can be used in the frame-based CL servers. This is why the frame-based forwarding scheme is considered to be an appropriate means for interconnecting ATM LANs. The frame-based scheme is less restricted in processing time. It can avoid useless transmissions because if a cell in the frame is lost the CL servers can detect the loss and can drop the whole frame. However, the frame-based scheme has the drawback that it requires large reassembly buffer and causes processing delays.

5.7.2 LAN Emulation

LAN emulation means that the point-to-point ATM switch provides the function of a virtual shared medium. From the protocol stack's point of view, the ATM layer behaves like an IEEE 802 MAC protocol underlying the LLC. The key attribute of the shared medium connection is that communication is done as a broadcast. That is, every station in a LAN receives all the packets from all other stations, and filters out the packets destined to itself. This feature of broadcast can be emulated in ATM networks using broadcast servers even though ATM is originally connection oriented.

LAN emulation service architecture is based on a client-server model. As shown in Figure 5.60, the clients of a LAN emulation may be ATM stations, ATM/LAN bridges, or routers. The servers of a LAN emulation are a *LAN emulation server* (LES), a *LAN emulation configuration server* (LECS), and a *broadcast and unknown server* (BUS). This architecture is only functional, so LAN emulation service configurations are not necessarily implemented in these three parts physically. In general, a client uses signaling virtual channel connections to exchange messages with servers, and data virtual channel connections to transfer connectionless data. The LES is responsible for registering MAC addresses to ATM addresses and resolving the addresses. The LECS locates the LES and provides configuration information for each emulated LAN segment. The BUS delivers broadcast or multicast frames. It is also responsible for

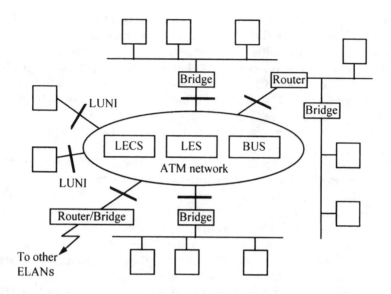

LECS: LAN Emulation Configuration Server
LES: LAN Emulation Server
BUS: Broadcast and Unknown Server
LUNI: LAN Emulation User to Network Interface

Figure 5.60 Sample configuration of ATM LAN using LAN emulation.

delivering the unicast frames whose destination address is either unregistered or unresolved yet.

The LAN emulation service function uses the following procedure: At first, a client contacts the LECS to locate the LES. Then an ATM connection is established to the LES, and the client registers its MAC address and ATM address to the LES. Each client maintains information mapping MAC addresses to ATM addresses for connection setup. But when a client does not have this information, it can obtain ATM addresses from MAC addresses using the LES. The LES makes a table for mapping MAC addresses to ATM addresses at the registration stage and updates it whenever a change occurs.

A client can send frames to the BUS even in the middle of the address resolution process. In this case the BUS broadcasts the frames to all clients, because it is the only method available to send frames to unregistered destination clients. In addition, for a delay-sensitive application, it is desirable to send frames using this method so that the service is unaffected by the delay that may occur during the address resolution and connection setup process.

To prevent the clients from abusing the broadcast channel, the number of broadcast frames that a client may send within a given time period is limited. If the BUS were capable of delivering the unresolved frames only to their destinations without broadcasting, then the traffic in the network would be much reduced. However it would result in complicated and costly BUS implementations.

Once the MAC address to ATM address resolution is completed, the client begins the connection setup process. After this process, unicast frames are transmitted directly to the destination client. The connection is automatically released after a fixed length of idle time. When transmitting multicast frames, the client takes a slightly different procedure. At first the client gets the address of the BUS from the LES, and then the client establishes a connection to the BUS to send multicast frames to the BUS. Then the BUS forwards the multicast frames to all destination clients by setting up point-to-multipoint or point-to-point connections.

Since a LAN emulation emulates the IEEE 802 MAC layer below the LLC it can support not only IP but also various network layer protocols such as SNA/APPN, IPX, and NetBios. This is in contrast to the IP over ATM approach, to be discussed next, which can support the IP protocol only. However, the LAN emulation has expansion limitations. While clients in an *emulated LAN* (ELAN) can communicate with each other directly, communications between ELANs are possible only through bridges or routers. So LAN emulation is suitable for small workgroup networks in a local area. ELANs interconnected by bridges and extended beyond a local area or small number of workgroups would be impractical. As the number of connected ELANs grows, the broadcast traffic passing over the bridges increases, and the bridges could become a bottleneck. To reduce broadcast traffic, routers can be used instead of bridges, and in order to reduce the number of interconnection devices, multiple ELANs can be interconnected by direct ATM connections.

5.7.3 IP Over ATM

While LAN emulation is a method suitable for networks for small workgroups, IETF's IP over ATM, which uses the native mode network layer protocols, is suitable for the interconnection of large-scale LANs and WANs across ATM. Because the IP over ATM specified in IETF's RFC1577 does not change the fundamental nature of the IP protocol, it is called *classical* IP over ATM. However, there are also other IP over ATM approaches that require more radical changes.

The classical IP over ATM approach is based on an IP subnetwork called *logical IP subnetwork* (LIS). An LIS consists of hosts and routers that have the same subnetwork mask and the same subnetwork address. Any two hosts in the same LIS communicate directly, but hosts in different LISs can communicate only through a router even if a direct ATM connection can be established between them. The size of *maximum transmission unit* (MTU) that a host can transmit over the ATM virtual channel is 9,180 octets. If all members in an LIS consent, this value can be changed.

To establish a direct virtual channel connection between two hosts in an LIS, mapping between ATM addresses and IP addresses is necessary. Based on this mapping, IP addresses are resolved to ATM addresses through the *ATM address resolution protocol* (ATMARP) and the inverse process is done through the *inverse ATMARP* (InATMARP). For address resolution, each host must register its IP address and ATM address to the ATMARP server in the same LIS. The ATMARP server can resolve the IP addresses in the same LIS.

Figure 5.61 shows an example of realizing the classical IP over ATM model using an ATMARP server. In the figure Host 1 (IP address A, ATM address X) wants to send some connectionless data packets to Host 2 (IP address B, ATM address Y). So Host 1 sends an ARP request to ARP server 1 in the same LIS. Then the ARP server 1 resolves IP address B to ATM address Y, and sends an ARP response to Host 1. Then Host 1 sets up a virtual channel connection (VCC) to Host 2 using the ATM address Y and transmits packets. Host 1 preserves the mapping information that IP address B is mapped to the ATM address Y for use in the next packet transmission. Because all these procedures are done on the IP over ATM layer, the user of the IP datagrams need not care about the ARP and connection setup procedures, and thus finds no difference from transmitting data over legacy IP protocols.

In encapsulating different network layer protocols, hosts can apply two

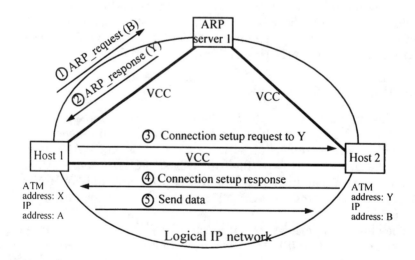

Figure 5.61 An example of data transfer using IP over ATM.

different methods. The first method is to multiplex multiple protocols in a single ATM virtual channel connection, and in this case the protocol of the carried PDU is identified by the LLC header and *subnetwork attaching point* (SNAP) header. The other method is to set up a VC for each protocol. The first method is suitable for the networks that support only PVCs, and the second method is suitable for the SVC networks in which VCCs can be created or removed flexibly.

The characteristics of the classical IP over ATM model can be summarized as follows. First, the MTU size is prespecified for all VCs in an LIS. Second, the LLC/SNAP encapsulation of IP packets is supported. Third, the end-to-end IP routing scheme remains the same. Fourth, IP addresses are resolved to ATM addresses by using the ATMARP service within the LIS. Fifth, each VC directly connects two IP members within the same LIS.

Although the classical IP over ATM method has the advantages of being conceptually simple and not requiring any change to existing systems, its performance is rather limited since communication among different subnetworks must be done through routers. This can cause a serious degradation of performance in an ATM network consisting of a large number of LISs. In ATM networks, hosts can communicate directly with each other without IP layer switching in routers, and this fact can be exploited in order to enhance performance by removing unnecessary relay nodes.

As a means to set up direct connections in *nonbroadcasting multiaccess* (NBMA) networks such as ATM, the IETF has introduced the *next hop resolution protocol* (NHRP), which relies on a new ARP server. The aim of the NHRP is to enable a source host to bypass all or some intermediate routers so as to establish a direct ATM connection to the destination host. In this protocol, some disjoint IP subnetworks are treated as one logical network called an NBMA network. In each NBMA network, there exists at least one IP-NBMA server that resolves IP addresses to NBMA addresses. The IP-NBMA server constructs an IP-NBMA address mapping table by utilizing the registration packets or by applying dynamic address learning mechanisms. NBMA networks are interconnected through IP routers.

The address resolving procedure of the NHRP is as follows: If the source host and the destination host are connected to the same subnetwork, the exiting address resolution procedure is applied. If the destination host is connected to a different subnetwork, an ARP query is sent to the IP-NBMA server. If the address entry is found, the IP-NBMA server returns the NBMA address; otherwise, the server forwards the ARP request to the next node toward the destination. To improve the performance of the protocol, responses are recorded at each node on the return path. If the destination address is not found within the NBMA network, a negative message is sent back to the source. An NHRP request never crosses the border of an NBMA network.

5.8 HIGH-SPEED AND REAL-TIME COMMUNICATION PROTOCOLS

In contrast to the existing communication networks with limited bandwidth and high error rates, emerging communication networks offer large bandwidth ranging from a few hundred Mbps to several Gbps, low error rates, and high reliability. Network services have been evolving from text-oriented data services to real-time multimedia (e.g., audio, video, graphic, and data) services such as video-on-demand, video conference, distributed processing, and virtual reality services, which require highly advanced communication functions such as multicast, synchronization, complex call control, and so on. Existing transport protocols such as TCP and OSI TP, however, are not suitable for this new communication environment because they cannot utilize high bandwidth due to the low processing speed, and cannot guarantee the quality of service required for real-time services. Therefore, various high-speed transport protocols have been proposed in the search for improved processing rates. In addition, new real-time and multicast communication protocols have been actively studied.

In this section, several typical examples of protocols that have been developed for such new communication environments are examined in terms of two categories: high-speed transport protocols and real-time communication protocols.

5.8.1 High-Speed Transport Protocols

The performance of existing transport protocol degrades in environments where the product of the transfer rate and the round-trip delay is large, namely, long and wide transport pipes. A network containing this kind of path may be called a *"long and fat network"* (LFN). Examples of the LFN can be found in the DS1 satellite channel and DS3 terrestrial long-distance fiber-optic links for which the bandwidth delay product is more than 1 Mbit. The fundamental problems of the existing transport protocols in LFN, taking TCP for example, are as follows.

The first problem is that the processing rate is restricted by the window size. The TCP header uses a 16-bit field to report the receiver window size to the transmitter, and hence the largest amount of data that can be sent without acknowledgment is limited to 65 Kbytes. But it is known that small windows can limit the packet processing in a high-speed network. Therefore, to circumvent this problem, a new TCP that can support large window sizes must be defined.

The second problem is that packet losses in an LFN can cause a critical effect on the throughput. The TCPs that have been implemented so far drained the transmission pipeline whenever packet loss occurred, and a slow-start scheme was taken as an action to recover the normal operation state. Recently, fast retransmission and recovery algorithms have been introduced that enable fast recovery without draining the transmission pipeline if a single packet loss occurs. However, if the number of packet losses exceeds one, a retransmission timeout occurs and the pipeline drainage occurs as well. On the other hand, expanding the window size to match the capacity of an LFN results in a corresponding increase in the probability of losing more than one packet per window. This could have a devastating effect on the throughput of TCP over an LFN. If the gateways on the LFN employ congestion control mechanisms based on some form of random dropping, the probability of dropping more than one packet per window becomes very high, and TCP performance seriously deteriorates.

As a means to react to multiple packet losses, *selective acknowledgments* (SACK) can be employed. Unlike the normal cumulative acknowledgments of TCP, selective acknowledgments give the transmitter complete information on the segments that have not been received by the receiver.[15]

The third problem is how to determine the round-trip time accurately. TCP is designed to provide reliable data delivery by retransmitting segments that are not acknowledged within some prespecified *retransmission timeout* (RTO) interval. As such, it is essential for TCP to be able to determine the RTO accurately in dynamic fashion. In general, RTO is determined by estimating the mean and variance of the measured *round-trip time* (RTT).

To solve the problems described, various high-speed transport protocols have been developed. They include XTP (eXpress Transfer Protocol), NETBLT (NETwork BLock Transfer), VMTP (Versatile Message Transfer Protocol), HSTP (High-Speed Transport Protocol), TP++, and so on. Among them, XTP and TP++ are described in the following subsections.

XTP

The XTP protocol provides the services of two layers by merging the network layer and the transport layer. This protocol, which is tailored for implementations in VLSI chips, provides the functions of a high-speed transport layer perfectly, and has been adopted by the U.S. Navy, IBM, and the French Army.

Systems such as distributed systems, transaction systems, and multimedia systems that are in use or will be used in the future require some unique features for their own purposes. For example, distributed systems that depend on the *remote procedure call* (RPC) require a reliable datagram method that can

---

15. Some evidence has been published in favor of selective acknowledgments, and selective acknowledgments have been included in, or proposed for, a number of experimental Internet protocols such as VMTP, NETBLT, RTP, and OSI TP4. IETF's RFC-1072 defined a new TCP "SACK" option to send a selective acknowledgment. However, there are important technical issues to be worked out concerning both the format and semantics of the SACK option.

support rapid connection setup and transactions, and real-time systems should be able to control delays. In addition, multimedia systems should be able to handle problems such as synchronization and bandwidth allocation according to the traffic having various characteristics, and video conference systems should be equipped with multicast capability. So that XTP can be used as a transport protocol suitable for the applications described, it is designed to provide various function as discussed in the following paragraphs.

To begin with, XTP has a large *sequence number* field so that it can support transmission of large data. If the sequence number field is small, then the amount of data on the transmission line is limited, and thus it is not suitable for environments in which the transmission speed is high.

To support various communication services, XTP has the *sort* field, which enables it to transmit data in the order of their importance. By using two types of packets—information packets and control packets—XTP can transmit control information and data streams separately. But in order to make processing simple, two types of packets are arranged to use the same format in the packet header and trailer, and both header and trailer are set to fixed sizes. If XTP operates in the no-error mode, it does not retransmit errored packets. This operation is applied in real-time communication services for which retransmission of errored packets is useless. XTP supports the multicast function that ensures end-to-end reliability, which is important in group communications.

XTP employs new technology in implementation. XTP is implemented in VLSI chips so that it can operate at a gigabit rate, which becomes possible by using fixed-length packets in multiples of four bytes. It processes, in parallel fashion, various functions such as address change, context generation, flow control, error control, transmission rate control, and host system interface, using the information in the header and trailer. Also it minimizes the use of timers that hinder high-speed protocol processing, so that the receiver uses only one timer.

To efficiently process the transactions that happen frequently in a distributed system, XTP employs a fast and simple connection setup method. To establish connection quickly, it uses an implicit setup method in which the transmitter transmits data immediately after sending the FIRST packet without waiting for the receiver's confirmation, and the receiver receives data only if it has an active context, as illustrated in Figure 5.62. Such an implicit connection setup method helps to reduce the connection setup delay that otherwise would last until two systems complete exchanging the FIRST message. After transmitting the FIRST packet, it continues transmitting data, and finally releases the connection by applying the control packets WCLOSE and RCLOSE three times (see Figure 5.62).

To support various formats of addresses, XTP uses the address segment that is composed of a fixed-size descriptor field, which describes the address format, and a variable-size field, which contains the real address. This

Host A   **Host B**

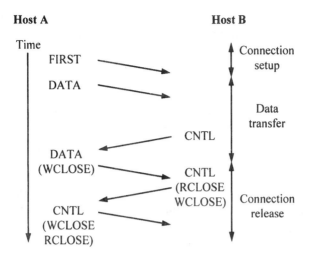

Figure 5.62   Connection setup, data transfer, and release procedure in XTP.

addressing scheme enables support of both the TCP and OSI address schemes, and supports multicast services as well. The address segment is included in the FIRST packet only, and other packets use the KEY field, which is an information index saved in a cache during the connection setup.

XTP employs not only the go-back-*N* scheme, which is used in TCP and TP4, but also the selective repeat scheme for error control. The receiver maintains the "spans" that indicate the portion transmitted without error, and notifies the sender so that the sender can retransmit the data not included in the "spans."

The flow control scheme of XTP is basically an end-to-end window scheme. For flow control, XTP uses parameters such as RSEQ, DSEQ, and ALLOC: The sender can continue transmitting packets without an ACK message until the transmitted number reaches the limitation parameter ALLOC. When reaching this limit point, the sender transmits the last packet with a request message. The receiver that receives the message transmits an ACK message to the sender, then the sender resumes transmitting packets.

A transmission rate control scheme is used in addition to the above flow control scheme, because the flow control itself cannot guarantee effective and error-free data transmission. The transmission rate control scheme is designed in such a way that the sender can transmit data fast regardless of the processing speed of the receiver as long as the receiver has enough buffer space. In this transmission rate control scheme, not only the transmitter and the receiver but also the intermediate routers cooperate to control flow all the way through the route. In fact, this is the scheme that controls the transmission rate by controlling the burst size and the time interval of the transmitter's data.

TP++

TP++ is a transport layer protocol that is developed by Bellcore for use in the LFN. Assuming that the network layer performs congestion control, but not error control, TP++ provides end-to-end error control. So TP++ can effectively handle the packet disorder induced by multiple-path transmissions and packet losses caused by network congestion.

TP++ was developed for real-time applications that require a constrained quality of service such as real-time packet delivery and high throughput. To meet such constraints, it was designed via a *top-down approach.* In addition, to simplify the protocol specification, validation, and development, the transport protocol was designed to be application independent. Connection setup is accomplished via timer-based scheme. Flow control is nonexistent except for the function to set the QoS factor of connection in consideration of the network condition. Error control is separated from flow control and uses both ARQ and FEC schemes.

Connection control of TP++ relies on the timer-based method, which has a shorter delay and smaller overhead than the handshake method. Connection is set up immediately without a handshake, and if the connection is not used for a certain period of time, it is released automatically. The connection is basically unidirectional, and the connection for the ACK message of the reverse direction is maintained independently. Connection for bidirection or multicast is set up by using multiple unidirectional connections. Each connection has a 16-bit *connection identifier* (CID) that uniquely identifies the connection, and each data packet in one connection has a continuous *connection sequence number* (CSN).

TP++ can use the time stamp method for data synchronization since the transmitter and the receiver operate in synchronized clocks. The transmitter creates a time stamp when transmitting each *transport PDU* (TPDU). The receiver checks the stamp and accepts the TPDU if it has arrived within its lifetime and discards it otherwise. When the receiver transmits an ACK message that also contains a time stamp, it uses the same CID that the transmitter had.

For error control, TP++ uses both the FEC and the ARQ schemes. TP++ uses the selective confirmation method in such a way that the ACK message can be issued only for the packet whose CSN value is the largest. The transmitter, after transmitting TPDUs, calculates the loss probability of the transmitted packets based on the network delay and the loss rate, and retransmits the packet whose loss probability exceeds the threshold. However, such an ARQ scheme is not adequate for the applications that require strict time delay. So in this case, TP++ applies the FEC scheme, which transmits the error correction information along with the corresponding user data such that the receiver can use it to correct errors. In this case each TPDU is divided into multiple FEC blocks and the corresponding error correction information is attached to each block.

5.8.2 Protocols for Real-Time Communication

Recently, distributed real-time multimedia applications such as remote video conferencing, virtual reality, and remote games have been developed for the Internet. But the existing service models, which are multipoint-to-multipoint communication oriented, do not provide the proper means to support distributed real-time multimedia applications that require guaranteed service quality. So the network structures and service models that can satisfy the requirements of those new applications have been sought, and various structures and models have been proposed, among which RSVP and RTP are considered in the following.

RSVP

RSVP (resource ReSerVation Protocol) is an Internet protocol for resource reservation that supports unidirectional resource reservation in a way similar to that of the ST-II protocol of the Internet. RSVP is a receiver-initiated resource reservation protocol, in which the receiver is responsible for initiating the resource reservation. This reservation scheme enables heterogeneous receivers within a multicast group to be accommodated. Specifically, each receiver can reserve the amount of resource that best fits its own environment, and can select, and change, the information streams that it wants. RSVP also provides various different reservation types and allows for each application process to specify how to combine reservation resources for the multicast group aggregated at the intermediate switches. This feature brings forth an efficient utilization of network resources. RSVP also employs the *soft-state* concept, which enables easy updating of the switch state, to support dynamic membership changes and automatic adaptation to routing changes. This feature enables RSVP to deal effectively with dynamic reconfiguration of multicast groups.

The procedure of RSVP resource reservation is illustrated in Figure 5.63(a) [Zhang93,Zhang95]. In the figure H1 and H2 are data sources, H3, H4, and H5 are receivers, and S1, S2, S3, and S4 are switches where resources are reserved. The lines depict the multicast routing *tree* for this multicast communication. Before or when each data source starts transmitting, it sends a path message containing the flow specification of the data source to the target multicast receiver group along the routing tree. The switch that receives a routing message checks if the routing information of the multicast destinations is readily existing, and creates one if it is nonexistent. The routing information includes the

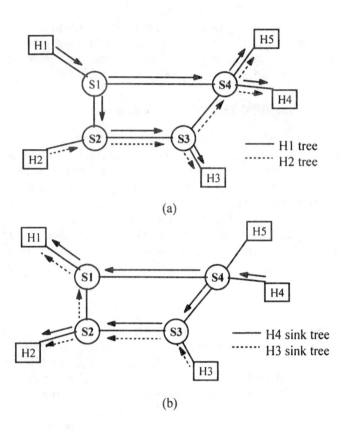

Figure 5.63 Example of RSVP setup: (a) path message flow and (b) setup message flow.

basic information such as the incoming and outgoing links for the sender and the receiver. In Figure 5.63(a), H1 and H2 send a path message to H3, H4, and H5 along the multicast routing trees depicted as solid and dotted lines, respectively. When a receiver receives a routing message from a source that wants to reserve resources, the receiver issues a resource reservation message based on the flow specification in the routing message. This reservation message is transmitted along the *sink tree*, which is the reverse route of the path message, to reach the data source. In Figure 5.63(b), the dotted lines depict the sink tree of H3 and the solid lines the sink tree of H4. Any switch on the route, if it cannot meet the resource reservation request, sends an RSVP reject message back to the data receiver and discards the reservation message [Zhang93].

The switches on the sink trees merge the reservation requests for the same multicast group by pruning those that carry a request for reserving resources smaller than, or equal to, the amount of resources of some previous request. For example, if H1 is a video source, and H4 has reserved enough bandwidth to receive the full video data stream while H5 wants to receive only low-resolution video data, if the reservation request from H4 is already granted, then the requested reservation of H5 is discarded as a result of these readily reserved sufficient resources. Once the reservation is established, the receiver periodically sends a reservation refresh message.

RTP

RTP (real-time transport protocol) is a protocol that provides end-to-end delivery real-time services such as interactive audio, video, and simulation data. RTP provides the functions such as payload type identification, sequence number verification, internal time stamp transfer, and transmission data monitoring. RTP transfers the data required for the receiver to reconstruct the data, rather than performing the various functions for real-time data transfer. In other words, RTP complements the functions of a lower layer network in the aspects of application programs, rather than perform independent transport layer functions. In general, application protocols are executed on the UDP level, exploiting the multicast function and the error control function provided by the UDP. In this context, transport layer services are provided by the combined contributions of RTP and UDP.

For the control of RTP, *real-time transport control protocol* (RTCP) is employed. RTCP monitors the quality of service and conveys information about the session participants. It periodically distributes control packets containing that information to all session participants, through the same distribution mechanisms as for the data packets.

Figure 5.64 shows the format of the RTP header. In the format, V denotes the version of RTP, P a padding bit, X the extension flag, CC the CSRC count (which contains the number of CSRC identifiers), M a marker whose interpretation is defined by a profile, and PT the payload type (which identifies

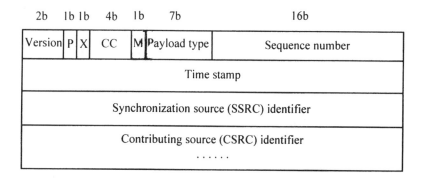

Figure 5.64 RTP header format.

the format of the RTP payload and determines its interpretation by the application).

SSRC is the *synchronization source* identifier, which identifies a stream of RTP packets that does not depend on the network address. Examples of SSRC include a microphone, camera, or RTP mixer. The receiver can classify received packets depending on their synchronization sources and can reorganize the packets for reconstruction. CSRC is the *contributing source* identifier, which identifies a stream of RTP packets that has contributed to the combined stream produced by RTP mixers. An RTP mixer refers to an intermediate system that receives RTP packets from one or more sources, possibly converts the data format, combines the information into new packets, and then forwards new RTP packets. Since the timing among multiple input streams is not synchronized in general, a PTR mixer makes timing adjustments among the streams and generates its own timing for the mixed stream.

To illustrate the use of RTP, Figure 5.65 depicts the configuration for a simple multicast audio conference. The network in the figure consists of host systems H1, H2, and H3, which participate in the conference, and RTP mixers M1 and M2. Hosts H1, H2, and H3 transmit encoded audio data, which are prefixed by RTP headers. The RTP header indicates what type of audio encoding is employed for each packet so that senders can change the encoding scheme during the conference. The SSRC identifier for each audio stream is randomly chosen so that a unique number can be assigned to each of the streams within a particular RTP session. The binding of the SSRC identifiers is provided through RTCP. Mixer M1 receives audio streams S1 and S2 in RTP data packets sent from H1 and H2. It then combines those streams, possibly changes the data format, and generates a new combined audio stream S4 based on the RTP header information such as time stamp, data format, and so on. The mixer adjusts the timing of S1 and S2 and generates its own timing for the mixed stream S4. The SSRC identifier of S4 is newly generated in mixer M1, and the CSRC identifier list consists of the SSRC identifiers of H1 and H2. In a similar manner, mixer M2 combines streams S3 and S4, and generates new mixed stream S5 of RTP data packets.

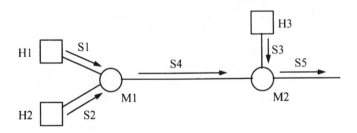

Figure 5.65 Example of RTP.

5.9 INTERNET

The Internet is the largest interconnected computer network in the world that provides information exchange. All the computers connected to the Internet can easily communicate with one another using the TCP/IP protocol, which enables connection among computers produced by different vendors.

The backbone network of the Internet was initiated by the ARPANet, which was sponsored by the U.S. Department of Defense in 1983. After ARPANet disappeared, NSFNET, a network connecting supercomputer centers, assumed the role of the Internet backbone in 1986. Later, the Internet evolved to become a national research computer network in the United States that connected university research centers, academic institutions, and nonprofit organizations. In the late 1980s the Internet began to expand all over the world.

Since 1983, the Internet has shown a remarkable growth rate. The number of networks connected to the Internet increased to about 100 in 1985, to about 200 in 1987, and to more than 500 in 1989. The growth rate accelerated in the 1990s: The number increased to about 2,000 in 1990, to about 5,000 in 1992, and to about 60,000 in early 1996 with about 5 million hosts and 40 million user stations connected to them. This exponential increment is expected to continue to connect about 10 million hosts and 100 million user stations by 2000. The Internet is also expected to become one of the major contributors in constructing information superhighways for the future.

In this section, the Internet organization is considered first and is followed by a discussion of the protocol suite employed by the Internet. Next, several popular Internet applications such as MBONE, WWW, and MIME are discussed.

5.9.1 Internet Organization

The *Internet Society* (ISOC), which is a nonprofit organization that looks after the prospects of the Internet, makes high-level decisions on the direction of Internet developments. For technical aspects of the Internet, ISOC relies on the IAB (Internet Architecture Board), which is composed of IETF (Internet Engineering Task Force) and IRTF (Internet Research Task Force). IETF develops Internet standards, and IRTF is in charge of research and development for future network techniques.

IETF is a informal organization consisting of network designers, network operators, network equipment manufacturers, users, and researchers in the related fields. IETF provides an open forum for discussion and has no limitation on participation, in either affiliation or format. One can either attend the meetings in person, or participate in discussions via e-mail. IETF carries out technical works related to the standard Internet protocols.

Because the Internet protocols include areas with very wide scopes, the standard work is divided into several areas: User Service (USV), Application

(APP), Transport Service (TSP), Internet (INT), IP Next Generation (IPng), Routing (RTG), Network Management (MGT), Operational Requirement (OPS), Security (SEC), and General (GEN) areas.

In developing Internet standards, only protocols that are well recognized and widely accepted are adopted as standards. Every standard proposal should go through practical implementation and performance verification tests in small- and large-scale networks before being selected as a standard. Such a standardization procedure helps to achieve the goal of the Internet standards such as technical excellence, preimplementation and verification, openness, fairness, and timeliness.

Actual standards-related activities are carried out *working groups* (WGs). There are about 60 active WGs, most of which have been assigned some specific topics for research or for standardization. Working groups are operational until the assignment is completed.

Each WG consists of a chairman and some number of individuals who are interested in the subject of the group. The most important step for establishing a WG is preparing its charter, which describes the objective and the work plan of the group. After examining this charter, IESG (Internet Engineering Steering Group) decides whether or not to approve the group. Once approved, the working group is given the right to use the file that records Internet drafts, and can call for meetings and conferences for fulfillment of its objectives. The chair of each working group is responsible for summarizing and recording all conferences and discussions.

Internet Drafts and *Request for Comments* (RFC) are two important documents. Internet Draft is an informal document prepared by working groups or individuals to report the progress of study and intermediate study results. In contrast, the RFC is a formal document that describes the results of studies on Internet-related technology and topics. It originates from a series of documents issued in 1969 related to the ARPANet project. An RFC is prepared out of the standard proposals after conducting a review process among IESG members and RFC editors. The standard proposals for consideration for RFC can be made by anyone, but Internet Drafts of working groups are the major source of them.

An RFC can be divided into the following four categories depending on its contents: prototype RFC, empirical RFC, informative RFC, and historic RFC.

5.9.2 Internet Protocol

As previously stated, the TCP/IP suite plays the role of an adhesive in seamlessly connecting the heterogeneous networks and computer and communication equipment of different vendors. The TCP/IP suite refers to a set of about 100 protocols that are necessary to connect computers and communication equipment to form one network. The most important protocols among them are

TCP and IP. The TCP/IP suite also includes various application programs such as *telnet* and *ftp* (refer to Section 5.3 for a detailed description of TCP/IP).

As high-speed networks emerge and the number of users explodes, and as multimedia services become available, TCP/IP has revealed its limitations. To adapt to new environments, IETF has initiated research for new Internet protocols in the early 1990s, along with the related standardization. This standardization process has been performed at various protocol levels. In the IP level, basic standards of the *next generation IP* (IPng) have been completed. Most distinctive characteristics of the next-generation IP protocol are the increased address field, which can support an increased number of hosts, and the simplified packet header, which can help increase packet processing capacity.

While the next-generation IP has reached its final stage of standardization, the TCP protocols and upper layers are still in the basic study or experimental stages. For TCP, research is in progress that increases the processing capacity by expanding the window size according to the characteristics of a high-speed network in which the propagation delay is relatively large compared to the transmission delay. In addition to the network and transport layer protocols, protocols that can support high-speed real-time multimedia communication service have been actively investigated also. For this, IETF has studied high-level protocols operating on TCP or UDP, such as RSVP, RTP, and RTCP. Because these protocols were already treated in the previous section, we discuss only the next-generation IP in the following.

Next-Generation IP

IP, which is the basic internetworking protocol of the Internet, has revealed the serious problem that the address field with its four-byte length would be exhausted soon in the 21st century. This means that a new IP with a layer address space must be developed before the address space becomes exhausted. IETF decided to take the responsibility of developing next-generation IP recommendations and formed an ad hoc study area named IPng in July 1993.[16]

Part of the technical criteria for the next-generation IP included the following requirements: First, it is required to support 10^9 or more networks and 10^{12} or more terminals, and is recommended to support 10^{12} or more networks and 10^{15} or more terminals. It must be able to support the existing routing schemes as well as mechanisms for a smooth transition from the existing IP (e.g., IPv4). It should be adaptable to network topology change, expandable, and

---

16. As candidates for the new IP, four proposals were discussed: TUBA (TCP and UDP, with Bigger Address), SIP (Simple Internet Protocol), Pip (P Internet Protocol), and CATNIP (Common Architecture for Next-generation Internet Protocol). Afterward, SIP and Pip were combined to form SIPP (Simple Internet Protocol Plus), which became the working document for IP, version 6.

independent of a physical medium. It is required to provide datagram service and to support the service class concept. It should support autoconfiguration, multicast, and host mobility functions, and should include control protocols for testing and debugging networks. It must support private networks and include network security functions such as authentication and key distribution.

Characteristics of IPv6

In July 1994, IETF decided to adopt SIPP as the working document for the IPng standardization. This work was assigned the version number 6, and hence the term IPv6 has been used to represent the next-generation IP. In the following, some features of IPv6 are discussed in comparison with the existing IPv4.

1. *Expanded addressing capability:* The IP address size is increased from 32 bits to 128 bits so that it can support a much larger number of addressable nodes, more levels of addressing hierarchy, and simple autoconfiguration. The address includes the *scope* field in the multicast address to increase the expandability and efficiency. The IP address can support a new type of address, called a *cluster address* that identifies a particular region rather than an individual node.

2. *Simplified header format:* The IPv6 has eliminated, or has changed the use of, some of the IPv4 header fields. This helps to simplify packet processing and increase the throughput. Figure 5.66 depicts the header format of IPv6.

3. *Improved options and extension support:* If some specialized packet processings are required, IPv6 utilizes the extension header in addition to the basic header. The version 6 header separation approach contrasts with IPv4 in that the extension header is included in the basic header. This helps to improve the processing efficiency because the extension header is added only when it is necessary. Also, this renders a flexible packet format, and yields very easy extension information modifications and additions. Figure 5.67 depicts the extension header structure of IPv6.

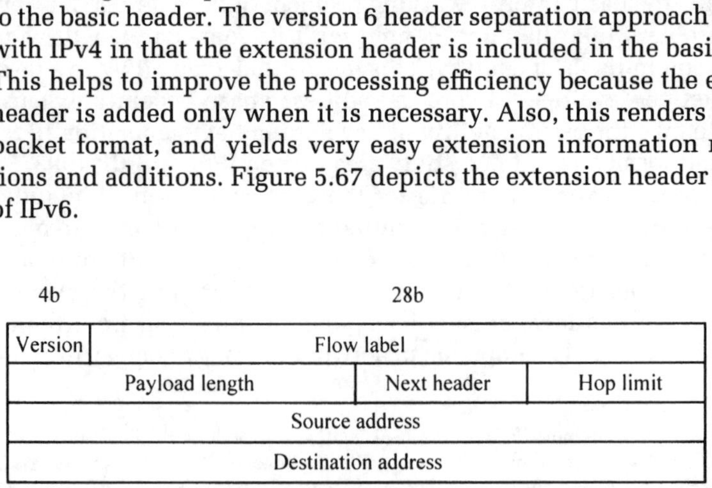

4b		28b		
Version	Flow label			
Payload length			Next header	Hop limit
Source address				
Destination address				

Figure 5.66   Header format of IPng.

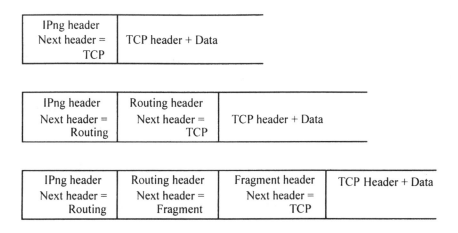

Figure 5.67 Extended header format of IPng.

4. *Flow labeling capability:* IPv6 can tag labels on the packet flows that require special treatment, such as real-time services and nondefault services, to meet the quality of service required for the flows.

5. *Authentication and security functions:* IPv6 supports security functions such as authentication and data integrity.

IPv6 Addressing

An IPv6 address is a 128-bit identifier for an interface or a set of interfaces. There are three types of IPv6 addresses:

- *Unicast:* An identifier for a single interface. A packet sent to a unicast address is delivered to the interface identified by that address.
- *Anycast:* An identifier for a set of interfaces (typically belonging to different nodes). A packet sent to an anycast address is delivered to one of the interfaces identified by that address. In this case the interface nearest to the transmitter according to the routing protocol is selected.
- *Multicast:* An identifier for a set of interfaces (typically belonging to different nodes). A packet sent to a multicast address is delivered to all the interfaces identified by that address.

IPv6 address is represented as a text string like X:X:X:X:X:X:X:X, where X represents the hexadecimal value of each 16-bit segment of the 128-bit address. For example:

1. FEDC:BA98:7564:3210:FEDC:BA98:7654:3210,

2. 1080:0:0:0:8:800:200C:417A,

respectively, represent IPv6 addresses. When an IPv6 address contains many consecutive 0's, the representation can be compressed. For example, the address in item 2 can be abbreviated to 1080::8:200:200C:417A.

In the environment where IPv4 and IPv6 nodes are mixed, it is convenient to use an alternative form such as X:X:X:X:X:X:D.D.D.D, where D represents the decimal value of an 8-bit segment of the 32-bit IPv4 address. For example:

3. 0:0:0:0:0:0:13.1.68.3,
4. 0:0:0:0:0:FFFF:129.144.52.38,

are the mixed representation, and their compressed forms are

3. ::13.1.68.3,
4. ::FFFF:129.144.52.38.

5.9.3 MBone

MBone (multicast backbone) is a virtual network designed to broadcast in real-time the IETF meetings by multicasting voice and moving image through the computer network. MBone demonstrated that it is possible to support real-time multicast video conferencing over the Internet.

The reason why MBone can provide real-time services is because it employs RTP, which makes it possible to reduce network delay and error through the time control and sequencing functions (see Sec. 5.8.2). As a consequence, MBones that employ RTP can support real-time services such as voice and video.

MBone is not a physical network but a virtual network constructed over the Internet. If a router that supports the multicast function is installed in the existing Internet, then a set of hosts capable of multicast functions is formed. The resulting subnetwork is called an *MBone island*, and if MBone islands are interconnected through the Internet, then the result is the MBone network. MBone uses the so-called *tunneling* method to construct the virtual multicast network. When transmitting multicast packets from an MBone island to another over the Internet, an MBone router encapsulates IP headers based on unicast routing schemes on top of MBone packets. Each host in an MBone island notifies to the multicast router that it wants to join the multicast group by using IGMP (Internet Group Management Protocol), and then the router routes multicast packets based on this information.

Figure 5.68 helps to illustrate multicasting through tunneling. In the figure, router R4 is assumed not to support multicast, and host H1 takes the role of a multicast router for subnetworks S1 and S2. So, multicast IP packets are

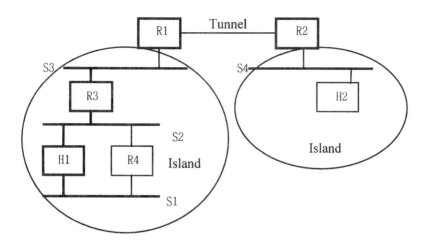

Figure 5.68 Example of multicasting in MBone.

transmitted to and from S1 and S2 via H1, and are delivered to subnetwork S3 via router R3, which has multicasting capabilities. The MBone island consisting of S1-H1-S2-R3-S3 is connected to the other island through the *tunnel*, which refers to a virtual path that conveys multicast packets between a pair of multicast routers. Router R1 on one end of the tunnel encapsulates multicast IP packets and then transmits them to the multicast router R2 on the other side of the tunnel. Receiving the encapsulated packets, R2 strips off the encapsulation header and forwards the original multicast packets to the destination hosts within the MBone island.

5.9.4 WWW

The WWW (World Wide Web) is a large-scale information retrieval system that manages information widely distributed within the Internet by employing hypermedia technology. The WWW has the following features:

1. *Integrated interface support:* WWW provides an integrated interface to each protocol, data format, address system, and others so that various services and databases that are supplied in Internet can be directly accessed.
2. *Convenient user environment:* WWW transparently supports most existing Internet applications such as services (i.e., *telnet, gopher, anonymous ftp, archie, usenet news, finger, wais,* and *whois*). This service combination provides the integrated user interface with a plethora of

Internet applications. Therefore, even an Internet novice can retrieve various hypermedia information only using the mouse, without memorizing complex commands.

3. *Easy expansion:* WWW server can easily expand its capabilities by supporting the standardized *common gateway interface* (CGI) between the WWW server and exterior applications. The CGI program also enables dynamically changing information to be retrieved in real-time in addition to static query information.

4. *Versatile platform:* WWW enables anyone in any system to build a WWW server and provide hypermedia information to others, since it is not designed to support any particular system.

WWW employs the client-server model as illustrated in Figure 5.69. It also uses standardized information expression, information transfer, and information naming methods so as to be able to process and transfer widely distributed information in a systematic manner.

Distributed information is stored in various WWW servers, and users access the servers using browsers. A server prepares structured files using *hypertext markup language* (HTML) as illustrated in Figure 5.70. The term *hypertext* refers to a set of texts linked in such a way that the related texts can be retrieved along the link. Hypermedia is similar to hypertext except that it is intended to deal with various media such as text, audio, image, and video.

WWW employs various standards to achieve consistency in producing and transferring information. WWW employs HTML as a standard for information production, and it uses *hypertext transfer protocol* (HTTP) for its information

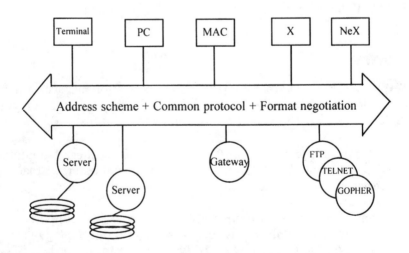

Figure 5.69   Client-server structure of the WWW.

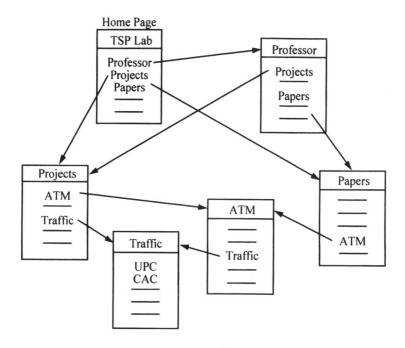

Figure 5.70 An example of the hypertext structure of WWW.

transfer standard. In addition, WWW uses the naming standard called *universal resource locator* (URL) to facilitate universal information retrieval in distributed servers. In the following, URL, HTTP, and HTML are discussed further.

URL

A URL is used to indicate the location of information distributed in a WWW server. Users can access the server storing the desired information by describing its location in terms of its URL. A URL expressions takes the following form:

Protocol://Internet address[: port number]/[directory]filename

The *protocol* part in the URL includes *http* and various other protocols such as *ftp, gopher, mailto, news, nntp,* and *telnet.* Table 5.10 lists some typical protocols used in WWW.

HTTP

HTTP is a protocol designed to transfer hypermedia between servers and clients. In addition to the object information, HTTP also transmits metadata

Table 5.10
Types of Protocols and URIs

Protocol Types	Usage	Examples of URIs
ftp	File transfer	ftp://internet.address.edu/path.splunge.txt
gopher	Menu-based information	gopher://domain.name.edu/Select_string
mailto	Send and receive messages	mailto:mail_address
news	Discussion group	news.group
telnet	Connect to a remote host	telnet://telnet_address
rlogin	Connect to a remote host	rlogin://usename@flober.rodent.edu
wais	Search indexed databases	wais://wais.server.edu/database

information over the bit space in the header. This HTTP header is an extended form of *multipurpose internet mail extensions* (MIMEs). Through this extension, the HTTP can carry binary information and nonstandard information that is agreed on by server and client. In general, delay occurs when the server and client make various negotiations before information transfer. Because the length of this delay is independent of the length of information to transfer, the resulting delay-induced overhead becomes relatively large in the case of short information. HTTP is a simple protocol designed to reduce such delays. HTTP is a stateless protocol in the sense that the server processes each of the client's requests individually, regardless of previous requests and connection states. HTTP uses eight-bit character code to be able to transfer all possible types of data.

Connection of HTTP is made through the four stages indicated in Figure 5.71:

1. *Connection setup:* A client accesses a server by using an Internet address and port number. The default port number is 80.
2. *Request:* The client sends a message to the server to request service. The request message includes, in its HTTP header, information on transaction methods and client's capabilities.
3. *Response:* The server sends a response to the client after finishing the client's request. The response message includes information on transaction state and the requested data.
4. *Connection release:* The client releases the connection.

A server can provide, using the CGI programs, the services that general application programs provide. For example, if a database is constructed as shown in Figure 5.72, it is possible to retrieve information in the database using a browser. The procedure for CGI program is as follows:

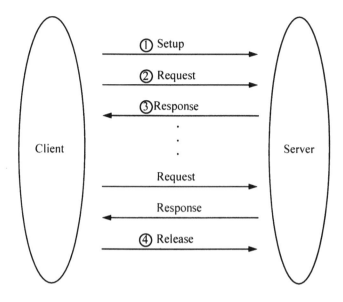

Figure 5.71   Transaction between client and server using HTTP.

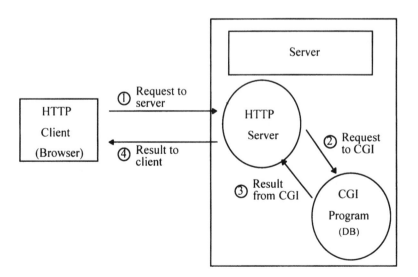

Figure 5.72   Example of CGI script.

1. A client sends requests to an HTTP server.
2. The HTTP server executes the CGI program, and transfers the requests to the program.
3. The CGI program returns the result to the HTTP server.
4. The HTTP server returns the result to the client.

HTML

To be able to retrieve distributed information in an integrated manner, it is necessary to express information in standard forms. HTML is the language employed by the WWW for a consistent expression of information. A hypermedia document includes text, images, and other types information. It can also link to information in different documents. HTML language is independent of the platform, so it can be used without restriction to any particular types of hardware or software. HTML describes a logical structure such as contents and paragraphs, with the job of presenting the document left for browsers. For instance, the HTML document in the Figure 5.73(a) looks like Figure 5.73(b) if viewed through a web browser. HTML is described in terms of the *standard generalized markup language* (SGML) terminology defined in ISO 8879. In fact, SGML is very complex and includes much more than is necessary for the WWW, so just essential parts of it are used to form HTML. The early version of HTML used to be very restrictive in expressing documents, but new, diverse functions have been added in newer versions such as HTML +.

5.9.5 MIME

To overcome the limitation of text-oriented e-mail, Internet researchers developed the MIME (Multipurpose Internet Mail Extension) protocol, which can transmit binary data along with text data, overcoming the limitation of the original text-oriented e-mail programs. The type of e-mail generally in use on the Internet is the one standardized in 1982 as RFC 822. Since this standard was designed to process text only, it could not handle multimedia messages such as audio and image. As a consequence, when transmitting nontext data by e-mail, it had to encode the data to a text format.

The limitation of RFC 822 becomes apparent when exchanging e-mail messages between RFC 822 hosts and X.400 hosts. That is, X.400 provides means to include nontext data to Internet e-mail, but SMTP, the e-mail transmitting protocol in the Internet, can transmit only seven-bit data. For the exchange of nontext e-mail, the transmitter must encode the data using *uuencode* and the receiver must decode it using *uudecode*. MIME has solved these problems, while keeping compatibility with the present RFC 822. The types of contents that MIME supports are listed in Table 5.11.

```
<html>
<title>
TSP Lab's Homepage
</title>

<body>
<body bgcolor="#8f8fbd" text="#ffffff" link="#ffff00" alink="#c0d9d9"
   vlink="#c0d9d9">

<center>
<a href="http://taebaek.snu.ac.kr:8000/intro/intro.html"> [Introduction] </a>
<a href="http://taebaek.snu.ac.kr:8000/professor/professor.html"> [Professor] </a>
<a href="http://taebaek.snu.ac.kr:8000/students/students.html"> [Students] </a>
<a href="http://taebaek.snu.ac.kr:8000/research/research.html"> [Research] </a>
<a href="http://taebaek.snu.ac.kr:8000/paper/paper.html"> [Paper] </a>
<a href="http://taebaek.snu.ac.kr:8000/e-mail/e-mail.html"> [e-mail] </a>
</center>

<center> <b> <h1> The Homepage of the TSP Lab</h1> </b> </center>

Welcome to the homepage of the TSP Lab! <br>
TSP Lab is the abbreviation of <h2> <i> Telecommunication and Signal Processing
Laboratory. </i> </h2>
The following figure is composed of clickable images.
Click your mouse botton where you want to go on the image. Enjoy our homepage,
and e-mail <a href="mailto:syh@hanla.snu.ac.kr">me</a>
if you find an error or have a suggestion. Thanks, good bye.<p>


<center>
<a href="/img/tsp-map/tsp-over.map">
<img src="/tsp-gif/tsp-over-1.gif" border=3 ismap> </a>
<br>
<hr size=5>
</center>

</body>
</html>
```

(a)

Figure 5.73   Example of HTML script and its display on a browser: (a) HTML script and (b) display on a browser. (Illustration continued on next page.)

Figure 5.73 (continued)

Table 5.11
Types of Contents Supported by MIME

Type	Subtype	Remarks
Text	Plain	Plain (unformatted) text
	Richtext	Formatted text data
Image	GIF	Image in GIF format
	JPEG	ISO JPEG standard
Audio	Basic	8kHz μ-law audio data
Video	MPEG	ISO MPEG standard
Message	Partial	To allow large objects to be delivered in several separate pieces of mail
	External-body	The actual body data are not included, but merely referenced
Multipart	Mixed	The body parts are independent and intended to be displayed serially
	Alternative	Each of the body parts is an alternative version of the same information
	Parallel	All of the parts intended to be presented in parallel, i.e., simultaneously
	Digest	To allow for a more readable digest format that is largely compatible with RFC 934
Application	Postscript	Uninterpreted binary data or information to be processed
	ODA, etc.	by a mail-based application

5.10 INFORMATION INFRASTRUCTURE AND INFORMATION SUPERHIGHWAY

In 1993, the United States announced the *National Information Infrastructure* (NII) initiative, and Japan announced their plan to build an "intellectual society." Both are examples of the keen interest that nations around the world have in building information infrastructures that can support the multimedia communications of the future. Furthermore, in the following year, plans for a *Global Information Infrastructure* (GII) and *Asia Pacific Information Infrastructure* (APII) were announced as complements to the national information infrastructures of individual nations.

Information infrastructure refers to the foundation required for production, processing, storage, and transfer of information for the benefit of people, society, and industry, and *information superhighway* is the surname given to the high-speed broadband communication network that is a key component of the information infrastructure.

This section looks at the perspective and the bases of the information infrastructure, and the information superhighways. First, the background for building an information infrastructure is considered, and the structure of the information infrastructure is discussed. Next, the services to be provided over information superhighways are surveyed and the relevant technologies examined. Finally, scenarios for building the information superhighway considered.

5.10.1 Background

The NII initiative of the United States was announced as a general and forward-looking information technology policy aimed at strengthening national competitiveness and maintaining technological and industrial leadership in the upcoming information age. The plan is intended to connect all public offices, companies, schools, libraries, hospitals, and homes across the nation through a broadband high-speed communication network so that anyone anywhere may easily access multimedia information services.

Building an "intellectual society," where human knowledge and its related activities become the most important resources, was believed by Japanese leaders to be the only way to overcome the limits of the current mass-consumption and mass-production society. So under the so-called new social infrastructure plan Japan intends to construct a nationwide all-fiber-optic communication network as a basis for building a national information infrastructure aimed at maintaining industrial leadership and solving various seemingly intractable problems such as the society's aging problem.

The *European Union* (EU), which is the largest single economic block in the world, has also come to regard the information infrastructure as being of the utmost importance for the future due to the consensus that the formation of a single unified European economic market, with the resultant increased employment, growth, and industrial competitiveness, can only be achieved if a *European information infrastructure* (EII) is in place. As a beginning, the EU is in the process of forming a *Trans European Network* (TEN) to support expedited unification of European markets and consolidation of European societies and economy.

It is natural to wonder about the reasons that drive this competitive movement toward the building of an information infrastructure or information superhighway. The main driving force is the idea that the country that controls information and communications in the twenty-first century will control the world. At a more concrete level, there appear to be three practical reasons: first, expectations of direct economic effects such as increased employment and economic growth; second, expectations of an increase in the efficiency of all social activities and the competitiveness of industries throughout the widespread utilization of information; and, third, the improvement of the citizen's quality of life through better social and governmental services.

The direct economic effects of the building of an information infrastructure include greater production, increased employment, and new market development. However, the indirect effects such as the increase in industrial competitiveness are expected to be much more important. This is based on the reasoning that technology is expected to be much more important than the traditional "three elements of manufacturing"—(capital, land and labor—(in the future, and the development, acquiring, and application of technology itself will critically depend on the information infrastructure. No less important than these economic effects is the increase in the overall standards of living. By using the information superhighways, public services in all areas of medicine, education, government, and transportation will improve and be more efficient. Also by enabling telecommuting, home shopping, and televoting, citizens should be able to enjoy a much more convenient lifestyle. In fact, the information infrastructure may bring about revolutionary changes in the social, cultural, economic, and political fabric and also the living patterns of individuals in the future.

5.10.2 Information Infrastructure

The information infrastructure has a service aspect and a technology aspect. The technology aspect may be viewed to be the *hard* infrastructure with the service aspect viewed as the *soft* infrastructure. That is, the information services are the *soft* infrastructure that can be built on the *hard* foundation of information technology.

To put information services to a practical use, some sort of mediating functions are necessary that will enable a socially, culturally, and economically harmonious introduction of information services. This requires us to divide the service aspect into two categories: first, the *information society layer,* which is related to the social, cultural, and human factors that influence the introduction of information services, and, further, the direction of the overall information infrastructure, and second, the *information application layer,* which is related to the actual provision and application of information services to individuals, society, and industry.

The technology aspect of the information infrastructure can also be divided into two categories: first, the *information processing layer,* which is related to information processing and storage retrieval functions, and second, the *information transfer layer,* which is related to various information transfer functions. Therefore, the information infrastructure can be regarded as being composed of four layers as shown in Figure 5.74.[17] In the following the important elements of each layer are considered.

First, in the information society layer, it is important to establish social and environment rules and regulations, taking into account social values, cultures, lifestyles, human rights, and so on, so that information services can be applied widely, harmoniously, and actively.

In the information application layer, it is essential to actively develop versatile applied services that will help to promote industrial productivity and improve social welfare. As a means for offering these services, it is important to develop various application software that can digitalize and systematically store all available information. Resurrecting all existing knowledge in universally accessible digital information is what will make the twenty-first century the knowledge-based society, and the layer in which this be carried out is the information application layer.

The information processing layer processes, stores, and retrieves information in an appropriate manner for use in high-speed multimedia environments. For this to happen, it is important to develop user interfaces having diverse expressive and understanding abilities. In addition, basic server, multimedia database, and hypermedia retrieval technologies are also important. Because the information processing layer pertains to various types of information processing, its scope and capability depend heavily on computer technology.

The information transfer layer is the layer that will actually transport the information to the user. The functions of this layer include the networking function that will enable smooth interworking of communication networks, and the transport function for the physical transfer and switching of information signals. As such, the information transfer layer relies heavily on the communications technology, and the term *information superhighway* itself refers to this information transfer layer.

This layering concept also offers a guideline for classifying the businesses involved in the information infrastructure: The information society layer may be related to information users, information application layer to information providers, the information processing layer to information businesses, and the information transport layer to network providers. The information user may be an individual, institute, or group that receives various information services through communication devices. The information providers, such as

---

17. This four-layer structure based on the functions, services, and related technology of the information infrastructure is the authors' private opinion. As such, other layering models are also possible. For example, the NII of the United States consider a three-layer model consisting of application, enabling services, and bitway layers, and these three layers are closely related to the information application, information processing, and information transfer layers, respectively.

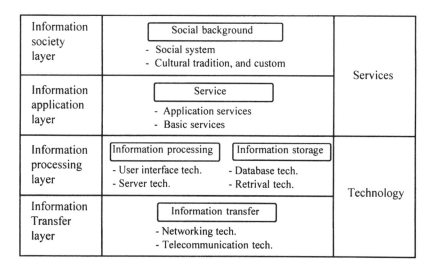

Figure 5.74   Layers of the information infrastructure.

universities, libraries, museums, hospitals, media companies, financial institutes, public offices, and broadcasters, actually produce raw information materials or services. The information businesses are the driving force behind the information superhighways in that they actually process and store each type of multimedia information and sell the end products or services to information users. The network providers operate the actual physical network infrastructure for transporting high-speed multimedia information among information users, businesses and providers.

5.10.3 Information Services

Once the information infrastructure is constructed, countless new services may be offered. In the following some of typical information services that will become available due to construction of information infrastructure are reviewed (refer to Table 5.12). They are the applied services that differ from the basic services discussed in Section 4.1 in that the applied services are for practical applications, whereas the basic services are for technological categorizations (refer to Section 6.2.1 for the definition of several basic and applied services).

Educational Area

Once the information infrastructure network is complete a revolution is expected in the field of education. Teachers will be able to use multimedia texts

Table 5.12
Applied Services in Information Infrastructure

Area	Applied Services
Education	Distance learning, electronic education, digital libraries, digital museums
Living	Distance diagnosis and treatment, electronic newspapers, electronic publishing, video-on-demand (VOD), home-shopping
Industry/business and public	Electronic document service, telecommuting, CALS

to increase the effectiveness of education, connect to digital libraries, and show information or related data on displays during class, use a videophone to connect to distant experts for opinions, or connect to other classrooms for video conferences. Students who could not attend the lectures will be able to rent videotapes or access on-line databases storing lecture material and study at any convenient time.

It will be also possible to offer educational services to people at home or offices through the use of distance learning tools. Distance learning enables the learners to access various developed electronic studying tools such as CD-ROM and study the material by themselves in an interactive fashion. Distance learning over the information superhighways will be even more effective for areas such as medical education, which require high-quality color pictures or images.

The digital library is probably the most important database in the information infrastructure. A digital library combines the traditional functions of existing libraries with new information systems. A digital library stores the record indices and texts of all books in traditional libraries, and also audio, graphic, images, and video data in a digital format. Digital libraries enable users to access the desired databases at any time and at any place through the information superhighways. Thus digital libraries will eliminate the spatial and temporal limitations of the existing libraries, and hence bring forth "open libraries" or "libraries without walls."

Living Area

From a social welfare point of view, telemedicine is one of the most typical services that can be offered over the information superhighways. Hospitals and health centers can connect to each other to exchange patient records, and doctors will be able to diagnose and treat patients remotely. Even a patient living in the countryside or in a remote place can get treatment from expert doctors through the information superhighways. This concept in the end could lead to home treatment.

Just as *video-on-demand* (VOD) is expected to become a dominant movie entertainment means, *news-on-demand* (NOD) is expected to become a predominant news access means in the future. Once newspapers start to be delivered or retrieved in an electronic form, the problem of newspaper delivery delays is solved and, further, personalized 24-hour newspapers can be created. Also, the advantages of digital media can be fully exploited to include audio narrations and visual images in the electronic newspaper.

The technologies used in the electronic newspaper, when applied to general publishing, will lead to electronic publishing services. With the rapid spread of computer desktop typesetting, printing via traditional typesetting has been disappearing. But much more than this, complete electronic publishing will also become possible. Not only will computer applications such as word processors be used to write and transfer electronic manuscripts between writers and publishers, but the publishing of books as electronic files will take place. Also books that had been stored in physical form at traditional libraries will be stored in an electronic form in digital libraries.

Another item that cannot be left out among the effects that information infrastructure will bring to home life is electronic commerce at home, examples of which include home banking and teleshopping (or home shopping). Existing 24-hour banking services and on-line transactions between banks, while helpful, have still not done away with the need for people to visit their banks. Once home banking becomes possible, assisted by a relevant electronic signature technology, it will become possible to handle all financial matters from the user's home. A prerequisite for a complete teleshopping service is high-resolution image services as well as remote payment transaction capability.

Industry/Business and Public Area

Electronic document service will be an important service in all industry/business and government sectors. For this service a single complete system can be established over the network such that the user just requests a petition and all the necessary paperwork is done automatically and delivered to the relevant people or institutes, followed by a message notification sent to the user on the transaction result. This can be extended for the processing and exchange of various administrative papers among various branches of government organizations, resulting in increased administrative efficiency.

In the business sector a tremendous amount of paperwork is generated in business activities such as material ordering, inventory, and exports/imports, and electronically maintaining and exchanging these documents would lead to a considerable amount of savings. By extending this to include all types of industrial, trade, distribution, standards, and marketing information, efficiency and competitiveness could be significantly enhanced, and, in this connection, there has been a recent surge of interest in CALS.[18] The effort to connect EDI and CAD systems into a single international information system also reflects these ideas.

Just as the EDI and CALS services solve the problem of delivering and exchanging documents for business, video conferencing and video telephony services can solve the problem of business traveling. Video telephones for two people or a full-scale video conferencing system for a group can help to resolve the problems of wasted time, transportation, and even family estrangement that are caused by business trips. In the end video conferencing and telephony along with electronic document exchange are the key tools for telecommuting.

5.10.4 Information Technology

Information means "news" in a narrow sense and "knowledge" in a broad sense, and if its interpretation is broadened further it can include all nonmaterialistic products. Nonetheless, information may be treated the same as any other materialistic product: That is, it is processed, stored, and then transported. This procedure as a whole can be regarded as defining the area of information technology. To repeat, information technology can be divided into information *processing technology*, information *storage technology*, and information *transfer technology*.

Information Processing Technology

Information processing refers to the process of changing the content and presentation method of information. Through this process, the interfaces between the user and the information device, between information devices, and between the network and information device become smooth. For example, audio and video signal compression and decompression technology enables the connection of information machines over limited bandwidth transmission paths, and also enables the storage of multimedia data in storage devices of finite capacity. Diverse user interface technologies enable an organic link between the user and the information device, so that any user can easily and efficiently access multimedia information. Also encryption technology and synchronization technology can be regarded as part of the information processing technologies. Explicit examples of information processing technologies are listed in Table 5.13.

---

18. CALS, *continuous acquisition and life cycle support* or *computer-aided acquisition and logistic support,* was originally a system used by the U.S. Department of Defense for logistical support. But as its use has spread to business and commercial areas, the abbreviation has also been interpreted as *commerce at light speed.*

Information Storage Technology

In part, information storage is a function used in all user devices, computers, and servers, but information storage is primarily related to databases. For information infrastructure services to be realized, large real-time multimedia databases must be installed in various places in the network. Each multimedia database must be capable of storing, searching, and retrieving at high speed a huge amount of multimedia information. Also, it must be furnished with automatic recovery mechanisms so that it can maintain the stored data safely and reliably. Information storage technology can be divided into the actual storing technology for saving data, retrieval technology for finding information, and server technology to serve large numbers of simultaneous user clients. Explicit examples of these information storage technologies are given in Table 5.14.

Information Transfer Technology

The information transfer function refers to the actual transport of information between users, and between the user and the end system. This is the traditional

Table 5.13
Information Processing Technologies

Subcategories	Information Processing Technologies
Information compression, decomposition	Image, video, voice compression/decompression, 3-D imaging
User interface	GUI, agent interface, speech recognition, natural language processing
Information protection	Information security, user authentication, information integrity
Synchronization	Intermedia (multimedia) synchronization, intramedia synchronization

Table 5.14
Information Storage Technologies

Subcategories	Information Storage Technologies
Database	Distributed DB, object-oriented DB, real-time DB, parallel DB, hypermedia DB
Information retrieval	Multimedia characteristics extracting and indexing, multimedia real-time parallel DB search
Server	Supercomputer, massively parallel processing computer, video-on-demand server

area of communications engineering. While information processing and storage are functions that developed relatively recently with the invention of computers, information transfer is a basic function that started with the invention of the telegraph and telephone more than a century ago. Therefore, information transfer technology can be thought of as the telecommunications technology, whose description is the fundamental intent of this book. To offer information infrastructure services, basic telecommunication technologies such as transmission, switching, signaling, and networking are needed. Fiber-optic transmission technology, wireless transmission technology, ATM switching, optical switching, user-network interface, network host access control, traffic control, optical media, and microwave technology can all be seen as basic elements of communications technology.

To transport high-bandwidth multimedia data effectively, various networking technologies must be added on top of this telecommunication technology base. For example, high-speed transport protocol, high-speed media access control, and multimedia communications protocol technologies are needed. In addition, high-bandwidth data transport, real-time data transport, network management, and an operating system technology are needed. Explicit examples of information transfer technology are listed in Table 5.15.

5.10.5 Building the Information Superhighways

Information superhighways, or the high-speed, broadband communication network, that carry out the actual transfer of information are the primary component of the information infrastructure and they are in the lowest layer of the information infrastructure. Only after building the information highways can various information processing and storage functions be installed to provide versatile applied information services to users who are distributed all over the network.

Currently, there are three different approaches toward building the

Table 5.15
Information Transfer Technologies

Subcategories	Information Transfer Technologies
Telecommunication	Optical/wireless transmission, ATM/optical switching, access and traffic control, optical media and devices, microwave and devices, intelligent networks
Networking	High-speed transport protocol, multimedia communication protocol, protocol conversion, large data transport, multipoint communication, network management

information superhighways. The first is to rely on, and to speed up, the existing telephone network, the second is to use the Internet, and the third is to use the cable and satellite network.

ISDN is basically the amalgamation of services on top of the existing telephony network, and BISDN is the extension of this idea to broadband services. Therefore, building information superhighways on the basis of existing telephone networks implies nothing but building the BISDN.

This approach to building information can be regarded as being idealistic. Because all interoffice and access networks will be based on the basic 155-Mbps high-speed ATM communications network, all types of applied information services in the living, industrial, and public areas can be supported. But connecting all access networks with optical fiber is something that will require a huge amount of investment and time, and it may become possible only if the government leads the investment. Examples of government-driven investment plans can be found in Japan, where investment of 53 trillion yen is projected by 2010, and Korea, where investment of 45 trillion won is projected by 2015.

In contrast to the telephone network-based approach, the approach based on the Internet is economical and gradual. Because building an internet does not mean building a new network, but connecting existing networks, it is relatively cheap and has a lot of flexibility. For these reasons, the Internet has been expanding rapidly and is becoming a plausible basis for the information infrastructure, without any overall investment plan, and based only on the spontaneous participation of its members. But the basic limit of the Internet is that it is a low-speed data network. Also in interconnecting various networks, multiple layers of protocols must be processed, resulting in large and highly variable end-to-end delays. These delays represent the potential bottleneck of the Internet in offering real-time services. The IETF is studying various methods to solve these problems (refer to Section 5.9).

The approach based on cable and satellite communications is regarded as the most practical method to provide broadband real-time services, because CATV networks are readily deployed in many places. If interactive functions can be added to CATV networks so that it can be used as a high-speed access network, and if it is connected to the backbone BISDN network, it will become possible to rapidly build the information superhighways. Table 5.16 lists comparisons of the three different approaches to building information superhighways in terms of network, terminal, service, user, and conventional paradigm.

From the viewpoint of the user access network, the CATV network-based approach appears to be in direct competition with the telephone network-based approach, but the two become complementary in that the BISDN can provide a backbone network for the interconnection of CATV networks. It will be also possible to interconnect CATV networks directly via satellite networks in the future without crossing over the BISDN, but it will take a long period of time to make the access network fully optical.

Table 5.16
Comparison of Approaches to Building Information Superhighways

	Telephone Network-Based Approach	*Internet-Based Approach*	*Cable/Satellite-Based Approach*
Network	BISDN	Internet	CATV/satellite network
Terminal	Telephone (broadband) + network termination (NT)	PC/workstation + modem	Television + set-top box
Service (examples)	Video dial tone (VDT)	MBone	Video-on-demand (VOD)
Major user	Home, business, public	Business, R&D institutes	Home
Conventional paradigm	Communication (communication)	Information (data processing)	Broadcasting (entertainment)

On the other hand, the approach based on CATV and satellite networks appears to be in direct competition with the approach based on the Internet. One side is a television-based approach, while the other is a computer-based approach. While the penetration of television is much larger, the growth of home computers is also exponential. From the functional point of view, a set-top box and television pair cannot be compared with the rapidly developing computers of today. But in the area of real-time video service, which is a common area of interest to all users, the television still has an advantage over the computer. Simply said, it is not possible currently to assert which method is better, but the approach that can better understand the user's demand in regard to the information superhighways' services and can provide faster, cheaper, and more user-friendly services will eventually win the competition.

Selected Bibliography

ANSI T1.606-1990, "Telecommunication-frame relay bearer service-architectural framework and service description," 1990.

ANSI T1.606add, "Addendum to T1.606," (T1X1/90-175), 1990.

ANSI T1.6ca, "Core aspects of frame protocol for use with frame relay bearer service," (T1S1/90-214), 1990.

ATM Forum, "LAN emulation over ATM specification": version 1.0, January 1995.

ATM Forum, "Network compatible ATM for local network applications," phase 1, version 1.0, April 1992.

DEC, Northern Telecom, Stratacom, Cisco, "Frame relay specification with extension based on proposed T1S1 standards," Revision 1.0, 1990.

ETSI ETS300211-217 and ETS300268-278, "Connectionless broadband dataservice, superset of bellcore SMDS."

Frame Relay Forum FRF.1, "User to network interface implementation agreement," version 2.0.

Frame Relay Forum FRF.2, "Frame relay network-to-network interface IA."

Frame Relay Forum FRF.3, "Multiprotocol encapsulation implementation agreement."

Frame Relay Forum FRF.4, "Switched virtual circuit, pending ratification."

Frame Relay Forum FRFTC93.67, "Frame relay MIB architecture and requirements."

Frame Relay Forum FRFTC93.67, "Frame relay multicast service description."

Frame Relay Forum FRFTC93.67, "Frame relay/ATM IA."

Frame Relay Forum FRFTC93.67, "Switched virtual circuit IA."

IAB & IESG, "Charter of the IAB", RFC1602, March 1994.

IAB & IESG, "The internet standards process—revision," RFC 1602, March 1994.

IEEE P802.12/D5, "Local and metropolitan area networks—part 12: demand-priority access method and physical layer specifications," Draft 5.0, 1994.

IEEE P802.3u/D4, "MAC parameters, physical layer, medium attachment units and repeater for 100 Mbps operation," 1995.

IEEE P802.6, "DQDB subnetwork of a metropolitan area network," 1991.

ISO-IEC 9314-1,-2,-3, "Fiber distributed data interface physical layer protocol, MAC and system management-100 Mbit/s optical fiber ring."

ITU-T Rec. I.122, "Framework for frame mode bearer service," 1993.

ITU-T Rec. I.233, "Frame mode bearer services," 1992.

ITU-T Rec. I.364, "Support of broadband connectionless data service on B-ISDN," 1993.

ITU-T Rec. I.365.1, "Frame relaying service specific convergence sublayer(FR-SSCS)," 1993.

ITU-T Rec. I.370, "Congestion management for the ISDN frame relaying bearer service," 1991.

ITU-T Rec. I.372, "Frame relaying bearer service network-to-network interface requirements," 1993.

ITU-T Rec. I.430, "Basic user-network interface—Layer 1 specification," 1993 (revision).

ITU-T Rec. I.431, "Primary rate user-network interface—Layer 1 specification," 1993 (revision).

ITU-T Rec. I.555, "Frame relaying bearer service interworking," 1993.

ITU-T Rec. Q.921(I.441), "ISDN user-network interface, data layer specification," 1988.

ITU-T Rec. Q.922, "ISDN data link layer specification for frame mode bearer services," 1992.

ITU-T Rec. Q.933, "DSS1 signaling specification for frame mode bearer service," 1991.

TA-TSY-000772, "Generic system requirement in support of SMDS," Issue 3, Bellcore, 1989.

Adam, J. F., H. H. Houh, M. Ismert, and D. L. Tennenhouse, "Media-intensive data communications in a "desk-area" network," *IEEE Commun. Mag.*, August 1994, pp. 60–67.

Ahmad, R., and F. Halsall, "Interconnecting high-speed LANs and backbones," *IEEE Network*, September 1994, pp. 36–43.

Ali, M. J., "Frame relay in public networks," *IEEE Commun. Mag.*, Vol. 30, No. 3, March 1992, pp. 72–80.

Armitage, G. J., and K. A. Adams, "How inefficient is IP over ATM anyway?," *IEEE Network*, January/February 1995, pp. 18–26.

Babson, M., D. Buster, G. De Val, and S. Xavier, "ATM switching and CPE adaptation in the North Carolina information highway," *IEEE Network*, November/December 1994, pp. 40–46.

Bantz, D., and F. Bauchot, "Wireless LAN design alternatives," *IEEE Network*, Vol. 8, No. 2, March/April 1994, pp. 43–53.

Bates, R. J., *Wireless Networked Communications*, New York: McGraw-Hill, 1994.

Berenbaum, A., J. Dixon, A. Iyenger, and S. Keshav, "A flexible ATM-host interface for XUNET II," *IEEE Network*, July 1993, pp. 18–23.

Bracket, C. A., "Dense WDM networks: Principle and applications," *IEEE JSAC*, Vol. JSAC-8, No. 6, August 1990, pp. 948–964.

Byrne, W. R., et al., "Evolution of metropolitan area networks to broadband ISDN," *IEEE Commun. Mag.*, Vol. 29, No. 1, January 1991, pp. 69–82.

Capell, R. L., D. A. Kettler, L. Corn, and R. Morris, "Evolution of the North Carolina information highway," *IEEE Network*, November/December 1994, pp. 54–70.

Carol, M. J., and R. D. Gitlin, "High performance optical local metropolitan area networks: Enhancements of FDDI and 802.6 DQDB," *IEEE JSAC*, Vol. JSAC-8, October 1990, pp. 1439–1448.

Canset, S., "RTP: A transport protocol for real-time applications," Internet Draft, November 1995.

Cavanagh, J. P., "Applying the frame relay interface to private networks," *IEEE Commun. Mag.*, March 1992, pp. 48–65.

Chao, J. H., et al., "IP on ATM local area networks," *IEEE Commun. Mag.*, August 1994, pp. 52–59.

Chen, K., "Medium access control of wireless LANs for mobile computing," *IEEE Network*, Vol. 8, No. 5, September/October 1994, pp. 50–63.

Cheriton, D. R., and C. L. Williamson, "VMTP as the transport layer for high-performance distributed systems," *IEEE Commun. Mag.*, June 1989, pp. 37–44.

Chesson, G., "The protocol engine project," *Unix Review*, September 1987.

Cheung, N. K., "The infrastructure for gigabit computer networks," *IEEE Commun. Mag.*, April 1992, pp. 67–68.

Clapp, G. H., "LAN interconnection across SMDS," *IEEE Network*, Vol. 5, No. 5, September 1991, pp. 25–32.

Clark, D. D., M. L. Lambert, and L. Zhang, "NETBLT: A bulk data transfer protocol," RFC 998, 1987.

Comer, D. E., *Internetworking with TCP/IP*, Vol. 1, Englewood Cliffs, NJ: Prentice Hall, 1991.

Damodaram, R., et al., "Network management for the NCIH," *IEEE Network*, November/December 1994, pp. 48–54.

Davidson, R. P., and N. J. Muller, *Interworking LANs: Operation, Design and Management*, Norwood, MA: Artech House, 1992.

Deering, S., "Host extensions for IP multicasting," RFD-1112, 1989.

Deering, S., "An architecture for IPv6 unicast address allocation," Internet Draft, March 1995(a).

Deering, S., "Internet protocol, version 6 (IPv6) specification," Internet Draft, June 1995(b).

Doeringer, W. A., et al., "A survey of light-weight transport protocols for high-speed networks," *IEEE Trans. Commun.*, November 1990, pp. 2025–2039.

Droms, R., "Dynamic host configuration protocol," RFC-1541, October 1993.

Fernandes, J. J. G., P. A. Watson, and J. C. Neves, "Wireless LANs: Physical properties of infra-red systems vs. MMW systems," *IEEE Commun. Mag.*, August 1994, pp. 68–73.

Fink, R. L.,and R. E. Ross, "Following the fiber distributed data interface," *IEEE Commun. Mag.*, March 1992, pp. 50–55.

Fischer, W., et al., "Data communications using ATM: Architectures, protocols, and resource management," *IEEE Commun. Mag.*, August 1994, pp. 24–33.

Graham, Ian S., *THE HTML Source Book*, New York: John Wiley & Sons.

Guarneri, R., and C. J. M. Lanting, "Frame relaying as a commun access to N-ISDN and B-ISDN data services," *IEEE Commun. Mag.*, June 1994, pp. 39–43.

Halsall, F., *Data Communications, Computer Networks and Open System*, Reading, MA: Addison-Wesley, 1992.

Handel, E.,and M. N. Huber, *Integrated Broadband Networks: An Introduction to ATM-Based Networks*, Reading, MA: Addison-Wesley, 1991.

Haner, M., et al., "Broadband fiber loops with wireless access," *European Conference on Optical Communication*, 1995.

Heinanen, J., "Multiprotocol encapsulation over ATM adaption layer 5," RFC-1483, July 1993.

Hemrick, C. F., R. W. Klessig, and J. M. McRoberts, "Switched multi-megabit data service and early availability via MAN technology," *IEEE Commun. Mag.*, April 1988, pp. 9–14.

Hornig, C., "Standard for transmission of IP datagrams over ethernet networks," RFC-894, 1984.

Hughes, J. P., and W. R. Franta, "Geographic extension of HIPPI channel via high speed SONET," *IEEE Network*, May/June 1994, pp. 42–53.

Huitema, Christian, *Internet Routing*, Englewood Cliffs, NJ: Prentice Hall, 1995.

Hunt, C., *TCP/IP Network Administration*, O'Reilly & Associates Inc., 1992.

Inoue, Y., and N. Terada, "Granulated broadband network," *IEEE Commun. Mag.*, April 1994, pp. 56–62.

Jabbari, B., et al., "Network issues for wireless communications," *IEEE Commun. Mag.*, Vol. 33, No. 1, January 1995, pp. 88–99.

Jacobson, V., "Congestion avoidance and control," *Computer Communications Review*, Vol. 18, No. 4, August 1988, pp. 314–329.

Jacobson, V., R. T. Braden, and D. A. Borman, "TCP extensions for high performance," RFC-1323, 1992.

Jain, R., "FDDI: Circuit issue and future plans," *IEEE Commun. Mag.*, September 1993, pp. 98–105.

Johnson, J. T., "Coping with public frame relay: A delicate balance," *Data Communication*, January 1992(a), pp. 31–38.

Johnson, J. T., "Frame relay products," *Data Communication*, May 1992(b).

Johnston, C. A., "Architecture and performance of HIPPI-ATM-SONET terminal adapters," *IEEE Commun. Mag.*, April 1995, pp. 46–51.

Kasahara, H., K. Imai, N. Morita, and T. Ito, "Distrubuted ATM ring-based switching architecture for MAN and BISDN access networks," in *Proceeding of Workshop on Broadband Communication*, January 1992, Estoril, Portugal, IFIP Technical Committee 6.

Katz, D., and P. S. Ford, "TUBA: Replacing IP with CLNP," *IEEE Network*, Vol. 7 No. 3, May 1993, pp. 38–47.

Kavak, N., "Data communication in ATM networks," *IEEE Network*, May/June 1995, pp. 28–37.

Kung, H. T., "Gigabit local area networks: A system perspective," *IEEE Commun. Mag.*, April 1992, pp. 70–78.

La Porta, T. F., and M. Schwartz, "Architectures, features and implementation of high-speed transport protocols," *IEEE Network*, May 1991, pp. 14–23.

Laubach, M., "Classical IP over ATM," RFC-1577, Hewlett-Packard, December 1993.

Lynch, D., and M. Rose, *Internet System Handbook*, Reading, MA: Addison-Wesley, 1993.

Minoli, D., *Enterprise Networking: Fractional T1 to SONET, Frame Relay to BISDN*, Norwood, MA: Artech House, 1992.

Modiri, N., "The ISO reference model entities," *IEEE Network*, Vol. 5, No. 4, July 1991, pp. 24–33.

Mollenaeur, J. F., "Networking for greater metropolitan areas," *Data Communications*, Vol. 17, No. 2, February 1988(a), pp. 155–178.

Mollenaeur, J. F., "Standards for metropolitan area networks," *IEEE Commun. Mag.*, April 1988(b), pp. 15–19.

Mukherjee, B., "WDM-based local lightwave networks part I: Single-hop systems," *IEEE Network*, May 1992(a), pp. 12–27.

Mukherjee, B., "WDM-based local lightwave networks part II: Multihop systems," *IEEE Network*, July 1992(b), pp. 20–33.

Muller, N. J., and R. P. Davidson, *LANs to WANs: Network Management in the 1990s*, Norwood, MA: Artech House, 1990.

Neufeld, G. W., et al., "Parallel host interface for an ATM network," *IEEE Network*, July 1994, pp. 24–34.

Newman, P., "ATM local area networks," *IEEE Commun. Mag.*, March 1994(a), pp. 86–98.

Newman, P., "Traffic management for ATM local area networks," *IEEE Commun. Mag.*, August 1994(b), pp. 44–50.

Newman, R. M., Z. L. Budrikis, and J. L. Hullett, "The QPSX MAN," *IEEE Commun. Mag.*, April 1988, pp. 20–28.

Pahlavan, K., and A. Levesque, "Wireless data communications," *Proc. of IEEE*, Vol. 82, No. 9, September 1994, pp. 1398–1430.

Pahlavan, K., T. Probert, and M. Chase, "Trends in local wireless networks," *IEEE Commun. Mag.*, Vol. 33, No. 3, March 1995, pp. 88–95.

Pandya, R., "Emerging mobile and personal communication systems," *IEEE Commun. Mag.*, Vol. 33, No. 6, June 1995, pp. 44–52.

Patterson, J. S., and W. L. Smith, "The North Carolina information highway," *IEEE Network,* November 1994, pp. 12–17.

Pehrson, B., P. Gunningberg, and S. Pink, "Distributed multimedia applications on gigabit networks," *IEEE Network,* Vol. 6, No. 1, January 1992, pp. 26–35.

Perlman, R., *Interconnections,* Reading, MA: Addison-Wesley, 1992.

Plummer, D. C., "An Ethernet address resolution protocol," RFC-826, 1982.

Postel, J., "Introduction to RFC authors," RFC-1543, October 1993.

Postel, J. B., and J. K. Reynolds, "Standard for transmission of IP datagrams over IEEE 802 networks," RFC-1042, 1988.

Rahnema, M., "Frame relaying and the fast packet switching concepts and issued," *IEEE Network,* Vol. 5, No. 4, July 1991, pp. 10–17.

Rannsom, M. N., and D. S. Spears, "Applications of public gigabit networks," *IEEE Network,* Vol. 6, No. 2, March 1992, pp. 30–41.

Saltzer, J. H., D. P. Reed, and D. D. Clark, "End to end arguments in system designs," *ACM Trans. on Computer Systems,* Vol. 2, No. 4, November 1984, pp. 277–288.

Sanders, R. M., and A. C. Weaver, "The express transfer protocol (XTP)—a tutorial," *ACM SIGCOMM,* 1991.

Schneiderman, R., *Wireless Personal Communications,* New York: IEEE Press, 1994.

Sequiun, H., "Optical fiber local network for distribution and interactive service," *Cable Television Engineering,* Vol. 14, No. 15, December 1988, pp. 637–638.

Sher, P. J. S., et al., "Service concept of the switched multi-megabit data service," in *Proc. GLOBECOM'88,* 1988, pp. 12.6.1–12.6.6.

Stallings, W., *Local and Metropolitan Area Networks,* New York: Macmillan, 1993.

Stallings, W., *Data and Computer Communications,* New York: Macmillan, 1994.

Stevens, R., *TCP/IP Illustrated,* Vol. 1, Reading, MA: Addison-Wesley, 1994.

Strayer, W. T., B. J. Dempsey, and A. C. Weaver, *XTP: The Xpress Transfer Protocol,* Reading, MA: Addison-Wesley, 1992.

Takashima, S., "Network," *NTT R&D,* Vol. 38, No. 4, 1989, pp. 441–458.

Tannebaum, A. S., *Computer Networks,* Englewood Cliffs, NJ: Prentice-Hall, 1988.

Thomson, S., "IPv6 address autoconfiguration," Internet Draft, February 1995.

Truong, H. L., W. W. Ellington, Jr., J. Y. Le Boudec, et al., "LAN emulation on an ATM network," *IEEE Commun. Mag.,* May 1995, pp. 70–85.

Vetter, R. J., D. H. C. Du, and A. E. Kleitz, "Networking supercomputing: High performance parallel interface [HIPPI]," *IEEE Network,* May 1992, pp. 38–44.

Vickers, B. J., and T. Suda, "Connectionless service for public ATM networks," *IEEE Commun. Mag.,* August 1994, pp. 34–42.

Watson, G., et al., "The demand priority MAC protocol," *IEEE Network,* Vol. 9, No. 1, January/ February 1995, pp. 36–43.

Zhang, L., and S. Deering, "RSVP: A new resource reservation protocol," *IEEE Network,* Vol. 7, No. 5, September 1993, pp. 8–18.

Zhang, L., R. Braden, and D. Estrin, "Resource reservation protocol—version 1 functional specification," Internet Draft, January 1995.

Broadband Video Services and Technology

6

Strengthened by rapidly developing optical communication, image processing, and VLSI technologies, the field of video communications has emerged into the limelight as an economically and technologically viable field. In fact, most services, excluding the high-quality video services, can already be provided in a limited way through the existing local data networks (i.e., LAN), public telephone networks (i.e., PSTN), and the narrowband integrated information networks (i.e., NISDN). Thus, it can be stated that the prerequisite to widespread residential deployment of the BISDN, which demands a comprehensive reorganization of existing networks, is the provision of high-quality video communications services.

From the standpoint of utilization format, the area of video services can be categorized into the distributive services (e.g., CATV, HDTV), which have distributed service characteristics; the video data services (e.g., videotex, video information retrieval, CD-ROM application), which have interactive service characteristics; the field of conversational bidirectional communication (video telephone/conferencing), which requires real-time transmission and switching, and the message services (e.g., store and forward services). In addition, there are multimedia services that provide various services such as video, audio, and data in a composite and integrated manner. There are challenging areas on the road toward an efficient dissemination of these various video communication applications. The most typical ones are the compression and efficient transmission of broadband video information, the communication procedures for simultaneous accommodation of multiple users, and the reliability and economic feasibility of various types of video terminals.

In its early stages, a video communication network was a separate network of the synchronous mode, independent of voice or data networks. However, due to the heavy cost of constructing several types of communication networks with the introduction of diverse sets of services, the concept of the asynchronous network emerged, which can provide integrated services of image, voice, and data through a single integrated network. However, the communication procedures and call connection of the asynchronous mode are more complex than those of the synchronous mode. Furthermore, the conversational video conferencing services must transmit various types of graphics and data that occur during the conference; hence, the associated communication and call connection procedures are much more complex compared to those of simple retrieval services such as a video database. Consequently, when an asynchronous conversational video service is to be provided, all these issues should be carefully considered in detail.

Figure 6.1 depicts the overall flow of information for video services, which in most cases include the audio and data components as well. The video, audio, and data signals in support of a video service are A/D-converted, coded, packetized on individual bases, and then multiplexed together. Depending on the applications the packetizing and/or multiplexing processes (i.e., the dotted blocks) may be omitted. Such processed information is transmitted across networks or stored in storage media. The received or retrieved information then passes through the reverse processing and is finally displayed.

In this chapter, broadband video services and technologies are discussed with emphasis on the communication applications in packetized form. To begin with, the history behind video services, video coding, and packetized video communications is given. Then broadband video services are defined with the most representative services such as video conferencing and video-on-demand described in detail. Next, image compression techniques are examined, and then we follow up with a detailed description of the MPEG-2 video coding system, an integrated embodiment of all existing video processing technologies. Next, video communications in high-speed packet networks are discussed, with an emphasis on video transmission over the ATM network. Finally, the status and trends of HDTV technology, whose market potential is huge not only for consumer electronics but also for communications and other industrial applications, are considered.

6.1 INTRODUCTION

Broadband video technology can be divided into coding technology for compressing tremendous amounts of video information, transmission technology for the efficient transmission of compressed video information, decoding technology for restoring compressed video information, and terminal technology, which integrates the first three. Usually, a broadband transmission medium is required for the transmission of a video signal; hence, video signal compression and restoration technologies are essential for the effective use of the transmission line. Video compression technology adopts various video/image coding techniques to eliminate redundant information from an image signal within the range that its characteristics do not change perceptibly.

For the economical provision of various communication services with

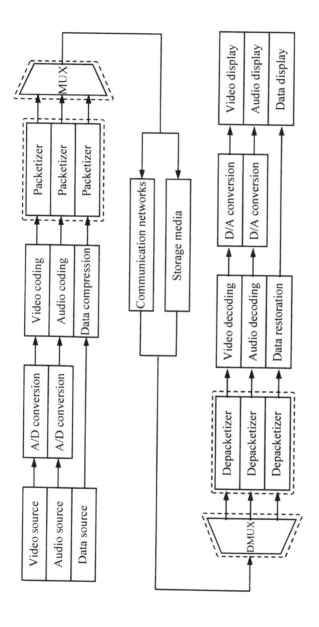

Figure 6.1 Overall flow of information for video services.

different characteristics, a communication network is required that can accommodate diverse forms of information media and at the same time maintain a unified access structure and transmission mode. From this viewpoint, it is difficult to provide a broad spectrum of video communication services through the existing PSTN or the NISDN. An alternative approach is the BISDN based on ATM. In the BISDN, the ATM transmission technique is adopted, which divides the information to be transported into ATM cells of a fixed size. The ATM transmission technique offers several advantages from various standpoints, as it can accommodate diverse types of services with a single type of network.

Terminal technology has a close relationship with the attributes of the services to be provided. For example, in distributive broadcasting services, the video encoder is far more complex than the video decoder. However, in conversational services the encoder and the decoder must coexist inside the same device; hence, for an economical implementation, the two parts should have about the same degree of complexity. The worldwide development trend in terminals is to adopt common coding schemes for different use among different service terminals.

In this introductory section, we introduce video services, and video compression and transmission technologies are described along with the related standards. Detailed discussions, however, follow in Section 6.2 for video services, in Sections 6.3, 6.4, and 6.6 for video compression and related subjects, and in Section 6.5 for video transmission.

6.1.1 Broadband Video Services

Most broadband services are associated with video services. To paraphrase, communication services take broadband characteristics when the video factor gets involved. Broadband video services include the basic services such as video telephony, video response service [i.e., *video-on-demand* (VOD)], video conferencing, broadcasting services (i.e., CATV, HDTV), transaction services, and so on, and the applied services such as *news-on-demand* (NOD), *near VOD* (NVOD), teleshopping, home banking, telegame, telework [i.e., computer-supported cooperative work (CSCW)], telemedicine, distance learning, digital library service, and so on. Aside from the video services listed, which are all provided through communications networks, there are also video services for storage media, whose typical examples are the CD-ROM and *digital video disk* (DVD).

The video services can be classified according to the required bandwidth, service characteristics, bit pattern, time relation, connection relation, and combinations of these. In terms of bandwidth, video services can be classified into narrowband services and broadband services; in terms of service characteristics, into distributive services and interactive services, which can be further subclassified into conversational services, message services, and retrieval services; in terms of bit pattern, into CBR services and VBR services; in terms of time relation, into real-time services and non-real-time services; and in terms of connection relation, into CO services and CL services. The last three classifications can be combined to yield the classes A, B, C, D for use in AAL classification (refer to Sections 4.1.1 and 4.7.1).

As a variation of the classification based on the service characteristics, video services may be categorized into *conversational services*, *on-demand services*, and *distribution services* for practical purposes, depending on the degree of interaction. Then the videophony, video conferencing, transaction, telework, distance learning, and telemedicine belong to the conversational services; the VOD, NOD, teleshopping, home banking, telegame, and digital library services belong to the on-demand services; and broadcasting and NVOD services belong to the distribution services. Conversational services and on-demand services may both be regarded as interactive services, with the information flow balanced for the former and unbalanced for the latter. On the other hand, on-demand services may be regarded as distribution services with a strong user controllability.

Each of the preceding broadband services is defined in Section 6.2, with the most representative services such as video conferencing and VOD services described in great detail.

6.1.2 Video Compression Technologies and Standards

All the video services are common in that they consist of tremendous amounts of information. For cost-effective transmission or storage, therefore, video data must be compressed within such a range that the effect of compression is imperceptible. As such, research in the area video communications and storage has concentrated on video compression during the past few decades.

Among the huge amount of data for video services, a significant amount of redundancy is also included. Because a video signal is formed by scanning several hundred lines (e.g., 525 lines for the NTSC TV signal) per picture (or frame), and 24 to 30 frames (e.g., 30 frames for the NTSC TV signal) per second, it is possible to scan the same information repeatedly. Further, since most changes within a frame appear in a gradual manner, a considerable amount of redundancy exists even within the same frame. Such redundancies in the temporal and spatial domain are targeted for elimination by introducing compression techniques.

The video compression (or video coding) techniques that can remove the spatial-domain redundancy are several, including *differential pulse code modulation* (DPCM), transform coding, and subband coding. The video compression technique that can dispose of the temporal domain redundancy is the motion estimation/compensation method. The former technique is called *intraframe coding* and the latter is called *interframe coding*. In general video compression, these intraframe and interframe coding techniques are used in conjunction and, depending on the applications, supplementary coding techniques such as *run-length coding* (RLC) and *variable-length coding* (VLC) are additionally applied in the end. Section 6.3 provides a detailed description of various intraframe and interframe coding techniques.

The video compression techniques mentioned here have been researched and standardized during the past decades. The most representative video coding standards effective to date are JPEG, H.261, MPEG-1, MPEG-2, and digital HDTV standards. JPEG is the standard for the compression of still images developed by the *Joint Photographic Experts Group* (JPEG) of the ITU-T, the IEC, and the ISO in the late 1980s. H.261 refers to the ITU-T Recommendation H.261 standard, which was developed about the same time for the compression of moving images. The target output rate of the H.261 compression is $p \times 64$ Kbps for $p = 1$–30, whose intended application was NISDN video services such as narrowband videotelephony and video conferencing.

MPEG-1 is the 1.5- Mbps video compression standard developed in 1990 by the *Moving Picture Experts Group* (MPEG) of the ISO and the IEC for use in digital storage devices. The MPEG-2 standard was standardized in 1994 by the same experts group, and its intended application encompasses all video signals up to the HDTV signals. MPEG-2 is a family of standards composed of the system video and audio standards. MPEG-2 is a flexible and scaleable standard that can be employed for all video applications, including digital broadcasting services by satellite, HDTV, DVD, VOD, and other services. In fact, the digital HDTV video coding standards are special cases of MPEG-2. For example, *advanced television* (ATV) of the United States uses the Main Profile at High Level (MP @ HL) MPEG-2 standard, and the digital video broadcasting (DVB) video coding standard of Europe uses the Spatially Scaleable Profile at High 1440 Level (SSP @ H1440) MPEG-2 standard.

MPEG-2 video compression and system standards are described in detail in Section 6.4, and the status and trends of HDTV standardization are discussed in Section 6.6.

6.1.3   Video Transmission

The characteristics of the video services discussed so far range from one extreme to another—transmission channels from a few kilobits per second to more than several hundred megabits per second, and service durations ranging from a few seconds to several hours, depending on the service type. Therefore, it is difficult to accommodate all of the various needs by either the circuit mode alone or the packet mode alone. Even conversational services demand real-time processing capability. To resolve such problems all together, it is necessary to enhance service efficiency through the employment of ATM technology.

ATM networks are packet networks in which the ATM cells, fixed-size packets, convey information. In ATM networks compressed signals are mapped into ATM cells and the relevant headers are attached for transmission. Virtual channels are established to guide the ATM cells to their destinations but they do not exclusively occupy true channels. So delays are not avoidable due to queuing within the networks and also due to the processing required for packetizing and depacketizing the video signals. Further, if congestion occurs in the network, the delay could grow exceedingly large, obstructing the real-time services.

Passing through the multiplexers/demultiplexers and switches in the ATM network, cells can experience jitter and loss, and the degree of jitter and loss depends on the traffic condition and the performance quality of the transmission media. Cell losses, in general, occur either due to limited buffer space in multiplexers/demultiplexers and switches or due to bit errors that occurred in the address field of the header.

As discussed so far, there are three important problems in video communication over ATM networks: time delay, cell jitter, and cell loss. ATM networks, therefore, must be designed, operated, and maintained in such a way that the effect of these three critical problems can be minimized to the levels that are acceptable for conversational services. Unless effective resolutions or compensations to those problems are devised, the merits of the ATM video transmission such as constant image quality, efficient bandwidth usage, and flexible accommodation of versatile services cannot be achieved.

Detailed discussions on the issues of packet video transmission are given in Section 6.5, together with the compensation methods for packet jitter and packet loss.

6.2 BROADBAND VIDEO SERVICES

In terms of the long history of communication services, broadband video services have emerged on the scene only recently as practically realizable services. This emergence owes much to the advances in optical communication supporting devices and image processing technologies, because they enabled the barrier of tremendous information overload of the video services to be overcome. Especially, the video processing technology has contributed in various ways such that the video service signals can be compressed and flexibly manipulated to match the desired service applications. Consequently, versatile video services have been designed, implemented, and practiced recently, and conventional analog-type video services have been converted digitally.

Among the various ways of classifying video services discussed in Sections 4.1.1 and 4.7.1, the simplified classification based on the service characteristics as introduced in Section 6.1.1 is most practical. So, according to this classification, video services are described here in the context of conversational services, on-demand services, and distribution services. Various video services belonging to the three classes are defined first, and then the most representative conversational and on-demand services, namely, video conferencing and VOD services, are described in detail. Finally, the architectures of video communication service terminals are considered in terms of related standards.

6.2.1 Definition of Video Services

The characteristics and services of each video service can be best represented when it is defined in terms of objective wording. So, this section is dedicated to the definition of the video services of common interest today [DAVIC 1.0].

Conversational Services

Videotelephony is a conversational service involving two users at separate locations with one party optionally being the service provider. The user initiates and controls the conversation, which is supported by real-time, bidirectional exchange of audio, video, and data.

Video conferencing is a general form of conversational service involving one or more people at geographically separate sites, with a multipoint service provider possibly involved. The user announces, establishes, controls, and moderates the conference, which is supported by real-time, bidirectional exchange of audio, video, and data.

Transaction services are used to present information to a user, who then reacts to the information, and these actions then alter information in a database. In this case, at least one owner of the information must be involved in the transaction.

Telework service (or CSCW) is a conversational service involving a user (i.e., teleworker), a service provider, and possibly one or more collaborators and a database. The user establishes a session via the service provider, activates and controls local and distant applications, and communicates with collaborators through audio, video, and data.

Distance learning is a conversational service involving users (i.e., students), instructors, and educational institutions who are not at the same geographic location. This service aims at providing "the virtual classroom" by making classes available regardless of the locations of the instructors and students. Users can schedule classes, browse through classes, join or leave classes, or end the class. During the class, functionalities such as camera control, document display, display control, and student's interaction with instructor must be provided.

Telemedicine service is similar to the combination of the video conferencing and retrieval services. X-ray images may be retrieved by the end user or distributed by the end user to other end users for consultation or further evaluation. In addition, real surgical procedures can be broadcast to end users (or students) or other consultants.

On-Demand Services

VOD is a network-delivered video service that provides the functionality of the home *video cassette recorder* (VCR) at the user's request and control. The user can control the service in terms of the following features: select/cancel, start/stop, pause, fast forward/reverse, scan forward/reverse, setting/resetting memory markers, showing counters, and jumping to different scenes.

NOD is a network-delivered news service involving a user, service providers, and network providers. The user obtains information on news items, which also includes summaries and headlines, in an interactive manner. The user can choose news items to view, the presentation level, and the presentation means (i.e., text only, or also moving video). Seamless integration of navigation and contents are important. NOD is similar to VOD in general, but differs in the following aspects: The target service information (i. e., news) is time varying, time limited, and, in many cases, live, so the database contents need to be updated frequently, with an update notification given to the user. The contents also need to be dynamically managed for easy access, diversified classification, and prioritized storage. NOD may be extended to include the capabilities to collect news from different sources based on user queries, and to generate and use a personal profile to change the presented news items and the way of presentation. This extension leads to the so-called "personalized news" service.

Teleshopping is an application service that allows the user to browse video catalogs or virtual shops to purchase products and services. The user can select items after getting enough information through video, text, audio, and graphics presentations, can learn about the charges and other conditions, and can order the product or service, with the delivery and payment methods determined between the provider and the user.

Home banking is an application service that provides the user with electronic access to offerings available in a bank, which may include retrieving account balances, moving money between accounts, making payment to third parties, applying for loans, and browsing through bank offerings.

Telegame is an application service that allows a user to play a game via a network-delivered service. The user will be presented with a menu of games that is available from their service provider. Once the user selects a game to play, then the game program is loaded into either the *subscriber's terminal unit* (STU) or a game machine located at the service provider, and game play is then started. The user can alter the state of the game, with the change shown in graphic/video stream for the user's viewing, and can play the game to the end or stop in the middle.

Digital library service is a variation of the VOD service with the variation appearing in the offered service formats, which include text, image, video, audio, graphics, and other data. In this context, digital library service may be called a "multimedia-on-demand" service. This service involves the user, service providers (i.e., digital library), and possibly network providers. The user can access via electronic means the digital library, which is a multimedia database storing all bibliographical and other multimedia data in digital formats, possibly in a linked manner, and can search and retrieve the desired information.

Distribution Services

Broadcast is a typical distribution service that provides multiple users with immediate, real-time access to multiple sources of television/radio/data programs. Interactivity for broadcasting can be provided between the user and the local processor of the STU as well as between the user and the service/network/content provider.

NVOD is a specific video broadcast service where the same audiovisual content is broadcast during specific time slots in order to provide the user with a permanent link to the chosen program. It is a distribution service that improves the availability of the audiovisual content without requiring a dedicated point-to-point connection to each user. Basic NVOD covers pure broadcast of video with no real interactivity between the user and the service/network provider. The user merely selects the best suitable start time for the interested program among the titles broadcast continuously, and gets the effect of a pause by choosing the program that started that much later. Intelligent NVOD offers the user a more effective handling of the pause feature, and the intelligence for this improved service may reside in the STU or in the service/network provider.

6.2.2 Representative Video Services

Among the various video services defined in the previous subsection, video conferencing may be selected as the most representative conversational video service, because other major conversational services such as video telephony, distance learning, and telemedicine services can be provided in a similar manner. Likewise, the VOD may be singled out as the most representative on-demand service since other major on-demand services such as NOD, telegame, and digital library services can be provided in almost the same way. Therefore, in the following, video conferencing and VOD services are described in detail.

Video Conferencing

Video conferencing, which in fact includes voice, audio, and other data as well to form a multimedia conference, helps to save time and expense that otherwise might be incurred due to travel and other related arrangements. Video conferencing is a conference that can provide virtual meeting environments to remotely located users through communication means. Video conferencing also helps to overcome the limitations of audio-based teleconferencing through visual effects.

A full-featured video conference system consists of a video subsystem, audio subsystem, control subsystem, transmission subsystem, and other peripheral units as shown in Figure 6.2, which is installed in a dedicated video conference room. The video subsystem consists of cameras, monitors, and a video control unit: Cameras include a close-up camera, wide-view camera, multifunction camera, and document camera; monitors include input video display and output video display; and the video control unit consists of a video switching device and a camera control device. The audio subsystem consists of telephones, microphones, speakers, and an audio control unit, and performs such functions as detection of camera control signal (by way of voice), echo cancellation, equalization, and mixing of several audio signals. The control subsystem consists of a main control device and a *multiprocessor unit* (MPU), and performs the control and OAM functions for the overall system. The signal processor unit compresses video, audio, and data signals, and multiplexes and transmits them through the transmission line. It also performs the reverse processes, and possesses error correction and self-diagnosis capabilities. Peripheral devices include remote control units, electronic blackboard, facsimile, VCR, copier, data terminal, and printer.

The described complete video conference system is useful for formal or official use, but is beyond the scope of average users. So, handy desktop video conferencing systems have been introduced recently for average users. This video conference system is built by furnishing video conferencing functions to regular personal computers. Although currently very limited in its video quality, speed, features and functionality, the desktop video conference system will improve in the future in accordance with the advancement of multimedia processing technology and the escalation of network speed.

The functionalities required in general for a video conference system are as follows:

1. *Support of multiconnection communications*: Video conferencing requires multiple connections for different media, where each connection is individually controllable.
2. *Support of multiparty connection*: Video conferencing requires participation of multiple users, among which information exchange is possible in multipoint-to-multipoint fashion. The type of media used may be different among all participants, and early and late leaving or joining are both allowed.

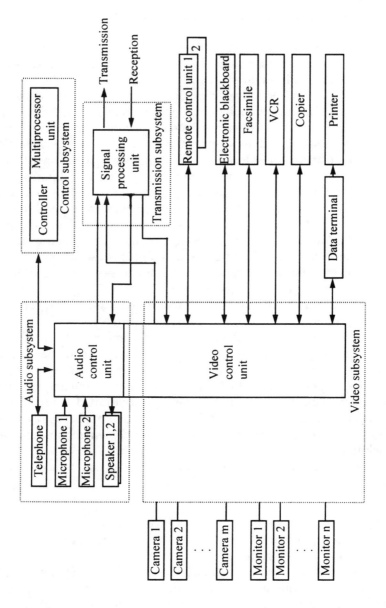

Figure 6.2 Block diagram of a video conferencing system.

3. *Guarantee of intermedia synchronization*: For example, lip synchronization is necessary between the voice and image media.
4. *Control of the right to speak*: The right to speak needs to be controllable whenever necessary, although it may be desirable not to control the right if a natural flow of conversation is desired.
5. *Support of real-time data transmission*: Supporting data or document transmission needs to be done in real time so it does not interfere with the smooth flow of the real-time video conference.

Video conference systems may be controlled through either a centralized or distributed control mechanism, as depicted in Figure 6.3. The centralized control mechanism employs the multipoint server as the control center, which receives information from every participating terminal, carries out various processes such as voice mixing and video switching, and finally returns the processed results to the terminals. This mechanism has the advantage that call structure is simple and network resource utilization is low. However, it has the drawback that the network delay becomes large because all information is transferred via the multipoint server. In contrast, in the distributed control system, call structure becomes complicated and control is intricate, but network delay is small.

Video-on-Demand

VOD service is beneficial to individual users because it can provide various video information together with the accompanying audio and data information on the user's request. The fact that control is on the user's side distinguishes VOD service from usual distribution services, and it may be the key to the success of VOD. VOD service is also called *movie-on-demand* (MOD) service and is similar to *video-dial-tone* (VDT) service.

VOD service may be divided into two classes depending on the allowed level of user's control: One is the *near VOD* (NVOD) service, which is basically a variation of broadcasting service with the variation being that a video program is distributed in multiple sessions, each of which has a different start time. The user can get the effect of pausing the video play but only by consulting the program schedule. The other is the *true VOD* service, which is an interactive service with each session established on a per-user basis. The user has the same degree of control as for his/her own video player. True VOD, however, requires substantially improved server systems and network facility.

The ATM-based network structure for the VOD service is as depicted in Figure 6.4. The VOD service network consists of video server, video archive, service operation center, service gateway, and *subscriber terminal unit* (STU).

The *video server* is the core element of the VOD service that provides the video and other programs upon the user's request. It stores information in

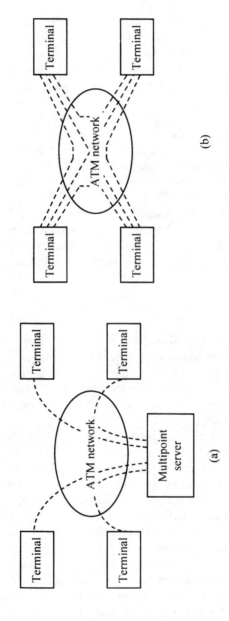

Figure 6.3   Multiparty mechanisms: (a) centralized mechanism and (b) distributed mechanism.

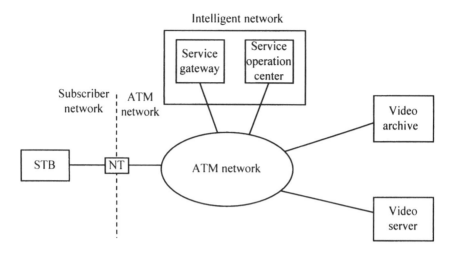

Figure 6.4 VOD service network architecture.

random accessible storage media such as hard disks and semiconductor memories. Since the video signal may be stored in the MPEG format, the video server must be capable of MPEG video decoding. The video server provides diverse functions such as a coding format conversion function for server-STU compatibility, a directory function for information retrieval, a statistical information processing function for various statistical analyses, a service management function for billing, and so forth. The video server consists of digital storage devices, disk controllers, program controllers, server controller, switch, interim memories, and network interfaces, as shown in Figure 6.5(a). The program controllers control the program stored in the storage devices, and the interim memory is to buffer the video data during interim periods.

The *video archive* stores video programs that the video server cannot store due to capacity limit. It stores the video program in compressed format in non-real-time storage media such as videotape and downloads them to the video server whenever necessary.

The *service operator center* enables the VOD service provider to manage its own set of program materials or to manage the user group for billing, subscriptions, and so on. For small networks the service operator center may be integrated with the video server.

The *service gateway* provides its interfaces for the user to locate and set up connections to the desired video servers. For small networks, the service gateway may be integrated with the video server or with the switching machine.

The *STU*, in general, consists of a set-top box and a remote controller, but in some cases a personal computer furnished with an appropriate network interface function and video processing capability is used as the STU. The STU

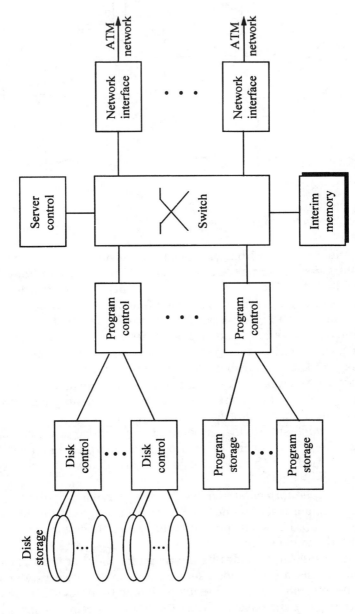

Figure 6.5 Configurations of (a) video server and (b) set-top box.

(b)

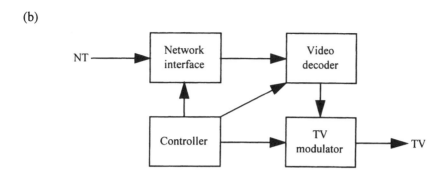

Figure 6.5 (continued)

needs to support versatile service providers, applications, and user interfaces. Internally the STU consists of the network interface, video decoder, TV modulator, and controller, as shown in Figure 6.5(b). The video decoder decodes the compressed video signals, and the TV modulator converts the digital images to analog TV signals.

The overall VOD service network is operated and managed by the operation systems, which communicate with various VOD network elements through standard interfaces and information models, conforming to the TMN standards.

6.2.3 Video Communication Terminal Equipment

To provide video services, the communication terminal equipment needs to have a video processing capability in addition to other audio, data, and signaling functions. If all of these functions are taken into account, the resulting terminal equipment becomes a general-purpose multimedia terminal. Further, if the network-dependent components are taken into consideration as well, the integrated reference model shown in Figure 6.6 results. Therefore, any video terminal equipment, if it is constructed on the basis of this model, can be used as the terminal for conversational services, on-demand services, and other video services.

The functions required for a general-purpose video (or multimedia) terminal equipment, as indicated in the Figure 6.6 terminal reference model, can be divided into two categories: video (or multimedia) processing functions, and communication and control functions. The video processing functions include audio/video input/output, telematic, audio/video codec, and multimedia MUX/ DMUX functions. The communication and control functions include system control, call control, service signal control, end-to-end signaling, AAL, ATM, and physical layer functions. Functional requirements for each of the functions are listed in Table 6.1.

The video codec includes the H.261 codec for conversational video

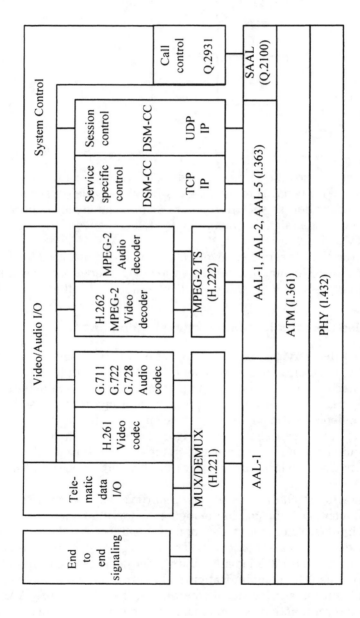

Figure 6.6 Video communication terminal reference model.

Table 6.1
Functional Requirements for Video Communications Terminal Reference Model

Category	Details
Multimedia processing	Video/audio codec
	MPEG-1/2 video decoder, MPEG-1 audio decoder
	H.261 video codec, G.711, G.722, G.728 audio codec
	Multimedia multiplexing
	H.221 MUX/DMUX (conversational video service/message service)
	H.222 DMUX (on-demand distribution/retrieval service)
	Multimedia input/output
	Video, audio I/O
	Telematic I/O
Communication and control	Service signal control
	DSM-CC, TCP/UDP/IP, etc. (on-demand distribution/retrieval service)
	H.242, H.230, H.221 (conversational video service/message service)
	Call control
	Q.2931 (user-network call control)
	AAL processing (I.363)
	AAL-1/2 processing (conversational video communication)
	AAL-2 processing (bidirectional video communication)
	AAL-5/2 processing (on-demand distribution/retrieval/message service)
	ATM layer processing (I.361)
	Physical layer processing (I.432)

communications, the G.711, G.722, and G.728 audio codecs, and the MPEG-1/2 video/audio decoders for on-demand and distribution services. The multimedia multiplexing function includes H.221 MUX/DMUX for conversational services, and H.222 DMUX for on-demand and distribution services. In terms of media (or multimedia) data flows, H.261 video codec, G.711, G.722, and G.728 audio codec, and H.221 MUX/DMUX are for *conversational media flow* (CMF), and MPEG-1/2 video/audio codec and H.222 DMUX are for *distributional media flow* (DMF).

The service signal flows can be divided into two types in a similar manner: One is *conversational service-signal flow* (CSF) for conversational services and the other is *distributional service-signal flow* (DSF) for on-demand distribution services. The CSF possesses the H.242, H.230, and H.221 protocol processing capabilities as prescribed in H.321, and is multiplexed by the H.221 MUX/DMUX, together with other multimedia data. The DSF forms an independent flow separately from multimedia data, and for this it employs the DSM-CC (see

Section 6.4.3), TCP/IP (see Section 5.3), and other protocols as regulated by ISO/IEC and DAVIC. Session control flow, which is used mainly for on-demand distribution services, uses the DSM-CC and UDP/IP protocols as regulated by ISO/IEC and DAVIC.

Call control uses the Q.2931 protocol as recommended by ITU-T and DAVIC, and is transported over the *signaling AAL* (SAAL). The Q.2931 and SAAL protocols are described in Section 4.8.

The AAL function includes AAL-1, AAL-2, and AAL-5 functions, as recommended in ITU-T Recommendation I.363, among which AAL-1/2 are for conversational services and AAL-5/2 are for on-demand, distribution, retrieval, and message services (AAL-2 standard is yet to be defined). The ATM layer function and the physical layer function, respectively, follow the ITU-T I.361 and I.432 recommendations. As the physical layer transport technology, SDH/ SONET, FTTC, HFC, and ADSL technologies can be used.

There are two representative communication terminal models that fall within the scope of the preceding terminal reference model. They are the H.321 and H.310 terminals, which are both recommended by ITU-T for use in the BISDN environment. The two differ in that H.321 is for adaptation of NISDN video terminals in the BISDN environment, whereas H.310 was designed for BISDN services. In the following the functions of the two terminals are briefly considered.

The H.321 terminal takes the video/audio codec, multimedia multiplexing, multimedia input/output, and service signal control functions of the H.320 NISDN terminal, and merely modifies the communication interface part to match the BISDN environment. The functional architecture of the H.321 terminal is as shown in Figure 6.7. It can easily be recognized that the H.321 terminal is an implementation of the conversational service related part (i.e., the left half) of the Figure 6.6 terminal reference model.

The H.310 terminal determines a unidirectional terminal (i.e., receive only or transmit only) or a bidirectional terminal, and can be used for conversational, on-demand, distribution, and message services. The functional architecture of the H.310 terminal is shown in Figure 6.8, where it can be seen that the H.310 is a mixed implementation of the Figure 6.6 reference model. Depending on the option of functions, either a conversational service terminal (by selecting H.261, H.221, etc.) or an on-demand distribution terminal (by selecting H.262, H.222, etc.) or other service-specific terminals can be realized from it.

6.3 VIDEO COMPRESSION TECHNIQUES

The dramatic advances in signal processing and VLSI technologies during the past decade have brought about significant progress in the development of compression technology for video signals at various transmission speeds. So video coders that were once regarded as technically impossible or economically

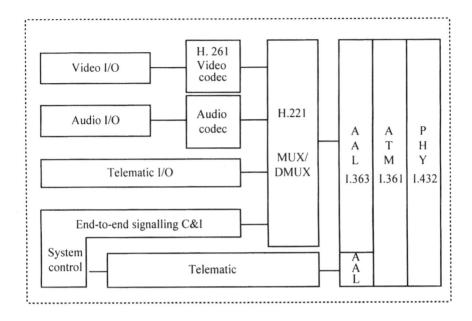

Figure 6.7 Functional architecture of H.321 terminal.

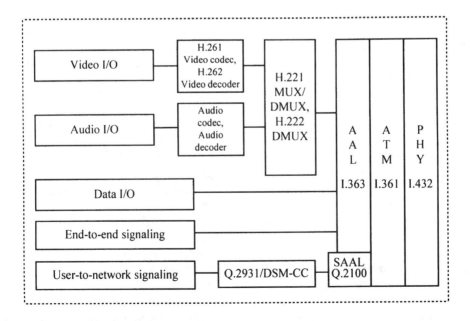

Figure 6.8 Functional architecture of H.310 terminal.

unfeasible in the past have emerged into practical use. Such progress in video compression techniques has enabled the video communication services considered in the previous section to be realized. In fact, those video services require the transmission of tremendous quantities of information, so for cost-effective transmission, video data must be compressed within a range such that the effect of compression remains imperceptible.

Video, or moving image, information is provided to viewers in a series of still images or frames, and the effect of movement is achieved through small, continuous changes in the frames. Since images are provided at the speed of 25 or 30 frames/s in the case of television and 24 frames/s in the case of movie film, continuous changes between frames appear as natural moving images to the human eye. Video signals are composed of spatial- and temporal-domain information. Spatial-domain information is provided in each frame, and temporal-domain information is provided by images that change with time (i.e., by differences between frames). Because changes between neighboring frames are minute, objects appear to move smoothly.

In digital video systems, each frame is sampled in units of *pixels*, or picture elements. Sample values for pixel luminance are quantized with eight bits per pixel in the case of *black and white* (BW) images. In the case of color images, each pixel maintains the associated color information also; for instance, the three pieces of luminance information designated as *red, green, blue* (RGB) are quantized to eight bits, respectively. Video information composed in this way possesses a tremendous amount of information; hence, for transmission or storage, the image compression (or coding) technique is required, which eliminates redundant information, mainly in the spatial and temporal domains.

In general, redundancies in the spatial domain are a result of the small differences between neighboring pixels of a given frame, and those in the temporal domain are due to the minute differences in contiguous frames caused by the movement of an object. The method of eliminating redundancies in the spatial domain is called *intraframe coding*, which can be divided into DPCM, transform coding, and subband coding. On the other hand, time-domain redundancies can be eliminated by the *interframe coding* method, which also includes the motion estimation/compensation method, which compensates for motion through estimation. In intraframe coding and interframe coding, run-length coding and variable-length coding are also used, and they exploit the statistical characteristics of data to further compress data without any loss of important information. For visual communication very high compression ratios, in the range of several tens to one reduction, are required. To meet this high-tier requirement, hybrid coding methods that combine multiple compression techniques are usually employed.

In this section various image compression techniques are explained. Spatial compression through intraframe coding is considered first, followed by temporal compression through interframe coding. Then, various intermediate image formats are examined. A whole section, Section 6.4, is dedicated to the MPEG system, which is an integrated embodiment of the above image processing technologies.

6.3.1 Intraframe Coding

Intraframe coding uses only the spatial information existing in each video frame. Since this type of coding does not use any temporal information, it can also be used for still-image coding, where real-time implementation does not matter. Intraframe coding of video signals can be realized rather simply and does not require memory that stores preframes or postframes. In general, the intraframe method can be categorized into three types: predictive coding, transform coding, and subband coding. Since each method has its own set of merits, a combination of two or more methods is often used. Each of these three intraframe coding techniques is discussed next.

Predictive Coding

Predictive coding is one of the oldest image compression techniques and is based on the fact that prediction errors are very small when the present pixel is predicted from neighboring pixels. The DPCM technique encodes the quantized value of the difference between present pixel value and predicted value (i.e., prediction error). The use of a great number of neighboring pixels for prediction can decrease prediction error and raise performance. But because the merits of using a large number of pixels do not outweigh the accompanying complexity, the number of neighboring pixels used for prediction is generally not more than four. Figure 6.9 depicts the DPCM technique.

Image degradations in prediction coding are due to granular noise, slope overload, and edge business. Granular noise results when the quantization step size is too big, whereas slope overload results when the quantization step size is too small. Granular noise and slope overload cause noise and degradations at the boundaries of the restored image. Edge business results when an image signal is displayed continuously in time, since pixels at the boundary of the object are quantized differently in neighboring frames. Therefore, a technique that gives a good-quality image for still images does not necessarily provide good-quality image for moving images.

To alleviate these degradations, quantizers can be contrived to reflect human visual characteristics, a noise reduction filter can be applied, and different adaptive prediction and quantization schemes can be used according to image contents. For example, the boundaries of objects can be treated differently from the flat parts. Much research has been based on this concept, but most of it is not very attractive because the hardware complexity increases too much compared to the corresponding improvement in the image quality. From the

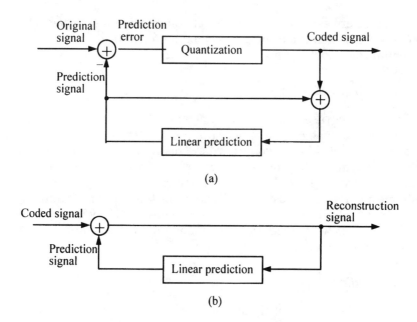

Figure 6.9   Prediction coding of DPCM: (a) encoder and (b) decoder.

performance standpoint, the predictive coding scheme, though it can be easily implemented, is not as effective as transform coding in reducing redundant information. Transform coding has a better compression ratio than predictive coding.

Transform Coding

As a result of extensive research in the past decades, transform coding has been employed in the world standards such as JPEG, H.261, and MPEG for still-image and moving image compression. The basic concept of transform coding is to obtain a high compression ratio by eliminating redundancies through orthogonal transforms.

Under the assumption that statistical characteristics of the image data are stationary, the *Karhunen-Loeve transform* (KLT) is known to be the best transform from the standpoint of mean square error. But due to the fact that basis functions must be sent to the decoder because basis functions of KLT are data dependent, and due to the lack of high-speed computation algorithms, it is impractical to employ KLT in real-time applications. Therefore, a fixed-basis orthogonal transform, which can be easily implemented while maintaining a similar level of performance, is demanded. In processing image signals, which are in general close to the first-order Markov signals whose correlation between adjacent pixels is very high (higher than 0.95), DCT is the right alternative that provides a comparable performance. In fact, DCT manifests good performance even when no assumption has been made about the statistical characteristics of the image data. DCT performs transforms in real numbers and fast computing algorithms are already in existence. Moreover, since VLSI-implemented products that operate in a wide range of transmission speeds are available, DCT is widely used for various image compression applications. The basic principle of DCT transform coding is depicted in Figure 6.10.

Input image is divided into blocks of $N \times N$ pixels in DCT, and block size is chosen by considering the requisite compression efficiency and picture quality. In general, the bigger the block size the greater the compression ratio, since more pixels are used to reduce redundancies. But when block size is too large design complexity increases, and the complexity overshadows the compression gain when the block outgrows the 8×8 size. According to experimental results, the 8×8 block size is known to be the most effective. After dividing an image into blocks, DCT is applied to each block.

The two-dimensional DCT and inverse DCT are defined as follows:

$$c_{u,\,v} = \frac{1}{4}\, e(u)e(v) \sum_{i=0}^{N-1}\sum_{j=0}^{N-1} b_{i,\,j} \cos\!\left(\frac{\pi u(2i+1)}{16}\right) \cos\!\left(\frac{\pi v(2j+1)}{16}\right) \quad (6.1)$$

$$b_{i,\,j} = \frac{1}{4}\sum_{u=0}^{N-1}\sum_{v=0}^{N-1} e(u)e(v)c_{u,\,v} \cos\!\left(\frac{\pi u(2i+1)}{16}\right) \cos\!\left(\frac{\pi v(2j+1)}{16}\right) \quad (6.2)$$

In the equations, $b_{i,\,j}$ is the $(i,\,j)$th pixel of each block, and $c_{u,\,v}$ is the transform coefficient corresponding to each frequency. Weighting factor $e(u)$ is $1/\sqrt{2}$ when $u = 0$, and 1 otherwise. The coefficient $c_{0,0}$ which designates the mean value of pixels of a specified block, is sometimes called the *DC* (or constant) *component*.

In this manner, the $b_{i,j}$'s are first transformed into $c_{u,v}$'s and then compressed through the coding steps shown in Figure 6.11. Transform coefficients $c_{u,v}$'s are quantized after thresholding to create as many 0 coefficient values as possible within the range for which degradations in picture quality do not occur, since with more 0's there are more chances for a greater compression ratio. To guarantee continuity between mean values of different blocks, DC components are excluded from thresholding and the values are quantized by a finer step size. Last, coefficients arranged in two dimensions are rearranged into coefficients arranged in one dimension using zigzag scanning, and then run-length coding is applied to both nonzero coefficients and the length of zero runs. Since long sequences of 0s occur when zigzag scanning is employed to change coefficients arranged in two dimensions to one dimension, the coding efficiency is

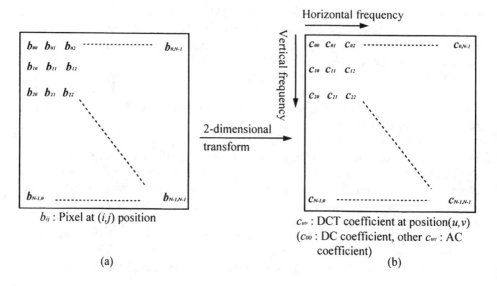

Figure 6.10   DCT block transform: (a) *N* × *N* original pixel block and (b) *N* × *N* DCT transform coefficient block.

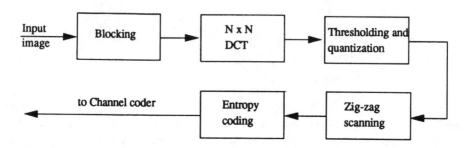

Figure 6.11   Block diagram of basic DCT transform coder.

improved correspondingly. Nonzero coefficients and zero-run lengths are coded using a variable-length code book designed on the basis of data statistics.

Image degradations due to the DCT coding scheme result from the ringing effects of blocks, including abrupt boundaries and traces of block boundaries in flat regions. These visual degradations can be reduced using a DCT coefficient quantizer that is based on human visual characteristics. In this case, a larger quantizer step size is used for those DCT high-frequency coefficients that are not susceptible to human eyes, and a smaller step size for the low-frequency coefficients that are susceptible to human eyes.

The thresholding processing and quantization in the basic DCT coding algorithm can apply the same technique to all the blocks regardless of the contents of the image data, and fine thresholding and quantization step size are required for restoration of high-quality pictures. But then the requisite bit rate increases too much for complex images; hence, an adaptive coding scheme should be considered for a trade-off between quality and bit rate. Some of the DCT coding techniques categorize blocks into several characteristic models according to block characteristics and handle them according to the properties of each model.

As described earlier, transform coding uses the orthogonal transform to reduce redundancies by eliminating data correlations, and its performance improves when there are fewer activities in a block. Using this fact, an adaptation method can be derived that divides image data into blocks of variable sizes according to the degree of data activities. That is, after dividing data into blocks of appropriate sizes, DCT coding can be applied to each block. Such an adaptive scheme divides highly active parts into very small blocks or low-activity parts into large blocks, and is thus suitable for the coding of text, drawings, and graphic images.

Subband Coding

Although the basic concept of subband coding is simple, progress in subband coding for image compression has only been achieved in recent decades. Subband coding is composed of two major steps. The first step is the subband filtering step, which splits an image signal into its constituent frequency components, and the second step is the coding step, which compresses each frequency band according to its respective characteristics. The two subband coding steps are separately discussed next.

Subband coding is accompanied by an analysis filter bank at the encoder and a synthesis filter bank at the decoder, respectively. The analysis filter bank splits the input into several different bands using a different sampling rate for each band. In contrast, the synthesis filter bank combines several band signals of different rates to synthesize the desired signal. Subband coding requires less processing time for each band, but requires many processors, say, one for each band. A simple example of QMF bank for subband coding of one-dimensional signals is shown in Figure 6.12.

After decomposing an image into several bands using the analysis filter bank, a different coding scheme can be applied to each band that is most appropriate for the given band. Since data characteristics of each band vary widely and human visual sensitivity to degradation also varies from band to band, better performance is obtained when each of the bands is processed according to its own set of characteristics.

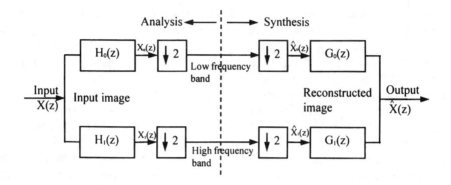

Figure 6.12   Block diagram of QMF bank for subband coding of one-dimensional signals.

The intraframe subband DCT coding scheme is currently widely used for subband coding, and is briefly described here.

As shown in Figure 6.13, each frame can be decomposed into four bands (LL, LH, HL, HH) by applying analysis filtering in the horizontal direction and then in the vertical direction.

The LL band includes most of the important data except the edges and boundaries; hence, it is necessary to minimize losses associated with coding of this particular band. Therefore, DCT coding is widely used for the coding of the LL band.

High-frequency bands (LH, HL, HH) mainly contain information about the edges of objects, backgrounds, and the boundaries, and pixel values are generally smaller than those of the LL band, so total information contained in these bands is in most cases smaller than that of the LL band. Also, human eyes are not sensitive to small changes in pixel values of the three high-frequency bands. Therefore, a simple nonuniform quantizer with a dead zone can be used that converts small pixel values to zero without any significantly perceptible degradation. A high compression ratio can be achieved by additionally applying the run-length coding to nonzero values and zero-run lengths. Since the quantizer converts many of the pixel values in the dead zone into zeros, longer lengths of zero runs are obtained.

Since sample rates decrease (e.g., to a fourth in the case of the four-subband coding mentioned earlier) after the band splitting, the subband coding technique is widely used for high-speed coding processing such as HDTV, which is difficult to perform before the band splitting procedure. Another effective application is packet image transmission, which exploits the fact that the importance of image information varies with each band (LL > HL, LH > HH). This is discussed in detail in Section 6.5.

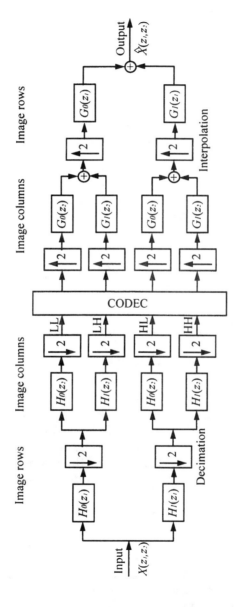

Figure 6.13 Block diagram for intraframe coding that combines subband analysis and transform coding.

6.3.2 Interframe Coding

As mentioned previously, numerous information redundancies exist between continuous image frames; hence, most information on the present frame can be determined from previous frames. For example, in most cases there is a great possibility that the same objects could occur among continuous frames, and if only the information related to motion is known, then data associated with the object can be coded logically in a single step. This concept also applies to backgrounds, and in order to achieve further compression of image information, such time axis redundancies should be eliminated as well. In general, the fast moving portion in a frame, even in TV programs or movies, is less than 5% of the frame; hence, motion estimation is the key to minimizing redundancies in time.

Figure 6.14 depicts the configuration of a general interframe coder employing motion estimation and compensation. This basic configuration consists of two stages: the first for performing motion estimation and compensation, and the second for compression. The motion of an object is estimated by calculating relative displacement between the previous frame and its corresponding image data, generally in units of blocks. The difference between present data and motion-compensated past data is coded to be compressed. Motion compensation is used to reduce time redundancies effectively, and is quite similar to the predictive coding mentioned earlier, which predicts the present pixel from neighboring pixels of the given frame. In the following some of the most frequently used motion estimation methods are discussed.

Motion Estimation

The motion estimation method consists of the pel (which is another acronym for picture element) recursive algorithm, which estimates pel-to-pel motion recursively, the *block matching algorithm* (BMA), which estimates block-to-block motion, and the block recursive matching algorithm, which is a mixture of the first two. In general, extensive computation time is required for motion estimation, so the block matching algorithm is widely used, since it is implementable in real time.

BMA estimates motion on a block basis. Since all pixels in a block are assumed to move in one direction in this algorithm, computation and the associated hardware are simple. The operation of BMA is illustrated in Figure 6.15. Each frame is first divided into $N \times N$ size blocks, and motion is estimated between the present frame and the previous frame. The reference for motion estimation can be normal mean square error or absolute difference error, and the block with the minimum error is picked for the decision of motion vector. The searching area of the previous frame is prespecified, so motion estimation is done on all the blocks within this searching area.

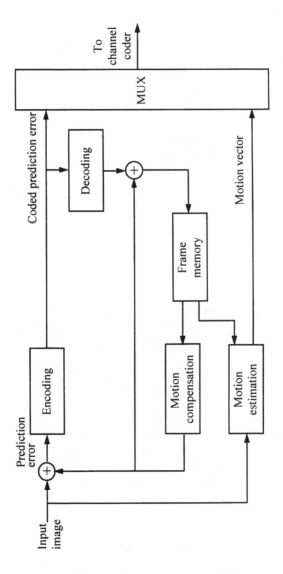

Figure 6.14 Interframe encoder employing motion estimation and compensation.

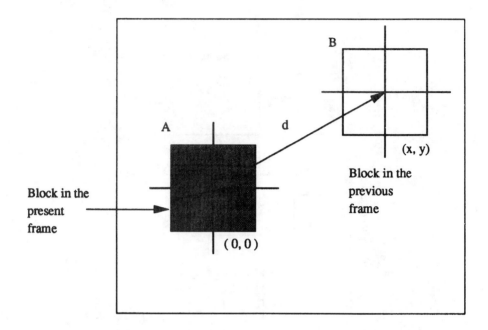

Figure 6.15 Illustration of motion vector estimation.

Since motion estimation in BMA is simple, it has already been implemented on a real-time-processing VLSI chip and is actually in use for image coders. A VLSI architecture with adjustable block size and searching area has also been developed.

Compression Coding After Motion Estimation

The purpose of motion estimation is to estimate present image data (or block) from previous or neighboring frames in order to reduce time redundancies. The most widely used technique is the motion compensation predictive coding scheme. In this scheme, prediction error, which is the difference between the present block and the motion compensated block (from previous frame), is coded. Through precise prediction of the present block from the previous frame, it can reduce prediction error and thus raise the compression ratio. This motion-compensated prediction coding scheme can be regarded as a method of the DPCM type that reduces time redundancies.

In general, the performance of the motion-compensated coding scheme relies on several factors. They are the maximum size of motion displacement (or the size of motion search area), the precision of the motion compensation method for estimating motion, and the adaptability of motion estimation to variations in time and spatial resolution with different buffer control schemes.

The simplest scheme for coding prediction error between the block estimated through motion estimation is the direct quantization method.

For interframe coding in general, most of the coding schemes described earlier can be adopted for the interframe coding of prediction errors. Among the schemes, DCT coding is the most widely employed in international standards such as H.261 and MPEG. Figure 6.16 shows the block diagram of the H.261 motion-compensated DCT predictive coder and decoder.

To further improve performance, human visual characteristics can be incorporated into the encoder design. Performance can also be enhanced by employing the aforementioned adaptive coding schemes.

Motion-compensated prediction coding can be used in combination with subband coding. In this case, each frame is band-split first, then the motion-compensated prediction coding scheme is applied to each of the bands. This scheme is suitable for packet image coding provided by the ATM network (refer to Section 6.5 for packet video).

6.3.3 Standard Intermediate Image Formats

Standardization of communication procedures, signal formats, and coding schemes is very important in order to exchange image information precisely and effectively. However, the television, which has been the most popular means for image distribution so far, has not yet achieved worldwide standardization and has different signal formats (e.g., NTSC, PAL, and SECAM) in various regions around the world. HDTV has experienced similar hardship in international standardization. However, worldwide standardization has been realized for video phone and video conference systems whose representative example is ITU-T's H.261. Trends in worldwide standardization of image processing in general can be summarized as follows.

Image processing consists of three steps: first, the preprocessing step, which preprocesses input signals from various video sources; second, the step for converting the image into a standard intermediate format; and third, the compression step. The compressed images are transmitted through digital transmission line and the receiver receives them and reconverts them into a standard intermediate format and displays the reconstructed images after the postprocessing step. Figure 6.17 is a depiction of the three stages of image processing.

The image signal source can be in various forms, such as video phone/conference, TV signals (NTSC, PAL, SECAM), HDTV signals (1,050-1,125 scanning lines, etc.), VTR tape signal (VHS, S-VHS), video films, and so on. Also, format and the resolution of each signal source and the bandwidth required for transmission have different characteristics, even within the same application. In the past, attempts have been made to unify these sources with different characteristics into a single standard video source according to each application;

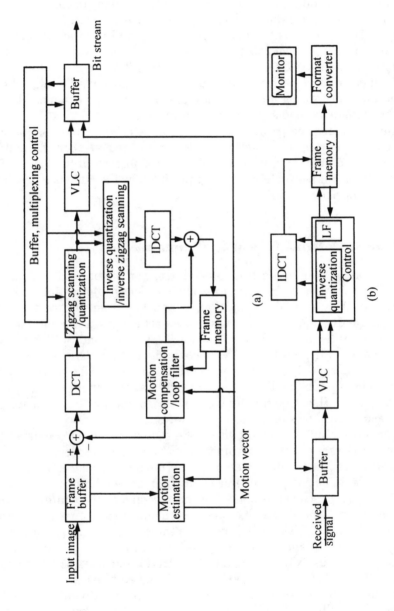

Figure 6.16 Block diagram of H.261 video codec: (a) video coder and (b) video decoder.

but now standardization of image signals within the same application has been virtually given up because of sharp conflicts among companies and among countries. Instead, the trend is toward providing compatibility among different image sources by creating common intermediate formats according to several of the resolution requirements through a preprocessing procedure.

The common formats are digital image formats created from different signal sources through preprocessing that enable communication between different equipment. The common formats have a hierarchical structure according to resolution, as shown in Figure 6.18. The minimum common format is *quarter CIF* (QCIF) with 176 × 120 pixels, which is the image format of video phone. *Common intermediate format* (CIF), which is four times the size of QCIF, is used for multimedia services that employ video phone and video conference. As indicated in Table 6.2, CIF has been standardized by ITU-T H.261, and the *source input format* (SIF) by ISO's MPEG-1. The ITU-R 601 format (704 × 480) of MPEG-2 Main Level, which is currently a digital format for analog CATV and TV signals, is four times the size of SIF and will be employed for the digital transmission of TV signals, digital VTR, high-quality video phone/conference, and multimedia services.

The all-digital HDTV system compresses the HDTV signal composed of video, audio, graphics, and data to a digital signal of about 20 Mbps, and then modulates it to a 6-MHz analog signal using a *vestigial sideband* (VSB) modulation scheme such as 8-VSB. For the HDTV signals, the SMPTE 274M format is used, and for its video coding MPEG-2 High Level is used (refer to Section 6.6 for HDTV).

Figure 6.17 Three stages of image preprocessing.

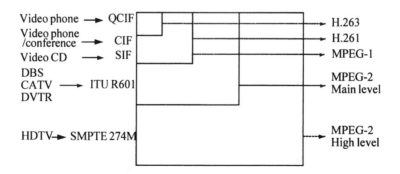

Figure 6.18 Hierarchical common intermediate format.

Table 6.2
Standardization of Image Data Compression

Organization		Application Area	Input Format (525 TV line)	Target Transmission Bit Rate	Remarks
ITU-T	H.261	Video phone/ teleconference	QCIF (176 × 120) CIF (352 × 240)	$p \times 64$ Kbps $p = 1, \ldots, 30$	
ISO/ITU-T	JPEG	Still image	Resolution $< 2^{16}$ (horizontal and vertical)	0.25–2.0 bit/pixel	
ISO	MPEG-1	Digital storage media (CD-ROM)	SIF (352 × 240, or 352 × 288)	1–1.5 Mbps	VHS
	MPEG-2 (Main Level)	Digital storage media (CD ROM DVTR, digital broadcasting)	ITU-T (704 × 480, or 704 × 576)	4–9 Mbps	S-VHS
FCC (U.S.)	ACATS	HDTV broadcasting (6-MHz channel)	Progressive (1280 × 720) or interlaced (1920 × 1080)	15–20 Mbps	

When the same video compression technique is applied among different image signals by adopting the common intermediate format, essential components of codecs can be maximally shared among different applications through a standardization of compression schemes. Then the mass production of video codec components becomes possible, which can reduce codec prices dramatically.

6.4 MPEG-2 VIDEO TECHNOLOGY

In the late 1980s, two international standards for image data compression were established. One is H.261, which is the video compression standard for video phone and video conferencing applications via ISDN at the bit rate of $p \times 64$ Kbps for $p = 1 - 30$. The other is JPEG, which is the still-image compression standard for use in computers, color facsimile, electronic still image cameras, and so on. After completion of these two standards, ISO and IEC collaborated to establish other moving picture compression standards for versatile applications. MPEG-1 (ISO/IEC 11172) is the result designed for storing moving pictures at 1.5 Mbps maximum on digital storage media such as compact disks.

MPEG-2 (ISO/IEC 13818) is the descendent of MPEG-1 and its standardization began in 1990 when the technical specification of MPEG-1 was finalized. The motivation of MPEG-2 was to establish a higher performance compression standard operating at higher bit rates. In contrast to MPEG-1, whose quality is approximately the same as that of a VHS tape due to the rather limited bit rate of 1.5 Mbps, MPEG-2 provides up to HDTV quality depending on the available bit rate. Initially, the committee proposed a two-stage plan: the first stage, called "MPEG-2" for realizing television quality at 5 to 10 Mbps, and the second stage called "MPEG-3" for HDTV applications. But the activity of the United States toward HDTV standardization at that time influenced the MPEG committee to merge MPEG-3 into MPEG-2, in 1992, so that MPEG-2 can be widely accepted in industry for versatile applications including HDTV.

Starting from 30 proposals submitted for evaluation, a test model was built first, and then the model was updated over and over again, incorporating new tools and improving existing ones. The technical specification was finalized in November 1993 at a meeting in Seoul, Korea, and it was approved as the international standard in November 1994 at a Singapore standards meeting. The final product is the MPEG-2 (ISO/IEC 13818) standard, which consists of three parts: system standard (13818–1), video standard (13818–2, also later adopted by ITU-T as the H.262 standard), and audio standard (13818–3). These three parts are depicted in Figure 6.19.

Due to its flexibility and outstanding compression performance, MPEG-2 has been adopted in various applications including *direct broadcasting by satellite (DBS)*, HDTV, *digital video disk* (DVD), VOD, and others (see Table 6.3).

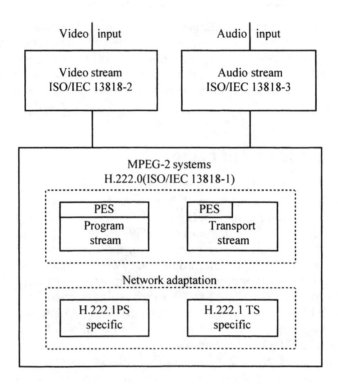

Figure 6.19 MPEG-2 Standards.

In this section, the MPEG-2 standard is examined in accordance with the image processing techniques described in Section 6.3. The features of the MPEG-2 video are discussed first, followed by a detailed description on the MPEG-2 video compression algorithm. Finally the MPEG-2 system standard is investigated.

6.4.1 Features of MPEG-2 Video

The MPEG-2 video standard is, in some sense, a full integration of existing image compression techniques that have been developed during the past three decades. It is a generic coding algorithm that can be adapted to all visual communication applications, even including HDTV distribution. For this reason, many distinctive features are incorporated in the MPEG-2 video standard.

Compatibility

The MPEG-2 video includes the MPEG-1 video syntactically; therefore *forward compatibility* is maintained between the two but backward compatibility is not.

Table 6.3
Application Areas of MPEG-2 Standard

Abbreviation	*Application Area*
CATV	Cable television or community antenna television
CDAD	Cable digital audio distribution
DAB	Digital audio broadcast
DTTB	Digital terrestrial television broadcast
EC	Electronic cinema
ENG	Electronic news gathering
HTT	Home television theater
IPC	Interpersonal communication
ISM	Interactive storage media
NCA	News and current affairs
NDB	Network database service
RVS	Remote video service
SSM	Serial storage media
STV	Satellite television broadcast
TTV	Terrestrial television broadcast

In other words, an MPEG-1 video bit stream can be decoded by an MPEG-2 video decoder but the opposite does not hold. This contrasts with the MPEG audio case where both forward and backward compatibilities are maintained.

Bit Rate

Major applications of the MPEG-2 video include digital compression of current and high-definition television signals. The current television quality is achieved at 3 to 9 Mbps, whereas HDTV quality is 17 to 30 Mbps. The bit rate is decided depending on channel capacity and required picture quality. For instance, 4 to 7 Mbps is used in digital DBS, 17 Mbps in U.S. HDTV, and 3 to 4 Mbps in ADSL-based VOD and in DVD.

Improvement over MPEG-1 Video

The MPEG-2 video is a generic video compression algorithm that outperforms the MPEG-1 video in many aspects. For better performance, the MPEG-2 video reexamined and enhanced many compression tools of the MPEG-1 video, and incorporated many new tools as well. The improved or newly added tools include manipulation of interlaced video, versatile motion estimation/compensation, nonuniform quantization of DCT coefficients, alternate scanning of quantized DCT coefficients, separate Huffman coding for intracoded macroblocks, and so on. Table 6.4 summarizes the difference between the MPEG-1 and the MPEG-2 video.

Profiles and Levels

MPEG-2 has a wide range of applications and each application requires a specific resolution and functionalities. The MPEG-2 video is classified into several categories depending on resolution and functionality (see Table 6.5).

The picture resolution of the MPEG-2 video is classified into four *levels*. The "low" level is the smallest picture resolution and also the target resolution of MPEG-1 video. The "main" level is the current television resolution and is the most widely used one. The "high-1440" and the "high" levels are for European and U.S. HDTV applications, respectively.

The functionality of MPEG-2 video is classified into five *profiles*. The "simple" profile is the one that excludes B pictures for simpler implementation. The "main" profile can have B pictures (i.e., bidirectional-predictive picture frames) and is the profile used in most applications. The "spatial scaleable," "SNR scaleable," and "high" profiles have a hierarchical coding structure and extended functionalities.

Some practical examples of the MPEG-2 video are as follows: The "main profile/main level (MP@ML)" is adopted in digital DBS and DVD; the "main profile/high level (MP@HL)" is used in the U.S. HDTV standard; the "spatial scaleable profile/high-1440 level (SSP@H1440)" is considered in European HDTV; and the "simple profile/main level (SP@ML)" is adopted in U.S. digital cable systems. Table 6.6 provides the primary specifications of the MP@ML standard.

Scanning

The MPEG-1 video deals with *progressive scan,* which is adopted in computer display. The MPEG-2 video, however, treats *interlaced scan* as well, which is used in television display. In MPEG-2, for efficient processing of interlaced video, both *frame structure* and *field structure* are allowed. Scenes that have plenty of motion should be processed by field structure since even fields and odd fields are relatively less correlated; whereas scenes that have little motion would be better processed by frame structure because a high correlation exists between even and odd fields. Furthermore, even in frame structure, field processing is allowed on a macroblock-by-macroblock basis for an efficient compression of local area with rich motion.

New Tools

A new motion estimation scheme adopted in MPEG-2 is the so-called dual prime scheme. It is basically a field-by-field motion estimation/compensation

Table 6.4
Comparison of MPEG-1 and MPEG-2 Specifications

Option	MPEG-2 (MP@ML)	MPEG-1
Number of pixels	720 × 480	360 × 240
Transmission rate	Below 15 Mbps	Below 1.5 Mbps
Picture processing mode	Frame mode/field mode	Frame mode
Prediction method of motion vector	Interframe prediction/interfield prediction	Interframe prediction
IDCT mismatch control	After inverse quantization, reverse the last bit when the sum of the DCT coefficients is even	After inverse quantization, decrease the absolute value by one when the DCT coefficients are odd
Escape syntax	Use the escape code (6 bit) + FLC (18 bit) to the (RUN, LEVEL) value, which cannot be coded by VLC	Use the escape code (6 bit) + FLC (14 bit) or FLC (22 bit) to the (RUN, LEVEL) value, which cannot be coded by VLC
Escape usage	The (RUN, LEVEL) value, which can be coded by VLC, is not allowed to use the escape format	The (RUN, LEVEL) value, which can be coded by VLC, is allowed to use the escape format
Horizontal position of chrominance signal	Horizontal sampling point of chrominance signal is equal to that of the luminance signal	Horizontal sampling point of chrominance signal is located in the middle of the samples of the luminance signal
Slice	Slice must start and end in the same line on MB	Slice need not to start and end in the same line on MB
D picture	Not used	One of the picture formats
Motion vector precision	Forward/backward motion vector flag must be "0" (i.e., only half pixel motion vector is used)	Possible to use pixel unit motion vector when the flag is assigned to 1
Aspect ratio information	Aspect ratio of 4 bits in header is display aspect ratio. Pixel-to-aspect ratio is obtained from it as well as frame size and display size	Pixel-to-aspect ratio of 4 bits is represented in header
Forward/backward f_code	f_code needed for decoding motion vector is forward/backward and horizontal/vertical code in picture header	f_code needed for decoding motion vector is forward/backward code in picture header
Limited variable flag and maximum horizontal size	The concept of the limited variable flag changed to the concept of profile and level; the horizontal size is below 720 pixels in MPEG-2 MP@ML	The limitation conditions are satisfied and the horizontal size is below 768 when the limited variable flag is assigned to 1

Table 6.5
MPEG-2 Video Profiles and Levels

Level/Profile	Simple	Main	SNR	Scalable	High
HIGH		MP@HL			HP@HL
HIGH 1440		MP@H1440		SSP@H1440	HP@H1440
MAIN	SP@ML	MP@ML	SNP@ML		HP@ML
LOW		MP@LL	SNP@LL		

Table 6.6
Primary Specification of MPEG-2 MP@ML

Category	Details
Frame format	720 × 480 (at 29.97 Hz), 720 × 576 (at 25 Hz)
Coding rate	Maximum 15 Mbps
Image format	4: 2: 0
Picture types	I (intra), P (predictive), B (bidirectional) picture
Coding unit	Frame structure/field structure
Motion vector quantization	
Frame structure	Frame (16 × 16)/field (16 × 8)/double prime (in case of $M = 1$)
Field structure	Field (16 × 16)/field (16 × 8)/double prime (in case of $M = 1$)
Motion vector search range	-128 pixel $\sim +127.5$ pixel, semipixel unit
Buffer size	1.75 Mbit (1,835,008 bit)
Compatibility	Forward compatibility for MPEG-1
DCT DC precision	8 / 9 / 10 / 11 bits
Varable-length coding table	MPEG-1 VLC table
DCT coefficient scanning	MPEG-1 scanning (zigzag)/alternative scanning
Variable bit rate	Applicable

scheme, but the number of bits required to transmit the motion vectors is drastically reduced compared with the ordinary field-by-field motion estimation/compensation. This method is especially useful in maintaining good picture quality when B frames need to be avoided in order to reduce coding delay and system complexity.

For the DCT of a macroblock, even in a frame structure, the frame DCT and the field DCT can be adaptively selected. Therefore, the frame DCT is usually applied to the areas with little motion, whereas the field DCT is usually used in the areas with rich motion. Note that in the field structure, the field DCT is the only choice.

In quantizing the DCT coefficients, a uniform quantizer was used in MPEG-1 where the quantization step size is constant regardless of the magnitude of the transform coefficients. In MPEG-2, on the contrary, a *nonuniform quantizer* can also be optionally used where the quantization step size increases along with the magnitude of the transform coefficients. The use of this nonuniform quantizer yields further data reduction at the expense of increased complexity.

In scanning the quantized transform coefficients prior to Huffman coding, only the zigzag scan was used in MPEG-1. In MPEG-2, however, the alternate scan can be optionally used, which scans the vertical frequency components relatively early for effective handling of interlaced video. This scheme is especially effective for interlaced video with rich motion where the vertical correlation is weak.

In Huffman coding of quantized DCT coefficients, a single two-dimensional (2-D) VLC table was used in MPEG-1. In MPEG-2, however, an optional *intra VLC table* is added for handling intra-coded macroblocks. Note that intra-coded macroblocks tend to produce large transform coefficients and have quite different statistics of (RUN, LEVEL) symbols. The intra VLC table is usually, but not always, better suited for coding intra-coded macroblocks.

Scalability

The concept of *scalability* is newly adopted in MPEG-2. There are three scalabilities in MPEG-2: spatial scalability, temporal scalability, and SNR scalability.

Spatial scalability handles the original high-resolution image (e.g., HDTV image) by splitting it into the low-resolution image (e.g., SDTV image), called the base layer, and the high-resolution error image, called the enhancement layer. The base layer is obtained by decimating the original image by two, both horizontally and vertically. The base layer is coded first using MPEG-2, reconstructed at the encoder, and then interpolated by two to produce the prediction image at the enhancement layer. At the enhancement layer, the error image is obtained by taking the difference between the original and the prediction images. The error image is then coded using MPEG-2 as well. The base layer bit stream and the enhancement layer bit stream are multiplexed and transmitted together. At the receiver, the base layer decoder decodes the base layer bit stream only to reconstruct the low-resolution pictures, while the high-resolution decoder decodes both bit streams and mixes them up to reconstruct the high-resolution pictures. In this way, the high- and low-resolution encoder and decoder can maintain full compatibility (upward and downward). European HDTV systems are likely to adopt this spatial scaleable scheme to maintain the compatibility between the HDTV and the SDTV. However, the U.S. HDTV system adopts the simulcast for downward compatibility, that is, the current NTSC television viewer can view the HDTV program or an NTSC version of the same program, which is also broadcast.

Temporal scalability and *SNR scalability* also split the high-resolution image into two layers, and the coded bit streams of the two layers are multiplexed and transmitted together. The temporal scalability splits the image in the temporal direction, whereas the SNR scalability splits the image in terms of the bit-per-pixel resolution.

6.4.2 MPEG-2 Video Compression Algorithm

The MPEG-2 video employs the video coder and decoder whose structures are as shown in Figure 6.20. The video compression algorithm is composed of motion estimation/compensation, DCT, scalar quantization, RLC, and Huffman coding functions. It is a hybrid algorithm that combines the intraframe and interframe coding schemes in a way similar to that shown in Figure 6.16.

The MPEG-2 video syntax has a hierarchical structure comprising the sequence layer, *group of picture* (GOP) layer, picture layer, slice layer, macroblock layer, and block layer. Among these, the *GOP layer* is designed for random access and recovery from transmission errors, and the *slice layer* is for resynchronization at the decoder in the case of transmission errors. A macroblock is the unit for motion estimation/compensation and a block is the unit for DCT. Therefore, even when a slice is lost at the decoder due to transmission errors in the channel, the next slice can be received after resynchronization at the start of the slice. When the decoder is turned on during the transmission of the bit stream or when a channel change occurs in the broadcasting environment, the pictures are reconstructed and presented from the next I picture (i. e., intra picture). Figure 6.21 illustrates the hierarchical structure of the MPEG-2 video.

Redundancy Reduction

The essence of video compression is to remove the redundancy inherent in the video. A list follows of redundancies and associated compression methods.

1. *Spectral redundancy*: The RGB (red, green, blue) signals coming from video cameras, for example, are highly correlated and take on large bandwidth. To decrease the amount of video sample data based on human perception, the RGB color space is converted to YCrCb color space. The Y (*luminance*) has the full bandwidth and is very sensitive to human perception. The Cr and Cb (*chrominance*) components have a narrower bandwidth and are less discernible by human eyes. Therefore, the chrominance components are usually decimated by two, both horizontally and vertically, which results in a reduced number of samples.

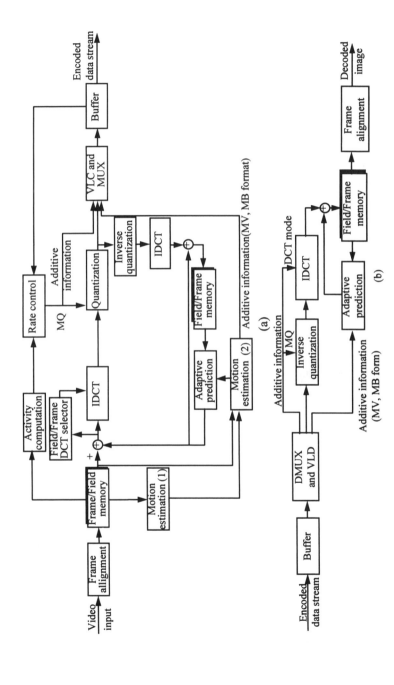

Figure 6.20 Structure of MPEG-2 video encoder and decoder: (a) encoder and (b) decoder.

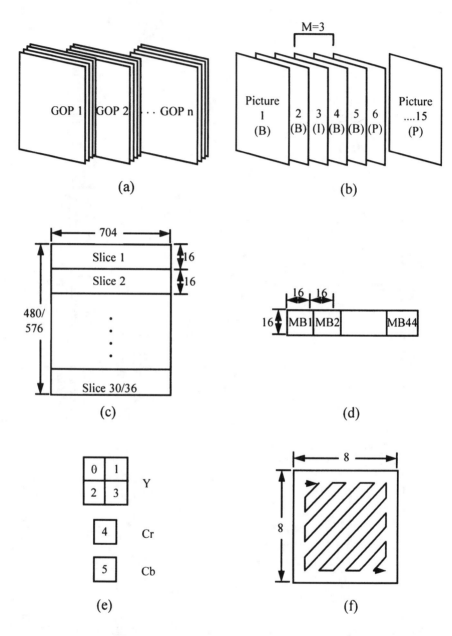

Figure 6.21 Layer structure of video data: (a) data stream layer; (b) group of picture (GOP) layer; (c) picture layer; (d) slice layer; (e) macroblock (MB) layer; and (f) block layer (zigzag scan).

2. *Spatial redundancy*: The input image is partitioned into 8×8 pixel units, called *blocks*. Each block goes through the DCT process, which is known to be the most efficient among known block transforms in terms of energy compaction and decorrelation. Its performance is close to that of the optimal KLT for the first-order Markov sequences with a very high correlation coefficient. Most natural images can be well modeled using the first-order Markov sequences and here the DCT provides a good compression means. For synthetic images, however, the DCT shows relatively poor performance. A number of efficient algorithms and architectures for fast computation of DCT/IDCT have been reported, including [Lee84, Hou87, Lee89, Yip88, Cho91].

After each block is transformed into the frequency domain through the 2-D DCT, the 64 real-valued DCT coefficients of the block are quantized to be represented by an integer. The quantization is a lossy process, but a high compression ratio can be obtained as the reward.

In MPEG-2, the *human visual system* (HVS) is incorporated into the quantization process. Human eyes are less sensitive to quantization noise at higher frequencies, and the quantization at higher frequencies is usually coarser for balanced human perception. For this purpose, two quantization tables with 64 entries each are often used (one for luminance, the other for chrominance) and the same tables are used at the encoder and the decoder.

3. *Temporal redundancy*: Adjacent pictures in an image sequence are highly correlated. The input picture is partitioned into 16×16 pixels, called *macroblocks*, and then each macroblock goes through motion estimation/compensation to remove the temporal redundancy. Each macroblock is coded based on interframe or intraframe mode. The decision is made on the basis of prediction errors. For instance, a picture right after a scene change is intraframe coded since there is no correlation between the current picture and the previous one.

4. *Statistical redundancy*: The DCT coefficients are quantized into integer values. Due to the energy compaction (into low frequencies) property of the DCT and the ascending property of the quantization matrix, most of the high-frequency coefficients become zero. By exploiting this statistical characteristic, further data compression can be achieved through entropy coding. This step is a lossless process and reversible. In MPEG-2, considering that most nonzero values tend to appear in the low-frequency region, the quantized coefficients are run-length coded just after using, for example, the zigzag scan. This results in a sequence of (RUN, LEVEL) symbols, where RUN is the number of successive zeros and LEVEL is the nonzero coefficient following RUN. The (RUN, LEVEL) symbols are then 2-D Huffman coded. When the last nonzero

value is reached, the *end of block* (EOB) mark is transmitted to avoid an unnecessary scan for the remaining coefficients.

Picture Types

There are three kinds of pictures in MPEG-2 video: I, P, and B pictures, which are categorized based on the method of motion estimation/compensation. A group of pictures starting with an I picture and delimited by the next I picture forms a GOP. The size of a GOP is therefore identical to the period of I pictures.

1. *Intra (I) picture*: This type of picture is intra-frame coded. Temporal redundancy therefore still remains and the compression ratio is relatively low. The I pictures are inserted periodically for the purpose of blocking propagation of errors caused by interframe coding and for the purpose of realizing random access in broadcasting or storage media environments. The I picture affects the total image quality substantially, so it must be quantized more finely than the P and B pictures.
2. *Predictive (P) picture*: This type of picture is coded using motion estimation/compensation from the anchor (or reference) frame. The anchor frame is selected from the previous I or P frames, depending on the number of B pictures. The P picture propagates coding errors to the following P and B pictures so it needs to be quantized more finely than B pictures.
3. *Bidirectionally predictive (B) picture*: This type of picture is coded using both forward and backward motion estimation/compensation from the previous and the following pictures. The reference pictures for the motion estimation/compensation are the nearest two I/P pictures. A coding error in this type of picture does not propagate and the quantization step size is controlled to be relatively large.

If N denotes the size of GOP, and M the number of consecutive B pictures plus one, then every GOP contains one I picture, $N/M - 1$ P pictures, and $N - N/M$ B pictures. For instance, if $M = 1$, then there is no B picture and the GOP structure becomes IPP..IPP.. . If $M = 3$ and $N = 9$, then the GOP structure becomes IBBPBBPBBIBB... . As the parameters M and N become larger, the coding delay as well as the random access time become longer, and the compression performance becomes higher.

Image Format

The MPEG-2 video has three different color image formats: 4:2:0, 4:2:2, and 4:4:4. The 4:4:4 format does not decimate chrominance samples, that is, the color components Cr and Cb have the same number of samples as the luminance component Y. The 4:2:2 format decimates Cr and Cb by two, horizontally, whereas the 4:2:0 format decimates both color components by two, horizontally and vertically. The decimation of color components provides a significant data reduction with little image degradation. Therefore, the 4:2:0 format is used in most applications including the Main Profile@Main Level. The 4:2:2 format is used for professional use in production studios or in broadcasting stations. The 4:4:4 format is rarely used due to its excessive amount of samples.

Motion Estimation and Compensation

In MPEG-2, there are three kinds of *motion estimation/motion compensation* (ME/MC) schemes: frame ME/MC, field ME/MC, and dual prime ME/MC. MPEG-2 requires that all ME/MC be performed in half pixel accuracy.

1. *Frame ME/MC:* This is also used in MPEG-1. The ME/MC is performed on the basis of a frame, that is, there is no discrimination between even fields and odd fields. The macroblock in the current frame is compared with every macroblock within the motion search area of the previous frame, and then the position of the macroblock that best matches the current macroblock is chosen for the motion vector evaluation (see Figure 6.15). More specifically, the cost function for the comparison is the mean absolute error or the mean square error, and the motion vector is the vector pointing to the position that gives the minimum cost.

   Practically, the picture data are given only in the pixel positions. The half pixel motion estimation is therefore performed in two steps: The first step is to find the integer pixel motion vector using, for example, a full search within the search area. The second step is to extend the accuracy to a half pixel unit by using half pixel interpolation and a nine-point search around the integer-pixel motion vector. A P picture requires one motion vector per macroblock, and a B picture two motion vectors per macroblock.

2. *Field ME/MC:* Two fields in a macroblock undergo two separate ME/MC processes. Field ME/MC is therefore carried out on the basis of a half macroblock whose size is 16×8 pixels. A P picture requires two motion vectors per macroblock, and a B picture four motion vectors per macroblock.

   In MPEG-2, the adaptive field/frame mode is provided, where, for each macroblock, the frame ME/MC and the field ME/MC are both performed and the one that gives the best result is chosen. Of course, the information about the selected mode should be transmitted to the decoder together with the motion vector.

3. *Dual Prime ME/MC:* This method is a compromise between the frame ME/MC and the field ME/MC. As mentioned, the field ME/MC scheme

has a comparatively heavy overhead for representing the motion vectors, whereas the frame ME/MC scheme has comparatively poor performance in the presence of rich motion. The dual prime scheme is designed to use a *full motion vector* plus a *delta motion vector* per macroblock, which results in a boost to performance comparable to that of field ME/MC. To avoid system complexity, the scheme is applied only to the cases when B frames are not present in the sequence.

The process of determining the full motion vector and the delta motion vector is very complicated. There are four candidates of full motion vectors and nine candidates of delta motion vector around each full motion candidate, yielding 36 combinations. The encoder should select one out of these 36 combinations on the basis of prediction performance. The decoding is, however, much simpler, and most off-the-shelf MPEG-2 decoder chips or systems can support the dual prime ME/MC scheme.

DCT and Quantization

The DCT is known to be the most effective transform in image compression in terms of energy compaction and decorrelation, and thus has been adopted in the international standards of H.261, JPEG, and MPEG. The size of DCT is decided to be 8 × 8 as a trade-off between the energy compaction efficiency and the complexity of the transform.

The efficiency of DCT is high for intraframe coded blocks due to high correlation among image samples but low for interframe coded blocks due to low correlation among samples in motion-compensated prediction error signals.

Because MPEG-2 video is designed to deal with interlaced video, the DCT can be applied either in the frame mode or field mode. In other words, the field DCT is more effective in interlaced video with rich motion.

Quantization is a process of representing the real-valued DCT coefficients by a finite number of bits. The DC and AC components are quantized separately. The DC component is more important in image reconstruction and is thus quantized using 8 to 11 bits. The AC coefficients are quantized using the quantization matrix and the quantizer scale factor. The quantization matrix incorporates the human visual system, and the scale factor controls the quantizer step size and resulting number of bits generated. The values of the quantization matrix are larger for higher frequencies; therefore, the high-frequency components, whose values tend to be very small in nature, usually become zero after quantization, which eventually leads to high data compression.

Variable-Length Coding

Variable-length coding reduces the average code length by assigning a longer code word to a less frequent symbol and a shorter code word to a more frequent one. *Huffman coding* is a typical variable-length coding method whose objective is to yield the average code length as close as the theoretical limit, called *entropy*. Huffman coding is used for variable-length coding of quantized DCT coefficients, differential motion vectors, macroblock types, coded block patterns, macroblock address, and so on.

In MPEG-2 image compression, the run-length coding and Huffman coding are combined to variable-length code the quantized DCT coefficients. Recall that a large number of zeros usually occurs in the quantized DCT coefficients. For primary data reduction in this situation, the quantized coefficients are zigzag scanned, as shown in Figure 6.21(f), to produce the (RUN, LEVEL) symbols previously mentioned. The (RUN, LEVEL) symbols have different statistics and a 2-D Huffman coding is adopted here instead of using two separate Huffman tables for RUN and LEVEL symbols.

In the case of motion vectors, the difference between the current motion vector and the previous one is taken first, and then it is Huffman coded. The horizontal component and the vertical one are coded separately. In the P pictures, the forward motion vector is transmitted; in the B pictures, the forward, the backward, or both motion vectors are transmitted, depending on the prediction performances.

In coding the macroblock information, the *macroblock address* (MBA), *macroblock type* (MBtype), and *coded block pattern* (CBP) are Huffman coded.

Video Buffer Verifier

The *video buffer verifier* (VBV) is a hypothetical decoder connected at the output of the encoder. The encoder monitors the buffer status of the hypothetical decoder and controls the bit rate such that the overflow or underflow can be avoided at the encoder/decoder in the CBR environment. In the case of VBR applications, the system target decoder, defined in the MPEG-2 system standard, applies.

Rate Control

In MPEG-2 compression, the three different picture types produce quite a fluctuating number of bits. Even in the same type of pictures, the number of generated bits can be significantly different depending on the variance of motion and scene complexity. Therefore, an appropriate bit allocation is needed that can determine the target bit budget of a GOP and assign bits to each frame within the GOP. Even in a frame, one may incorporate the human visual system in deciding the quantizer step size. For instance, a complex scene, like a forest, to whose quantization noise human eyes are insensitive, can be quantized using coarser step size, whereas a smooth scene, like a human face, to whose quantization noise human eyes are very sensitive, can be more finely quantized. The

MPEG-2 standard does not specify how to control the number of bits generated from the encoder. The encoder should be carefully designed so that no overflow or underflow occurs at the output buffer and the picture quality is globally constant to human perception.

6.4.3 MPEG-2 System

The MPEG-2 standard comprises three main parts: system (part 1), video (part 2), and audio (part 3). In MPEG-2-based compression systems, video and audio are compressed using the MPEG-2 video and audio standards, respectively. For practical application, the compressed video and audio bit streams need to be merged into a single bit stream so that it can be stored in digital storage media or transmitted through broadcasting/communication channels. A simple and naive way of accomplishing this goal is to segment video and audio bit streams appropriately and put them into predefined time slots. This scheme is frequently used in a telecommunication system where, for instance, 24 (or 30) PCM voice channels are multiplexed into fixed slot formats to form the DS-1 (or DS-1E) signal. This kind of approach, however, has little flexibility and does not meet the versatile requirements of a wide range of applications. Further, such time-division multiplexing does not fit with the ATM network. An alternative to this fixed-format multiplexing can be found in the packet or ATM-cell-based multiplexing. The MPEG-2 system layer is what enables the packet or cell-based transmission and storage of compressed video and audio signals. In the following, the functions and structures of the MPEG-2 system layer are outlined.

MPEG-2 Data Streams

The MPEG-2 system layer provides multiplexing and media synchronization functions that are necessary to store or transmit the MPEG-2 data streams. The MPEG-2 system layer can also provide synchronization of MPEG-2 decoders to the MPEG-2 program sources as necessary. An MPEG *program* refers to a set of video, audio, and data information that is multiplexed into a simple data stream for storage and playback or for transmission across a network. Such individual MPEG program streams, which are called *elementary streams*, are packetized to large, variable-size packets to form *packetized elementary streams* (PES).

Combining one or more streams of PES packets that have a common time base into a single stream yields a *program stream* (PS). If the elementary streams need to be in separate streams that are not multiplexed, each elementary stream can be encoded as a separate PS. The PS is designed for use in an error-free environment and is suitable for applications that may involve software processing of system information. The PS is segmented into large, variable-size packets.

Combining one or more programs that have one or more independent time bases into a single stream produces a *transport streams* (TS). The TS is designed for use in error-probable environments such as storage or transmission in lossy or noisy media and in the cases when multiple programs need to be multiplexed into a simple data stream. TS is segmented into fixed-size 188-byte packets, and TSs can be multiplexed into a new TS.

As such, PSs and TSs are designed for different applications, and their definitions do not follow a strict layer model. One is not a subset of the other, and either one can be converted into the other. In this conversion, the PES is taken as the intermediate data stream. Because all fields necessary for a stream are not always contained in the other stream, some fields must be derived for the conversion.

Figure 6.22 shows the flow of various information streams: Video data and audio data are separately processed to form the video PES and audio PES, respectively. The video and audio PESs are multiplexed in the PS and the TS multiplexers to generate the PS and the TS, respectively. The packetizers and the PS and TS multiplexers belong to the category of the MPEG-2 systems standard (ISO/IEC 13818–1), whereas the video encoder and the audio encoder belong to the categories of the MPEG-2 video standard (ISO/IEC 13818–2) and the MPEG-2 audio standard (ISO/IEC 13818–3), respectively.

Packetized Elementary Stream (PES)

The PES is formed out of MPEG-2 video and audio program streams called elementary streams. The PES packet is large and of variable size, and consists

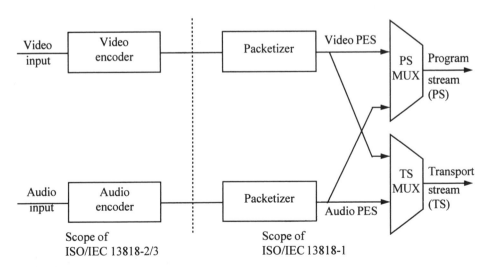

Figure 6.22 MPEG-2 system structure.

of the PES packet header and payload as shown in Figure 6.23(a). In general, PES packets are much larger than the TS packet.

PS and TS are logically constructed from PES packets. Also they can be converted to each other by way of PES packets. But PES streams do not contain some necessary information contained in PSs and TSs, such as the pack header, as can be seen in Figure 6.23.

A continuous sequence of PES packets whose payloads consist of data from a single elementary stream and have the same stream identification may construct a PES stream. In this case the optional PES header of the PES packet includes the *elementary stream clock reference* (ESCR) field, from which the decoders of the PES streams can derive timing information. The PES streams contain the bytes from the elementary stream contiguously in their original order.

Program Stream (PS)

The PS is connected by combining one or more of PES packet streams that have a common time base. The PS packet size is variable and may be very large. The PS is a contiguous stream of the PS packets, each of which consists of the pack header and the packs, with each pack including one or multiple PES packets, as shown in Figure 6.23(b). In general PS packets are much larger than the size of the TS packet.

The PS is designed for storage or communication in the environment where errors are very unlikely and where software-based processing is a major consideration. A typical application for the PS could be storage using *digital storage media* (DSM).

The PS, as well as the constituent elementary streams, may have either a fixed or variable rate. In either case, the syntax and semantics constraints on the stream are identical. The rate of the PS is defined by the values and locations of the *system clock reference* (SCR) and the program-mux-rate fields in the PS pack header. The SCR is the time stamp from which the decoders of the PS can derive timing information.

The prototypical decoder for the PS, when a composite video and audio PES streams are assumed to form the PS stream, takes the configuration shown in Figure 6.24(a). The channel-specific decoder in the front end is necessary in the case when the PS is stored or transported in the same channel-specific format. The PS demultiplexer in the center demultiplexes the PS stream into video and audio elementary streams, which are fed, respectively, to the video decoder and the audio decoder for final restoration of the video and audio signals. The PS decoder also extracts timing information using the SCR and distributes it to the video and audio decoders to aid their synchronized restoration. In the case of the DSM decoder, a *DSM command and control* (DSM-CC) functional block is additionally needed for an integrated control of the decoder.

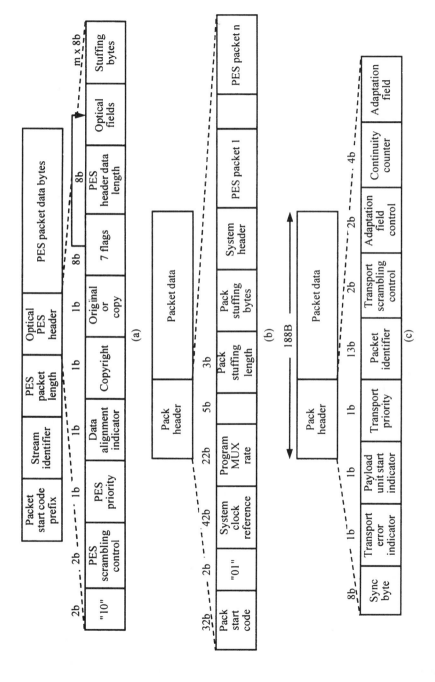

Figure 6.23 MPEG-2 system packet formats: (a) PES packet format; (b) PS packet format; and (c) TS packet format.

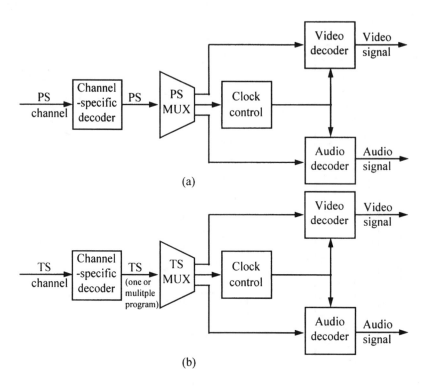

Figure 6.24 Prototypical MPEG-2 decoder system configurations: (a) PS decoder system and (b) TS decoder system.

Transport Stream (TS)

The TS is constructed in various ways: It may be constructed by combining one or more programs from elementary coded data streams that have one or more independent time bases, or it may be constructed from PSs or from other TSs that may contain one or more programs. The TS is a contiguous stream of TS packets that consists of the header and payload. The TS packet has a fixed size of 188 bytes and the format shown in Figure 6.23(c).

The TS is tailored for storage or communication in the environment where errors are likely, with the error occurring in the form of bit value errors or loss of packets. A typical example of the TS application may be the transport of video signals over the ATM network (see Section 6.5.5).

As for the case of the PS, the TS, as well as its constituent elementary streams, may be either fixed or variable rate, and the syntax and semantics constraints on the stream are identical in either case. The rate of the TS is defined by the values and the locations of the *program clock reference* (PCR) fields, which are separate for each program, in general. The PCR is the time stamp from which the decoders of the TS can derive timing information.

The prototypical decoder for the TS, if a composite video and audio PES streams are assumed to form the TS stream, has the configuration shown in Figure 6.24(b). It is basically the same as that for the PS decoder shown in part (a) except that the TS demultiplexer needs to handle the TS that contains multiple programs. As in the case of the prototypical PS decoder, the TS demultiplexer demultiplexes the TS stream into video and audio elementary streams and extracts timing information using the PCR and distributes it to video and audio decoders. The channel-specific decoder, if it detects errors in the incoming signal, can notify error messages to the video and audio decoders in various ways through some communication channels not shown in the figure.

Synchronization of Streams

The MPEG-2 system assumes a constant delay model in which the time delay of the entire system, from the input of the encoder to the output of the decoder, is constant. This time delay includes all of the time segments required for coding, coder buffering, multiplexing, storage or transmission, decoder buffering, decoding, and presentation.

Since the end-to-end delay through the entire system is constant, the audio and video presentations can be exactly synchronized. For this the decoder needs to have a system clock whose frequency and instantaneous value match those of the encoder. The time information necessary to convey the encoder system's clock to the decoder is the SCR for the PS systems and the PCR for the TS systems.

In the encoder system, there is a single system clock, and it is needed as the reference in generating the time stamps that indicate the correct decoding and presentation time of the video and audio signals. The time stamps that indicate the decoding time are called the *decoding time stamps* (DTS), and those that indicate the presentation time are called the *presentation time stamp* (PTS). The time stamps that indicate the system clock in the PS are the SCR, and those in the TS are the PCR.

In the decoder system, the decoding and presentation time can be derived based on the SCR and PCR. The resulting decoder system clock enables video and audio signals to be precisely synchronized to each other, thus forming a constant-delayed replica of the encoder input signal. However, it is also possible to construct decoders that do not have constant delay. In this case, the synchronization between the presented video and audio may not be precise, and the behavior of the buffer in the decoder may not follow the reference pattern of the constant delay model. Therefore, the buffer needs to be carefully handled such that overflow does not occur since it can cause data loss and a significant adverse effect on the decoding process.

6.5 PACKET VIDEO TRANSMISSION

The recently developed packet network technology and the advances in ATM technology necessitate processing and transfer of image information in packet form. The packet-mode transfer of image information differs from the conventional circuit-mode transfer in various aspects. The most important difference is that the packet-mode transfer allows VBR transmission with a constant image quality, whereas the circuit mode requires CBR transmission to match a specific bandwidth.

As discussed in Section 6.3, image information compressed by image coding has a bandwidth that varies with time depending on the amount of redundant information. Therefore, in order to transmit an image signal through a circuit-mode network, it is necessary to control the degree of image compression to generate a constant rate stream for the compressed image signal. For this purpose, a controllable buffer is employed, as shown in Figure 6.25(a). The state of this buffer is monitored and fed back to the image quality control blocks, which control the degree of image compression. That is, if the buffer fills, the compression ratio of the encoder is increased; if the buffer drains, the ratio is decreased and/or redundant data are stuffed. As such, the circuit-mode transfer of an image has the drawbacks that the image quality varies according to the buffer state and that it cannot utilize the given bandwidth efficiently.

In contrast, packet-mode transfer allows VBRs, and hence no buffer control is needed to maintain a constant bandwidth. Therefore, it has the virtue that the design of the encoder is simple and picture quality can be kept constant with time. Furthermore, under the requirement of the same picture quality,

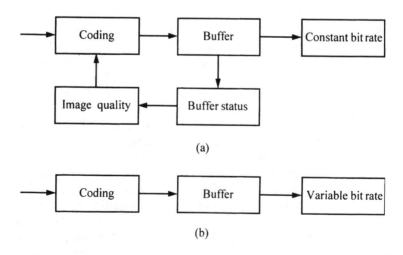

(a)

(b)

Figure 6.25 Video codecs: (a) for constant bit rate and (b) for constant image quality.

statistical multiplexing of several video signals can help to achieve far better utilization of bandwidth in packet networks than in circuit-switching networks. Based on these results, packet-mode transfer appears to be a highly proficient scheme that satisfies the need for both constant picture quality and efficient bandwidth usage.

From the network's point of view, the packet mode has a more fundamental virtue: It can accommodate efficient integration of complex information such as voice, data, image, video, and text through a single network, and it can adapt easily to ever-changing user demands and service environments. In fact, this is one of the core factors that influenced the adoption of ATM, which is a modification of packet communication, as the basic mode of transmission and switching for the BISDN. Note, however, that packet transmission raises a new set of problems to be solved, including packet jitter and packet loss, which are not problems in circuit-switching networks.

In this section, various problems associated with the packet communication of video information are discussed, as well as their solutions. Problems and requirements of video transmission across a packet network are described first. The compensation of packet jitter and clock synchronization in VBR packet transmission, and the problems due to packet loss and bit error control, are discussed. Next, hierarchical video coding appropriate for packet video transmission will be considered, and then packet video transmission in the ATM network is examined.

6.5.1 Problems of Video Transmission in Packet Networks

Processing video signal transmission in broadband packet networks is illustrated in Figure 6.26, in which 53-byte-long ATM cells are assumed for the video packets (5 bytes for the header and 48 bytes for the user information space). In the figure, we can see that an analog video signal is converted into a digital signal and compressed through video coding. Although it depends on the coding scheme used, the compressed video signal in most cases manifests a VBR flow whose bandwidth varies with time.

A compressed video signal is mapped into packets with the addition of a header using a packetizer, and delay is introduced in this process. The amount of delay varies depending on the degree of change in the bandwidth of the compressed video signal, but if the average bandwidth of the compressed video signal is around 1 to 100 Mbps, the associated time delay is about 5 to 400 ms on average. As a reference point, coding in circuit-switching networks introduces 10 to 100 ms of time delay at the buffer in converting a VBR signal to a CBR signal.

After the video packets have been formed, they arrive at the receiver terminal via a large number of packet multiplexers and packet-switching networks.

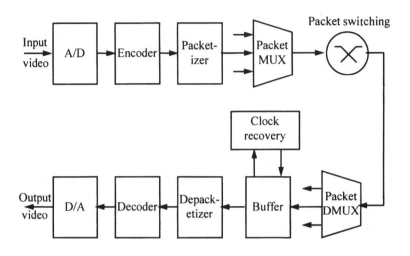

Figure 6.26 Image processing of broadband packet network.

Here, packet jitter and packet loss can occur depending on the traffic condition of the network, and they must be compensated for at the receiving terminal.

Causes of packet loss in the network can be divided into two types. One is the packet loss due to limited buffer space in packet switches and multiplexers, and the other is the packet loss that occurs when packets do not arrive at the receiver terminal because of bit errors in the address of the header.

When packets arrive at the receiving terminal, it separates the misdelivered packets from the readily arrived packets, compensates for the lost packets, and then stores them in the buffer memory. The buffer of the receiving terminal eliminates packet jitter; that is, it adds compensating delay to the varying delays of each packet in order to make the overall delay constant.

Afterwards, packets are read from the buffer memory by the video decoder's clock, and additional packet losses could occur at this stage if the encoding clock of the transmitter is different from the decoding clock of the receiver. Therefore, clock recovery and synchronization is required at the receiving terminal to prevent additional packet losses.

As has been discussed to this point, three main problems (i.e., time delay, packet jitter, and packet loss) have to be resolved for video transmission in a broadband packet network. Here, time delay in the packet networks is not anticipated to be much longer than the video transmission delay in the existing circuit-switching networks, since no bandwidth adjustment buffer is needed other than the packetizing and depacketizing process. Consequently, packet jitter, the associated clock regeneration at the receiving terminal, and packet loss remain the main problems of video transmission in packet networks. In other words, without effective compensation for these problems, the merits of packet video transmission (i.e., constant image quality and efficient bandwidth usage) cannot be achieved. To maintain constant picture quality in a packet network, compressed video signal packets must arrive at the receiver terminal without any packet losses within a specified time. To achieve such a result, however, inefficient bandwidth (almost the maximum bandwidth of packet switching) must be allocated, as is the case in circuit-switching networks. On the other hand, to achieve efficient bandwidth usage, packet loss and jitter due to packet multiplexing and the resultant delay are inevitable.

In the subsequent two sections we describe packet jitter compensation and receiver clock regeneration, which are the main problems of video transmission in broadband packet networks. Then we consider packet loss compensation methods, which provide a possible way to satisfy both of the two conflicting requirements mentioned above.

6.5.2 Packet Jitter Compensation and Clock Regeneration

Packet jitter refers to the fluctuations in intervals between packets caused by the variations in waiting time of packets at buffers of the packet multiplexers and switches according to the traffic condition of the network. In other words, packet jitter is attributable to the variable delay of each packet from transmitter to receiver.

If the time it takes for the i'th packet to be formed, leave the transmitter, and arrive at the receiving terminal is defined as the waiting delay, $Dq(i)$, then the delay that varies from packet to packet (i.e., packet jitter) can be written as follows:

$$J(i) = \begin{cases} 0, & i = 1, \\ Dq(i) - Dq(i - 1), & i = 2, 3, \dots . \end{cases} \tag{6.3}$$

This varying packet jitter changes to fixed delay by adding compensation delay $Dr(i)$, and in order to achieve perfect packet jitter compensation, the following relation must be satisfied:

$$Dq(i) + Dr(i) = \max\{Dq(j)\}, \ i = 1, 2, \dots . \tag{6.4}$$

Figure 6.27 illustrates packet jitter and compensated delay. Figures 6.27(a–c) represent image compression and packet generation; Figure 6.27(d) represents delay at the receiver terminal; and Figure 6.27(e) represents delay compensation at the receiver buffer until the packet is read by the receiver clock. Table 6.7 lists the degree of associated queuing delay, jitter, and compensation delay of each packet.

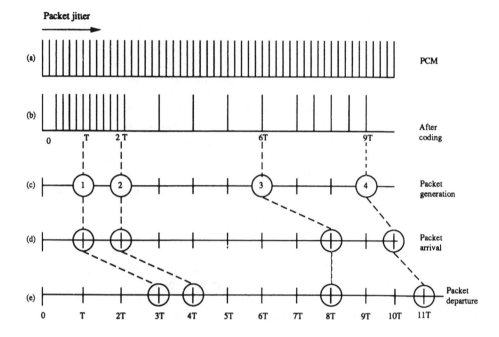

Figure 6.27   Illustration of packet jitter and delay compensation.

Table 6.7
The Degree of Queuing Delay, Jitter, and Compensation
Delay Associated with Figure 6.27

Packet, i	1	2	3	4
Queuing delay, $Dq(i)$	0	0	2T	T
Jitter, $J(i)$	0	0	2T	$-$T
Compensated delay, $Dr(i)$	2T	2T	0	T

The general solution for preventing packet loss due to packet jitter is to delay the packet arriving first at the receiver's buffer as much as the maximum packet delay, under the assumption that the receiver's clock is the same as the transmitter's clock.

To depacketize video packets that have arrived, synchronization is required between the transmitting terminal and the receiving terminal. Therefore, the clock of the transmitting terminal must be recovered at the receiving terminal. Since the clock cannot be sent separately, the receiver clock must be regenerated from the arriving packets. But since the packet arrival rate is not constant due to jitter and the occasional packet losses, the clock synchronization problem becomes complicated. For clock regeneration, the receiver first checks for the presence of any packet losses and substitutes any lost packets with other packets stored in the buffer memory of the receiving terminal. Ample buffer memory is required to prevent extra packet losses and to accommodate the maximum jitter generated in the packet network as well. In the end, the transmitter clock must be regenerated from the packets that contain jitter, and therefore clock synchronization regeneration methods in VBR transmission are different from those in CBR transmission.

In case a common reference clock exists in the network, as is the case for SDH-based BISDN, a synchronous clock recovery method can be employed. In this method, the transmitter writes down the difference information between the reference clock and the transmitter clock on a packet, then the receiver recovers the transmitter's clock using this transmitted clock information and the reference clock. The SFET, TS, and SRTS schemes discussed in an earlier section are typical examples of the synchronous clock recovery methods. ITU-T recommended the SRTS as the source clock recovery scheme for the AAL-1, together with the adaptive clock method to be discussed later. An illustration was given in Figure 4.42 on applying SRTS to VBR services. However, in most packet networks, the reference clock is not available, so synchronous clock recovery is not practically applicable.

In packet networks where a common reference clock is not available, an asynchronous clock method (or an adaptive clock method) is used. In this method, the receiver recovers the transmitter's clock taking advantage of the fact that the amount of data transmitted during a fixed amount of time is proportional to the transmitter's clock frequency. This, of course, applies to the case of CBR transmission. However, if video packets are transmitted in VBR, clock information cannot be extracted directly from the buffer memory state, since it does not reflect the actual transmitter clock. Therefore, the clock information is generally included in the transmitted packets. Here, clock information should have a particular pattern that can be easily identified by the receiving terminal. For example, a particular clock pattern can be sent in every video scanning line period (i.e., 63.6 μs). Although this clock information is inserted periodically at the transmitting side, the periodicity might become unclear as the packets arrive at the receiving terminal, due to packet jitter. Therefore, a clock generation method is required that can minimize jitter effects. Figure 6.28 shows the time average method that generates the transmitter's clock by detecting clock patterns from the received packets and connecting them to PLL. It calculates the number of clock patterns detected from the arrived packets over a fixed duration of time and compares it to the number of clock patterns generated by the receiver's clock during the same duration. The receiver's clock

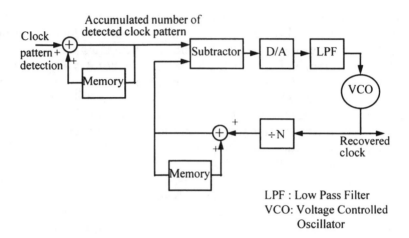

Figure 6.28 Time average clock recovery method.

is then adjusted by lowpass-filtering the difference. In the figure N denotes the period of the clock pattern divided by the period of clock.

As an example, we consider the case when the difference between the transmitter clock and the receiver clock amounts to 10 ppm (i.e., 10^{-5}). If jitter is not considered, and one clock pattern is inserted in every video scanning period, one clock pattern difference appears every 6.36 s, which is 100,000 times 63.6 μs. For the time unit that calculates the difference of the number of clock patterns, 1 minute is adequate. When this time unit becomes larger, the difference of the number of patterns approaches the difference in transmitter/receiver clocks excluding jitter effects. But if it becomes too long, the buffer may overflow or underflow, and hence the time unit should not be lengthened infinitely.

In the time average method shown in Figure 6.28, the mean jitter value approaches zero over a long period of time while the transmitter/receiver clock difference accumulates. The time average method uses this difference to control the receiver's clock and causes it to converge with the transmitter's clock. The memory in the figure continues to add up the difference between clock patterns. Even though jitter effects remain in the receiver's clock for the first few periods (about 1 min in this case), as time passes the memory plays the role of making it converge with the transmitter's clock.[1]

---

1. It is also possible to recover the transmitter's clock by extracting clock information out of the time intervals of packet arrival, not out of the difference of the clock pattern numbers. In this case the converging speed can be increased, but implementation becomes more complicated.

6.5.3 Packet Loss Compensation and Error Correction

The extent of picture quality degradation of video information caused by packet losses or bit errors varies depending on the coding scheme used. In general, the bigger the compression ratio, the worse the picture degradation. In the worst case, a single bit error or a single packet loss could cause loss of synchronization in the decoding stage, thus causing severe picture degradation until synchronization is recovered.

Table 6.8 shows the mean time between bit errors and packet losses associated with network error and packet loss for various video transmission rates. According to the table, in the case of a high-quality TV signal coded at around 155 Mbps, in order to guarantee more than 2 hours without error or packet loss, the bit error rate and packet loss rate should be 10^{-12} and 10^{-10}, respectively. But it is very difficult to meet this requirement at the network level. Therefore, in order to achieve such bit error and packet loss rates, an error compensation method must be considered at the transmitter/receiver terminal level.

In general, two methods are used to combat bit errors and packet losses: *error concealment* and *error correction.* Error concealment refers to the subjective compensation method that conceals the errored or lost fraction of an image from human eyes in the receiver using the image signals in the neighborhood of the errored part or using the prestored image signals. The error concealment method does not require any pretreatment in the transmitter and this does not increase the transmission rate, but the concealed image is not correct and the decoder implementation becomes complex. Error correction is an objective compensation method that transmits prearranged error-correcting information in addition to the pure image data so that the receiver can correct errors and recover the original image correctly.

Figure 6.29 is a block diagram of a system that simultaneously corrects bit errors and regenerates packet losses by employing the *Reed-Solomon* (RS) error correction coding scheme. The transmitter puts compressed video information

Table 6.8
Average Interval Due to Bit Error and Cell Loss

Average Bit Rate		*64 Kbps*	*1.5 Mbps*	*45 Mbps*	*135 Mbps*
Bit error rate	10^{-6}	16 seconds	0.7 seconds	22 ms	7.4 ms
	10^{-9}	4.3 hours	11 minutes	22 seconds	7.4 seconds
	10^{-12}	6 months	7.7 days	6.2 hours	2.1 hours
Cell loss rate	10^{-6}	1.7 hours	4.3 minutes	8.5 seconds	2.8 seconds
	10^{-8}	6.9 days	7.1 hours	14 minutes	4.7 minutes
	10^{-10}	1.9 years	1 month	1 day	7.9 days

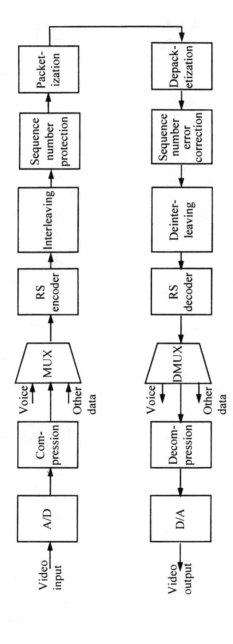

Figure 6.29 Image processing using RS encoder and interleaving.

and other information to be transmitted into the buffer and then takes them out in K-byte blocks and converts them into N-byte blocks by appending parity bytes after RS coding.

RS coding can regenerate up to $N - K$ deleted bytes among N bytes when there is no bit error. When bit errors exist, they can be corrected as well, but the capability to regenerate deleted bytes decreases. This phenomenon can be mathematically expressed as

$$2e + E < N - K + 1 \tag{6.5}$$

where e designates the number of errored bytes caused by bit errors and E designates the number of deleted bytes. For example, in the case of RS coding with $N = 32$ and $K = 28$, with no error occurring in each byte, up to 4 deleted bytes among the 32 bytes can be regenerated, and with no deleted bytes, up to 2 byte errors can be corrected. (For a more detailed discussion on the structuring and interleaving of the RS coded data, refer to Section 4.7.3.)

6.5.4 Hierarchical Video Coding

Careful selection of a video compression coding scheme is another way of minimizing picture quality degradation caused by packet losses in packet networks. First, hierarchical coding can be considered, whose most representative example is subband coding. In hierarchical coding, the video signal is decomposed into several parallel signals, which are compressed separately. Here, each parallel signal has different hierarchical characteristics with a different priority. Among the various hierarchical coding schemes, the subband coding presented is one of the most versatile and widely applicable coding schemes in use.

The virtue of subband coding lies in the fact that the coding method for each band can be freely chosen, since the derivation procedure and compression procedure of each parallel signal are independent of those of other signals. Also, since the subband itself naturally converts video information into different hierarchical resolutions, in case terminals with different resolutions (e.g., NTSC, EDTV, HDTV) communicate with each other, just the required bandwidth can be used by adding and subtracting parallel signals. As described earlier, the image signal is first split into different frequency bands via the filter bank, mainly into frequency bands of two-dimensional space. Here, the higher the frequency band, the lower the importance of each frequency band is for image regeneration.

Figures 6.30 and 6.31 depict, for example, a four-band rectangular subband and a three-band rhombic subband, respectively. We observe from the figures

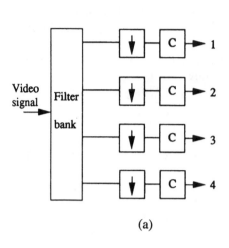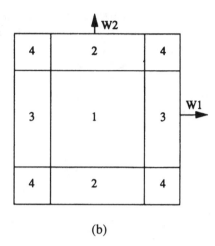

Figure 6.30 Two-dimensional four-channel square subband decomposition: (a) processing and (b) spectrum.

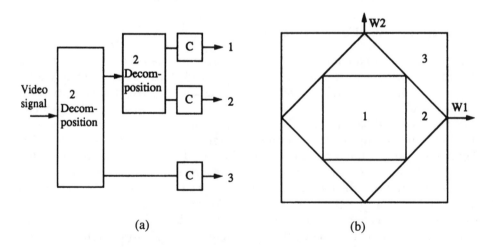

Figure 6.31 Two-dimensional three-channel rhombic subband decomposition: (a) processing and (b) spectrum.

that as the number indicating each parallel signal increases, the importance of that band for regenerating the original image decreases. Note that the overall information capacity of the subband parallel signal is the same as the original information capacity.

Since the bandwidth allocated to individual parallel signals has decreased, the sampling rate of each parallel signal can be lowered in proportion to the reduced bandwidth. This is very important for high-speed circuit implementation, because a more complex coding method that could not be employed for the original high-bit-rate video signal can now be used for the new low-rate signal. For these reasons, subband coding is widely used as an effective coding scheme independent of network structure.

To demonstrate how effective the subband coding is in packet networks, we consider the images shown in Figure 6.32. Figure 6.32(a) is the original image, and Figure 6.32(b,c) are the reconstructed images. Figure 6.32(b) is the image reconstructed using only frequency bands 1 and 2 after the rectangular band-splitting depicted in Figure 6.30, and Figure 6.32(c) has been reconstructed from frequency bands 1 and 2 after the rhombic band splitting scheme of Figure 6.31.

Comparing Figures 6.32(b,c), we can see that since most image information is contained in the low-frequency bands, the difference in the degree of picture quality degradation is not as great as the difference in the bandwidths. In fact, rhombic band-splitting is closer to the actual behavior of human eyes, but it requires more complex procedures in general.

Therefore, if different packets are formed for different frequency band signals, the influence on picture quality due to packet losses of high-frequency bands can be minimized. Low-frequency band packets still require minimal packet loss rate. But since the information capacity protected by band-splitting is limited to low-frequency band packets, implementing of the error correction encoder could become easier and more bandwidth efficient.

As has been described so far, hierarchical coding that encompasses subband coding is an extremely effective coding method for video communication in packet networks where packet losses are inevitable. Another major advantage of hierarchical coding is that it can be used for minimizing the overall information capacity to be processed by the network in multicast communication like video conference and CATV. Furthermore, the process of assembling various parts of different images into a single image is very simple in this multicast environment.

6.5.5 Video Transmission Over ATM Network

Since the ATM network transports signals in the form of ATM cells, video transmission over the ATM network becomes a type of packet video transmission. Among the four AALs—AAL-1, AAL-2, AAL-3/4, and AAL-5—AAL-3/4 is more adequate for data transmission and the other three are all capable of video signal transport.

ATM video transmission can be carried out by mapping a video signal to the payload space of the ATM cell. The video signal to be mapped may take any arbitrary format, but it is most appropriate to use a video signal that is in

Figure 6.32 Images for illustration of band-splitting: (a) original image; (b) reconstructed image using frequency bands 1 and 2 after rectangular band-splitting (2D, separable QMF filter); and (c) reconstructed image using frequency bands 1 and 2 after rhombic band-splitting (2D, nonspecific QMF filter).

MPEG-2 format. Therefore, in the following the cases in which MPEG-2 video signals are mapped into AAL-1, AAL-2, and AAL-5 for ATM, video transmission will be considered.[2]

For MPEG-2 video transmission, a lossless, constant-delay connection is needed between the source multiplexer and the receiver demultiplexer. Because MPEG-2 video data are compressed data, a small amount of data loss can cause a wide range of damage. Because a small delay variation can cause overflow or underflow of the buffer in the MPEG-2 system, constant delay is critical. Once such a loss-free constant-delay connection is set up, the MPEG-2 decoder can synchronize its clock using the SCR or PCR conveyed along the video data.

From Figures 6.6 and 6.19, we can see that for MPEG-2 video transmission over ATM there are two available MPEG-2 streams—the PS packet stream and the TS packet streams—and three available AAL types—AAL-1, AAL-2, and AAL-5. Which combination to select among the six possible combinations for ATM video transmission is up to the QoS requirement and the available transmission environment.

To begin with, since AAL-1 is designed to support circuit emulation services across ATM networks, it provides constant delay for all connections, and it also has support for FEC. Therefore, AAL-1 can satisfactorily provide functions required for MPEG-2 video transmission. However, the fact that AAL-1 can support only CBR services may become a drawback in the long run. Further, for the end-user equipment that is readily capable of handling AAL-5 in support of nonisochronous data and signaling, adding an AAL-1 capability could be an additional burden.

The AAL-2, by nature, is supposed to support effective VBR MPEG-2 video transmission, but its functionality is yet to be defined. In practice, the applicability of AAL-2 would depend on how well its functionality could take into account the economic and performance considerations.

Since AAL-5 is designed to support high-speed VBR data services, it can possibly be a very effective method to transport MPEG-2 video signals if the required service QoS is not stringent. In particular, for end-user equipment that is readily furnished with the AAL-5 capability for data and signaling treatments, MPEG-2 video transmission can be supported without additional cost. However, AAL-5 has several problems that need to be resolved before it can be useful for a wider set of applications: First, it may not provide the desired QoS when a large amount of VBR traffic is loaded. Second, it is not equipped with the FEC capability in its common part AAL.[3] Third, it can potentially introduce large amounts of delay variations across the network connection.

---

2. With regard to video coding and AAL mapping, the ITU-T recommended H.261 video codec and AAL-1 for visual communication purposes, and the ATM Forum and DAVIC recommended MPEG-2 video codec and AAL-5 for VOD service purposes.
3. Establishing an additional *video-audio service-specific convergence sublayer* (VASSCS) on top of the common part AAL-5 has been studied.

If the PS and the TS are compared to each other, the PS is devised on the assumption that the transport medium is error-free, while the TS is designed for use in a noisy environment where errors are likely to occur. Therefore, for ATM video transmission the TS is much more adequate than the PS, in general. Further, it is also more desirable, in this case, to store video signals in TS packets than in PS packets.

6.6 HIGH-DEFINITION TELEVISION

High-definition television (HDTV) is a system that transfers high-resolution images through band-limited channels. HDTV has more than twice the horizontal and vertical resolution of existing color TV, and provides wide screen images with an aspect ratio of 16:9 and CD-quality multichannel sounds.

Since the HDTV study project was first proposed to CCIR (the predecessor of ITU-R) by Japan in 1972, several countries throughout the world have started to develop HDTV systems. In particular, Japan developed a HDTV prototype employing the *multiple sub-Nyquist sampling encoding* (MUSE) format in 1984 for the first time in the world, and started experimental broadcasting through the broadcasting satellite BS-3b in November 1991. Europe adopted the analog *high-definition multiplexed analog components* (HD-MAC) system as the EU standard system in 1986, but has given up developing the system. Europe has resumed the HDTV standardization work together with the *European project for digital video broadcasting* (EP-DVB), and completed the DVB specification, a digital television standard, in 1995. The DVB specification is designed such that it can include the HDTV specification of Europe. The United States also started to develop an *advanced television* (ATV) system and organized an *Advisory Committee on Advanced Television Service* (ACATS) in support of this under the *Federal Communications Commission* (FCC) in 1987. In February 1993, the United States decided to standardize digital HDTV in a manner totally different from that of the MUSE system of Japan and the HD-MAC system of Europe. In November 1993, the *grand alliance* (GA) system, which unified the specifications of four different digital systems was approved by FCC, and its standardization was completed in November 1995.

Until 1980, which may be called the first era of HDTV research, not only Japan but also the United States and Europe were interested in the analog HDTV system. However, the fast advances in digital signal processing techniques turned the direction of development toward digital system in the late 1980s. Since the first digital system proposal of DigiCipher, the United States has led the so-called digital TV era by turning the direction of HDTV development toward the digital ATV, which has the same bandwidth as that of existing TV. Nowadays, standardization and transition toward a fully digital system is nearing completion.

The international standardization of the MPEG-2 system and the development of LSI technologies contributed to the development of digital HDTV. Initially MPEG-3 was planned separately for the purpose of HDTV, but it was decided in 1993 to expand MPEG-2 to include the HDTV video coding. The U.S. ATV specification has adopted the MP@HL (Main Profile at High Level) of MPEG-2, and the digital TV standard of Europe has adopted SSP@H1440 (Spatially Scaleable Profile at High 1440 Level) of MPEG-2.

Fully digital HDTV specifications adopt the digital format in both the signal compression and transmission stages, so it is possible to process and transmit images, voice, and data all together. Therefore, future HDTV is expected to be not just a simple broadcasting receiver system but an integrated multimedia terminal.

In this section, the studio and production specifications of HDTV are reviewed first. Then the standardization process of HDTV and standard specifications of the United States, Europe, and Japan are examined. Finally, the HDTV coding and transmission schemes are investigated.

6.6.1 Studio and Production Specifications of HDTV

Studio specifications must precede the establishment of international specifications on HDTV, and transmission specifications must be defined in conformance with them. The HDTV committee organized in 1974 within the ITU-R carried out standardization of the international specifications for unifying program production. The standards on the basic parameters of HDTV, subjective evaluation of image quality, and provisions about international exchange of programs were thus obtained and prescribed in ITU-R Recommendations 709, 710, and 714. They can be summarized as follows.

1. ITU-R HDTV specifications for studio and international exchange of programs prescribe various parameters for the photoelectricity transformation, screen characteristic, scanning characteristic, signal format, analog representation, digital representation, and so on. The details are given in Table 6.9.

2. The subjective evaluation of the image quality of HDTV is the provisions of the observation condition for evaluation, and the details are as follows: sight distance of 3H, maximum luminance value of 150 to 200 cd/m, chromaticity of the background of D (white), disposition of an observer within a 30-degree horizontal from the center of the screen, screen size of 1.4m (55 in.), and so on.

3. The international exchange of the electronic programs made by HDTV is recommended for live broadcast or the use of videotapes is encouraged to maintain the best quality.

Table 6.9
HDTV Standard Specification of ITU-R

Category	Subcategory	Details
Optoelectronic conversion	Optoelectronic transfer characteristics before nonlinear precorrection	Assumed to be linear
	Overall optoelectronic transfer characteristics at the source	A function of the luminance of the image (γ = 0.45)
	Chromaticity coordinates	Coefficients on X-Y chromaticity coordinates R(0.640,0.330), G(0.300,0.600), B(0.150,0.060)
Picture characteristic	Aspect ratio	16:9
	Sample per active line	1,920
	Lattice structure of the pixel	Square pixel
	Number of active lines per frame	1,080 (GA, MPEG-2)
Picture scanning characteristic	Order of sample scanning	From left to right and from top to bottom
	Method of scanning	Currently 2:1 (interlaced scanning) Eventually 1:1 (progressive scanning)
	Picture rate	Take readily used field/frame rates
Signal format	Conceptual nonlinear precorrection of primary signals	Correction coefficient γ = 0.45
	Derivation of luminance signal	$Y = 0.2125\,R + 0.7154G + 0.0721B$
	Derivation of color-difference signals (analog coding, $B - Y / R - Y$)	$P_B = 0.6389{\cdot}(B - Y)$ $P_R = 0.6349{\cdot}(R - Y)$
Analog representation	Nominal level (R, G, B, Y)	Reference black 0, and reference white 700 mV
	Format of synchronizing signals	Trilevel bipolar
	Timing reference	Center of the horizontal; synchronization timing reference
	Synchronization level	350 mV; every signal component has timing signal
	Horizontal, vertical blanking interval	To be induced from parameters of picture characteristics and picture scanning characteristics

(continued)

Table 6.9
HDTV Standard Specification of ITU-R (Continued)

Digital representation	Coded signals	R, G, B or Y, C_1, C_2
	Sampling lattice (R, G, B, Y)	Orthogonal
	Sampling lattice (C_1, C_2)	Cosituated with each other and with alternative luminance samples
	Sampling frequency (R, G, B, Y)	Integer multiple of 2.25 MHz
	Sampling frequency (C_1, C_2)	Half the luminance sampling frequency
	Number of samples per scanning line	1,920
	Signal coding	8-bit uniform quantization
	Nominal level	Luminance (black level 16, white level 235); chrominance (225 level)
	Synchronization method	Use SAV (start of actual video) and EAV (end of actual video)
	Bit rate	0.8 to 1.2 Gbps for current implementation, 2.0 to 3.0 Gbps for some future implementations

6.6.2 HDTV Standards

The existence of a single "international standard" for HDTV is extremely desirable, so that the entire world can use the same HDTV format. However, this has not happened for real HDTV standardization. So, in the following subsections, the three most representative HDTV/digital TV standards—ATV in the United States, DVB in Europe, and MUSE in Japan—are considered individually.

ATV Standard of the United States

In 1987 ACATS began to standardize ATV with the goal of introducing HDTV in terrestrial broadcasting. About 20 formats for ATV were proposed, including the DigiCipher system by GI. The United States began to seriously consider adopting digital HDTV broadcasting systems in 1990, and various digital HDTV formats were proposed including those by Zenith/AT&T group, GI/MIT group, and ATRC group (North American Philips, Thomson, DSRC, and NBC). In 1992, the FCC decided to introduce digital HDTV (i.e., ATV) in the terrestrial 6-MHz

TV band and requested the groups to prepare a unified system. The Grand Alliance (GA) was formed in answer to this request in May 1993 and, after two years of development and field tests, drafted the ATV standard jointly with ACATS in November 1995. The FCC was scheduled to approve the ATV standards in 1996.

The U.S. activities related to the development of HDTV are listed in chronological order in Table 6.10.

The principal specifications of the GA system are summarized in Table 6.11. To be compatible with computers every studio format should have a square pixel structure, and every HDTV screen larger than 34 inches should be able to scan progressively with more than 797.5 scanning lines and display 60 or more frames per second. Also, movies and other materials with the frame rate of 24/30 frames/s should be scanned progressively. There are four studio signal formats for the GA system, two for HDTV and two for *standard definition television* (SDTV), as listed in Table 6.11.

The 8-VSB modulation scheme, developed by Zenith, was adopted as the standard for HDTV transmission. In 1995 the *coded orthogonal frequency-division multiplexing* (COFDM) standard developed in Europe was also revisited because it was known to possess such desirable properties as single-frequency network realizability, superior crosstalk protection capability in fading channels, and high preference among broadcasting stations. However, ACATS

Table 6.10
HDTV-Related Activities in the United States

Year	Event
1977	SMPTE organized the HDTV research group.
1981	SMPTE gave the first trial performance of HDTV.
1982	ATSC was established for the purpose of HDTV and ATV.
1987	FCC organized ACATS and started the development of ATV.
1988	FCC announced trial decision on ATV and new research guidelines.
1990	GI proposed fully digital DigiCipher system.
1990	FCC announced new test schedule for the six proposed formats.
1991	*Advanced Television Test Center* (ATTC) executed the test on the six formats.
1993	February, FCC decided to adopt digital format, excluding the narrowband MUSE system.
1993	May 24, Grand Alliance was established.
1993	November, FCC approved the GA system.
1994	As a transmission scheme, 8 VSB was adopted for terrestrial broadcasting, and 16 VSB for CATV broadcasting.
1995	September, the field test of the GA system was reported.
1995	September, ATSC adopted GA system.
1995	November 28, ACATS recommended the final format to FCC.
1996	FCC will approve the GA system and announce the ATV standard for the United States.

Table 6.11
Basic Specifications of U.S. GA System

Category	Subcategory	Details			
Studio signal format	Frame format (width × height)	1,920 × 1,080 square pixel	1,280 × 720 square pixel	704 × 480 ITU-R 601	640 × 480 square pixel
	Aspect ratio	16:9	16:9	16:9 or 4:3	4:3
	Frame rate (or field rate)	23.976/24 Hz progressive scanning	23.976/24 Hz progressive scanning	23.976/24 Hz progressive scanning	23.976/24 Hz progressive scanning
		29.97/30 Hz progressive scanning	29.97/30 Hz progressive scanning	29.97/30 Hz progressive scanning	29.97/30 Hz progressive scanning
		59.94/60 Hz interlaced scanning	59.94/60 Hz progressive scanning	59.94/60 Hz interlaced scanning	59.94/60 Hz interlaced scanning
		—	—	59.94/60 Hz progressive scanning	59.94/60 Hz progressive scanning
Compression scheme of the image signal		MPEG-2 MP@HL (Main Profile@High Level)			
Voice signal		Dolby AC-3			
Multiple access scheme		MPEG-2 system			
Modulation scheme		For terrestrial broadcasting: 8-VSB For cable transmission: 16-VSB			

decided not to adopt the COFDM scheme as the FCC standards because the COFDM format did not prove to be superior to the VSB format.

HDTV and Digital Television Standards in Europe

Since 1986 Europe has proceeded with the development of HD-MAC, which is an analog system with a MAC format. However, because this system turned out to be inferior to the digital system of the United States, and not superior to the already existing analog system of Japan, MUSE, Europe abandoned the system in 1993 and founded the EP-DVB to study the digital television broadcasting specifications of Europe. The EP-DVB prepared specifications for terrestrial broadcasting, satellite broadcasting, and cable broadcasting, based on which the

European Telecommunication Standards Institute (ETSI) has completed the specifications for DVB.

The DVB specification is, in fact, an outcome of various digital system development projects. The DVB specification considers the maximized compatibility among media by unifying the basic structures of digital broadcasting formats for satellite, terrestrial, and cable broadcasting. The DVB specification has been also adopted as the basic structure in the DAVIC specification for the VOD area and is expected to be a unified European format.

The specifications for satellite and CATV broadcasting were completed by ETSI at the end of 1994, and the specification draft on terrestrial broadcasting was completed in 1995. These specifications adopt MPEG-2 as the standard compression format of image and voice, and adopt the *quadrature phase shift keying* (QPSK) modulation scheme for satellite broadcasting and the *quadrature amplitude modulation* (QAM) scheme for CATV transmission, respectively. They also adopt the *orthogonal frequency-division multiplexing* (OFDM) scheme for terrestrial digital broadcasting, because it is suitable for realizing a *single-frequency network* (SFN) for high-quality mobile services. These three specifications are different only in transmission parts depending on the media, but they share the same basic structures in other parts such as image/voice coding and internal signal processing, thus accomplishing maximum compatibility among different media. Apparently these specifications are for digital TV in the SDTV grade, but they can be easily converted to HDTV specifications by increasing the bit rate or bandwidth. All the terrestrial TV formats adopt the OFDM as the modulation scheme in Europe, which is different from that of the United States.

The DVB specifications are summarized in Table 6.12.

MUSE System and Digital Television Standards of Japan

Japan's NHK started developing HDTV in 1964 with the support of the Japanese government. In the 1980s Japan completed the MUSE system, the first HDTV

Table 6.12
Basic Specifications of European DVB System

Category	Details
Compression scheme of the image signal	MPEG-2 SSP@H1440 (Spatially Scalable Profile at High 1440 Level)
Multiple access scheme	MPEG-2 system
Modulation scheme	For terrestrial broadcasting: COFDM For cable transmission: QAM For satellite broadcasting: QPSK

broadcasting system, with an analog transmission format, and tested satellite broadcasting in 1991. Even if somewhat late compared to the United States, Japan has also started to develop the terrestrial and satellite digital broadcasting systems inspired by the U.S. efforts on digital systems.

The Japanese activities on HDTV development are listed in Table 6.13 in chronological order.

The MUSE system was developed for HDTV broadcasting through satellites. It is very flexible since it uses *vestigial sideband* (VSB) systems and amplitude modulation of package media such as CATV, VCR, and video disk. The MUSE system specifications are summarized in Table 6.14.

Japan plans to digitize communication satellite broadcasting by 1996, broadcast satellite broadcasting by 2008, and the terrestrial microwave TV broadcasting between 2000 and 2009, respectively. Japan also plans to develop *ultra definition television* (UDTV) by 2005, which is an even higher definition digital television.

6.6.3 Coding Methods of HDTV Signals

The HDTV studio specification requires a very large bandwidth to provide high-resolution images, high-quality voice, and additional information. Although it varies depending on the specific scheme, 20 MHz of analog bandwidths is required for each of the R, G, B signals, which amounts to 50 to 70 MHz total even

Table 6.13
HDTV-Related Activities in Japan

Year	Event
1964	NHK started developing HDTV.
1972	NHK submitted the study plan to CCIR for the development of HDTV.
1972	NHK announced Hi-Vision camera for testing and display.
1984	NHK announced MUSE format for satellite HDTV broadcasting.
1985	NHK developed overall Hi-Vision system, and tested broadcasting in MUSE format.
1986	NHK proposed the 1125/60 HDTV format as the international standard, but was rejected.
1988	NHK broadcasted Seoul Olympic games through HDTV.
1989	Test broadcasting of Hi-Vision for an hour per day from June 1.
1990	Commercial HDTV product became available on the market.
1991	Test broadcasting was started for eight hours per day through channel 9 of the satellite BS-3b from November.
1992	Rapid progress in size and cost reductions of HDTV set. (Number of MUSE LSI IC reduced to five, a third of original number.)
1994	Broadcasting time was increased to nine hours per day from January, and ten hours per day from October.
1996	Plans advance to start digital broadcasting through a communication satellite.

after conversion into the composite signal of a luminance signal and two chrominance signals. In contrast, the bandwidth allocated for HDTV transmission is limited to 6 MHz for the terrestrial channel and 27 MHz for the satellite channel, as indicated in Table 6.15. When a simple PCM scheme is used as the preprocessing for digital transmission, the HDTV signal has about a 1.2-Gbps rate, which contrasts with the bit rate of about 15 to 20 Mbps (using 8-VSB modulation) through the 6-MHz terrestrial channel, 20 Mbps (using QPSK modulation) through the satellite, and 45 Mbps to a few hundred Mbps through the optical cable. Therefore, it is essential to compress the signal bandwidth by employing various video processing techniques for transmitting the originally

Table 6.14
Basic Specifications of Japanese MUSE System

Category	Subcategory	Details
Studio signal format	Frame format (width × height)	1,920 × 1,035
	Aspect ratio	16:9
	Frame rate (or field rate)	59.94/60 Hz, interlaced scanning
Compression scheme of the image signal		Motion compensated multiple subsampling
Voice signal		Four-channel stereo, PCM
Transmission scheme		FM modulation through satellite

Table 6.15
Bandwidth of Transmission Media

Parameters	Terrestrial TV VHF/UHF	Satellite	Analog Coaxial Cable	Analog Optical Fiber	Digital Optical Fiber
Total bandwidth	VHF 34 + 42 MHz, UHF 420 MHz	C and Ku bands 500 MHz	Less than 1,000 MHz	350 ~ 2,200 MHz	560 ~ 2,400 Mbps (single mode)
Bandwidth per channel	6 MHz (M,N) 7 MHz (B) 8 MHz (D, I, K, L)	Narrow band RF 24,27,36 MHz Wideband RF 54, 72 MHz	6–12 MHz	6–55 MHz	n × 155 Mbps (SDH)

wideband signals over the band-limited channels. Most HDTV systems reduce the signal bandwidth by employing motion compensation on one hand and the transform coding technique on the other hand in order to reduce the temporal and spatial redundancies, respectively. In the following, the HDTV coding schemes are examined for the cases of analog and digital systems separately.

Analog HDTV System

The representative analog transmission system is the MUSE system of Japan, which was devised for satellite broadcasting. Because this system adopts FM modulation for satellite transmission, the available bandwidth of the coded signal is about 8 MHz, which implies that about a 10:1 bandwidth reduction is necessary. To achieve this, the fact that human eyes have a low resolution for moving areas is exploited. That is, time-spatial subsampling is used as the basic compression scheme, with the type of subsampling varying depending on the amount of motion. Consequently, the encoder performs interframe coding for stationary areas (temporal subsampling), and intraframe coding for moving areas (spatial subsampling). This system is generally classified as an analog system considering that the analog modulation scheme is adopted as the transmission scheme, even though it partially uses the basic digital signal processing techniques such as digitization of pixel values and subsampling. The final compressed signal is FM modulated with the bandwidth tripled, and is finally transmitted to subscribers through DBS. Since the bandwidth decreases to 6 MHz in the case of terrestrial broadcasting, the resolution of images or bandwidth of the original signals should be reduced by reducing the number of scanning lines or by performing other processes. The analog system has the drawbacks that it is not compatible with multimedia (or BISDN) systems and that the picture quality degrades for each retransmission. Figure 6.33 depicts the MUSE encoder and decoder systems.

Digital HDTV System

The representative full-digital systems are DVB of Europe and ATV of the United States, both of which employ the digital image processing techniques for video compression, and the digital modulation techniques for transmission. Since both the systems employ MPEG-2-based coding schemes for video compression, the basic video structures for video coding are very similar to each other (i.e., MPEG-2 MP@HL in ATV, and MPEG-2 SSP@H1440 in DVB; refer to Section 6.4 for MPEG-2). Hence in the following the video compression scheme of the ATV system is briefly considered.

In the ATV system with terrestrial broadcasting capability, 8-level VSB modulated signals are transmitted through the existing channel with 6-MHz bandwidth. In this case 20- to 25-Mbps digital transmission is possible. However since a 5- to 7-Mbps error correction code is necessary for reliable

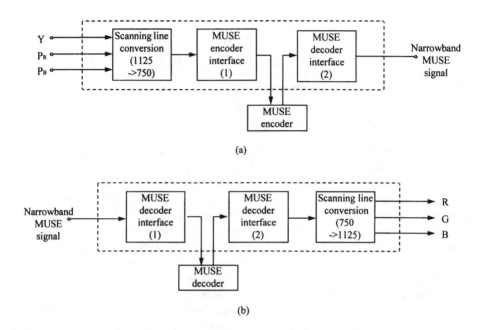

Figure 6.33   Block diagram of narrowband-MUSE system: (a) encoder and (b) decoder.

transmission even under unfavorable conditions, the transmission rate for a pure image signal is only 15 to 20 Mbps. Therefore 50:1 to 100:1 data compression is needed in order to transmit a 1-Gbps HDTV signal at such a low bit rate. In the ATV system interframe motion-compensated prediction and DCT coding are used for video compression. The encoding and decoding structures of the ATV system are depicted in Figure 6.34.

6.6.4 Transmission of HDTV Signal

To achieve efficient transmission of HDTV signals, various conditions must be considered such as analog bandwidth, digital transmission rate, transmission medium, transmission system, and applications. As examined in Section 6.6.3, when analog bandwidths of the signal and channel are taken into account, the available digital transmission rate over terrestrial and satellite links is much lower than that usable in wired links. If coaxial cable or optical cable is employed as the transmission medium, the available transmission rate becomes large and flexible. In particular, in the ATM network, constructed on the SDH/SONET fiber physical layer, an almost constant picture quality can be achieved through VBR transmission. In the ATM network, bit rates in multiple of 155 Mbps (e.g., 622 Mbps, 2.5 Gbps) and fractions of 155 Mbps can be flexibly

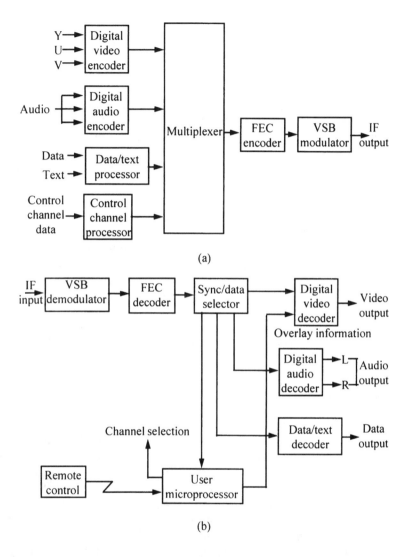

Figure 6.34 Block diagram of ATV system: (a) encoder and (b) decoder.

used for HDTV signal transmission. In terms of transmission error, the error rate is high for terrestrial or satellite channels due to instabilities in the atmosphere. It is low for optical-cable-based ATM network, but cell losses due to buffer overflow and misdelivery must be taken into account in this case.

HDTV signal modulation schemes can be classified into digital modulation and analog modulation. As for analog modulation, *frequency modulation* (FM) is the most typical scheme, which is now being used for the satellite communication in the MUSE system. As for digital modulation, several schemes are available that depend on the transmission medium. Digital modulation refers to the processing that loads the digital data onto the carrier signal in order to transmit the data streams through the bandpass channel. The parameters of the carrier available for the digital modulation are magnitude, phase, and frequency, which respectively render the digital modulation schemes *amplitude shift keying* (ASK), *phase shift keying* (PSK), and *frequency shift keying* (FSK). If the phase information of the transmitted signal is utilized for signal detection at the receiver, it is called *coherent modulation;* otherwise, it is termed *noncoherent modulation.* A typical example of the noncoherent modulation scheme is *differential PSK* (DPSK), in which the information is carried through the phase difference between successively transmitted symbols.

The choice of modulation schemes depends on the target transmission medium. In the case of terrestrial and wired communications whose medium characteristics are linear, the multilevel ASK is widely used, while in the case of satellite communications whose carrier frequency is tens of gigahertz, the PSK scheme with a constant modulus is used due to the nonlinear characteristics of the amplifier in this high-frequency band.

In the HDTV and digital television standards, various modulation schemes such as VSB, COFDM, QAM, and QPSK are used. Since QAM and QPSK are frequently used in digital communications, only VSB and COFDM are considered further in the following.

VSB Modulation in ATV

The VSB modulation seen in the NTSC system is also used in the ATV system for terrestrial broadcasting and wired transmission. A digital M-VSB modulator has M-level data inputs composed of various information, a synchronization signal, and a pilot signal. In the case of 8-VSB for terrestrial broadcasting, the data signal must be spectrum-shaped to lie in the 6-MHz bandwidth. The shaping filter characteristic is flat in the passband except for the boundaries, and the characteristics of the boundaries at both ends are made the same, even though they need not be, because of the *vestigial sideband* (VSB) characteristics of the transmitted signal. Figures 6.35(a,b) depict the HDTV and the typical NTSC TV signal spectrums, respectively, and Figure 6.35(c) depicts the frequency spectrum of the filter used in the 8-VSB scheme.

The VSB filter processing in the transmitter generates the stable in-phase and quadrature-phase signals through the complex filtering. In this filter processing, the *root-raised cosine Nyquist* (RRCN) filtering and the sinc (i.e., sin x/x) filtering (for the compensation of D/A converter) are performed. The baseband signals whose phases are orthogonal with each other are converted to the analog waveforms through the D/A converter, and then multiplied by the carrier waveforms with orthogonal phase relations, resulting in the *intermediate*

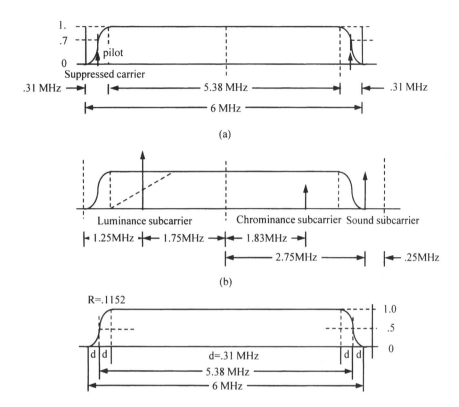

Figure 6.35 Spectrum of VSB modulation: (a) HDTV spectrum; (b) NTSC spectrum; and (c) amplitude response of 8-VSB filter.

frequency (IF) VSB signal. The frequency of the IF carrier is 46.69 MHz, which is obtained by adding the center frequency (44.0 MHz) of the IF signal to a quarter of the symbol rate (i.e., 10.76 MHz / 4 = 2.6905 MHz). On the other hand, *Reed-Solomon* (RS) coding is used to enhance the error correction capability, and *trellis-coded modulation* (TCM) and interleaving are used to cope with random errors and burst errors, respectively.

COFDM Modulation in DVB

Coded orthogonal frequency-division multiplexing (COFDM), the standard scheme for terrestrial broadcasting in Europe, is directly applicable to the MPEG-2 coded system. It is designed to avoid *cochannel interference* (CCI) and *adjacent channel interference* (ACI), and to make the best use of the spectrum resources at the same time. It also enables a single-frequency network to be built.

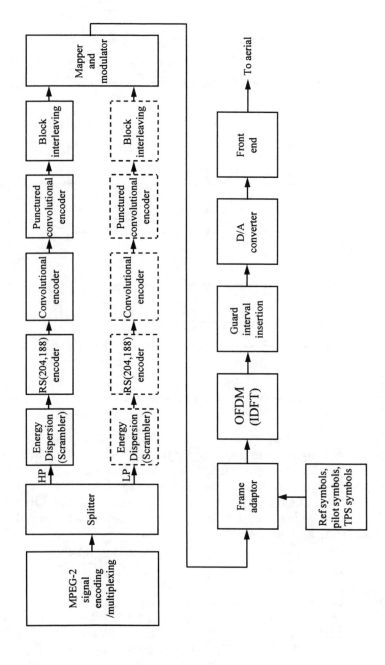

Figure 6.36 Block diagram of COFDM system in DVB format.

Figure 6.36 depicts the block diagram of the COFDM scheme adopted in the DVB specification. A packet received from the MPEG-2 multiplexer is composed of 188 bytes, which include a 1-byte synchronization signal. The splitter in the figure splits the input signal into *high-priority* (HP) and *low-priority* (LP) signals, and for each split signal an internal code and an external code are applied to enhance the error correcting capability in the terrestrial transmission. The RS code is used for the external code, and the convolutional interleaving is also used for the external coded signals to withstand the burst errors. The punctured code based on the 1/2 convolutional code with constraint length 7 is used for the internal code, and a bit-unit interleaving based on the block is used for the internal interleaving. The processed HP and LP signals are then mapped and modulated through QPSK, 16-QAM, 64-QAM, and so on. The combined signal is transmitted in the form of an OFDM frame based on 6,785 subcarriers, and the OFDM signal can be generated by the IFFT process. The *transmission parameter signaling* (TPS) symbol in the figure is intended to convey transmission-related information such as the QAM constellation value, inner code rates, and guard interval.

In the COFDM modulation for terrestrial broadcasting, the external code and external interleaving are the same as those for satellite and wired broadcasting so as to achieve maximum compatibility, and the internal code (i.e., the punctured convolutional code) is the same as that of satellite broadcasting. The QAM modulation level and the code rate in COFDM can be freely determined and two-level hierarchical channel coding and modulation are also possible.

Selected Bibliography

ATRC, "System description of advanced digital television," February 1991.

Datapro Reports on Telecom., "Video conferencing systems," November 1990.

DAVIC1.0, "Description of DAVIC functionalities," Part 1, Rev. 4.0, 1995.

First International Workshop on Packet Video, Workshop Notes, New York, May 1987.

Second International Workshop of Packet Video, Workshop Notes, Torino, Italy, September 1988.

Third International Workshop of Packet Video, Workshop Notes, Torino, Italy, March 1990.

Fourth International Workshop on Packet Video, Workshop Notes, Kyoto, Japan, August 1991.

ISO 10918(JPEG), "Digital compression and coding of continuous-tone still images," 1991.

ISO 11154, "Progressive bilevel image compression," 1994.

ISO-IEC/JTC1/SC2/WG12(MHEG), "Information processing, coded representation of multimedia and hypermedia information objects," 1994.

ISO-IEC/JTC1/SC2/WG8, "Coded representation of picture and audio information," JPEG Draft, January 1990.

ISO/IEC 10918-1, "Digital compression and coding of continuous-tone still image," part 1: Requirement and Guidelines, 1993.

ISO/IEC 11172-1, "Information technology—generic coding of moving pictures and associated audio, part 1: systems," March 1994.

ISO/IEC 11172-2, "Information technology—generic coding of moving pictures and associated audio, part 2: video," March 1994.

ISO/IEC 11172-3, "Information technology—generic coding of moving pictures and associated audio, part 3: audio," March 1994.

ISO/IEC 13818–1, "Information technology—generic coding of moving pictures and associated audio, part 1: systems, recommendation," March 1994.

ISO/IEC 13818–2, "Information technology—generic coding of moving pictures and associated audio, part 2: video, recommendation," March 1994.

ISO/IEC 13818–3, "Information technology—generic coding of moving pictures and associated audio, part 3: audio, recommendation," March 1994.

ISO/IEC CD-11172, "Information technology—coding of moving pictures and associated audio for digital storage," 1993.

ITU-R Rec. 601, "Encoding parameters of digital television for studios," 1982.

ITU-R Rec. 656, "Interfaces for digital component video signals in 525-line and 625-line television systems," 1982.

ITU-R Rec. 709, "Basic parameter values for the HDTV standard for the studio and for international programme exchange," 1990.

ITU-R Rec. 710, "Subjective assessment methods for image quality in high-definition television," 1990.

ITU-R Rec. 714, "International exchange of programme electronically produced by means of high-definition television," 1990.

ITU-T Rec. G.726, "40, 32, 24, 16 kbits/s adaptive differential pulse code modulation (ADPCM)," 1990.

ITU-T Rec. H.221, "Frame structure for a 64 to 1920 kbit/s channel in audiovisual teleservices," 1995 (revision).

ITU-T Rec. H.230, "Frame-synchronous control and indication signals for audiovisual systems," 1995 (revision).

ITU-T Rec. H.242, "System for establishing communication between audiovisual terminals using digital channels up to 2 Mbit/s," 1993 (revision).

ITU-T Rec. H.261, "Video coder for audio visual services at 64 kbit/s," 1993 (revision).

ITU-T Rec. H.263, "Video coding for low-bit rate communications," 1995.

ITU-T Rec. H.320, "Narrow-band visual telephone systems and terminal equipment," 1993 (revision).

ITU-T Rec. H.324, "Terminal for low-bit rate multimedia communications," 1995.

Zenith, AT&T, "Technical description of digital spectrum compatible HDTV," February 1991.

Brotman, S. N., *Telephone Company and Television Competition*, Norwood, MA: Artech House, 1990.

Chang, Y., et al., "An open-systems approach to video on demand," *IEEE Commun. Mag.*, Vol. 31, May 1994, pp. 68–80.

Chiang, T., and D. Anastassiou, "Hierarchical coding of digital television," *IEEE Commun. Mag.*, Vol. 32, No. 5, May 1995, pp. 38–45.

Cho, N. I., and S. U. Lee, "Fast algorithm and implementation of 2-D discrete cosine transform," *IEEE Trans. Circuits, Syst.*, Vol. 38, No. 3, 1991, pp. 297–305.

Cuevas, F. A., M. Bertran, F. Oller, and M. Selga, "Voice synchronization in packet switching networks," *IEEE Network*, September 1993, pp. 20–25.

Deloddere, D., W. Verbiest, and H. Verhill, "Interactive video on demand," *IEEE Commun. Mag.*, Vol. 32, No. 5, May 1995, pp. 82–88.

DePryker, M., and A. D. Gelmen, "Video on demand," *IEEE Commun. Mag.*, Vol. 32, No. 5, May 1995, p. 67.

Dixit, S., and P. Skelly, "MPEG-2 over ATM for video dial tone networks: Issues and strategies," *IEEE Network*, September 1995(a), pp. 30–40.

Dixit, S., "A look at video-digital tone network," *IEEE Spectrum*, Vol. 32, April 1995(b), pp. 64–65.

Fleisher, P. E., et al., "Digital transport of HDTV on optical fiber," *IEEE Commun. Mag.*, Vol. 29, No. 8, August 1991, pp. 36–41.

Fox, J. R., "Cable television network options in the U.K. for the 1990's," *IEEE J. of Lightwave Commun. Syst.*, February 1990.

Haghiri, M. R., and F. W. P. Vreeswijk, "HDMAC coding for MAC Compatible broadcasting of HDTV signals," *IEEE Trans. on Broad.*, Vol. 36, No. 4, December 1990.

Hou, H. S., "A fast recursive algorithm for computing the discrete cosine transform," *IEEE Trans. Acoust., Speech, Signal Process.*, Vol. ASSP-35, October 1985, pp. 1455–1461.

Kang, M., and I. Lee, "Optical CATV technology development direction," *J. of Korea Inst. of Telematics and Electronics*, Vol. 16, No. 16, December 1989.

Karlsson, G., and M. Vetterly, "Subband coding of video for packet networks," *SPIE, Optical Eng.*, Vol. 27, July 1988.

Kishimoto, R., and I. Yamashita, "HDTV communication systems in broadband communication networks," *IEEE Commun. Mag.*, Vol. 29, No. 8, August 1991, pp. 28–35.

Lacert, C., et al., "The first fully digital broadband switched FTTH pilot trial," in *Proc. of Broadband '90*, 1990.

Large, D., "Tapped fiber vs. fiber reinforced coaxial CATV systems," *IEEE J. of Lightwave Tech.*, February 1990.

Lee, B. G., "A new algorithm to compute the discrete cosine transform." *IEEE Trans. Acoust., Speech, Signal Process.*, vol. ASSP-32, pp. 1243–1245, Dec. 1984.

Lee, B. G., "Input and output index mappings for a prime factor decomposed computation of discrete cosine transform," *IEEE Trans. Acoust., Speech, Signal Process.*, vol. ASSP-37, no. 2, pp. 237–244, Feb. 1989.

Legall, D. J., "MPEG: A video compression standard for multimedia applications," *Communications of the ACM*, Vol. 34, No. 4, April 1991.

Leger, F., "The prospect optical cable to the home," *Cable Com. Mag.*, January 1989.

Lukacs, M. E., and D. G. Boyer, "A universal broadband multipoint teleconferencing service for the 21st century," *IEEE Commun. Mag.*, November 1995, pp. 36–43.

Maglaris, B., et al., "Performance models of statistical multiplexing in packet video communications," *IEEE Trans. on Commun.*, Vol. 36, July 1988, pp. 834–844.

Netravali, A., and B. Haskell, *Digital Pictures-Representation and Compression*, New York: Plenum Press, 1988.

Ninomiya, Y., et al., "An HDTV broadcasting system using a bandwidth compression technique-MUSE," *IEEE Trans. on Broad.*, Vol. 33, No. 4, December 1987.

Paik, W., "Digicipher-all digital, channel compatible, HDTV broadcast system," *IEEE Trans. on Broad.*, Vol. 36, No. 4, December 1990.

Pancha, P., and E. M. Zarki, "MPEG coding for variable bit rate video transmission," *IEEE Commun. Mag.*, May 1994, pp. 54–66.

Pangrac, D. M., "Application of optical fiber transmission technology to existing CATV networks," *Cable Television Eng.*, Vol. 14, No. 6, June/September 1989.

Perry, T. S., "HDTV and the new digital television," *IEEE Spectrum*, Vol. 32, No. 4, April 1995, pp. 34–35.

Prycker, M. D., and A. D. Gelman, "Video on demand," *IEEE Commun. Mag.*, Vol. 32, No. 5, May 1994, pp. 67–109.

Sari, H., G. Karam, and I. Jeanclaude, "Transmission techniques for digital terrestrial TV broadcasting," *IEEE Commun. Mag.*, February 1995, pp. 100–109.

Verbiest, W., and L. Pinnoo, "A variable bit rate video codec for asynchronous transfer mode networks," *IEEE JSAC*, Vol. 7, June 1989, pp. 761–770.

Yang, P. P. N., and M. J. Narashima, "Prime factor decomposition of the discrete cosine transform," in *Proc. ICASSP '85*, March 1985, pp. 772–775.

Yates, R. K., N. Mahe, and J. Masson, *Fiber Optics and CATV Business Strategy*, Norwood, MA: Artech House, 1990.

Yip, P., and K. R. Rao, "The decimation-in-frequency algorithms for a family of discrete sine and cosine transforms," *Circuit, Systems, and Signal Process.*, 1988, pp. 4–19.

Appendix A
Related Standards

Broadband and Optical Subscriber Network

1. ANSI T1.413, Asymmetric Digital Subscriber Spec.
2. ANSI X3T9.5 TP/PMD, "Performance of 125 Mb/s 32-CAP transceiver," 1992.
3. DAVIC1.0, "Delivery system architectures and APIs," DAVIC1.0 Spec, Part 4, Rev. 4.0, 1995.
4. ETSI 300–324–1 & 300–347–1, "Signalling protocol and switching: V interface at the digital local exchange and V5.2 interface for the support of access network," part 1: V5.2 Spec., September 1994.
5. ETSI, TM3 RG12, "Asymmetric Digital Subscriber Spec."
6. ITU-T Rec. G.650, "Definition and test methods for the revalent parameters of single mode fibers," 1993.
7. ITU-T Rec. G.651, "Characteristics of a 50/125-μm multimode graded index optical fiber cable," 1993 (revision).
8. ITU-T Rec. G.652, "Characteristics of single-mode optical fiber cable," 1993 (revision).
9. ITU-T Rec. G.653, "Characteristics of dispersion-shifted single-mode optical fiber cable," 1993 (revision).
10. ITU-T Rec. G.654, "Characteristics of a 1550-nm wavelength loss-minimized single-mode optical fiber cable," 1993 (revision).
11. TA-NWT-001209, "Generic requirements for fiber optic branching components," Bellcore, 1991.
12. TA-TSV-001294, "Generic requirements for element management layer(EML) functionality and architecture," Issue 1, Bellcore, December 1992.
13. TR-NWT-000909, "Generic requirements and objectives for fiber in the loop systems," Issue 1, Bellcore, December 1991.
14. TR-TSY-000303, "Integrated digital loop carrier system generic requirements, objectives, and interface," Bellcore, 1990.

Synchronous Digital Transmission

1. ANSI T1.102, "Digital hierarchy interfaces."
2. ANSI T1.105–1988, "American national standard for telecommunications-digital hierarchy-optical interface rates and formats specification," 1988.
3. ANSI T1.105–1991, "Digital hierarchy-optical interface rates and formats specifications (SONET)," 1991.
4. ANSI T1.105a-1991, "Supplement to T1.105," 1991.
5. ANSI T1.106–1998, "American national standards for telecommunications-digital hierarchy-optical interface specifications (single mode)," 1988.
6. ANSI T1.117, "Digital hierarchy-optical interface specifications (SONET) (single mode-short reach)," 1991.
7. ITU-T Rec. G.703, "Physical/electrical characteristics of hierarchical digital interfaces," 1991.
8. ITU-T Rec. G.707, "Synchronous digital hierarchy bit rates," 1995 (revision).
9. ITU-T Rec. G.708, "Network node interface for the synchronous digital hierarchy," 1993 (revision).
10. ITU-T Rec. G.709, "Synchronous multiplexing structure," 1993 (revision).
11. ITU-T Rec. G.774, "Synchronous digital hierarchy (SDH) management information model for the network element view," 1992.
12. ITU-T Rec. G.781, "Structure of recommendations on equipment for the synchronous digital hierarchy (SDH)," 1994 (revision).
13. ITU-T Rec. G.782, "Types and general characteristics of synchronous digital hierarchy (SDH) equipment," 1994 (revision).
14. ITU-T Rec. G.783, "Characteristics of synchronous digital hierarchy (SDH) equipment functional blocks," 1994 (revision).
15. ITU-T Rec. G.784, "Synchronous digital hierarchy (SDH) management," 1994 (revision).
16. ITU-T Rec. G.803, "Architecture of transport networks based on the synchronous digital hierarchy (SDH)," 1993 (revision).
17. ITU-T Rec. G.804, "ATM cell mapping into plesiochronous digital hierarchy (PDH)," 1994.
18. ITU-T Rec. G.825, "The control of jitter and wander within digital network which are based on the synchronous digital hierarchy (SDH)," 1993.
19. ITU-T Rec. G.831, "Management capabilities of transport networks based on the synchronous digital hierarchy (SDH)," 1993.
20. ITU-T Rec. G.957, "Optical interfaces for equipments and systems relating to the synchronous digital hierarchy," 1995 (revision).
21. ITU-T Rec. G.958, "Digital line systems based on the synchronous digital hierarchy for use on optical fiber cables," 1994 (revision).
22. ITU-T Rec. G.SHR-1 (Draft), "SDH protection: rings and other architectures," May 1994.

23. ITU-U Rec. G.702, "Digital hierarchy bit rates," 1988.
24. SR-NWT-001756, "Automatic protection switching for SONET," Issue 1, Bellcore, 1990.
25. SR-NWT-002224, "SONET synchronization planning guidelines," Issue 1, Bellcore, 1992.
26. TA-NWT-001042, "Generic requirements for operations interfaces using OSI tools: SONET path switched ring information model," Issue 3, Bellcore, 1992.
27. TA-NWT-001250, "Generic requirements for synchronous optical network (SONET) file transfer," Issue 2, Bellcore, 1992.
28. TN-NWT-001042, "Generic requirements for operations interfaces using OSI tools: synchronous optical network (SONET) transport information model," Issue 1, Bellcore, 1992.
29. TR-NWT-000253, "Synchronous optical network (SONET) transport systems: common generic," Issue 2, Bellcore, 1991.
30. TR-NWT-001230, "SONET bidirectional line switched ring equipment generic criteria," Issue 2, Bellcore, 1992.
31. TR-TSP-000496, "SONET add/drop multiplex equipment (SONET ADM) generic criteria," Issue 3, Bellcore, 1992.
32. TR-TSY-00023, "Wideband and broadband digital cross-connect generic requirements and objectives," Issue 2, Bellcore, 1989.
33. TR-TSY-000303, "Integrated digital loop carrier system generic requirements, objectives, and interface," Issue 1, Revision 3, Bellcore, 1990.

BISDN and ATM Technology

1. ATM Forum 93–215 (R8), "Broadband interconnection interface specification document."
2. ATM Forum 93–590, "Data exchange interface specification document."
3. ATM Forum, "ATM user-network interface specification version 2.0," 1992.
4. ATM Forum, *ATM user-network interface specification:* Version 3.0, Prentice Hall, 1993.
5. ATM Forum, "Network compatible ATM for local network applications," Phase 1, Version 1.0, 1992.
6. ITU-T Rec. F.811, "Broadband connection-oriented bearer services," 1992.
7. ITU-T Rec. F.812, "Broadband connectionless data bearer service," 1992.
8. ITU-T Rec. I.113, "Vocabulary terms for broadband aspects of ISDN," 1993 (revision).
9. ITU-T Rec. I.120, "Integrated services digital networks (ISDN)," 1993 (revision).
10. ITU-T Rec. I.121, "Broadband aspects of ISDN," 1991 (revision).
11. ITU-T Rec. I.140, "Attribute technique for the characterization of the

telecommunication services supported by an ISDN and network capability of an ISDN," 1992 (revision).

12. ITU-T Rec. I.150, "BISDN asynchronous transfer mode functional characteristics," 1993 (revision).
13. ITU-T Rec. I.211, "BISDN service aspects," 1993 (revision).
14. ITU-T Rec. I.311, "BISDN general network aspects," 1993 (revision).
15. ITU-T Rec. I.321, "BISDN protocol reference model and its application," 1991.
16. ITU-T Rec. I.327, "BISDN functional architecture," 1993.
17. ITU-T Rec. I.356, "BISDN ATM layer cell transfer performance," 1993.
18. ITU-T Rec. I.361, "BISDN ATM layer specification," 1993 (revision).
19. ITU-T Rec. I.362, "BISDN ATM adaptation layer (AAL) functional description," 1993 (revision).
20. ITU-T Rec. I.363, "BISDN ATM adaptation layer (AAL) specification," 1993 (revision).
21. ITU-T Rec. I.364, "Support of broadband connectionless data service on BISDN," 1993.
22. ITU-T Rec. I.371, "Traffic control and congestion control in BISDN," 1993.
23. ITU-T Rec. I.374, "Framework recommendation on network capabilities to support multimedia services," 1993.
24. ITU-T Rec. I.413, "BISDN user-network interface," 1993 (revision).
25. ITU-T Rec. I.432, "BISDN user-network interface—physical layer specification," 1993 (revision).
26. ITU-T Rec. I.610, "BISDN operation and maintenance principles and functions," 1993 (revision).

High-Speed Data Networks and Services

1. ANSI T1.606–1990, "Telecommunication-frame relay bearer service-architectural framework and service description," 1990.
2. ANSI T1.606add, "Addendum to T1.606," (T1X1/90–175), 1990.
3. ANSI T1.6ca, "Core aspects of frame protocol for use with frame relay bearer service," (T1S1/ 90–214), 1990.
4. ANSI X3.183, "High performance parallel interface (HIPPI)."
5. ANSI X3T9.3, "High-speed serial interface (HSSI)."
6. ANSI X3T9.5, "FDDI twisted pair physical medium dependent (known as CDDI or FDDI over twisted pair)."
7. ANSI/EIA, "RS 232C & DTE/DCE interface."
8. ATM Forum, "LAN emulation over ATM specification: version 1.0," January 1995.
9. ATM Forum, "Network compatible ATM for local network applications," phase 1, version 1.0, April 1992.

10. DEC, Northern Telecom, Stratacom, Cisco, "Frame relay specification with extension based on proposed T1S1 standards," Revision 1.0, 1990.
11. ETSI ETS300211–217 & ETS300268–278, "Connectionless broadband dataservice, superset of Bellcore SMDS."
12. Frame Relay Forum FRF.1, "User to network interface implementation agreement," version 2.0.
13. Frame Relay Forum FRF.2, "Frame relay network-to-network interface IA."
14. Frame Relay Forum FRF.3, "Multiprotocol encapsulation implementation agreement."
15. Frame Relay Forum FRF.4, "Switched virtual circuit, pending ratification."
16. Frame Relay Forum FRFTC93.67, "Frame relay MIB architecture and requirements."
17. Frame Relay Forum FRFTC93.67, "Frame relay multicast service description."
18. Frame Relay Forum FRFTC93.67, "Frame relay/ATM IA."
19. Frame Relay Forum FRFTC93.67, "Switched virtual circuit IA."
20. IAB & IESG, "Charter of the IAB," RFC1602, March 1994.
21. IAB & IESG, "The internet standards process—revision," RFC 1602, March 1994.
22. IEEE 802.11, "Wireless LAN, between 2400MHz & 2500MHz (2400MHz to 2483.5MHz in United States), total equivalent bandwidth 8Mbits/s in 1Mbits/s hops."
23. IEEE 802.12, "100 Base-VG 100Mbits/s Demand priority," (in process).
24. IEEE 802.3, "Carrier sense multiple access/collision detection, access method and physical layer specification, includes 10Base2, 10Base5, 10Base-T-Ethernet LAN."
25. IEEE 802.3J, "10BaseF:optical fiber passive and active star."
26. IEEE 802.4, "Token-passing bus."
27. IEEE 802.5, "4Mbits/s and 16Mbits/s specification for token ring."
28. IEEE 802.7, "Slotted ring."
29. IEEE 802.X (802.13), "100Base-T 100Mbits/s CSMA/CD, (100Mbits/s Ethernet)" (in process).
30. IEEE P802.12/D5, "Local and metropolitan area networks-part 12: demand-priority access method and physical layer specifications," Draft 5.0, 1994.
31. IEEE P802.3u/D4, "MAC parameters, physical layer, medium attachment units and repeater for 100Mbps operation," 1995.
32. IEEE P802.6, "DQDB subnetwork of a metropolitan area network," 1991.
33. Internet RFC 1094, "Network file system (NFS)."
34. Internet RFC 1098, "Simple network management protocol (SNMP)."
35. Internet RFC 768, "User datagram protocol (UDP)."
36. Internet RFC 791, "Internet protocol (IP)."
37. Internet RFC 792, "Internet control message protocol (ICMP)."
38. Internet RFC 793, "Transmission control protocol (TCP)."

39. Internet RFC 826, "Ethernet address resolution protocol (ARP)."
40. Internet RFC 959, "File transfer protocol (FTP)."
41. ISO-IEC 9314–1,-2,-3, "Fiber distributed data interface physical layer protocol, MAC and system management-100 Mbit/s optical fiber ring."
42. ITU-T Rec. I.122, "Framework for frame mode bearer service," 1993.
43. ITU-T Rec. I.233, "Frame mode bearer services," 1992.
44. ITU-T Rec. I.364, "Support of broadband connectionless data service on B-ISDN," 1993.
45. ITU-T Rec. I.365.1, "Frame relaying service specific convergence sublayer (FR-SSCS)," 1993.
46. ITU-T Rec. I.370, "Congestion management for the ISDN frame relaying bearer service," 1991.
47. ITU-T Rec. I.372, "Frame relaying bearer service network-to-network interface requirements," 1993.
48. ITU-T Rec. I.430, "Basic user-network interface—layer 1 specification," 1993 (revision).
49. ITU-T Rec. I.431, "Primary rate user-network interface—layer 1 specification," 1993 (revision).
50. ITU-T Rec. I.555, "Frame relaying bearer service interworking," 1993.
51. ITU-T Rec. Q.921 (I.441), "ISDN user-network interface, data layer specification," 1988.
52. ITU-T Rec. Q.922, "ISDN data link layer specification for frame mode bearer services," 1992.
53. ITU-T Rec. Q.933, "DSS1 signaling specification for frame mode bearer service," 1991.
54. TA-TSY-000772, "Generic system requirement in support of SMDS," Issue 3, Bellcore, 1989.

Broadband Video Services and Technology

1. ATRC, "System description of advanced digital television," February 1991.
2. Datapro Reports on Telecom., "Video conferencing systems," November 1990.
3. DAVIC1.0, "Description of DAVIC functionalities," Part 1, Revision 4.0, 1995.
4. First International Workshop on Packet Video, Workshop Notes, New York, May 1987.
5. Fourth International Workshop on Packet Video, Workshop Notes, Kyoto, Japan, August 1991.
6. ISO 10918 (JPEG), "Digital compression and coding of continuous-tone still images," 1991.
7. ISO 11154, "Progressive bilevel image compression," 1994.

8. ISO-IEC/JTC1/SC2/WG12 (MHEG), "Information processing, coded representation of multimedia and hypermedia information objects," 1994.
9. ISO-IEC/JTC1/SC2/WG8, "Coded representation of picture and audio information," JPEG Draft, January 1990.
10. ISO/IEC 10918–1, "Digital compression and coding of continuous-tone still image, part 1: requirement and guidelines," 1993.
11. ISO/IEC 11172–1, "Information technology—generic coding of moving pictures and associated audio, part 1: systems," March 1994.
12. ISO/IEC 11172–2, "Information technology—generic coding of moving pictures and associated audio, part 2: video," March 1994.
13. ISO/IEC 11172–3, "Information technology—generic coding of moving pictures and associated audio, part 3: audio," March 1994.
14. ISO/IEC 13818–1, "Information technology—generic coding of moving pictures and associated audio, part 1: systems, recommendation," March 1994.
15. ISO/IEC 13818–2, "Information technology—generic coding of moving pictures and associated audio, part 2: video, recommendation," March 1994.
16. ISO/IEC 13818–3, "Information technology—generic coding of moving pictures and associated audio, part 3: audio, recommendation," March 1994.
17. ISO/IEC CD-11172, "Information technology—coding of moving pictures and associated audio for digital storage," 1993.
18. ITU-R Rec. 601, "Encoding parameters of digital television for studios," 1982.
19. ITU-R Rec. 656, "Interfaces for digital component video signals in 525-line and 625-line television systems," 1982.
20. ITU-R Rec. 709, "Basic parameter values for the HDTV standard for the studio and for international programme exchange," 1990.
21. ITU-R Rec. 710, "Subjective assessment methods for image quality in high-definition television," 1990.
22. ITU-R Rec. 714, "International exchange of programme electronically produced by means of high-definition television," 1990.
23. ITU-T Rec. G.726, "40, 32, 24, 16kbits/s adaptive differential pulse code modulation (ADPCM)," 1990.
24. ITU-T Rec. H.221, "Frame structure for a 64 to 1920 kbit/s channel in audiovisual teleservices," 1995 (revision).
25. ITU-T Rec. H.230, "Frame-synchronous control and indication signals for audiovisual systems," 1995 (revision).
26. ITU-T Rec. H.242, "System for establishing communication between audiovisual terminals using digital channels up to 2 Mbit/s," 1993 (revision).
27. ITU-T Rec. H.261, "Video coder for audio visual services at 64kbit/s," 1993 (revision).
28. ITU-T Rec. H.263, "Video coding for low-bit rate communications," 1995.
29. ITU-T Rec. H.320, "Narrow-band visual telephone systems and terminal equipment," 1993 (revision).

30. ITU-T Rec. H.324, "Terminal for low-bit rate multimedia communications," 1995.
31. ITU-T Rec. T.122, "Multipoint communications service for audiographics and audiovisual conferencing service definition," 1993.
32. ITU-T Rec. T.123, "Protocol stacks for audiographic and audiovisual teleconference applications," 1994 (revision).
33. ITU-T Rec. T.124, "Generic conference control," 1995.
34. ITU-T Rec. T.125, "Multipoint communications service protocol specifications," 1994.
35. ITU-T Rec. T.170, "Audio-visual interactive (AVI) systems, introduction, principles, concept and models."
36. ITU-T Rec. T.171, "Coded representation of multimedia/hypermedia (integrated voice, data and video transmission) information objects (common text with ISO-SG13)."
37. ITU-T Rec. T.172, "Description of AVI scriptware (software structure underlying interactive capabilities) function (SG 13)."
38. ITU-T Rec. T.173, "Coded representation of AVI scriptware functions."
39. ITU-T Rec. T.175, "Protocols for browsing, downloading and executing AVI functions (protocol I)."
40. ITU-T Rec. T.176, "Protocols for synchronization of executed AVI applications (protocol B)."
41. Second International Workshop of Packet Video, Workshop Notes, Torino, Italy, September 1988.
42. Third International Workshop of Packet Video, Workshop Notes, Torino, Italy, March 1990.
43. Zenith, AT&T, "Technical description of digital spectrum compatible HDTV," February 1991.

Appendix B
List of Acronyms

AAL	ATM adaptation layer
ABR	Available bit rate
AC	Access control
ACATS	Advisory Committee on Advanced Television Service
ACI	Adjacent channel interference
ACF	Access control field
ACK	Acknowledgment
A/D	Analog-to-digital (converter)
ADC	Analog-to-digital conversion
ADM	Add-drop multiplexer
ADPCM	Adaptive differential pulse code modulation
ADS	Active double star
ADSL	Asymmetric digital subscriber line
ADTV	Advanced digital television
AF	Address filter
AIN	Advanced intelligent network
AIS	Alarm indication signal
AL	Alignment
AM	Asynchronous multiplexing
ANSI	American National Standard Institute
AP	Access point
APD	Avalanche photo diode
APId	Access point identifier
APII	Asia Pacific Information Infrastructure
APS	Automatic protection switching
ARP	Address resolution protocol
ARPA	Advanced Research Project Agency
ARPANet	ARPA network
ARQ	Automatic repeat request
ASE	Application service element
ASK	Amplitude shift keying
ASN	Abstract syntax notation
ASTA	Advanced software technology and algorithms

ATDM	Asynchronous time division multiplexing
ATM	Asynchronous transfer mode
ATMARP	ATM address resolution protocol
ATMR	ATM ring
AT&T	American Telephone and Telegraph
ATTC	Advanced Television Test Center
ATV	Advanced TV
AU	Access unit
AU	Administrative unit
AUG	Administrative unit group
B	Bidirectional (picture)
BA	Building automation
BAS	Building automation system
BAsize	Buffer allocation size
B-ICI	Broadband inter-carrier interface
B-NT	Broadband network termination
B-TA	Broadband terminal adapter
B-TE	Broadband terminal equipment
BBIDS	Broadband integrated distributed star
BBN	Broadcast banyan network
BBTG	Broadband task group
Bellcore	Bell Communication Research
B/Etag	Begin/End tag
BCN	Backward congestion notification
BDCS	Broadband digital cross-connect system
BECN	Backward explicit congestion notification
BER	Bit error rate
BGP	Border gateway protocol
BH	Buried heterostructure
BIA	Broadband integrated access
B-ICI	Broadband intercarrier interface
BIGFON	Breitbandiges Integriertes Glasfoser Fernmelde Ortsnetz Network
BIM	Byte interleaved multiplexing
BIP	Bit interleaved parity
BISDN	Broadband integrated services digital network
BISUP	BISDN user part
BMA	Block matching algorithm
BOM	Beginning of message
BPON	Broadband passive optical network
bps	bits per second (bits/s)
BRA	Base rate access
BRHR	Basic research and human resources
BSRF	Basic synchronous reference frequency

BT	British Telephone
BT	Burst tolerance
BTRL	British Telecom Research Laboratory
BUS	Broadcast and unknown server
BW	Black and white
BWB	Bandwidth balancing
C	Chooser
C	Container
C-MAC	Compatible multiplexed analog components
CA	Collision avoidance
CAC	Connection admission control
CAD	Computer-aided design
CALS	Commerce at light speed
CALS	Continuous acquisition and life-cycle support
CALS	Computer-aided acquisition and logistic support
CAM	Computer-aided manufacturing
CAM	Content addressable memory
CAP	Carrierless amplitude-modulation and phase-modulation
CAT	Category
CATV	Cable television or community antenna television
CBDS	Connectionless broadband data service
CBP	Coded block pattern
CBR	Constant bit rate
CC	Congestion control
CC	Command and control
CCI	Cochannel interference
CCIR	International Radiocommunications Consultative Committee
CCITT	International Telegraph and Telephone Consultative Committee
CCS	Common channel signaling
CD	Collision detection
CD	Count down
CDAD	Cable digital audio distribution
CDMA	Code division multiple access
CD-ROM	Compact disk read only memory
CDV	Cell delay variation
CE	Connection element
CEN	Comite Europeenne de Normalisation
CENELEC	Comite Europeenne de Normalisation Electrotechnique
CEPT	Conference Europeenne des Postes et Telecommunication
CEQ	Customer equipment
CG	Channel group
CGI	Common gateway interface
CH	Channel header

CID	Connection identifier
CIE	Commission Internationale de L'Eclairage
CIF	Common intermediate format
CIR	Cell insertion ratio
CIR	Command information rate
CL	Connection-less
CLNAP	Connectionless network access protocol
CLNIP	Connectionless network interface protocol
CLNP	Connectionless network protocol
CLP	Cell loss priority
CLR	Cell loss ratio
CLSF	Connectionless service function
CMF	Conversational media flow
CMISE	Common management information service element
CMOS	Complementary metal-oxide semiconductor
CMT	Connection management
CMTT	Comission Mixte pour des Transmissions Televisuelles et Sonones
CNET	Centre National d'Etudes des Telecommunications
CNM	Customer network management
CNRI	Corporation for National Research Initiatives
CO	Central office
CO	Connection-oriented
COCF	Connection-oriented convergence function
CODEC	Coder-decoder
COFDM	Coded orthogonal FDM
CO-LAN	Central office-local area network
COM	Continuation of message
COT	Central office terminal
CP	Call processor
CP	Cascade port
CP	Cell processor
CP	Connection point
CPCS	Common part convergence sublayer
CPE	Customer premises equipment
CPI	Common part indication
CPN	Customer premises network
C/R	Command and response
CRC	Cyclic redundancy check
$C_r C_b$	Chrominance component (symbol)
CRF	Connection related function
CRR	Class related rule
CS	Capability set
CS	Convergence sublayer

CS	Cyclic sequence
CSA	Carrier serving area
CSCW	Computer-supported cooperative work
CSDC	Circuit switched digital capability
CSDN	Circuit switched digital network
CSF	Conversational service-signal flow
CSI	Convergence sublayer indication
CSMA/CA	Carrier-sense multiple access with collision avoidance
CSMA/CD	Carrier-sense multiple access with collision detection
CS-MUX	Circuit-switched multiplexer
CSN	Connection sequence number
CS-PDU	Convergence sublayer protocol data unit
CSRC	Contributing source
CT	Central terminal
DA	Destination address
D/A	Digital-to-analog (converter)
DAB	Digital audio broadcast
DAC	Digital-to-analog conversion
DAE	Dummy address encoder
DAP	Data access protocol
DARPA	Defense Advanced Research Projects Agency
DAS	Dual-attachment station
DAVIC	Digital Audio-Visual Council
DB	Database
DB-IR	Direct beam infrared
DBP	Deutsche Bundes Post
DBS	Direct broadcast by satellite
DC	Distribution center
DCC	Data communication channels
DCN	Data communication network
DCS	Digital cross-connect system
DCT	Discrete cosine transform
DDM	Direct-division multiplexing
DDS	Digital data service
DE	Discard eligibility
DFB-LD	Distributed feedback laser diode
DF-IR	Diffused infrared
DFT	Discrete Fourier transform
DH	Double heterostructure
DL	Data link
DLC	Digital loop carrier
DLCI	Data link connection identifier
DMF	Distributional media flow

DMPDU	Differential MAC protocol data unit
DMT	Discrete multitone (modulation)
DMUX	Demultiplexer
DNS	Domain name service
DPA	Demand priority access
DPCM	Differential pulse code modulation
DPG	Data packet group
DPSK	Differential PSK
DQ	Distributed queue
DQDB	Distributed queue dual bus
DQSM	Distributed-queue state machine
DRAM	Dynamic random access memory
DS	Direct sequence
DSF	Distribution service-signal flow
DSL	Digital subscriber line
DSM	Digital storage medium
DSM-CC	DSM command and control
DS-n	Digital signal level n
DSS	Distributed sample scrambler
DSX	Digital system cross-connect
DTS	Decoding time stamp
DTTB	Digital terrestrial television broadcast
DVB	Digital video broadcasting
DVD	Digital video disk
DVTR	Digital video tape recorder
DWDM	Dense wavelength division multiplexing
DWMT	Discrete wavelet multitone (modulation)
DXC	Digital cross-connect
DXI	Data exchange interface
E	East
EA	Extended address
EBCN	Explicit backward congestion notification
EC	Electronic cinema
ECC	Embedded control network
ECL	Emitter-coupled logic
ECMA	European Computer Manufacturers Association
ECR	Errored cell ratio
ED	End delimiter
EDFA	Erbium-doped fiber amplifier
EDI	Electronic data interchange
EDTV	Extended definition television
EFCI	Explicit forward congestion indication
EFCN	Explicit forward congestion notification

EFS	End-of frame sequence
EGP	Exterior gateway protocol
EIA	Electronic Industry Association
EII	European Information Infrastructure
ELAN	Emulated local area network
ENG	Electronic news group
E/O	Electrical/optical
EOB	End of block
EOM	End of message
EP-DVB	European Project for Digital Video Broadcasting
EPRCA	Enhanced proportional rate control algorithm
EQTV	Extended quality television
ESCR	Elementary stream clock reference
ESnet	Energy Science Network
ESS	Electronic switching system
ET	Exchange termination
ETSI	European Telecommunication Standards Institute
EU	European Union
FA	Factory automation
FAS	Flexible access system
FAX	Facsimile
FC	Frame control
FCC	Federal Communications Commission
FCN	Forward congestion notification
FCS	Frame check sequence
FCS	Fast circuit switching
FCS	Frame control segment
FDDI	Fiber distributed digital interface
FDM	Frequency division multiplexing
FDMA	Frequency division multiple access
FEC	Forward error correction
FECN	Forward explicit congestion notification
FET	Field effect transistor
FFOL	FDDI follow-on LAN
FFT	Fast Fourier transform
FH	Frequency hopping
FIFO	First-in-first-out
FITL	Fiber-in-the-loop
FIX	Federal Internet Exchange
FLC	Fiber loop carrier
FM	Frequency modulation
FOA	First office application
FOH	Fixed overhead

FPLMTS	Future Public Land Mobile Telecommunications System
FPS	Fast packet switching
FR	Frame relay
FS	Frame state
FSK	Frequency shift keying
FSS	Frame synchronous scrambler
FTAM	File transfer, access, and management
FTP	File transfer protocol
FTTB	Fiber-to-the-building
FTTC	Fiber-to-the-curb
FTTH	Fiber-to-the-home
FTTO	Fiber-to-the-office
FTTZ	Fiber-to-the-zone
GA	Grand Alliance
GaAs	Gallium arsenide
GCRA	Generic cell rate algorithm
GFC	Generic flow control
GI	General Instrument
GIF	Graphics interchange format
GII	Global information infrastructure
GMT	Greenwich mean time
GoP	Group of picture
GOSIP	Government OSIP
Gbps	Gigabits per second (Gbits/s)
GRIN	Graded Index
H	Halt
H	Header
HBT	Heterojunction bipolar transistor
HCS	Header check sum
HDLC	High-level data link control
HD-MAC	High definition multiplexed analog components
HDSL	High-speed digital subscriber line
HDTV	High definition television
HE	Header extension
HEC	Head error control
HEL	Header extension length
HFC	Hybrid fiber/coax
Hi-OVIS	Highly optical visual information system
HIPPI	High performance parallel interface
HL	High level
HLF	High-level functions
HLI	High-level interface
HLPI	High-level protocol identifier

H-MUX	Hybrid multiplexer
HOL	Head-of-line
HP	Higher-order path
HP	High priority
HP	High profile
HPA	Higher-order path adaptation
HPC	Higher-order path connection
HPCC	High performance computing and communication
HPCS	High performance computing system
HPT	Higher-order path termination
HRC	Hybrid ring control
HSN	High-speed network
HSTP	High-speed transport protocol
HTML	HyperText markup language
HTT	Home television theater
HTTP	HyperText transfer protocol
HVS	Human visual system
I	Idle
I	Intra (picture)
IAB	Internet Architecture Board
IBS	Intelligent building system
IC	Integrated circuit
ICF	Isochronous convergence function
ICI	Inter-exchange carrier interface
ICI	Interface control information
ICI	Interchannel interface
ICMP	Internet control message protocol
IDCT	Inverse discrete cosine transform
IDFT	Inverse discrete Fourier transform
IDU	Interface data unit
IE	Inter-Exchange
IEC	International Electrotechnical Commission
IEEE	Institute of Electrical and Electronics Engineers
IESG	Internet Engineering Steering Group
IETF	Internet Engineering Task Force
IF	Intermediate frequency
IFFT	Inverse fast Fourier transform
IFU	Interface unit
IGMP	Internet group management, protocol
IGP	Interior gateway protocol
IHL	Internet header length
IITA	Information infrastructure technology and application
I-MAC	Isochronous MAC

IM/DD	Intensity modulation/direct detection
IMPDU	Initial MAC protocol data unit
IN	Intelligent network
InATMARP	Inverse ATM address resolution protocol
INN	INFAS and NTT network
INS	Information and network system
IP	Internet Protocol
IP	Internetworking protocol
IPC	Input port controller
IPC	Inter-personal communication
IPRM	Integrated protocol reference model
IPng	Internet protocol-new generation
IPv4	Internet protocol version 4
IPv6	Internet protocol version 6
IR	Infrared
IRP	Internal reference point
IRTF	Internet Research Task Force
ISDN	Integrated services digital network
ISLN	Integrated services local network
ISM	Industrial, scientific, and medical
ISM	Interface subscriber module
ISM	Interactive storage medium
ISO	International Standard Organization
ISOC	Internet Society
ISSI	Inter-switching system interface
ISUP	ISDN user part
IT	Information type
ITU	International Telecommunication Union
ITU-R	ITU Radio-Communications (Sector)
ITU-T	ITU Telecommunication (Sector)
IVD	Integrated voice and data
IWU	Interworking unit
JFET	Junction field effect transistor
JPEG	Joint Photographic Experts Group
Kbps	kilobits per second (Kbits/s)
KLT	Karhunen-Loeve transform
ksps	kilo symbols per second
LAB	Latency adjustment buffer
LAN	Local area network
LAP	Link access protocol
LATA	Local access and transport area
LC	Link connection
LCD	Loss of cell delineation

LCF-PMD	Low-cost fiber-PMD
LCN	Local communication network
LD	Laser diode
LE	Local exchange
LECS	LAN emulation configuration server
LED	Light emitting diode
LEO	Low Earth orbit
LES	LAN emulation server
LEX	Local exchange
LFN	Long and fat network
LI	Length indicator
LI	Line interface
LIS	Logical IP subnetwork
LL	Low level
LL	CLogical link control
LMDS	Local multipoint distribution service
LME	Layer management entity
LMI	Local management interface
LOC	Loop optical control
LOF	Loss of frame
LOP	Loss of pointer
LOS	Loss of signal
LOS	Line of sight
LP	Local port
LP	Lower-order path
LP	Low priority
LPA	Lower-order path adaptation
LPC	Lower-order path connection
LPF	Low-pass filter
LPT	Lower-order path termination
LS	Local switch
LSI	Large scale integrated circuit
LT	Line termination
LTE	Lightwave transmission equipment
LU	Line unit
LUT	Lookup table
LW	Lightwave
MA	Medium adapter
MAC	Medium access control
MAC	Multiple analog components
MAN	Metropolitan area network
MAP	Manufacturing automation protocol
MB	MacroBlock

MBA	MacroBlock address
Mbtype	MacroBlock type
MBone	Multicast backbone
Mbps	Megabits per second (Mbit/s)
MBS	Maximum burst size
MC	Motion compensation
MCF	MAC convergence function
MCF	Message communication function
MCM	Multicarrier modulation
MDF	Main distribution frame
ME	Mapping entity
ME	Motion estimation
MEO	Middle Earth orbit
MHS	Message handling service
MID	Message identifier
MID	Multiplexing identifier
MIME	Multipurpose Internet mail extension
MIN	Multistage interconnection network
MISFET	Metal-insulator-semiconductor field effect transistor
ML	Main level
MMDS	Multichannel (or Microwave) multipoint distribution service
MO	Management object
MOD	Movie on demand
MP	Main profile
MPDU	MAC protocol data unit
MPEG	Moving picture expert group
MPMP	Microwave point-to-multipoint
MPU	Multi processor unit
MQW	Multiple quantum well
MS	Multiplexer section
MSDU	MAC service data unit
MSOH	Multiplexer section overhead
MSP	Multiplexer section protection
MSS	MAN switching system
MST	Multiplexer section termination
MTS	Multiplexer timing source
MTU	Maximum transmission unit
MUSE	Multiple sub-Nyquist sampling encoding
MUX	Multiplexer
MVC	Maintenance voice channel
NAK	Negative acknowledgment
NBMA	Non-broadcast multiaccess
NC	Network connection

NCA	News and current affairs
NDB	Network database service
NDF	New data flag
NE	Network element
NETBLT	Network block transfer
NERN	National Research and Education Network
NFS	Network file system
NHRP	Next hop resolution protocol
NII	National Information Infrastructure
NISDN	Narrowband integrated services digital network
NJ	Negative justification
NNI	Network node interface
NOD	News on demand
NPC	Network parameter control
NPI	Null pointer indication
NS	Next station
NSFnet	National Science Foundation Network
NSI	NASA Science Internet
NSP	Network service protocol
NSS	Non-spread spectrum
NT	Network termination
NTM	Network traffic management
NTSC	National Television System Committee
NTT	Nippon Telephone and Telegraph
NVOD	Near video on demand
OA	Office automation
OAM	Operation and management
OAM&P	Operation administration maintenance & provisioning
OC-m	Optical carrier level m
ODF	Optical distribution frame
OEIC	Opto-electronic integrated circuit
OFDM	Optical frequency division multiplexing
OFDM	Orthogonal frequency division multiplexing
OIU	Office interface unit
OLU	Optical line unit
ONT	Optical network termination
ONU	Optical network unit
OOF	Out-of-frame
OPC	Output port controller
OS	Operating system
OSI	Open system interconnection
OSIE	Open systems interconnection environment
OSIP	Open systems interconnection profile

OSPF	Open shortest path first
OTDR	Optical time domain reflectometer
P	Padding
P	Parity
P	Pointer
P	Priority
P	Protection
P	Predictive (picture)
PA	Preamble
PA	Pre-arbitrated
PAD	Padding
PAL	Phase alteration by line
PBX	Private branch exchange
PC	Priority control
PC	Personal computer
PCC	Programmable cross-connect
PCI	Protocol control information
PCM	Pulse code modulation
PCN	Personal communication network
PCR	Peak cell rate
PCR	Program clock reference
PD	Photo diode
PDE	Power distribution enclosure
PDH	Plesiochronous digital hierarchy
PDU	Protocol data unit
PES	Packetized elementary stream
PHY	Physical (layer)
PI	Physical interface
PJ	Positive justification
PLCF	Physical layer convergence function
PLCP	Physical layer convergence procedure
PLL	Phase-locked loop
PM	Physical medium
P-MAC	Packet MAC
PMB	Programmable multiplexer bank
PMD	Physical medium dependent (layer)
PMTU	Path maximum transmission unit
POH	Path overhead
PON	Passive optical network
POS	Point of sales
POTS	Plain old telephone service
PPI	PDH physical interface
PPL	Passive photonic loop

PPM	Part per million
PRA	Primary rate access
PRBS	Pseudo-random binary sequence
PRCA	Proportional rate control algorithm
PRM	Protocol reference model
PS	Previous station
PS	Protection switching
PS	Program stream
PSDN	Packet switched data network
PSDN	Public switched data network
PSK	Phase shift keying
PSF	Protection switching failure
PSPDN	Packet-switched public data network
PSM	Protocol state machine
PSR	Previous slot release
PSTN	Public switched telephone network
PSU	Power supply unit
PT	Path trace
PT	Payload type
PTE	Path termination equipment
PTP	Point-to-point
PTMP	Point-to-multipoint
PTR	Pointer
PTS	Presentation time stamp
PVC	Permanent virtual circuit
P/Z/N	Positive/zero/negative
Q	Quiet
QA	Queue arbitrated
QAM	Quadrature amplitude modulation
QCIF	Quarter common intermediate format
QOS	Quality of service
QPSK	Quadrature phase shift keying
QPSX	Queued packet and synchronous exchange
R	Reset
RA	Running adder
RAI	Remote alarm indication signal
RARP	Reverse address resolution protocol
RBOC	Regional Bell operating company
RCLOSE	Read close
RDI	Remote defect indication
REG	Register
REI	Remote error indication
REQ	REQuest

RES	REServed
RF	Radio frequency
RF	Receiver failure
RFC	Request for comments
RFI	Remote failure indication
RG	Regenerator
RGB	Red green blue
RIP	Routing information protocol
RLC	Run length coding
RM	Resource management
RMAC	Repeater medium access control
RMN	Remote multiplexer
RMT	Ring management
RN	Remote node
ROLC	Routing over larger clouds
RQ	Request
RRCN	Root raised cosine nyquist
RS	Reed-solomon (code)
RSOH	Regenerator section overhead
RSP	Regenerator section protection
RST	Regenerator section termination
RP	Reservation protocol
RT	Remote terminal
RTCP	Real-time transport control protocol
R-TDMA	Reservation TDMA
RTO	Retransmission timeout
RTP	Real-time transport protocol
RTS	Residual time stamp
RTT	Round-trip time
RVS	Remote video service
S	Set
S	Sender
SA	Section adaptation
SA	Synchronous allocation
SA	Source address
SAAL	Signaling AAL
SACK	Selective acknowledgment
SAP	Service access point
SAPI	Service access point identifier
SAR	Segmentation and reassembly
SAS	Single-attachment station
SC	Sequence count
SCM	Subcarrier multiplexing

SCR	Sustainable cell rate
SCR	System clock reference
SD	Start delimiter
SDH	Synchronous digital hierarchy
SDT	Structured data transfer
SDTV	Standard definition television
SDU	Service data unit
SE	Switching element
SECAM	Sequential couleur avec memoire
SEMF	Synchronous equipment management function
SF	Split filter
SFET	Synchronous frequency encoding technique
SFN	Single frequency network
SHR	Self healing ring
SG	Study group
SGML	Standard generalized markup language
SIG	Signaling processor
SIN	Subscriber interface network
SIP	SMDS interface protocol
SIPP	Simple internet protocol plus
SIU	Subscriber interface unit
SLC	Subscriber loop carrier
SLM	Signal label mismatch
SM	Synchronous multiplexing
SMDS	Switched multi-megabit date service
SMF-PMD	Single mode fiber-PMD
SMN	SDH management network
SMT	Station management
SMTP	Simple mail transfer protocol
SN	Sequence number
SN	Subnetwork
SNA	Systems network architecture
SNAP	SubNetwork attaching point
SNI	Subscriber node interface
SNMP	Simple network management protocol
SNP	Sequence number protection
SNP	SNR profile
SOH	Section overhead
SONET	Synchronous optical network
SP	Signal processor
SPAG	Standard promotion and application group
SPC	Stored program control
SPE	Synchronous payload envelope

SPI	SDH physical interface
SPM-PMD	SONET physical layer mapping-PMD
SRTS	Synchronous residual time stamp
SS	Spread spectrum
SSCF	Service specific coordination function
SSCOP	Service specific connection oriented protocol
SSCS	Service specific convergence sublayer
SSM	Single segment message
SSM	Serial storage media
SSP	Spatially scaleable profile
SSRC	Synchronization source
SSS	Self synchronous scrambler
ST	Segment type
STB	Set-top box
STB	Satellite television broadcast
STG	Synchronous timing generation
STM-n	Synchronous transport module level n
STP	Shielded twisted pair
STP	Signaling transfer point
STS-m	Synchronous transport signal level m
STU	Set-top unit
STU	Subscriber's terminal unit
SU	Segment unit
SVC	Signaling virtual channel
SVC	Switched virtual channel
SW	Switch
SWAN	Sociocultural Welfare Advancement Network
T	Trailer
TA	Telecommunication automation
TA	Terminal adapter
TAT	Theoretical arrival time
TC	Transmission convergence
TCP	Termination connection point
TCP	Transmission control protocol
TCM	Trellis coded modulation
TCM	Time compression modulation
TDM	Time division multiplexing
TDMA	Time division multiple access
TE	Terminal equipment
TEI	Terminal endpoint identifier
TEID	Terminal equipment identification
TEN	Trans European network
TF	Transmitter failure

THT	Token holding time
TIM	Trace identification mismatch
TMN	Telecommunication management network
TNT	Trunk number translator
TOP	Technical and office protocol
TOS	Type of service
TP	Twisted pair
TPDU	Transport
TPE	Transmission path endpoint
TP-PMD	Twisted pair-PMD
TPON	Telephony over passive optical network
TPS	Transmission parameter signaling
TRT	Token rotation time
TS	Time stamp
TS	Transport stream
TSI	Time slot interchange
TT	Trail termination
TTB	Terrestrial television broadcast
TTL	Time to live
TTOSS	Totally transparent optical subscriber system
TTR	Timed token rotation
TTRT	Target token rotation time
TU	Tributary unit
TUG	Tributary unit group
TWT	Traveling wave tube
UDP	User datagram protocol
UDTV	Ultra-high definition television
UHF	Ultra high frequency
UI	Unit interval
UNEQ	Unequipped
UNI	User network interface
UPC	Usage parameter control
URL	Universal resource locator
U-SDU	User service data unit
UTP	Unshielded twisted pair
VBI	Vertical blanking interval
VBR	Variable bit rate
VBV	Video buffer verifier
VC	Virtual channel
VC	Virtual container
VCC	Virtual channel connection
VCCE	Virtual channel connection endpoint
VCI	Virtual channel identifier

VCO	Voltage controlled oscillator
VCR	Video cassette recorder
VDSL	Very-high-speed digital subscriber line
VDT	Video dial tone
VHF	Very high frequency
VLC	Variable length coding
VLSI	Very large scale integrated (circuit)
VMTP	Versatile message transfer protocol
VOD	Video on demand
VP	Virtual path
VPC	Virtual path connection
VPCE	Virtual path connection endpoint
VPI	Virtual path identifier
VPT	Virtual path termination
VQ	Vector quantization
VRS	Video response system
VSB	Vestigial sideBand
VT	Virtual tributary
VTR	Video tape recorder
W	West
W	Working
WAN	Wide area network
WBC	Wide band channel
WCLOSE	Write close
WDM	Wavelength division multiplexing
WG	Working group
WIM	Word interleaved multiplexing
WSP	Wideband service point
WTSC	World Telecommunication Standardization Conferences
WWW	World Wide Web
XC	Crossconnect
XTP	eXpress transfer protocol
Y	Luminance component (symbol)
2B1Q	2 Binary 1 quaternary
4B3T	4 Binary 3 ternary
4b/5b	4 binary 5 binary

About the Authors

Byeong Gi Lee received the B.S. and M.E. degrees in electronics engineering from Seoul National University, Seoul, Korea, and Kyungpook National University, Taegu, Korea, in 1974 and 1978, respectively, and the Ph.D. degree in electrical engineering from University of California, Los Angeles, in 1982. From 1974 to 1979 he was at the Department of Electronics Engineering of ROK Naval Academy, Chinhae, Korea, as an instructor and naval officer. From 1982 to 1984 he worked for Granger Associates, Santa Clara, CA, where he did R&D on applications of digital signal processing to digital transmission. During 1984–1986 he was a member of the technical staff at AT&T Bell Laboratories, North Andover, MA, where he worked on lightwave transmission system development and the related standard works. Since September 1986 he has been with the Department of Electronics Engineering, Seoul National University, where he is now a professor. His current fields of interest include digital transmission and integrated broadband networks, theory and applications of digital signal processing, and circuit theory. He is a coauthor of *Introduction to ISDN, Broadband Telecommunication Systems, Electronics Engineering Experiment Series (five volumes)*, and editor of the *HDTV Dictionary*, all in Korean; and a coauthor of *Scrambling Techniques for Digital Transmission* (Springer-Velag, 1994). He holds six U.S. patents with one more patent pending. Dr. Lee received the 1984 Myril B. Reed Best Paper Award and Exceptional Contribution Awards from AT&T Bell Laboratories. He is a senior member of IEEE; a board member of Korean Institute of Communication Sciences (KICS) and KSEETT; and a member of KITE, KISS, ASK, and Sigma Xi.

Minho Kang received a B.S. degree from Seoul National University, Seoul, Korea, an M.S. degree from the University of Missouri (Rolla), and a Ph.D. from the University of Texas at Austin in 1969, 1973, and 1977, respectively, all in electrical engineering. He worked for AT&T Bell Laboratories, Holmdel, NJ, in the optoelectronics research area from 1977 to 1978, and then worked for the Electronics and Telecommunications Research Institute (ETRI), Daejon, Korea, in the optical communications and transmission systems development area until January 1990. He was chief executive of the Research Center and Quality Assurance Center, and is presently an executive vice president in charge of the Overseas Business Group at Korea Telecom, the leading integrated telecommunications network operator in Korea. He was a lecturer at Seoul National University between 1979 and 1985 and served as the Electronics Research Coordinator at the Korean Ministry of Science and Technology between 1985 and 1988. He is a coauthor of *Introduction to ISDN, Introduction to Electrical Communication Technology, Principles of Optical Fiber Communication Technology, Laser Applications,* and *Broadband Telecommunication Systems,* all in Korean. He received the National Order of Merit, Dongbaeg-Jang, and the Grand Prize of New Industrial Technology Management. Dr. Kang is a senior member of IEEE; a member of the Optical Society of America; a board member of ICCC, KITE, KIEE, KICS, and KISS; and a member of Phi Kappa Phi and Eta Kappa Nu.

Jonghee Lee received a B.S. degree in electrical engineering from Seoul National University, Seoul, Korea, in 1971, and M.S. and Ph.D. degrees from the University of Pennsylvania in 1976 and 1980, respectively, both in systems engineering. From 1980 to 1985 he was with AT&T Bell Laboratories and then with Bell Communications Research, where he was responsible for product planning and development in digital terminals and crossconnects. From 1985 to 1990 he worked for Daeyoung Electronics Co., where he was the executive managing director responsible for R&D in communication systems. In 1990 he founded Dongjin Datacom, an EDI software company, and in 1991 he founded MODACOM Co., a telecommunications consulting and systems integration company. He is currently the president of MODACOM. He has been a lecturer at Seoul National University (1986, 1987) and Korea University (1993, 1994). He is the author of *Intelligent Building System Planning and Design,* and a coauthor of *Broadband Telecommunication Systems,* all in Korean. Dr. Lee is a board member of KICS; an advisory member of the KITE's Telecommunication Society; and a member of IEEE.

Index

The Artech House Telecommunications Library

Vinton G. Cerf, Series Editor